T0215140

Total Mean Curvature and Submanifolds of Finite Type

Second Edition

SERIES IN PURE MATHEMATICS*

ISSN: 1793-1185

Editor: C C Hsiung
Associate Editors: S Kobayashi, I Satake, Y-T Siu, W-T Wu and M Yamaguti

*To view the complete list of the published volumes in the series, please visit:
http://www.worldscientific.com/series/spm

Series in Pure Mathematics – Volume 27

Total Mean Curvature and Submanifolds of Finite Type

Second Edition

Bang-Yen Chen

Michigan State University, USA

World Scientific

NEW JERSEY · LONDON · SINGAPORE · BEIJING · SHANGHAI · HONG KONG · TAIPEI · CHENNAI

Published by

World Scientific Publishing Co. Pte. Ltd.

5 Toh Tuck Link, Singapore 596224

USA office: 27 Warren Street, Suite 401-402, Hackensack, NJ 07601

UK office: 57 Shelton Street, Covent Garden, London WC2H 9HE

Library of Congress Cataloging-in-Publication Data
Chen, Bang-yen.
    Total mean curvature and submanifolds of finite type / by Bang-Yen Chen (Michigan State University, USA). -- 2nd edition.
       pages cm. -- (Series in pure mathematics ; volume 27)
    Includes bibliographical references and index.
    ISBN 978-9814616683 (hardcover : alk. paper) -- ISBN 978-9814616690 (pbk. : alk. paper)
    1. Submanifolds. 2. Curvature. I. Title.
    QA649.C484 2014
    516.3'6--dc23

                                                                        2014036152

British Library Cataloguing-in-Publication Data
A catalogue record for this book is available from the British Library.

Printed in Singapore

To my wife Pi-Mei and
my children Bonny, Emery and Beatrice
and their families
in appreciation of their love and support

Foreword

The present book *"Total Mean Curvature and Submanifolds of Finite Type, Second Edition"* by MSU Distinguished Professor Bang-Yen Chen is the 2014 new edition of the 1984 first edition with the same title. Here follows a quote from the Bulletin of the London Mathematical Society's review of the 1984 book, written together with Paul Verheyen: *"Just as his previous books* ("Geometry of Submanifolds", Dekker, New York, 1973 and "Geometry of Submanifolds and Its Applications", Science University of Tokyo, 1981), *the author has written the present work as an invitation for further research on the topics that it deals with.* The World Scientific Publ. Co. could hardly have taken a better start for its new Series in Pure Mathematics than by editing Professor Bang-Yen Chen's beautiful new book as its Volume 1". It is a real joy in the new edition to see *how seriously indeed this invitation has been taken at heart by many geometers from all around the world* and *how the author's own thoughts on these topics have deepened and expanded in the mean time.* And let me be permitted to consider the 5-minimal curve that is shown on the cover, which has become the logo of PADGE right from the moment that it was "drawn" by Professors Bang-Yen Chen and Franki Dillen, as a token of the close co-operation of Professor Chen with many members of the KU Leuven and former KU Brussel's Section of Geometry over a period of about forty years now and also as a tribute to the memory of Professor Dillen.

The chapters 5 and 6 on *Total Mean Curvature* and on *Submanifolds of Finite Type* of the first edition now have turned into somewhat adapted new chapters 5 and 6 with the same title and now extra chapters 7, 8, 9 and 10 on *Biharmonic Submanifolds and Biharmonic Conjectures, λ-biharmonic and Null 2-type Submanifolds, Applications of Finite Type Theory* (a.o. including the presentation of a new variational principle of submanifolds for which the curve on the cover arose as a very particular example) and *Additional Topics in Finite Type Theory* were added in the new edition. The chapters 1, 2, 3 and 4 on *Differentiable Manifolds, Riemannian and Pseudo-Riemannian Manifolds, Hodge Theory and Spectral Geometry* and *Submanifolds* of the new edition are reworked and where appropriate expanded versions of the first four chapters of the first edition. And, from beginning to end, the author made the effort to include many historical notes and observations which are very welcome for a better comprehensive readability as such and for well allowing to put the presented materials in their wider mathematical and scientifical contexts.

Professor Bang-Yen Chen is one of the leading experts of *the general geometry of submanifolds* of our time. The new book very nicely presents the just listed really important topics in this field, whereas his previous book *"Pseudo-Riemannian Geometry, δ-Invariants and Applications"* (World Scientific Publ. Co., Singapore, 2011) similarly presented equally important such topics related to *the Chen curvatures*. And for a survey of this field as a whole, see Professor Chen's contribution on *"Riemannian Submanifolds"* in the *"Handbook of Differential Geometry, Vol. 1"* (Elsevier Science, Amsterdam, 2000, edited by Franki Dillen e.a.).

The first edition of the present book, over the years, for many geometers has been their first real meeting with basic differential geometry and in particular with the geometry of submanifolds. For the coming generations of mathematicians and of scientists and of engineers, (for the side of applications thinking on the relevance per se of the geometry of the position vector field and of the submanifolds' surface tensions a.o. and i.p. on the interest of various aspects of pseudo Riemannian geometry), I do hope that this would also be the case for the new edition. Anyway, for the professional mathematicians as well as for all other readers of this book, and luckily for them in a very small format, and, of course, only for whatever it may be worth, hereafter I would like to formulate a personal view on the geometry of submanifolds today.

From Jacob Bronowski's "The Origins of Knowledge and Imagination" come the following quotes: *"The place of sight in human evolution is cardinal"* and *"The world of science is wholly dominated by the sense of sight"*, and in the same author's "The Ascent of Man" one finds a discussion of *the classical theorem of Pythagoras* as wonderful connection between the two.

Euclid's *"Elements"* (~ -300) concerned the state of geometry at that time as science of our environment as experienced by our visual and motoric senses, presented in the axiomatic-deductive trend which aims for security of the mathematical activities, as protection against otherwise mostly intuitive and eventually too loose proceedings. Likely "because it involves a happening taking place at infinity and because our kind is not able to see so far", the parallel postulate of the Elements' planar Euclidean geometry inevitably, right away was pretty intriguing for many a scholar. Simon Stevin's *"De Thiende"* (1585) and *"De Beginselen der Weeghkonst"* (1586), respectively conclusively dealing with all operations involving real numbers in terms of calculations only involving natural numbers via the decimal system and introducing "the rule of the parallelogram" for the addition of vectors, paved the way for Descartes' *"Géométrie"* (1637) of which the programme was to build up the whole 2D Euclidean geometry based only on the determination of the distances of pairs of points by means of *the theorem of Pythagoras expressed in a 2D Cartesian co-ordinate system*: for co-ordinate axes x and y enclosing an angle θ, the distance Δs between points (x, y) and $(x + \Delta x, y + \Delta y)$ being given by $\Delta s^2 = \Delta x^2 + 2\cos\theta\,\Delta x\,\Delta y + \Delta y^2$; hereby the delicate role played by the axiomatical foundation of synthetic geometry basically was taken over by the geometrical foundation of the real number system, and moreover this programme works for all dimensions alike. Thereupon, the infinitesimal calculus could geometrically be developed and in turn this allowed Newton ($\sim$ 1670) to analytically determine *the curvature of the curves in a Euclidean plane* $\mathbb{E}^2$ at any of their points and Euler ($\sim$ 1760) to determine *the curvature behavior of the surfaces* M^2 *in a Euclidean space* $\mathbb{E}^3$ at any of their points in terms of the curvatures there of the normal sections in all tangent directions to M^2.

In his *"Disquisitiones generalis circa superficies curvas"* (1827), Gauss carried over Descartes' programme to the inner geometry of surfaces M^2 in $\mathbb{E}^3$. The surfaces are described by *curvilinear*, say (u, v) *co-ordinates* and *"a geometrical structure"* is defined on these surfaces M^2 by their *line element* ds as expressed by the infinitesimal distance function which is naturally induced on these surfaces M^2 from the standard theorem of Pythagoras' Euclidean distance function of the ambient space $\mathbb{E}^3$, i.e. via *a generalised theorem of Pythagoras* on M^2, namely, ds^2 is given by a general homogeneous quadratic polynomial in infinitesimal changes of the curvilinear co-ordinates: $ds^2 = E\,du^2 + 2F\,du\,dv + G\,dv^2$. And, Gauss could prove his "Krümmung" K to depend only on E, F and G, i.e. his *theorema egregium* stating the invariance of K under all surface isometries, after having shown that K equals the product of the principal Euler curvatures, whereupon he could make the fundamental distinction between *the intrinsic and the extrinsic geometries of surfaces* M^2 *in* $\mathbb{E}^3$. In this setting finally the parallel postulate problem could be resolved, since, locally, *the 2D hyperbolic geometry of Lobachevsky-Bolyai is realised as the intrinsic geometry on the pseudo-spheres or tractroids* which are "concrete" surfaces with constant $K < 0$ *in* $\mathbb{E}^3$. The intimate link between the validity of this non-Euclidean geometry on the one hand and the validity of Euclidean geometry itself on the other brought along the profound and quite revolutionary re-evaluation of the whole field of mathematical logic on which, till at present, we can pretty well base actual confidence when doing our mathematical doings.

And in *"Ueber die Hypothesen,* and respectively *Tatsachen, welche der Geometrie zu Grunde liegen"* (of 1854, published in 1866, and respectively 1868) Riemann, and respectively Helmholtz essentially followed the same programme to come for arbitrary dimensions n to Riemann-Finsler geometry, and respectively proper *Riemannian geometry.* They both started from n-fold extended "Mannigfaltigkeiten" M^n for their basic spaces, i.e. both used *systems of local co-ordinates* $(x^1, x^2, \ldots, x^n)$, and where Riemann then by hypothesis set off with a general Riemann-Finsler metric which, to get more explicit and to go for most possible simplicity in his further exposition, he then specialised to a *positive definite metric tensor* $g = g_{hk} dx^h dx^k$, Helmholtz straightforwardly came up with a squared line element ds^2 given by such a homogeneous quadratic polynomial in the infinitesimal changes of the co-ordinates, i.e. given by a generalised theorem of Pythagoras, because he found this to be the only factual possibility to allow for reasonable measurements of distances. The intrinsic geometry of surfaces M^2 in E^3 thus having been the inspiration for the wider 2D Riemannian geometry, cfr. e.g. Vincent Borrelli e.a.'s "tore plat en 3D", may show how human's geometrical experiences in "their space" actually have influenced the consideration of abstract Riemannian spaces (M^n, g) of arbitrary dimensions, Riemann and Helmholtz hereto moreover having been motivated by their reflections on physics and on human vision, respectively.

Next, in his "Raum und Zeit" (1908), Minkowski made the extension of our natural measure of distances in space (x, y, z) to a corresponding natural measure of distances in spacetime $(x, y, z; t)$ by *an indefinite revision of the classical theorem of Pythagoras,* as supported by the physical "Weltpostulat"as well as by our psychologically different appreciations of time and of location, namely, to the indefinite 4D *Minkowski metric* of index 1: $ds^2 = dx^2 + dy^2 + dz^2 - dt^2$; (likely hereby the three letter combination "ict" showed up for the first and best time, and as Minkowski put it: "Man kann danach das Wesen dieses Postulates mathematisch sehr prägnant in die magische Formel kleiden: 300000 km $=$ i sek"). And thus could begin the development of nD *pseudo* or *semi Riemannian geometry* along the lines of proper definite Riemannian geometry. The proper Euclidean, respectively the proper Riemannian spaces, in some sense conversely, then are the pseudo Euclidean, respectively the pseudo Riemannian spaces of index 0.

Finally, *the isometrical embedding theorems* of Nash (1956) and of Clarke and Greene (1970) state that *every abstract nD pseudo Riemannian manifold can be isometrically embedded into $(n + m)D$ pseudo Euclidean spaces with appropriately large co-dimensions m*, (Nash having done the embedding of Riemannian manifolds in Euclidean spaces). Thus, pseudo Riemannian geometry essentially is equivalent with the intrinsic geometry of the pseudo Riemannian submanifolds of pseudo Euclidean spaces, and, as such, is part of *the geometry of submanifolds of pseudo Euclidean spaces* which itself can be seen for arbitrary dimensions and co-dimensions to correspond in our natural imagination to the abstraction of our basic static and dynamic visual sense-experiences of "the real curves and the real surfaces that we do encounter in our real worlds".

In conclusion, the just sketched development of geometry may show it not to be so amiss to consider the general theory of submanifolds as the geometry of the human kind, from which perspective it moreover may be well in return to look back at Bronowski's quotes and discussion that were mentioned above by way of introduction to this sketch.

Leopold Verstraelen

The Research Center of
the Serbian Academy of Sciences and Arts
and the University of Kragujevac

31 - 05 - 2014

Preface

It has been known for a long time that the curvature of a closed surface is related to a topological invariant, namely, the Euler characteristic, via the well-known Gauss-Bonnet theorem. Since then it was well understood that integrals of curvature invariants play very important roles in many different aspects of Riemannian geometry, such as the index theorem, heat equation, volume of tubes and geodesic balls, submanifolds theory, spectral geometry, etc.

The concept of mean curvature was first introduced by Marie-Sophie Germain (cf. [Germain (1831)]). In the early 1820s, S. Germain (1776-1831) and S. D. Poisson (1781-1840) applied total mean curvature to describe elastic shells. Total mean curvature for surfaces in Euclidean 3-space was investigated again in the 1920s by Blaschke's school, and later by T. J. Willmore in the middle of the 1960s.

In order to understand the total mean curvature for general Euclidean submanifolds, the author introduced the notions of order and type for Euclidean submanifolds in the late 1970s. By applying such notions, he introduced the notions of finite type submanifolds and finite type maps. The study of finite type submanifolds and finite type maps provides a very natural way to relate the geometry of Riemannian manifolds with the spectral behaviors of the Riemannian manifolds via their immersions. Thus one often gets useful information on eigenvalues of a Riemannian manifold which can always be isometrically imbedded in a Euclidean space according to Nash's imbedding theorem. By showing that a certain submanifold is of k-type, one can, in principle, determine k eigenvalues of the Laplacian from the roots of its minimal polynomial.

A submanifold M of a (pseudo) Euclidean is called biharmonic if each (pseudo) Euclidean coordinate function of M is a biharmonic function.

The study of biharmonic submanifolds was initiated by the author in the middle of the 1980s in his program of understanding the theory of finite type submanifolds. Independently, biharmonic submanifolds as biharmonic maps were also investigated by G.-Y. Jiang in his study of Euler-Lagrange's equation of bienergy functional. A long standing conjecture of the author states that minimal submanifolds are the only biharmonic submanifolds in Euclidean spaces. The study of biharmonic submanifolds is nowadays a very active research subject. In particular, since 2000 biharmonic submanifolds have been receiving a growing attention and have become a popular research subject of study with many important progresses.

The major results, up to the early 1980s, on total mean curvature and finite type submanifolds were collected in the first edition of author's book *"Total Mean Curvature and Submanifolds of Finite Type"* published about 30 years ago. Since then there are numerous important developments on both subjects. In this second edition, we present numerous important new results on both subjects, including recent developments on biharmonic submanifolds and biharmonic conjectures as well as the recent solution of Willmore conjecture by F. C. Marques and A. Neves, developed after the publication of the first edition.

This book attempts to strike a balance between giving detailed proofs of basic results and stating many results whose proofs would take us too far afield. The author also made the effort to include many historical notes and observations for a better comprehensive readability. It is the author's hope that the readers will find this second edition a good introduction to both subjects of total mean curvature and finite type theory; providing the necessary background as well as a useful reference to recent and further research of both subjects.

In concluding the preface, the author would like to thank World Scientific Publishing for the invitation to undertake this project. He also would like to express his many appreciation to Professors David E. Blair, Ivko Dimitric, Oscar J. Garay, Shun Maeta, Stefano Montaldo, Ye-Lin Ou, Mira Petrović-Torgašev, Bogdan D. Suceavă, Joeri Van der Veken and Luc Vrancken for reading parts of the manuscript and offering many valuable suggestions. In particular, the author thanks Professor Leopold Verstraelen for writing an excellent foreword for this second edition.

May 31, 2014

B.-Y. Chen

Contents

Chapter 1

Differentiable Manifolds

A differentiable manifold is a type of manifold that is locally similar enough to a Euclidean space to allow one to do calculus. Any manifold can be described by a collection of charts. One may then apply ideas from calculus while working within the individual charts, since each chart lies within a linear space to which the usual rules of calculus apply. If the charts are suitably compatible, then computations done in one chart are valid in any other differentiable chart.

With his 1827 fundamental work

"Disquisitiones generales circa superficies curvas",

C. F. Gauss (1777-1855) demonstrated the existence of an intrinsic geometry of surfaces based only on the measurements of the lengths of arcs in these surfaces. As such, the 2-dimensional Euclidean geometry was generalized to a very large class of surfaces. Furthermore, for the first time Gauss made use of the so-called curvilinear coordinates to describe surfaces, which is the origin of a local chart around a point on a manifold.

In his famous habilitation lecture at Göttingen

"Über die Hypothesen welche der Geometrie zu Grunde liegen",

B. Riemann (1826-1866) discussed the foundations of geometry, introduced n-dimensional manifolds, formulated the concept of Riemannian manifolds and defined their curvature.

The works of physicist J. C. Maxwell (1831-1879) and mathematicians G. Ricci (1853-1925) and T. Levi-Civita (1873-1941) led to the development of tensor analysis and the notion of covariance, which identifies an intrinsic geometric property as one that is invariant with respect to coordinate transformations. These ideas had important applications in Einstein's theory of general relativity.

A modern definition of a 2-dimensional manifold was given by H. Weyl (1885-1955) in his 1913 book on Riemann surfaces [Weyl (1913)] and also by O. Veblen and J. H. C. Whitehead in [Veblen and Whitehead (1932)]. The widely accepted general definition of a manifold in terms of an atlas is due to H. Whitney (1907-1989) in [Whitney (1936)].

1.1   Tensors

Let U, V and W be vector spaces over a field $\mathbb{F}$ of dimensions m, n and r, respectively. A map f of $U \times V$ into W is called *bilinear* if it is linear in each variable separately, i.e., if

$$
\begin{aligned}
f(a_1 u_1 + a_2 u_2, b_1 v_1 + b_2 v_2) = {} & a_1 b_1 f(u_1, v_1) + a_1 b_2 f(u_1, v_2) \\
& + a_2 b_1 f(u_2, v_1) + a_2 b_2 f(u_2, v_2)
\end{aligned}
\tag{1.1}
$$

for all vectors $u_1, u_1 \in U$, $v_1, v_2 \in V$ and scalars $a_1, a_2, b_1, b_2 \in \mathbb{F}$.

More generally, let $V_1, \ldots, V_k$ be k finite dimensional vector spaces over a field $\mathbb{F}$. A map f of $V_1 \times \cdots \times V_k$ into a vector space W is called *multilinear* if it is linear in each of the variables separately.

We denote by $\mathrm{Hom}(V, W)$ the space of linear maps from V to W. The $\mathrm{Hom}(V, W)$ is a vector space over $\mathbb{F}$ of dimension nr. Let V^* denote the space of all linear functions on V, i.e., $V^* = \mathrm{Hom}(V, \mathbb{F})$. V^* is called the *dual space* of V.

We define the map $\varphi : V \times W \to \mathrm{Hom}(V^*, W)$ as follows: To each $(v, w) \in V \times W$, we assign a linear map $\varphi(v, w)$ by

$$
\varphi(v, w) v^* = v^*(v) w
\tag{1.2}
$$

for $v^* \in V^*$. It is easy to see that $\varphi : V \times W \to \mathrm{Hom}(V^*, W)$ is bilinear. Moreover, if $v_1, \ldots, v_n$ form a basis of V and $w_1, \ldots, w_r$ form a basis of W, then $\varphi(v_i, w_j)\,(i = 1, \ldots, n; j = 1, \ldots, r)$ form a basis of $\mathrm{Hom}(V^*, W)$.

Let f be a bilinear map of $U \times V$ into W. We denote a linear map

$$
\alpha_f : \mathrm{Hom}(U^*, V) \to W
\tag{1.3}
$$

by $\alpha_f(\varphi(u_i, v_j)) = f(u_i, v_j)$, where $u_1, \ldots, u_m$ is a basis of U and $v_1, \ldots, v_n$ is a basis of V. Then (1.3) defines the map α_f on the basis $\{\varphi(u_i, v_j)\}_{i,j}$ of $\mathrm{Hom}(U^*, V)$. We then extend α_f to all of $\mathrm{Hom}(U^*, V)$ so as to be linear. One may verify that the linear map α_f is in fact independent of the choice of basis $u_1, \ldots, u_m$ and $v_1, \ldots, v_n$. Consequently, to each bilinear map f from $U \times V$ into W, we have associated a linear map α_f of $\mathrm{Hom}(U^*, V)$ into W so that $f = \alpha_f \circ \varphi$.

Let $\bar{U}$ be the free vector space whose generators are the elements of $V_1 \times \cdots \times V_k$, i.e., $\bar{U}$ is the set of all finite linear combinations of symbols of the form $(v_1, \ldots, v_k)$ with $v_i \in V_i$.

Let N be the vector subspace of $\bar{U}$ spanned by elements of the form:

$$a(v_1, \ldots, v_k) - (v_1, \ldots, av_i, \ldots, v_k),$$

$$(v_1, \ldots, v_i + w_i, \ldots, v_k) - (v_1, \ldots, v_k) - (v_1, \ldots, w_i, \ldots, v_k).$$

We denote by $V_1 \otimes \cdots \otimes V_k$ the factor space $\bar{U}/N$. This vector space is called the *tensor product* of $V_1, \ldots, V_k$. We define a multilinear map φ of $V_1 \times \cdots \times V_k$ into $V_1 \otimes \cdots \otimes V_k$ by sending $(v_1, \ldots, v_k)$ into its coset *mod* N. We write

$$\varphi(v_1, \ldots, v_k) = v_1 \otimes \cdots \otimes v_k.$$

Let W be a vector space and $\psi : V_1 \times \cdots \times V_k \to W$ a multilinear map. We say that the pair (W, ψ) has the *universal factorization property* for $V_1 \times \cdots \times V_k$ if, for every vector space U and every multilinear map $f : V_1 \times \cdots \times V_k \to U$, there exists a unique linear map $h : W \to U$ such that $f = h \circ \psi$.

The following three propositions are well-known. For their proofs, see [Kobayashi and Nomizu (1963)].

Proposition 1.1. *The pair* $(V_1 \otimes \cdots \otimes V_k, \varphi)$ *has the universal factorization property for* $V_1 \times \cdots \times V_k$. *If a pair* (W, ψ) *has the universal factorization property for* $V_1 \times \cdots \times V_k$, *then* $(V_1 \otimes \cdots \otimes V_k, \varphi)$ *and* (W, ψ) *are isomorphic in the sense that there exists a linear isomorphism* $\sigma : V_1 \otimes \cdots \otimes V_k \to E$ *such that* $\psi = \sigma \circ \varphi$.

Proposition 1.2. *If* $\{e_{i_r, r}\}$ *is a basis of* V_r, $1 \leq i_r \leq \dim V_r$, *then*

$$\{e_{i_1, 1} \otimes \cdots \otimes e_{i_k, k}\}$$

is a basis for $V_1 \otimes \cdots \otimes V_k$.

Proposition 1.3. *We have the following:*

(i) *There is a unique isomorphism of* $U \otimes V$ *onto* $V \otimes U$ *which send* $u \otimes v$ *into* $v \otimes u$ *for all* $u \in U$ *and* $v \in V$.

(ii) *There is a unique isomorphism of* $(U \otimes V) \otimes W$ *onto* $U \otimes (V \otimes W)$ *which sends* $(u \otimes v) \otimes w$ *onto* $u \otimes (v \otimes w)$ *for all* $u \in U$, $v \in V$ *and* $w \in W$.

(iii) *If* $U_1 \oplus U_2$ *denotes the direct sum of* U_1 *and* U_2, *then*

$$(U_1 \oplus U_2) \otimes V = U_1 \otimes V \oplus U_2 \otimes V,$$

$$U \otimes (V_1 \oplus V_2) = U \otimes V_1 \oplus U \otimes V_2.$$

1.2 Tensor algebra

Let V be a finite dimensional vector space and V^* the dual space of V, i.e., the space of all linear functions on V. If $v \in V$ and $w \in V^*$, we put

$$\langle v, w^* \rangle = w^*(v).$$

Let $\{e_1, \ldots, e_n\}$ be a basis of V and $\{e^{1^*}, \ldots, e^{n^*}\}$ the corresponding basis for V^* so that

$$\langle e_i, e^{j^*} \rangle = \delta_i^j = \begin{cases} 1 & \text{if } i = j; \\ 0 & \text{if } i \neq j, \end{cases}$$

where δ_i^j are the Kronecker deltas.

We want to study those spaces of the form $v_1 \otimes \cdots \otimes v_k$ where each of V_i is either V or V^*. If there are r copies of V and s copies of V^*, then the space is called a *space of type* (r, s); r is the *contravariant degree* and s the *covariant degree*. Given two tensor spaces U of type (r, s) and V of type (p, q), the associative law for tensor products defines a tensor space of type $(r + p, s + q)$. We consider the ground field $\mathbb{F}$ as a tensor space of type $(0, 0)$. The tensor product defines a multiplicative structure on the weak direct sum of all tensor products of V and V^*. We denote this space by $T(V)$, i.e.,

$$T(V) = \mathbb{F} + V + V^* + V \otimes V + V \otimes V^* + V^* \otimes V$$
$$+ V^* \otimes V^* + \cdots .$$

The $T(V)$ with its multiplicative structure is called the *tensor algebra* of the vector space V.

We shall give the expressions of tensors with respect to a basis of V. Let $e_1, \ldots, e_n$ be a basis of V and let $e^{1^*}, \ldots, e^{n^*}$ be its dual basis. Then, by Proposition 1.2, $\{e_{i_1} \otimes \cdots \otimes e_{i_r}\}$ is a basis of $V \otimes \cdots \otimes V$ (r copies). We denote this space by V^r.

Every contravariant tensor K of degree r can be expressed uniquely as a linear combination:

$$K = \sum_{i_1 \cdots i_r} K^{i_1 \cdots i_r} e_{i_1} \otimes \cdots \otimes e_{i_r},$$

where $K^{i_1 \cdots i_r}$ are the components of K with respect to the basis $e_1, \ldots, e_n$ of V. Similarly, every covariant tensor L of degree s can be expressed as a linear combination:

$$L = \sum_{j_1 \cdots j_s} L_{j_1 \cdots j_s} e^{j_1^*} \otimes \cdots \otimes e^{j_s^*},$$

where $L_{j_1 \cdots j_s}$ are the components of L.

We define the notion of *contraction* as follows. Let $U = V_1 \otimes \cdots \otimes V_k$ with $V_i = V$ and $V_j = V^*$. Let U' be the tensor product of all the factors of U in the same order omitting v_i and v_j. The map of $V_1 \times \cdots \times V_k$ into U' defined by

$$(v_1, \ldots, v_k) \mapsto \langle v_i, v_j \rangle v_1 \otimes \cdots \otimes \hat{v}_i \otimes \cdots \otimes \hat{v}_j \otimes \cdots \otimes v_k$$

is bilinear. So it defines a map of U into U', which is called the *contraction* with respect to i and j.

The word *tensor* was first introduced by William Rowan Hamilton (1805-1865) in the middle of the nineteenth century to describe something different from what is now meant by a tensor. The contemporary usage of tensor was introduced by Woldemar Voigt (1850-1919) in [Voigt (1898)].

1.3   Exterior algebra

Let V be an n-dimensional vector space over a field $\mathbb{F}$. Denote by π_r the permutation group on r letters. Then π_r acts on $V^r = V \otimes \cdots \otimes V$ (r copies) as follows: Given any permutation $\sigma \in \pi_r$ and any tensor of the form $v_1 \otimes \cdots \otimes v_r$ in V^r, we define

$$\sigma(v_1 \otimes \cdots \otimes v_r) = v_{\sigma(1)} \otimes \cdots \otimes v_{\sigma(r)}.$$

We extend it by linearity to all of V^r.

A tensor K in V^r is called *symmetric* if $\sigma(K) = K$ for each permutation $\sigma \in \pi_r$. K is called *skew-symmetric* if $\sigma(K) = (\text{sign}\,\sigma)K$ for each $\sigma \in \pi_r$, where $\text{sign}\,\sigma$ is either 1 or -1 according to σ is an even permutation or odd permutation.

For any $K \in V^r$, we introduce the following two operations:

$$S_r(K) = \frac{1}{r!} \sum_{\sigma \in \pi_r} \sigma(K), \tag{1.4}$$

$$A_r(K) = \frac{1}{r!} \sum_{\sigma \in \pi_r} (\text{sign}\,\sigma)\sigma(K). \tag{1.5}$$

Since $S_r \circ \sigma = \sigma \circ S_r = S_r$ and $A_r \circ \sigma = \sigma \circ A_r = (\text{sign}\,\sigma)A_r$, $S_r(K)$ is a symmetric tensor and $A_r(K)$ is a skew-symmetric tensor. S_r is called the *symmetrization* and A_r the *alternation*.

It is easy to verify that the alternation $A_r : V^r \to V^r$ is linear. Let us denote the kernel of A_r by N^r. Then we have a natural isomorphism:

$$V^r / N^r \cong A_r(V^r).$$

We denote V^r/N^r by $\wedge^r(V)$. The elements of $\wedge^r(V)$ are called *r-vectors*. We define a multiplication on

$$\wedge(V) = \wedge^0(V) + \wedge^1(V) + \wedge^2(V) + \cdots$$

by $\alpha \wedge \beta = A_{r+s}(\alpha \otimes \beta)$ for $\alpha \in \wedge^r(V)$ and $\beta \in \wedge^s(V)$. Then $\alpha \wedge \beta \in \wedge^{r+s}(V)$. The sign of the permutation on $r+s$ letters which moves the first r letters pass the last s letters is $(-1)^{rs}$. Thus we have

$$\alpha \wedge \beta = (-1)^{rs}\beta \wedge \alpha. \tag{1.6}$$

The $\alpha \wedge \beta$ defined above is called the *wedge* (or *exterior*) *product* of α and β. It is direct to verify that $\wedge(V)$, with the wedge product, is an associative algebra, called the *exterior* or *Grassmann algebra* of V. This notion was introduced by Hermann Grassmann (1809-1877) in [Grassmann (1844)].

If $\{e_1,\ldots,e_n\}$ is a basis of V, then the exterior algebra $\wedge(V)$ is of dimension 2^n. Moreover, 1 and the elements

$$e_{i_1} \wedge \cdots \wedge e_{i_r}, \ \ 1 \le i_1 < \cdots < i_r \le n, \ \ r = 1,\ldots,n,$$

form a basis of $\wedge(V)$.

Proposition 1.4. *Let* $v_1,\ldots,v_r$ *be* r *vectors in a vector space* V. *Then* $v_1,\ldots,v_r$ *are linearly independent if and only if*

$$v_1 \wedge \cdots \wedge v_r \ne 0. \tag{1.7}$$

Proof. If $v_1,\ldots,v_r$ are linearly dependent, we can express one of them say v_r as $v_r = \sum_{i=1}^{r-1} a_i v_i$ for some scalars $a_1,\ldots,a_{r-1}$. Thus we have

$$v_1 \wedge \cdots \wedge v_r = v_1 \wedge \cdots \wedge v_{r-1} \wedge \sum_{i=1}^{r-1} a_i v_i = 0.$$

On the other hand, if $v_1,\ldots,v_r$ are linearly independent vectors, we can always find $v_{r+1},\ldots,v_n$ such that $v_1,\ldots,v_n$ form a basis of V. Hence we have $v_1 \wedge \cdots \wedge v_r \ne 0$. $\qquad\square$

We also need the following proposition for later use.

Proposition 1.5. (Cartan's Lemma) *Assume that* $v_1,\ldots,v_r$ *are linearly independent in* V *and* $w_1,\ldots,w_r$ *are* r *vectors in* V. *If we have*

$$\sum_{i=1}^{r} w_i \wedge v_i = 0, \tag{1.8}$$

then

$$w_i = \sum_{j=1}^{r} a_{ij}v_j, \ \ i = 1,\ldots,r, \tag{1.9}$$

for some scalars a_{ij} *satisfying* $a_{ij} = a_{ji}$.

Proof. Let $\{v_1, \ldots, v_n\}$ be a basis of V. We can write w_i as

$$w_i = \sum_{j=1}^{r} a_{ij} v_j + \sum_{j=r+1}^{n} b_{ij} v_j.$$

Thus, by (1.8), we find

$$\sum_{i<j\leq r} (a_{ij} - a_{ji}) v_i \wedge v_j + \sum_{i\leq r<j} b_{ij} v_i \wedge v_j = 0,$$

which imply $a_{ij} = a_{ji}$ and $b_{ij} = 0$. $\qquad\square$

From the definition of wedge product we obtain

$$(\theta \wedge \omega)(X, Y) = \frac{1}{2} \{\theta(X)\omega(Y) - \omega(X)\theta(Y)\} \qquad (1.10)$$

for $X, Y \in V$ and $\theta, \omega \in V^*$.

For each vector $X \in V$, we define the *interior product* ι_X with respect to X by

(a) $\iota_X a = 0$ for each $a \in \wedge^0(V)$ and

(b) $(\iota_X \omega)(Y_1, \ldots, Y_{r-1}) = r\omega(X, Y_1, \ldots, Y_{r-1})$ for $Y_1, \ldots, Y_{r-1} \in V$ and $\omega \in \wedge^r(V^*)$, where V^* is the dual space of V.

Remark 1.1. It is easy to verify that ι_X is a skew-derivation of $\wedge(V^*)$ into itself, i.e.,

$$\iota_X(\omega \wedge \omega') = (\iota_X \omega) \wedge \omega' + (-1)^r \omega \wedge \iota_X \omega'$$

for $\omega \in \wedge^r(V^*)$.

1.4 Differentiable manifolds

A topological space is called *separable* if it contains a countable dense subset, i.e., there exists a sequence $\{C_i\}_{i=1}^{\infty}$ of elements of the space such that every nonempty open subset of the space contains at least one element of the sequence. Let M be a separable topological space. We assume that M satisfies the *Hausdorff separation axiom* which states that any two distinct points can be separated by disjoint open sets. By an *open chart* on M we mean a pair (U, Φ), where U is an open subset of M and Φ is a homeomorphism of U onto an open subset of a Euclidean n-space $\mathbb{E}^n$.

A Hausdorff space M is said to have a *differentiable structure* of dimension n if there is a collection of open charts (U_α, Φ_α), where α belongs to some indexing set A such that the following three conditions are satisfied:

(a) $M = \cup_{\alpha \in A} U_\alpha$, i.e., $(U_\alpha)_{\alpha \in A}$ is an open covering of M.

(b) For any $\alpha, \beta \in A$, $\Phi_\beta \circ \Phi_\alpha^{-1}$ is a differentiable map of $\Phi_\alpha(U_\alpha \cap U_\beta)$ onto $\Phi_\beta(U_\alpha \cap U_\beta)$.

(c) The collection $(U_\alpha, \Phi_\alpha)_{\alpha \in A}$ is a maximal family of open charts which satisfy both conditions (a) and (b).

By "differentiable" in (b) we mean differentiable of class C^∞ unless mentioned otherwise.

By a *differentiable manifold* of dimension n we mean a Hausdorff space together with a differentiable structure of dimension n. For simplicity, we call a differentiable manifold simply a *manifold*. Let (U, Φ) be an open chart of a manifold M of dimension n. Denote by $x_1, \ldots, x_n$ a Euclidean coordinate system of $\mathbb{E}^n$. The system of functions $x_1 \circ \Phi, \ldots, x_n \circ \Phi$ is called a *local coordinate system* on U and U is called a *coordinate neighborhood*. In the definition, if $\mathbb{E}^n$ were replaced by the complex Euclidean n-space $\mathbb{C}^n$, then M is called a *complex manifold* of n complex dimension.

A map $\phi : M \to N$ between two manifolds is called *differentiable* if for every chart (U_α, Φ_α) on M and every chart (V_β, Ψ_β) on N with $\phi(U_\alpha) \subset V_\beta$, the map $\Psi_\beta \circ \phi \circ \Phi_\alpha^{-1}$ of $\Phi_\alpha(U_\alpha)$ into $\Psi_\beta(V_\beta)$ is differentiable.

Let $u_1, \ldots, u_n$ be a local coordinate system on U_α and $y_1, \ldots, y_m$ a local coordinate system on V_β. If ϕ is a differentiable map of M into N, then locally ϕ can be expressed by a set of differentiable functions:

$$y_1 = y_1(u_1, \ldots, u_n), \ldots, y_m = y_m(u_1, \ldots, u_n).$$

By a differentiable map of a closed interval $[a, b]$ into a manifold M, we mean the restriction of a differentiable map of an open interval $I \supset [a, b]$ into M. By a (differentiable) curve in M we mean a differentiable map of a closed interval into M.

Let $\mathcal{F}(p)$ be the algebra of differentiable functions in a neighborhood of a point p. Let $\gamma(t)$ be a differentiable curve in M with $\gamma(t_0) = p$. The vector tangent to the curve $\gamma(t)$ at p is a map $X : \mathcal{F}(p) \to \mathbb{R}$ defined by

$$Xf = \frac{df(\gamma(t))}{dt}\bigg|_{t=t_0}.$$

In other words, Xf is the derivative of f in the direction of the curve $\gamma(t)$ at $t = t_0$. The vector X satisfies the following two conditions:

(1) X is a linear map of $\mathcal{F}(p)$ into $\mathbb{R}$.

(2) $X(fg) = (Xf)g(p) + f(p)Xg$ for $f, g \in \mathcal{F}(p)$.

The set of maps X of $\mathcal{F}(p)$ into $\mathbb{R}$ satisfying these two conditions forms a real vector space. Let $u_1, \ldots, u_n$ be a local coordinate system defined in a coordinate neighborhood U of p. For each i, $(\partial/\partial u_i)_p$ is a map from $\mathcal{F}(p)$ into $\mathbb{R}$ satisfying conditions (1) and (2) above. We shall show that the set of vectors at p forms the vector space with basis $(\partial/\partial u_1)_p, \ldots, (\partial/\partial u_n)_p$.

Given a curve $\gamma(t)$ with $\gamma(t_0) = p$, let $u_i = \gamma_i(t)$ be its equations in terms of $u_1, \ldots, u_n$. Then we have

$$\left(\frac{df(\gamma(t))}{dt} \right)_{t_0} = \sum_{i=1}^{n} \left(\frac{\partial f}{\partial u_i} \right)_p \left(\frac{d\gamma_i(t)}{dt} \right)_{t_0}.$$

Thus every vector at p is a linear combination of $(\partial/\partial u_1)_p, \ldots, (\partial/\partial u_n)_p$.

Conversely, for a given linear combination $X = \sum_i a_i \, (\partial/\partial u_i)_p$, let us consider the curve defined by

$$u_i = u_i(p) + a_i t, \quad i = 1, \ldots, n.$$

Then X is the vector tangent to this curve at $t = 0$.

To prove the linear independence, we assume that $\sum_i a_i \, (\partial/\partial u_i)_p = 0$. Then we have

$$0 = \sum_i a_i \left(\frac{\partial u_j}{\partial u_i} \right)_p = a_j, \quad j = 1, \ldots, n.$$

Consequently, we obtain the following.

Proposition 1.6. *Let M be an n-dimensional manifold and $p \in M$. If $u_1, \ldots, u_n$ is a local coordinate system on a coordinate neighborhood containing p, then the set of vectors at p tangent to M is an n-dimensional vector space over $\mathbb{R}$ with basis $(\partial/\partial u_1)_p, \ldots, (\partial/\partial u_n)_p$.*

We denote by $T_p M$ the vector space tangent to M at p, which is called the *tangent space* of M at p. Its elements are called tangent vectors of M at the point p.

1.5   Vector fields and differential forms

Through out this book, by a *manifold* we mean a connected smooth manifold of dimension ≥ 2 unless mentioned otherwise. By a *closed manifold* we mean a compact manifold without boundary.

A *vector field* X on a manifold M is an assignment of a vector X_p to each point $p \in M$. If f is a differentiable function on M, then Xf is a function on M such that $(Xf)(p) = X_p f$. A vector field X is called *differentiable* if

Xf is differentiable for every differentiable function f on M. In terms of a local coordinate system $u_1, \ldots, u_n$, a differentiable vector field X can be expressed by

$$X = \sum_{i=1}^{n} X^i \frac{\partial}{\partial u_i},$$

where X^i are differentiable functions.

Let $\mathfrak{X}(M)$ be the set of all differentiable vector fields on M. Then $\mathfrak{X}(M)$ is a real vector space under the natural addition and scalar multiplication. Given two vector fields X, Y on M, we define the bracket $[X, Y]$ as a map from the ring of functions on M into itself by

$$[X, Y]f = X(Yf) - Y(Xf). \tag{1.11}$$

Then $[X, Y]$ is again a vector field on M. In terms of a local coordinate system $u_1, \ldots, u_n$, we write

$$X = \sum_{i=1}^{n} X^i \frac{\partial}{\partial u_i}, \quad Y = \sum_{j=1}^{n} X^j \frac{\partial}{\partial u_j}.$$

Then we have

$$[X, Y]f = \sum_{j,k=1}^{n} \left\{ X^k \frac{\partial Y^j}{\partial u_k} - Y^k \frac{\partial X^j}{\partial u_k} \right\} \frac{\partial}{\partial u_j}. \tag{1.12}$$

With respect to this bracket operation, $\mathfrak{X}(M)$ becomes a Lie algebra over $\mathbb{R}$ (of infinite dimension). In particular, we have the *Jacobi identity*:

$$[[X, Y], Z] + [[Y, Z], X] + [[Z, X], Y] = 0 \tag{1.13}$$

for $X, Y, Z \in \mathfrak{X}(M)$.

We may regard $\mathfrak{X}(M)$ as a module over the ring $\mathcal{F}(M)$ of differentiable functions on M as follows: If $f \in \mathcal{F}(M)$ and $X \in \mathfrak{X}(M)$, then we define fX by $(fX)_p = f(p)X_p$. We have

$$[fX, gY] = fg[X, Y] - f(Xg)Y - g(Yf)X$$

for $f, g \in \mathcal{F}(M)$ and $X, Y \in \mathfrak{X}(M)$.

For each point $p \in M$, we denote by $T_p^* M$ the dual space of the tangent space $T_p M$. Elements of $T_p^* M$ are called *covectors* at p. An assignment of a covector at each point $p \in M$ is called a 1-*form*.

For each $f \in \mathcal{F}(M)$, the *total differential df* of f is defined by

$$\langle (df)_p, X \rangle = Xf$$

for $X \in T_pM$. If $u_1, \ldots, u_n$ is a local coordinate system in M, then the total differentials $(du_1)_p, \ldots, (du_n)_p$ form a basis of T_p^*M. In fact, they form the dual basis of the basis $(\partial/\partial u_1)_p, \ldots, (\partial/\partial u_n)_p$ of T_pM.

In a coordinate neighborhood of $p \in M$, each 1-form ω can be expressed locally as

$$\omega = \sum_{i=1}^n \omega_i du_i.$$

The 1-form ω is called differentiable if its components ω_i are differentiable functions. It can be verified that this condition is independent of the choice of local coordinate system. Throughout this book we shall only consider forms which are differentiable. We denote by $\wedge^1(M)$ the space of all 1-forms on M.

Let $\wedge T_p^*(M)$ be the exterior algebra over $T_p^*(M)$. An r-form ω on M is an assignment of an element of degree r in $\wedge T_p^*(M)$ to each point $p \in M$. In terms of a local coordinate system $u_1, \ldots, u_n$, we have

$$\omega = \sum_{i_1 < i_2 < \cdots < i_r} \omega_{i_1 \cdots i_r} du_{i_1} \wedge \cdots \wedge du_{i_r}.$$

The r-form ω is called differentiable if its components $\omega_{i_1 \cdots i_r}$ are differentiable. By an r-from we shall always mean a differentiable r-form. We denote by $\wedge^r(M)$ the space of r-forms on M for $r = 0, 1, \ldots, n$. Each $\wedge^r(M)$ is a vector space over $\mathbb{R}$. We put $\wedge^0(M) = \mathcal{F}(M)$.

We set

$$\wedge(M) = \sum_{r=0}^n \wedge^r(M).$$

With respect to wedge product, $\wedge(M)$ is an algebra over $\mathbb{R}$.

Let d denote the *exterior differentiation*, which is characterized by the following four conditions:

(1) d is an $\mathbb{R}$-linear map of $\wedge(M)$ into itself with $d(\wedge^r(M)) \subset \wedge^{r+1}(M)$.

(2) For each $f \in \wedge^0(M)$, df is the total differential of f.

(3) If $\omega \in \wedge^r(M)$ and $\eta \in \wedge^s(M)$, then

$$d(\omega \wedge \eta) = d\omega \wedge \eta + (-1)^r \omega \wedge d\eta.$$

(4) $d^2 = 0$.

In terms of a local coordinate system $u_1, \ldots, u_n$ on M, if

$$\omega = \sum_{i_1 < i_2 < \cdots < i_r} \omega_{i_1 \cdots i_r} du_{i_1} \wedge \cdots \wedge du_{i_r},$$

then

$$d\omega = \sum_{i_1 < i_2 < \cdots < i_r} d\omega_{i_1 \cdots i_r} \wedge du_{i_1} \wedge \cdots \wedge du_{i_r}. \tag{1.14}$$

We mention the following result for later use. For its proof, see for instance, page 36 of [Kobayashi and Nomizu (1963)].

Proposition 1.7. *If ω is an r-form on M, then*

$$dw(X_0, X_1, \ldots, X_r) = \frac{1}{r+1} \sum_{i=0}^{r} (-1)^i X_i(\omega(X_0, \ldots, \hat{X}_i, \ldots, X_r))$$

$$+ \frac{1}{r+1} \sum_{0 \le i \le j \le r} (-1)^{i+j} \omega([X_i, X_j], X_0, \ldots, \hat{X}_i, \ldots, \hat{X}_j, \ldots, X_r). \tag{1.15}$$

In particular, if ω is a 1-form, we have

$$dw(X, Y) = \frac{1}{2}\{X\omega(Y) - Y\omega(X) - \omega([X, Y])\}. \tag{1.16}$$

For a given map ϕ from a manifold M into another manifold N, the *differential* of ϕ at a point $p \in M$ is the linear map $(\phi_*)_p$ of $T_p M$ into $T_{\phi(p)} N$ defined as follows:

For each $X \in T_p M$, we choose a curve $x(t)$ in M such that X is the vector tangent to $x(t)$ at $p = x(t_0)$. Then $(\phi_*)_p(X)$ is defined as the vector tangent to the curve $\phi(x(t))$ in N at $\phi(p) = \phi(x(t_0))$. Throughout this book we shall denote the differential of ϕ either by ϕ_* or by $d\phi$.

It is easy to verify that $(\phi_*)_p$ is independent of the choice of the curve and if g is a function in a neighborhood of $\phi(p)$, then $((\phi_*)_p X)g = X(g \circ \phi)$. The transpose of $(\phi_*)_p$ is a linear map of $T^*_{\phi(p)} N$ into $T^*_p M$.

For any r-form ω' on N, $\phi^* \omega'$ is an r-form on M defined by

$$(\phi^* \omega')(X_1, \ldots, X_r) = \omega'(\phi_* X_1, \ldots, \phi_* X_r) \tag{1.17}$$

for $X_1, \ldots, X_r \in T_p M$, $p \in M$. The exterior differentiation d commutes with ϕ^*, i.e.,

$$d \circ \phi^* = \phi^* \circ d. \tag{1.18}$$

Definition 1.1. A *submanifold* of a manifold M is a pair (N, ϕ), where ϕ is a differentiable map from a manifold N into M such that, for each point $p \in N$, $(\phi_*)_p$ is injective. In this case, ϕ is called an *immersion*. If, furthermore, ϕ is also injective, (N, ϕ) is called an *imbedded submanifold* of M and ϕ is an *imbedding*.

Definition 1.2. A map $\phi : N \to M$ is called *proper* if $\phi^{-1}(K)$ is compact for any compact subset K of M.

It is well-known that $\phi : N \to M$ is a proper imbedding if and only if there exists a covering of M by coordinate neighborhoods $\{U_\alpha\}$ such that $N \cap U_\alpha$ is defined by

$$y_1^\alpha(p) = \cdots = y_k^\alpha(p) = 0,$$

where $k = \dim M - \dim N$ and $y_1^\alpha, \ldots, y_n^\alpha$ are local coordinates on U_α.

1.6 Sard's theorem and Morse's inequalities

Given a map ϕ from an m-dimensional manifold M into an n-dimensional manifold N, the maximum rank that ϕ can have is the minimum of m and n. If $m < n$, the image of M is a lower dimensional object in N. In fact, no matter what values of m and n are, "in general" a point of N is not an image of a point of N when the rank of ϕ is less than n. The "generality" is in the sense of measure zero.

Definition 1.3. Let ϕ be a differentiable map of an m-dimensional manifold M into an n-dimensional manifold N. The points $p \in M$ where $\operatorname{rank}(\phi_*)_p < n$ are called the *critical points* of ϕ. All other points of M are called *regular points*. A point $q \in N$ such that $\phi^{-1}(q)$ contains at least one critical point is called a *critical value*. All other points of N are called *regular values*.

It is clear that if $\dim M < \dim N$, all points of M are critical. Moreover, if $q \in N$ does not lie in $\phi(M)$, it is a regular value.

We state Sard's theorem as follows.

Theorem 1.1. [Sard (1942)] *Let M and N be two manifolds of dimension m and n, respectively. If $\phi : M \to N$ is a differentiable map, then the critical values of ϕ form a set of measure zero.*

For the proof of Sard's theorem, see [Sard (1942); Sternberg (1964)].

Let M be an n-dimensional closed manifold and let f be a differentiable real-valued function on M. A point $p \in M$ is called a *critical point* of f if and only if $(f_*)_p = 0$. If we choose a local coordinate system $\{y_1, \ldots, y_n\}$ in M, this means that $(\partial f/\partial y_i)_p = 0$ for $i = 1, 2, \ldots, n$.

If p is a critical point of f, then the matrix

$$\left(\frac{\partial^2 f}{\partial y_j \partial y_i}\right)_p$$

represents a symmetric bilinear map f_{**} on the tangent space T_pM. A critical point p of f is called *non-degenerate* if f_{**} is non-degenerate at p. In this case, the dimension of a maximum dimensional subspace of T_pM on which f_{**} is negative-definite is called the *index* of the critical point p. A function f on M is called *non-degenerate* if all of its critical points are non-degenerate. A non-degenerate function is also called a *Morse function*. According to Sard's theorem almost all of functions on M are non-degenerate (except a set of measure zero).

We give the following notations:

$\Psi(M)$ = the set of non-degenerate functions on M;

$\beta_k(f)$ = the number of critical points of index k of f, $f \in \Psi(M)$;

$\beta(f) = \sum_{k=0}^n \beta_k(f)$, $n = \dim M$.

For any field $\mathbb{F}$, let $H_k(M; \mathbb{F})$ denote the k-th *homology group* of M with coefficients in $\mathbb{F}$. We put

$b_k(M; \mathbb{F}) = \dim H_k(M; \mathbb{F})$;

$b(M; \mathbb{F}) = \sum_{k=0}^n b_k(M; \mathbb{F})$;

$b(M) = \max\{b(M; \mathbb{F}) : \mathbb{F} \text{ fields}\}$.

The $b_k(M; \mathbb{F})$ is called k-*th Betti number* of M over the field $\mathbb{F}$ and $b(M)$ the *total Betti number*. Betti number was named by Henri Poincaré (1854-1912) after Enrico Betti (1823-1892).

We mention the following results for later use (cf. [Milnor (1963)]).

Theorem 1.2. (Weak Morse Inequalities) *Let M be a closed n-manifold. Then, for any field $\mathbb{F}$ and differentiable function f on M, we have*

$$\beta_k(f) \geq b_k(M; \mathbb{F}),$$

$$\sum_{k=0}^n (-1)^k \beta_k(f) = \sum_{k=0}^n (-1)^k b_k(M; \mathbb{F}) = \chi(M),$$

where $\chi(M)$ is the Euler characteristic of M.

Theorem 1.3. (Reeb's theorem) *Let M be a closed n-manifold. If there exists a differentiable function f on M with only two non-degenerate critical points, then M is homeomorphic to an n-sphere.*

Remark 1.2. Theorem 1.3 remains true even if the critical points are degenerate. However, it is not true that M must diffeomorphic to an n-sphere with its usual differentiable structure. In fact, J. W. Milnor constructed in [Milnor (1956)] a 7-sphere with non-standard differentiable structure which admits a differentiable function with two critical points.

1.7 Lie groups and Lie algebras

A *Lie group* G is a group which is at the same time a differentiable manifold such that the group operation $G \times G \ni (a, b) \mapsto ab^{-1} \in G$ is a differentiable map into G. A Lie group H is called a *Lie subgroup* of another Lie group G provided that H is both an abstract subgroup and an immersed submanifold of G.

Lie groups were named after Sophus Lie (1842-1899). The term *groupes de Lie* first appeared in 1893 in the thesis of Lie's student Arthur Tresse.

Definition 1.4. A *closed subgroup* H of a Lie group G is an abstract subgroup that is a closed subset of G.

The following is well-known.

Theorem 1.4. *If H is a closed subgroup of a Lie group G, then H is a Lie subgroup of G.*

Example 1.1. The set $\mathfrak{gl}(n; \mathbb{R})$ of all $n \times n$ real matrices is in a natural way a real vector space, and hence it is a manifold. The set $GL(n; \mathbb{R}) = \{g : \det g \neq 0\}$ of all invertible matrices in $\mathfrak{gl}(n; \mathbb{R})$ forms a group under matrix multiplication. The formula for the determinant of a matrix implies that the determinant function $\det : \mathfrak{gl}(n; \mathbb{R}) \to \mathbb{R}$ is a smooth map. Therefore $GL(n; \mathbb{R})$ is an open submanifold of $\mathfrak{gl}(n; \mathbb{R})$. The formulas for matrix multiplication and inverses show that for $GL(n; \mathbb{R})$, the group operation $G \times G \ni (a, b) \mapsto ab^{-1} \in G$ is a differentiable map. Hence $GL(n; \mathbb{R})$ is a Lie group, which is called the *general linear group*.

The determinant function $\det : GL(n; \mathbb{R}) \to \mathbb{R} \setminus \{0\}$ is a smooth homomorphism. It kernel $SL(n; \mathbb{R}) = \{a : \det a = 1\}$ is a closed subgroup of $GL(n; \mathbb{R})$, called the *special linear group*.

A *Lie algebra* over $\mathbb{R}$ is a real vector space $\mathfrak{g}$ equipped with a bilinear function $[\ ,\ ] : \mathfrak{g} \times \mathfrak{g} \to \mathfrak{g}$, called its *bracket operation* which satisfies

(1) $[X, Y] = -[Y, X]$ (skew-symmetry);

(2) $[[X, Y], Z] + [[Y, Z], X] + [[Z, X], Y] = 0$ (Jacobi identity).

For example, $\mathfrak{gl}(n; \mathbb{R})$ is a Lie algebra by defining
$$[x, y] = xy - yx,$$
where xy denotes the matrix multiplication of x and y.

If a is an element of a Lie group G, let us define $L_a : G \to G$ by $L_a(g) = ag$; and R_a by $R_a(g) = ga$ for all $g \in G$. Then both L_a and R_a are smooth maps. In fact, L_a is a diffeomorphism since $L_{a^{-1}}$ is its inverse. Similarly for R_a.

A vector field X on a Lie group G is called *left-invariant* if $dL_a(X) = X$ for all $a \in G$. Since $dL_a(X_g) = X_{ag}$ for all $a, g \in G$, left-multiplication merely permutes the tangent vectors constituting X. It is easy to verify that every left-invariant vector field is smooth.

Given a Lie group G, let $\mathfrak{g}$ be the set of all left-invariant vector fields on G. Then the usual addition of vector fields and scalar multiplication by real numbers make $\mathfrak{g}$ a real vector space. Moreover, $\mathfrak{g}$ is closed under the bracket operation. Hence $\mathfrak{g}$ is a Lie algebra, which is called the *Lie algebra of G*. Because the function $\mathfrak{g} \to T_e(G)$ sending each $X \in \mathfrak{g}$ to its value $X_e \in T_e(G)$ is a linear isomorphism, the Lie algebra $\mathfrak{g}$ of G and G have the same dimension. An important example is that the Lie algebra of the general linear group $GL(n; \mathbb{R})$ is the matrix Lie algebra $\mathfrak{gl}(n; \mathbb{R})$.

The term "Lie algebra" was introduced by H. Weyl in the 1930s.

1.8   Fibre bundles

We say that a Lie group G is a *Lie transformation group* on a manifold M or G *acts on M* if the following three conditions are satisfied:

(a) Every element $a \in G$ induces a transformation of M, denoted by
$$x \mapsto xa, \quad x \in M;$$

(b) $G \times M \ni (a, x) \mapsto xa \in M$ is a differentiable map;

(c) $(xa)b = x(ab)$ for all $a, b \in G$ and $x \in M$.

Let $e \in G$ denote the identity element of a Lie group G. We say that G acts *effectively* (resp., *freely*) on M if $xa = x$ for all $x \in M$ (resp., for some $x \in M$) implies that $a = e$.

Definition 1.5. Let M be a manifold and G a Lie group. A *fibre bundle* over M with the structure group G and (typical) fibre F consists of a manifold P and an effective action of G on P which satisfies the following conditions:

(1) There exist an open covering $(U_i)_{i \in A}$ of M and diffeomorphisms $h_i : \pi^{-1}(U_i) \to U_i \times F$ which map the fibre $\pi^{-1}(x)$ onto $\{x\} \times F$, where $\pi : P \to M$ is the projection;

(2) Define $\varphi_{i,x} : F \to \pi^{-1}(x)$, $x \in U_i$, by $\varphi_{i,x}(u) = h_i^{-1}(x, u)$, then

$$g_{ji}(x) = \varphi_{j,x}^{-1} \circ \varphi_{i,x} \in G, \quad x \in U_j \cap U_i;$$

(3) $g_{ji} : U_j \cap U_i \to G$ is differentiable.

The family of the maps g_{ji} are called the *transition functions* of the fibre bundle.

The notion of fibre spaces were introduced by Herbert K. J. Seifert (1907-1996) in [Seifert (1932)].

Definition 1.6. Let E and M be two manifolds and $\pi : E \to M$ a map. Then E is called an n-dimensional (real) *vector bundle* over M if the following two conditions are satisfied:

(1) $\pi^{-1}(x)$ is a (real) vector space for each $x \in M$;

(2) There exist an open covering $\{U_i\}$ of M and diffeomorphisms

$$h_i : \pi^{-1}(U_i) \to U_i \times \mathbb{R}^n$$

such that $\varphi_{i,x} : \mathbb{R}^n \to \pi^{-1}(x)$ are isomorphisms of vector spaces. In this case, $g_{ji}(x) = \varphi_{j,x}^{-1} \circ \varphi_{i,x} : \mathbb{R}^n \to \mathbb{R}^n$, $x \in U_j \cap U_i$, is an element of $GL(n; \mathbb{R})$.

Similarly, we may define complex vector bundles over a complex manifold M. Let $\pi : E \to M$ be the projection of a vector bundle over M. A map $s : M \to E$ is called a *cross section* if $\pi \circ s = id$, where id denotes the identity map. A similar definition applies to fibre bundles. We denote by $\Gamma(E)$ the space of all cross-sections of E.

Example 1.2. Let M be a manifold of dimension n. The set

$$TM = \cup_{p \in M} T_p M$$

of all tangent vectors to M has a differentiable structure defined as follows: Let U be a coordinate neighborhood in M with local coordinates $u_1, \ldots, u_n$

and $\bar{V}$ a real n-dimensional vector space with a fixed basis $\{e_1, \ldots, e_n\}$. We define a map

$$\Phi_U : U \times \bar{V} \to \cup_{p \in U} T_p M$$

by the condition that $\Phi_U(p, y) = \sum_{i=1}^{n} y^i \left(\frac{\partial}{\partial u_i}\right)_p$ for $y = \sum_{i=1}^{n} y^i e_i \in \bar{V}$. It is clear that Φ_U is one-to-one. We put $\Phi_{U,p} = \Phi_U(p, y)$. Let V be another coordinate neighborhood with local coordinate system $v_1, \ldots, v_n$. Suppose $U \cap V \neq \emptyset$. Then we define $\Phi_{V,p}$ to be the maps associated with V.

Let us put

$$g_{UV}(p) = \Phi_{U,p}^{-1} \circ \Phi_{V,p}, \quad p \in U \cap V. \tag{1.19}$$

Then $g_{UV}(p) : \bar{V} \to \bar{V}$ is one-to-one. In terms of local coordinate systems, we have the following: Let $y = \sum_{i=1}^{n} y^i e_i$. We put

$$y' = g_{UV}(p)(y). \tag{1.20}$$

We find

$$\Phi_{U,p}^{-1}(y') = \sum_{k=1}^{n} y'^k \left(\frac{\partial}{\partial u_k}\right)_p, \tag{1.21}$$

$$\Phi_{V,p}(y) = \sum_{i=1}^{n} y^i \left(\frac{\partial}{\partial v_i}\right)_p = \sum_{i,k=1}^{n} y^i \left(\frac{\partial u_k}{\partial v_i}\right)_p \left(\frac{\partial}{\partial u_k}\right)_p. \tag{1.22}$$

Thus (1.19), (1.20), (1.21) and (1.22) imply

$$y'^k = \sum_{i=1}^{n} y^i \left(\frac{\partial u_k}{\partial u_i}\right)_p. \tag{1.23}$$

These equation define $g_{U,V}(p)$ as a linear automorphism of $\bar{V}$. We give to $\bar{V}$ the topology and differentiable structure of the Euclidean n-space. Denote by $GL(\bar{V})$ the general linear group of $\bar{V}$. Then (1.23) shows that $g_{UV}(p)$ defines a map $g_{UV} : U \cap V \to GL(\bar{V})$ which is differentiable.

We take a covering of M by the coordinate neighborhoods $U, V, W, \cdots$ etc. Since the map g_{UV} is differentiable, it follows that $T(M)$ is a fibre bundle over M with Φ_U as coordinate functions. We call $T(M)$ the *tangent bundle* of M. This defines meanwhile a topology of $T(M)$ characterized by the condition that Φ_U maps open set of $U \cap \bar{V}$ into open sets in $T(M)$. This topology on $T(M)$ is in fact Hausdorff. A similar argument yields that all tensors of type (r, s) over a manifold M form a fibre bundle over M, called the *tensor bundle* of type (r, s).

Similarly, $T^*(M) = \cup_{p \in M} T_p^*(M)$ is a fibre bundle over M, which is called the *cotangent bundle* of M.

Denote by $\pi : T(M) \to M$ the projection map, i.e., the map which maps every element in T_pM onto p. A map s of M into $T(M)$ is called a *cross-section* if $\pi \circ s = id$. Similar definitions apply to other fibre bundles over M. Tangent bundle, cotangent bundle and tensor bundles of M are important kinds of vector bundle over M.

Example 1.3. By a linear frame u at a point $p \in M$ we mean an ordered basis $u = \{X_1, \ldots, X_n\}$ of the tangent space T_pM. Let $\mathcal{LF}(M)$ denote the set of all linear frames at all points of M and let π be the map $\mathcal{LF}(M) \to M$ which maps a linear frame u at p onto p. The general linear group $GL(N; \mathbb{R})$ acts on $\mathcal{LF}(M)$ on the right as follows:

Let $a = (a_j^i) \in GL(n; \mathbb{R})$ and $u = (X_1, \ldots, X_n)$ a linear frame at p. Then, by definition, ua is the linear frame $(X_1', \ldots, X_n')$ at p with $X_i' = \sum_{j=1}^n a_i^j X_j$. It is clear that $GL(N; \mathbb{R})$ acts freely on $\mathcal{LF}(M)$ and $\pi(u) = \pi(v)$ if and only if $v = ua$ for some $a \in GL(N; \mathbb{R})$.

To introduce a differentiable structure on $\mathcal{LF}(M)$, let $(u_1, \ldots, u_n)$ be a local coordinate system on a coordinate neighborhood U. Every linear frame v at $p \in U$ can be expressed uniquely in the form $v = (Y_1, \ldots, Y_n)$ with $Y_i = \sum_{j=1}^n Y_i^j \frac{\partial}{\partial u_j}$, where (Y_i^j) is a non-singular matrix. This shows that $\pi^{-1}(U)$ is one-to-one correspondent to $U \times GL(n; \mathbb{R})$. We can make $\mathcal{LF}(M)$ into a differentiable manifold by taking (u_i) and (Y_i^j) as a local coordinate system on $\pi^{-1}(U)$. From these we may verify that $\mathcal{LF}(M)$ is a fibre bundle over M with the structure group $GL(n; \mathbb{R})$. We call $\mathcal{LF}(M)$ the *linear frame bundle* of M.

Example 1.4. If M admits a Riemannian metric, then one may consider the space $\mathcal{O}(M)$ of all orthonormal frames on M. By similar consideration, $\mathcal{O}(M)$ becomes a manifold which is a fibre bundle over M with structure group $O(n)$. If one considers the set of all unit tangent vectors to M, then one obtains the *unit tangent bundle* $T^1(M)$ over M.

1.9 Integration of differential forms

In this section, we shall present the theory of integration of differential forms on a manifold.

Definition 1.7. A manifold M of dimension n is called *orientable* if there is a differentiable n-form which is nowhere zero. Two such forms define the same orientation if they differ from one other by a factor which is positive.

An orientable manifold has exactly two possible orientations. Let ω and ω' be two n-forms which determine an orientation of M. Put $\omega' = f\omega$, then f is either positive or negative everywhere. Thus the only possible orientations on M are given by ω and $-\omega$. The manifold M is called *oriented* if such an n-form ω is given.

Definition 1.8. The *support* of a real function f on a manifold M is the closure of the set of points of M at which f is not equal to zero. More generally, the support of a form ω is the closure of the set of points on M where ω is not equal to zero.

An open covering of a manifold M is called *locally finite* if any compact subset of M meets only a finite number of its elements.

Theorem 1.5. *Let B be a family of open sets of a manifold M which form a base for the topology of M. Then there is a locally finite open covering of M whose elements are in B.*

Proof. Since M is separable, there is a countable open covering $\{C_i\}$ of M such that each C_i has compact closure $\bar{C}_i$. We put

$$D_j = \cup_{1 \le i \le j} \bar{C}_i.$$

Then $\{D_j\}$ form a countable covering of M by compact sets D_j with $D_j \subset D_{j+1}$. We now construct compact sets E_j such that

$$D_j \subset E_j, \quad E_j \subset \text{Interior of } E_{j+1}.$$

We use the method of induction. Suppose that $E_1, \ldots, E_j$ are constructed. Because $E_j \cup D_{j+1}$ is compact, it has a finite covering by open sets with compact closures. We put E_{j+1} to be the union of these closures.

Let $S_j = \text{Interior of } E_j$ and $T_j = E_j \cap (M \setminus S_{j-1})$, where we assume that sets with negative indices are empty set. Since $E_{j-1} \subset S_{j-1}$ and $T_j \subset M \setminus S_{j-1}$, we have

$$T_j \cap E_{j-2} = \emptyset.$$

For $p \in T_j$, there is a set of B containing p which is contained in S_{j+1}, but it does not meet E_{j-2}. These sets, for all $p \in T_j$, form a covering of T_j.

On the other hand, T_j is a closed subset of a compact set E_j, it is compact. Therefore the above covering has a finite subcovering, which we call K_j. We denote by K' the family of the sets of K_j for all j. The sets of K' form a covering of K. Indeed, if $p \in M$, there is a positive j such that $p \in E_j$, and $p \notin E_{j-1}$. Hence p belongs to T_j and is covered by a set of K'.

Moreover, if $j \geq k + 2$, E_k meets no set of K_j. Since every compact set of M is contained in a certain E_k, it follows that the covering K' is locally finite. □

We need the following.

Theorem 1.6. (Partition of unity) *Let $(U_i)_{i \in A}$ be an open covering of a manifold M. Then there are functions (g_α) satisfying the following four conditions:*

(1) *Each g_α is differentiable and $0 \leq g_\alpha \leq 1$.*

(2) *The support of each g_α is compact and contained in one of the U_i.*

(3) *Every point of M has a neighborhood which meets only a finite number of the supports of $g_{\alpha'}$.*

(4) $\sum g_\alpha = 1$.

For the proof, see for instance page 272 of [Kobayashi and Nomizu (1963)].

Theorem 1.7. *Let M be an oriented manifold of dimension n. Then there is one and only one functional which assign to a differentiable n-form Φ with compact support, a real number called the integral of Φ over M, denoted by $\int \Phi$, such that*

(1) $\int \Phi_1 + \Phi_2 = \int \Phi_1 + \int \Phi_2$;

(2) *If the support of Φ is contained in a coordinate neighborhood U with coordinates $u_1, \ldots, u_n$ such that $du_1 \wedge \cdots \wedge du_n$ defines the orientation and $\Phi = \Phi(u_1, \ldots, u_n) du_1 \wedge \cdots \wedge du_n$, then*

$$\int \Phi = \int_U \Phi(u_1, \ldots, u_n) du_1 \wedge \cdots \wedge du_n,$$

where the right-hand side is a Riemannian integral.

Proof. Let Φ be an n-form with compact support S. We choose an open covering (U_i) of M such that each U_i is a coordinate neighborhood. Let g_i be a corresponding partition of unity. Then every point $p \in S$ has a neighborhood V_p which meets only a finite number of the supports of g_i. These V_p for all $p \in S$ form a covering of S. Because S is compact, it has a finite subcovering. Therefore there are only a finite number of $g_i \Phi$ which are not identically zero.

We define

$$\int \Phi = \sum_{\alpha} \int g_{\alpha} \Phi,$$

where the right-hand side is a finite sum. Because the differential n-form in each summand has a support lying in a coordinate neighborhood U_i, we can evaluate it according to the formula in condition (2).

We now prove that the above definition is independent of the various choices made. These are:

1) Choice of the neighborhood U_i which contains the support of $g_{\alpha}\Phi$;

2) Choice of the covering $\{U_i\}$ and the corresponding partition of unity g_{α}.

Suppose that the support of $g_{\alpha}\Phi$ be in two coordinate neighborhood U and V with local coordinates $u_1, \ldots, u_n$ and $v_1, \ldots, v_n$, respectively. We can take an open set W containing the support in which both coordinate system are valid. On W we have

$$g_{\alpha}\Phi = \Phi(u_1, \ldots, u_n)du_1 \wedge \cdots \wedge du_n$$
$$= \Psi(v_1, \ldots, v_n)dv_1 \wedge \cdots \wedge dv_n, \qquad (1.24)$$

where

$$\Psi(v_1, \ldots, v_n) = \Phi(u_1(v_1, \ldots, v_n), \ldots, u_n(v_1, \ldots, v_n))\frac{\partial(u_1, \ldots, u_n)}{\partial(v_1, \ldots, v_n)}.$$

We may assume that the Jacobian determinant is positive throughout W. The equation

$$\int \Phi = \int \Phi(u_1, \ldots, u)\frac{\partial(u_1, \ldots, u_n)}{\partial(v_1, \ldots, v_n)}dv_1 \wedge \cdots \wedge dv_n$$

is then exactly the formula for the transformation of multiple integrals. Therefore it follows that our definition is independent of the choice of the particular choice of the neighborhood U_i in the evaluation of the summands.

Next, we consider a second covering (V_j) by coordinate neighborhoods and let g' be a corresponding partition of unity. Then $(U_{\alpha} \cap V_{\beta})$ will be a covering of M with the functions $g_{\alpha}g'_{\beta}$ as a corresponding partition of unity. It follows that

$$\sum_{\alpha} \int g_{\alpha}\Phi = \sum_{\alpha,\beta} \int g_{\alpha}g'_{\beta}\Phi$$

and

$$\sum_\beta \int g'_\beta \Phi = \sum_{\alpha,\beta} \int g_\alpha g'_\beta \Phi.$$

This proves the independence of the integral of the choice of covering and the corresponding partition of unity.

The uniqueness is obvious. □

Let M be an oriented manifold of dimension n. A differential n-form ω is said to be > 0 or < 0 according to ω or $-\omega$ defines the orientation.

Definition 1.9. A *domain D with regular boundary* is a subset of M such that if, $p \in D$, either

(a) p has a neighborhood belonging entirely to D, or
(b) there is a coordinate neighborhood U of p with coordinates $u_1, \ldots, u_n$ such that $U \cap D$ is the set of all points $q \in D$ with $u_n(q) \geq u_n(p)$.

Points with property (a) are called *interior points* of D and points with property (b) are called *boundary points*. The set of all boundary points of D is called the *boundary* of D, which will be denoted by ∂D.

It is known that the boundary of a domain with regular boundary is a closed submanifold which is regularly imbedded. If D is orientable, so is the boundary ∂D.

We now consider a domain D with regular boundary ∂D and suppose it is compact. Define the characteristic function $h(p)$, $p \in M$, by

$$h(p) = \begin{cases} 1 & \text{if } p \in D, \\ 0 & \text{if } p \in M \setminus D. \end{cases}$$

We define the integral over D of an n-form Φ on M by

$$\int_D \Phi = \int h\Phi.$$

1.10 Stokes' theorem

The following is the well-known *Stokes theorem*.

Theorem 1.8. *Let M be an oriented manifold of dimension n and let ω be an $(n-1)$-form with compact support. Then for any domain D with regular boundary in M we have*

$$\int_D d\omega = \int_{\partial D} \omega. \tag{1.25}$$

Proof. Let (U_i) be an open covering of M by coordinate neighborhood such that for each U_i either $U_i \cap \partial D = \emptyset$ or U_i has property (b) of Definition 1.9. Let the functions (g_α) be a corresponding partition of unity.

Since ∂D and D are both compact, each of them meets only a finite number of the support of g_α. Thus we have

$$\int_{\partial D} \omega = \sum_\alpha \int_{\partial D} g_\alpha \omega$$

and

$$\int_D d\omega = \sum_\alpha \int_D d(g_\alpha \omega).$$

Therefore, it suffices to prove (1.25) for each summand of the above sums, i.e., under the assumption that the support of ω lies in a coordinate neighborhood U_i. Assume that $u_1, \ldots, u_n$ is a local coordinate system in U_i such that $du_1 \wedge \cdots \wedge du_n > 0$. Put

$$\omega = \sum_{j=1}^n (-1)^{j-1} a_j dU_1 \wedge \cdots \wedge \widehat{du_j} \wedge \cdots \wedge du_n.$$

Then we have

$$d\omega = \sum_{j=1}^n \frac{\partial a_j}{\partial u_j} du_1 \wedge \cdots \wedge du_n.$$

We first consider the case where $U_i \cap \partial D = \emptyset$. Then the right-hand side of (1.25) is zero. The set U_i either belongs to $M \setminus D$ or to the interior of D. If the first possibility holds, then $h = 0$, so (1.25) holds.

Next, suppose that U_i lies in the interior of D. Then we have $h = 1$ and the integral in the left-hand side of (1.25) is equal to

$$\int_D d\omega = \int_C \left(\sum_{j=1}^n \frac{\partial a_j}{\partial u_j} \right) du_1 \wedge \cdots \wedge du_n,$$

where C is a cube in the space of the coordinates u_j containing the support of ω in its interior. We may choose m sufficiently large so that C is defined by the inequalities $|u_j| \leq m$. This integral is now just an iterated integral in classical analysis. Thus we have

$$\int_C \frac{\partial a_j}{\partial u_j} du_1 \cdots du_n = \pm \int_{F_j} \{ a_j(u_1, \ldots, u_{j-1}, m, u_{j+1}, \ldots, u_n)$$

$$- a_j(u_1, \ldots, u_{j-1}, -m, u_{j+1}, \ldots, u_n) \} du_1 \cdots du_{j-1} du_{j+1} \cdots du_n,$$

where F_j is the union of the appropriate faces of the cube. Because the support of ω lies inside C, the above integral is equal to zero. Thus (1.25) holds. Now, we consider the case where U_i has property (b) in Definition 1.9. We assume that ∂D is contained in the subset defined by $u_n = 0$. When $u_n(p) \geq 0$, $h(p) = 1$, in the space of the coordinates $u_1, \ldots, u_n$ we take a cube defined by $|u_k| \leq m$, $k = 1, \ldots, n-1$; $0 \leq u_n \leq m$ such that the support of ω is contained in the union of its interior and the side $u_n = 0$. As in the above, we have

$$\int_C \frac{\partial a_k}{\partial u_k} du_1 \wedge \cdots \wedge du_n = 0$$

for $k = 1, \ldots, n-1$. Also we have

$$\int_C \frac{\partial a_n}{\partial u_n} du_1 \wedge \cdots \wedge du_n$$
$$= (-1)^{n-1} \int_{\partial D} a_n(u_1, \ldots, u_{n-1}, 0) du_1 \cdots du_{n-1}.$$

But the right-hand side of the last equation is equal to $\int_{\partial D} \omega$. Thus we have (1.25). $\square$

Remark 1.3. This modern form of Stokes' theorem is a vast generalization of a classical result first discovered by Lord Kelvin (1824-1907), who communicated it to George Stokes (1819-1903) in a letter dated July 2, 1850 (cf. [Katz (1979)]).

1.11 Homology, cohomology and de Rham's theorem

Homology theory started with the Euler polyhedron formula. Homology classes and relations were first defined rigorously by Henri Poincaré in [Poincaré (1895)]. Poincaré was also the first to consider chain complexes.

In this section we present homology and cohomology groups on a manifold M. We use Stokes' theorem to prove de Rham's theorem.

Notation 1.1. We denote by Δ_p the *simplex* in $\mathbb{R}^p$ defined by $0 \leq x_i \leq 1$ with $\sum_{i=1}^p x_i = 1$.

Thus Δ_0 is a point, Δ_1 is the unit interval $[0, 1]$ and Δ_2 is a triangle, etc.

On the simplex Δ_p we introduce the barycentric coordinates defined by

$$y_0 = 1 - \sum_{i=1}^p x_i, \quad y_j = x_j, \quad j = 1, \ldots, p,$$

so that $y_0 + y_1 + \cdots + y_n = 1$ and $0 \leq y_j \leq 1$.

The map $\delta_{p,i}$ $(i = 0, 1, \ldots, p+1)$ from Δ_p into Δ_{p+1} is defined by

$$\bar{y}_k(\delta_{p,i}(y_0, \ldots, y_p)) = \begin{cases} y_k & \text{if } k < 1; \\ 0 & \text{if } k = i; \\ y_{k-1} & \text{if } j > 1, \end{cases}$$

where $\bar{y}_k$ are the *barycentric coordinates* of Δ_{p+1}.

Lemma 1.1. *We have*

$$\delta_{p+1,i} \circ \delta_{p,j} = \delta_{p+1,j} \circ \delta_{p,i+1} \tag{1.26}$$

for $j \leq i$.

Proof. By comparing both sides of (1.26) acting on a p-simplex, we get the lemma. $\qquad\square$

Definition 1.10. A differentiable *singular p-simplex* on a manifold M is a map of Δ_p into M which can be extended to a differentiable map of a neighborhood of Δ_p in $\mathbb{R}^p$ into M. A differentiable *p-chain* is a finite formal linear combination with real coefficients of singular p-simplices.

The set $C_p(M)$ of all p-chains on M form a real vector space in a obvious way. If ϕ is a differentiable map of a manifold M into another manifold N, we define a linear map $\hat{\phi} : C_p(M) \to C_p(N)$ by

$$\hat{\phi}(s) = s \circ \phi \tag{1.27}$$

for simplices and extend it by linearity. If s is a p-simplex, then $s \circ \delta_{p-1,i}$ is a $(p-1)$-simplex. We define the *boundary* of s by

$$\partial(s) = \sum_{i=0}^{p} (-1)^i s \circ \delta_{p-1,i}. \tag{1.28}$$

We extend ∂ by linearity to a map of $C_p(M)$ into $C_{p-1}(M)$.

Lemma 1.2. *We have*

$$\partial \circ \partial = 0. \tag{1.29}$$

Proof. By linearity, it suffices to prove (1.29) for a simplex. Let s be a q-simplex. If $q \leq 1$, there is nothing to prove. If $q \geq 2$, we have

$$\partial^2 s = \partial(\partial s) = \sum_{i=0}^{q} \sum_{j=0}^{q-1} (-1)^i (-1)^j s \circ \delta_{q-1,i} \circ \delta_{q-2,j}$$

$$= \sum_{0 \leq j \leq i \leq q} (-1)^{i+j} s \circ \delta_{q-1,i} \circ \delta_{q-2,j} + \sum_{0 \leq i < j \leq q} (-1)^{i+j} s \circ \delta_{q-1,i} \circ \delta_{q-2,j}.$$

By Lemma 1.1 we have $s \circ \delta_{q-1,i} \circ \delta_{q-2,j} = s \circ \delta_{q-1,j} \circ \delta_{q-2,i+1}$ for $j \le i$. Thus we can rewrite the first sum as

$$\sum_{0 \le j \le i \le q} (-1)^{i+j} s \circ \delta_{q-1,j} \circ \delta_{q-2,i+1}.$$

Now, if we put $k = i+1$, this will cancel with the second sum which proves (1.29). $\qquad\square$

Definition 1.11. A p-chain c is called a *cycle* if $\partial c = 0$. The p-chains of the form $c = \partial d$ for some $(p+1)$-chains d are called *p-boundaries*.

Lemma 1.2 shows that the space $B_p(M)$ of p-boundaries is a subspace of the space $Z_p(M)$ of p-cycles.

Definition 1.12. The quotient space $Z_p(M)/B_p(M)$, denoted by $H_p(M)$, is called the *p-th homology group* of M.

Remark 1.4. If $\phi : M \to N$ is a differentiable map, then $\hat{\phi}$ commutes with ∂, i.e., $\partial\hat{\phi} = \hat{\phi}\partial$. Thus $\hat{\phi}$ maps cycles into cycles and boundaries into boundaries. Hence it induces a map of $H_p(M)$ into $H_p(N)$.

Definition 1.13. A differential form ω is called *closed* if $d\omega = 0$. It is *exact* if $\omega = d\eta$ for some form η. The quotient of the space of closed p-forms by the space of exact p-forms is called the *p-th de Rham cohomology group* of M, named after Georges de Rham (1903-1990), which is denoted by $H^p(M)$. The quotient of the space closed p-form with compact support over the space of exact form $d\eta$, where η is a $(p+1)$-form of compact support is denoted by $H_c^p(M)$.

When M is compact, we have $H^p(M) = H_c^p(M)$.

Let $c = \sum_s c_s$ be a p-chain and ω is a p-form of compact support. We define $\int_c \omega = \sum_s \int_{c_s} \omega$. It follows from Stokes' theorem that

$$\int_c d\omega = \int_{\partial(c)} \omega. \tag{1.30}$$

The following results are well-known.

Theorem 1.9. *The bilinear map of $C_p(M) \times \wedge^p(M) \to \mathbb{R}$ given by $\int_c \omega$ induces a bilinear map of $H_p(M) \times H^p(M) \to \mathbb{R}$.*

Proof. It follows from (1.30) that the integral of an exact form over a cycle and the integral of a closed form over a boundary both vanish. Thus the integral of a closed form over a cycle depends only the cohomology class of the closed form and the homology class of the cycle. $\qquad\square$

A fundamental results in algebraic topology asserts that this bilinear map is non-singular. Thus we have the following well-known de Rham's isomorphism.

Theorem 1.10. [de Rham (1931)] $H^p(M)$ *is the dual space of* $H_p(M)$. *In particular, we have the following isomorphism:*

$$H^p(M) \cong H_p(M). \tag{1.31}$$

The dimension of $H_p(M)$ is the p-th *Betti number* of M, which is denoted by $b_p(M)$.

Let α be a closed p-form and β a closed q-forms on M. Denote by $[\alpha]$ and $[\beta]$ the corresponding cohomology classes represented by α and β, respectively. It is direct to verify that the wedge product $\alpha \wedge \beta$ is a closed $(p+q)$-form on M.

Definition 1.14. Let c_α and c_β denote the corresponding homology classes in $H_p(M)$ and $H_q(M)$ associated with $[\alpha]$ and $[\beta]$ through the natural isomorphism (1.31). Then the homology class $c_{\alpha \wedge \beta}$ corresponding to $[\alpha \wedge \beta]$ is called the *cup product* of c_α and c_β, which is denoted by $c_\alpha \cup c_\beta$.

The notion of cup product was introduced by E. Čech and H. Whitney in the 1930s.

1.12   Frobenius' theorem

Let M be an n-dimensional manifold. An r-dimensional *distribution* on M is an assignment $\mathcal{D}$ defined on M which assigns to each point $p \in M$ an r-dimensional linear subspace $\mathcal{D}_p$ of T_pM.

An r-dimensional distribution $\mathcal{D}$ is called differentiable if there are r differentiable vector fields on a neighborhood of p which, for each point q in this neighborhood, form a basis of $\mathcal{D}_q$. Throughout this book, by a distribution, we shall mean a differentiable distribution.

A vector field X belongs to the distribution $\mathcal{D}$ if $X_p \in \mathcal{D}_p$ for each $p \in M$. In this case, we denote it by $X \in \mathcal{D}$. A set of r vector fields $X_1, \ldots, X_r$ in an r-dimensional distribution $\mathcal{D}$ is called a local basis of $\mathcal{D}$ if $X_1, \ldots, X_r$ are linearly independent at each point. A distribution $\mathcal{D}$ is called *involutive* if $[X, Y] \in \mathcal{D}$ for any $X, Y \in \mathcal{D}$.

An imbedded submanifold N of M is called an *integrable submanifold* of a distribution $\mathcal{D}$ if $\iota_*(T_p(N)) = \mathcal{D}_{\iota(p)}(M)$ for all $p \in N$, where ι is the inclusion map. An r-dimensional distribution $\mathcal{D}$ on M is called *completely* *integrable* if, for each point $p \in M$, there is a coordinate neighborhood U and local coordinates $y_1, \ldots, y_n$ on U such that all the submanifolds of U given by

$$y_{r+1} = const., \ldots, y_n = const.$$

are integrable submanifolds of $\mathcal{D}$.

The well-known theorem of Frobenius can be stated as follows.

Theorem 1.11. *A distribution $\mathcal{D}$ on a manifold M is completely integrable if and only if it is involutive.*

Proof. The necessity is obvious. To prove the sufficiency, we restrict our attention to local matters.

If $r = 1$, this reduces to existence theorem of ordinary differential equations. In fact, given a vector field X, by this existence theorem, we can introduce a local coordinate system $y_1, \ldots, y_n$ on M such that $X = \partial/\partial y_1$.

Now, we shall prove the theorem by induction on r. Let $p \in M$ and $X_1, \ldots, X_r$ be r independent vector fields on a coordinate neighborhood V of p. We introduce local coordinates $y_1, \ldots, y_r$ such that $X_1 = \partial/\partial y_1$ and $y_1(p) = 0$. We put

$$f_j = X_j y_1, \quad Y_1 = X_1, \quad Y_j = X_j - f_j X_1$$

on V for $j = 2, \ldots, r$. Then $Y_1, \ldots, Y_r$ are also linearly independent, which span the distribution $\mathcal{D}$ on V. Moreover, we have $Y_j y_1 = 0$ for $j = 2, \ldots, r$.

Let us consider

$$W = \{q \in V : y_1(q) = 0\}.$$

Then W is an $(r-1)$-dimensional submanifold of V and $Y_2, \ldots, Y_r$ are tangent to W, i.e., $Y_j(q)$ are tangent to W for $q \in W$. Let Z_j be the restriction of Y_j on W. The $\iota_* (\partial/\partial z_j)_q = (\partial/\partial y_j)_q$ for $q \in W$, where ι is the inclusion. If $Y_j = \sum_{k=2}^n Y_j^k \partial/\partial y_k$, then we have $Z_j = \sum_{k=2}^n Z_j^k \partial/\partial y_k$, where Z_j^k are the restriction of Y_j^k to W. Now, we want to claim that $Z_2, \ldots, Z_r$ span an involute $(r-1)$-dimensional distribution on W. In fact, if $[Z_i, Z_j]_q$ did not lie in the space spanned by $Z_2, \ldots, Z_r$, then $[Y_i, Y_j]_q$ would not lie in $\mathcal{D}_q$ since none of $Y_2, \ldots, Y_r$ has any component relative to $\partial/\partial y_1$. Therefore, by induction, we can find an integral submanifold of Span $\{Z_2, \ldots, Z_r\}$ through each point $p \in W$. Thus we can find local coordinates $u_1, \ldots, u_n$ in some neighborhood of $p \in W$ such that the distribution spanned by $Z_2, \ldots, Z_r$ has integral submanifolds given by

$$u_{r+1} = const., \ldots, u_n = const.$$

Now, we define local coordinates $v_1, \ldots, v_n$ about p as follows: Let $v_1 = y_1$ and $v_j(y_1, \ldots, y_n) = u_j(y_2, \ldots, y_n)$. The Jacobian determinant of the v's with respect to the y's does not vanish at p so that the v's do form a coordinate system about p. Moreover, because $\partial v_j / \partial y_1 = 0$ for $j = 2, \ldots, n$, we have $Y_1 = \partial / \partial y_1 = \partial / \partial v_1$. Since $Y_1 v_{r+k} = 0$, we have

$$\frac{\partial}{\partial v_1}(Y_j v_{r+k}) = [Y_j, Y_1] v_{r+k}.$$

Now, because $Y_1, \ldots, Y_r$ span an involutive distribution, we may find some function D^i_{jh} so that

$$[Y_i, Y_1] = D^1_{i1} Y_1 + \sum_{j=2}^{r} D^j_{i1} Y_j.$$

Thus we have

$$\frac{\partial}{\partial v_1}(Y_j v_{r+k}) = \sum_{i=2}^{r} D^i_{j1}(Y_i v_{r+k})$$

for $j = 2, \ldots, r$. Thus, for each fixed k, we may regard the function $Y_j v_{r+k}$ as solutions of the homogeneous linear differential system of $r-1$ differential equations with respect to the independent variable v_1,

On the other hand, if $v_1 = 0$, we have $Y_j v_{r+k} = Z_j v_{r+k} = 0$. Thus, by the uniqueness of the solutions, we find that $Y_j v_{r+k}$ vanishes identically. Therefore, $Y_1, \ldots, Y_r$ can be expressed in terms of $\partial / \partial v_1, \ldots, \partial / \partial v_r$. This proves that $\mathcal{D}$ is completely integrable.   □

Let $\mathcal{D}$ be an r-dimensional distribution on M. We put

$$\Omega = \{1\text{-forms } \omega \text{ on } M : \omega(X) = 0 \text{ for } X \in \mathcal{D}\}, \tag{1.32}$$

and let $I(\Omega)$ be the ideal generated by Ω in the ring of exterior polynomials on M.

Theorem 1.11 of Frobenius can also be rephrased as the following.

Theorem 1.12. *A distribution $\mathcal{D}$ on a manifold M is involutive if and only if $d\Omega$ is contained in the ideal $I(\Omega)$, where Ω is the differential system defined by* (1.32).

Remark 1.5. Despite being named after Frobenius, the theorem was first proven by F. Deahna in [Deahna (1840)] and A. Clebsch (1833-1872) in [Clebsch (1866)]. Deahna was the first to prove the sufficient conditions for the theorem and Clebsch developed the necessary conditions. G. Frobenius (1849-1917) in [Frobenius (1877)] is responsible for applying the theorem to Pfaffian systems, which paves the way for its usage in differential topology.

Chapter 2

Riemannian and Pseudo-Riemannian Manifolds

Riemannian geometry was first put forward in generality by B. Riemann in the middle of nineteenth century. In his famous inaugural lecture

"Über die Hypothesen welche der Geometrie zu Grunde liegen",

at Göttingen, Riemann discussed the foundations of geometry, introduced n-dimensional manifolds, formulated the concept of Riemannian manifolds and defined their curvature [Riemann (1854)]. Riemannian geometry, including the Euclidean geometry and the classical non-Euclidean geometries as most special particular cases, deals with a broad range of more general geometries whose metric properties vary from point to point.

Development of Riemannian geometry resulted in synthesis of diverse results concerning the geometry of surfaces and the behavior of geodesics on them, with techniques that can be applied to the study of differentiable manifolds of higher dimensions. It enabled Einstein's general relativity theory (1915), made profound impact on group theory and representation theory, as well as analysis, and spurred the development of algebraic and differential topology. Since every manifold admits a Riemannian metric, Riemannian geometry often helps to solve problems of differential topology. Most remarkably, by using Riemannian geometry, G. Y. Perelman proved in 2003 Thurston's geometrization conjecture; consequently solved in the affirmative famous Poincaré's conjecture posed in 1904.

Under the impetus of Einstein's Theory of General Relativity, the positiveness of the inner product induced from Riemannian metric was weakened to non-degeneracy. Consequently, one has the notion of pseudo-Riemannian manifolds. Furthermore, inspired by the theory of general relativity and string theory, mathematicians and physicists study not only Riemannian manifolds but pseudo-Riemannian manifolds and their submanifolds as well.

2.1   Symmetric bilinear forms and scalar products

A *symmetric bilinear form* on a finite-dimensional real vector space V is
a $\mathbb{R}$-bilinear function $B : V \times V \to \mathbb{R}$ such that $B(u,v) = B(v,u)$ for all
$u, v \in V$.

   A symmetric bilinear form B is said to be *positive definite* (resp. *positive
semi-definite*) if $B(v,v) > 0$ (resp. $B(v,v) \geq 0$) for all $v \neq 0$. Similarly, a
symmetric bilinear form B is called *negative definite* (resp. *negative semi-
definite*) if $B(v,v) < 0$ (resp. $B(v,v) \leq 0$) for all $v \neq 0$. B is said to be
non-degenerate whenever $B(u,v) = 0$ for all $u \in V$ implies $v = 0$.

Definition 2.1. The *index* of a symmetric bilinear form B on V is the
dimension of a largest subspace $W \subset V$ on which $B|_W$ is negative definite.

   If we choose a basis $v_1, \ldots, v_n$ of V, then the $n \times n$ matrix $(b_{ij}), b_{ij} = B(e_i, e_j)$, is called the *matrix of B with respect to* $e_1, \ldots, e_n$. Since B is
symmetric, the matrix (b_{ij}) is symmetric. A symmetric bilinear form is
non-degenerate if and only if the matrix of B with respect to one basis is
invertible.

Definition 2.2. A *scalar product* g on a finite-dimensional real vector space
is a non-degenerate symmetric bilinear form. An *inner product* is a positive
definite scalar product.

   By a *scalar product space* (V, g) we mean a vector space V equipped
with scalar product g. A subspace U of a scalar product space is called
non-degenerate if $g|_U$ is non-degenerate.

   Two vectors u, v of a scalar product space V are called *orthogonal*, which
are denoted by $u \perp v$, if $g(u,v) = 0$. Two subsets $P, Q \subset V$ are said to be
orthogonal, denoted by $P \perp Q$, if $g(u,w) = 0$ for all $u \in P$ and $w \in Q$.

   For a subspace $U \subset V$, put $U^{\perp} = \{v \in V : v \perp U\}$. Then $(U^{\perp})^{\perp} = U$.

Lemma 2.1. *A subspace U of a scalar product space V is non-degenerate
if and only if V is the direct sum of U and $U^{\perp}$.*

Proof. Since

$$\dim(U + U^{\perp}) + \dim(U \cap U^{\perp}) = \dim U + \dim U^{\perp} = \dim V,$$

$U + U^{\perp} = V$ holds if and only if $U \cap U^{\perp} = \{0\}$ holds. The later condition
means that U is non-degenerate. $\square$

On a scalar product space V, the *norm* $\|v\|$ of a vector v is defined to be $\sqrt{|g(v,v)|}$. A vector of norm one is called a *unit vector*. A set of mutually orthogonal unit vectors is called a *orthonormal set*. A set of n orthonormal vectors $e_1, \ldots, e_n$ of V is called an *orthonormal basis* whenever $n = \dim V$.

Lemma 2.2. *A scalar product space V of positive dimension admits an orthonormal basis.*

Proof. Since g is non-degenerate, there is a unit vector $e_1 \in V$. Let U_1 be the subspace spanned by e_1. Then $U_1^{\perp}$ is a non-degenerate subspace. Thus there is a unit vector $e_2 \in (U_1)^{\perp}$. The pair $\{e_1, e_2\}$ is an orthonormal basis of $\mathrm{Span}\{e_1, e_2\}$. By continuing this process $(n-1)$-times, we obtain an orthonormal basis $e_1, \ldots, e_n$ of V. $\qquad\square$

For an orthonormal basis $e_1, \ldots, e_n$ of a scalar product space V, we have

$$g(e_i, e_j) = \epsilon_i \delta_{ij}, \quad \epsilon_i = g(e_i, e_i) = \pm 1,$$

where δ_{ij} is the *Kronecker delta*, which is equal to 1 if $i = j$; and equal to 0 if $i \neq j$. Every vector $v \in V$ can be expressed in a unique way as

$$v = \sum_{i=1}^{n} \epsilon_i g(v, e_i) e_i.$$

For an orthonormal basis $e_1, \ldots, e_n$ of a scalar product space V, the number of negative signs in the signature $(\epsilon_1, \ldots, \epsilon_n)$ is the *index of V*.

A linear transformation $T : V \to W$ between two scalar product spaces is called a *linear isometry* if it preserves the scalar products. Two scalar product spaces are linear isometric if and only if they have the same dimension and the same index.

2.2 Riemannian and pseudo-Riemannian manifolds

A *(pseudo-Riemannian) metric tensor g* on a manifold M is a symmetric non-degenerate $(0,2)$ tensor field on M of constant index, i.e., g assigns to each point $p \in M$ a scalar product g_p on T_pM and the index of g_p is the same for all $p \in M$. Very often, we use $\langle \ , \ \rangle$ as an alternative notation for g. Thus we have $g(v, w) = \langle u, v \rangle$.

A *pseudo-Riemannian n-manifold* is by definition an n-dimensional manifold equipped with a (pseudo-Riemannian) metric tensor g. The common value $s, 0 \leq s \leq n$, of index on M is called the *index* of M. If $s = 0$,

M is called a *Riemannian manifold*. In this case, each g_p is a positive definite inner product on T_pM. A pseudo-Riemannian manifold (resp. metric) is also known as a *semi-Riemannian manifold* (resp. metric). A pseudo-Riemannian metric on an even-dimensional manifold M is called a *neutral metric* if its index is equal to $\frac{1}{2}\dim M$.

If the index of M is one, M is called a *Lorentz manifold* and the corresponding metric is called *Lorentzian*. A manifold of dimension ≥ 2 admits a Lorentzian metric if and only if it admits a 1-dimensional distribution. Lorentz manifolds named after the physicist Hendrik Lorentz (1853-1928).

A tangent vector v of a pseudo-Riemannian manifold M is called *spacelike* (resp. *timelike*) if $v = 0$ or $\langle v,v\rangle > 0$ (resp. $\langle v,v\rangle < 0$). A vector v is called *lightlike* or *null* if $\langle v,v\rangle = 0$ and $v \neq 0$.

The *light cone* $\mathcal{LC}$ of $\mathbb{E}_s^n$ is defined by $\mathcal{LC} = \{v \in \mathbb{E}_s^n : \langle v,v\rangle = 0\}$. A curve in a pseudo-Riemannian manifold is called a *null curve* if its velocity vector is a lightlike at each point.

For a timelike vector v in a Lorentz manifold, the set $\bar{C}(v)$ consists of all causal vectors w with $\langle v,w\rangle < 0$ is called the *time cone* (or the *causal cone*) containing v. The opposite time cone is the set $\bar{C}(-v)$ consists of all causal vectors w with $\langle v,w\rangle > 0$.

In each tangent space T_pM of a Lorentz manifold M, there exist two time cones and there is no intrinsic way to distinguish one from the other. To choose one of them is to time-orient T_pM. Let ζ be a function on M which assigns to each $p \in M$ a time cone ζ_p in T_pM. ζ is called smooth if for each $p \in M$, there is a smooth vector field V on some neighborhood U of p such that $V_q \in \zeta_q$ for each $q \in U$. Such a smooth function is called a *time-orientation* of M. To choose a specific time-orientation on M is to *time-orient* M.

A *spacetime* is a time oriented 4-dimensional Lorentz manifold. As with any time-oriented spacetime, the time-orientation is called the *future*, and its negative is the *past*. A tangent vector in a future time cone is called *future-pointing*. Similarly, a tangent vector in the pass time cone is called *pass-pointing*.

A vector in a Lorentzian vector space that is *non-spacelike* (i.e., either lightlike or timelike) is called *causal*. A *causal curve* in a spacetime is a curve whose velocity vectors are all non-spacelike.

Let $\{u_1,\ldots,u_n\}$ be a coordinate system on an open subset $U \subset M$, where $n = \dim M$. Then the components g_{ij} of the metric tensor g on U are given by

$$g_{ij} = \langle \partial_i, \partial_j\rangle, \quad 1 \leq i,j \leq n,$$

where $\partial_i = \partial/\partial u_i$, $i = 1, \ldots, n$. Since g is a symmetric $(0,2)$ tensor field, we have $g_{ij} = g_{ji}$ for $1 \leq i, j \leq n$. Hence the metric tensor on U can be written as

$$g = \sum_{i,j=1}^{n} g_{ij} du_i \otimes du_j. \tag{2.1}$$

At each point p in the Euclidean n-space $\mathbb{E}^n$, there exists a canonical linear isomorphism from $\mathbb{E}^n$ onto $T_p\mathbb{E}^n$. In terms of natural coordinates on $\mathbb{E}^n$, it sends a vector v to $v_p = \sum v^j \partial_j$. The inner product on $\mathbb{E}^n$ gives rise to a metric tensor on $\mathbb{E}^n$ with

$$\langle v_p, w_p \rangle = \sum_{j=1}^{n} v_j w_j \tag{2.2}$$

with $v = \sum_{j=1}^{n} v_j \partial_j$ and $w = \sum_{j=1}^{n} w_j \partial_j$.

For an integer $s \in [0, n]$, if we change the first s plus signs in (2.2) to minus sign, then it gives rise to a metric tensor

$$\langle v_p, w_p \rangle = -\sum_{j=1}^{s} v_j w_j + \sum_{k=s+1}^{n} v_k w_k \tag{2.3}$$

of index s. The resulting *pseudo-Euclidean space* is denoted by $\mathbb{E}^n_s$.

If $s = 0$, $\mathbb{E}^n_s$ reduces to the Euclidean n-space $\mathbb{E}^n$. The $\mathbb{E}^n_1$ is called a *Minkowski n-space*. When $n = 4$ and $s = 1$, it is the simplest example of a relativistic spacetime, known as the *Minkowski spacetime*, introduced in 1908 by one of Einstein's teachers, Hermann Minkowski (1864-1909). The Minkowski spacetime is the earliest treatment of space and time as two aspects of a unified whole, the essence of special relativity.

2.3   Levi-Civita connection

Let M be an n-manifold. Denote by $\mathcal{F}(M)$ the set of all smooth real-valued functions on M. If f_1, f_2 are smooth functions on M, so is their sum $f_1 + f_2$ and product $f_1 f_2$. The usual algebraic rules hold for these two operations, which make $\mathcal{F}(M)$ a commutative ring. We denote by $\mathfrak{X}(M)$ the set of all smooth vector fields on M.

For $V, W \in \mathfrak{X}(M)$, the bracket $[V, W]$ is defined by

$$[V, W]_p(f) = V_p(Wf) - W_p(Vf)$$

at each $p \in M$ and $f \in \mathcal{F}(M)$. The bracket operation $[\ ,\ ]$ on $\mathfrak{X}(M)$ is a $\mathbb{R}$-bilinear and skew-symmetric, which also satisfies the Jacobi identity:

$$[X, [Y, Z]] + [Y, [Z, X]] + [Z, [X, Y]] = 0.$$

These makes $\mathfrak{X}(M)$ an infinite-dimensional Lie algebra.

Definition 2.3. An *affine connection* ∇ on a manifold M is a function $\nabla : \mathfrak{X}(X) \times \mathfrak{X}(M) \to \mathfrak{X}(M)$ such that

(1) $\nabla_X Y$ is $\mathcal{F}(M)$-linear in X;
(2) $\nabla_X Y$ is $\mathbb{R}$-linear in Y;
(3) $\nabla_X(fY) = (Xf)Y + f\nabla_X Y$ for $f \in \mathcal{F}(M)$.

$\nabla_X Y$ is called the *covariant derivative* of Y with respect to X.

The *torsion tensor* T of an affine connection ∇ is a tensor of type $(1,2)$ defined by $T(X, Y) = \nabla_X Y - \nabla_Y X - [X, Y]$.

The following theorem shows that on a pseudo-Riemannian manifold there exists a unique connection sharing two further properties.

Theorem 2.1. *On a pseudo-Riemannian manifold M, there exists a unique affine connection ∇ such that*

(4) ∇ *is torsion free, i.e.,* $[Y, Z] = \nabla_Y Z - \nabla_Z Y$, *and*
(5) $X\langle Y, Z \rangle = \langle \nabla_X Y, Z \rangle + \langle Y, \nabla_X Z \rangle$

for $X, Y, Z \in \mathfrak{X}(M)$. This unique affine connection ∇ is called the Levi-Civita connection of M and it is characterized by the Koszul formula

$$2\langle \nabla_Y Z, X \rangle = Y\langle Z, X \rangle + Z\langle X, Y \rangle - X\langle Y, Z \rangle$$
$$- \langle Y, [Z, X] \rangle + \langle Z, [X, Y] \rangle + \langle X, [Y, Z] \rangle . \tag{2.4}$$

Proof. Assume that ∇ is an affine connection on M which satisfies both properties (4) and (5). Then after applying (4) and (5) on the right-hand side of (2.4) we get $2\langle \nabla_Y Z, X \rangle$. Hence ∇ satisfies the Koszul formula. Therefore, there exists only one affine connection on M which satisfies both properties (4) and (5). For the existence, let us define $F(Y, Z, X)$ to be the right-hand side of (2.4).

A direct computation shows that the function $X \mapsto F(Y, Z, X)$ is a $\mathcal{F}(M)$-linear for fixed Y, Z, So, it is a 1-form. Thus there is a unique vector field, denoted by $\nabla_Y Z$, such that $2\langle \nabla_Y Z, X \rangle = F(Y, Z, X)$ for all X. Thus the Koszul formula holds and from it we can deduce properties (1)-(5). $\square$

Koszul's formula is named after Jean-Louis Koszul (1921-). Throughout this book we shall use the Levi-Civita connection on pseudo-Riemannian manifolds. Let $\{u_1, \ldots, u_n\}$ be a local coordinate system on an open subset U of a pseudo-Riemannian n-manifold M. The *Christoffel symbols* for the coordinate system are the real-valued functions Γ_{ij}^k on U such that $\nabla_{\partial_i}\partial_j = \sum_{k=1}^n \Gamma_{ij}^k \partial_k$, $1 \leq i,j \leq n$.

Since the connection ∇ is not a tensor, the Christoffel symbols do not obey the usual tensor transformation rule under change of coordinates.

For the Christoffel symbols we have the following.

Proposition 2.1. *Let M be a pseudo-Riemannian n-manifold and let $\{u_1, \ldots, u_n\}$ be a coordinate system on an open subset $U \subset M$. Then*

(1) $\nabla_{\partial_i} Y = \sum_{k=1}^n \left\{ \dfrac{\partial Y_k}{\partial u_i} + \sum_{j=1}^n \Gamma_{ij}^k Y_j \right\} \partial_k$ *and*

(2) $\Gamma_{ij}^k = \sum_{t=1}^n \dfrac{g^{kt}}{2} \left\{ \dfrac{\partial g_{jt}}{\partial u_i} + \dfrac{\partial g_{it}}{\partial u_j} - \dfrac{\partial g_{ij}}{\partial u_t} \right\}$,

where $Y = \sum_{j=1}^n Y_j \partial_j$ and (g^{ij}) is the inverse matrix of (g_{ij}).

Proof. Statement (1) is an immediate consequence of property (3) given in Definition 2.3. To prove (2), let us put $X = \partial_t, Y = \partial_i, Z = \partial_j$ in the Koszul formula. Since the brackets are zero, it leaves

$$2\langle \nabla_{\partial_i}\partial_j, \partial_t \rangle = \frac{\partial g_{jt}}{\partial u_i} + \frac{\partial g_{it}}{\partial u_j} - \frac{\partial g_{ij}}{\partial u_t}.$$

But from the definition of Christoffel symbols we have

$$2\langle \nabla_{\partial_i}\partial_j, \partial_t \rangle = 2 \sum_{k=1}^n \Gamma_{ij}^k g_{kt}.$$

Attacking both equations with $\sum_t g^{tk}$ yields the required formula. $\qquad\square$

Remark 2.1. Christoffel symbols, named after Elwin Bruno Christoffel (1829-1900), are the coordinate expressions for the Levi-Civita connection derived from the metric tensor. The Levi-Civita connection is named after Tullio Levi-Civita, although originally discovered in [Christoffel (1869)]. Nevertheless, T. Levi-Civita and G. Ricci used Christoffel's symbols to define the notion of parallel transport.

2.4 Parallel transport

Let $\phi : N \to M$ be a smooth map between two manifolds. The *differential* at a point $p \in N$ is a linear map $\phi_{*p} : T_pN \to T_{\phi(p)}M$ defined as follows: For each $X \in T_pN$, $\phi_{*p}X$ is the tangent vector in $T_{\phi(p)}M$ such that

$$(\phi_{*p}X)f = X(f \circ \phi), \quad \forall f \in \mathcal{F}(N).$$

The *dual* of the differential ϕ_* is denote by ϕ^*.

For any q-form ω on M, define the q-form $\phi^*\omega$ on N by

$$(\phi^*\omega)(X_1, \ldots, X_q) = \omega(\phi_* X_1, \ldots, \phi_* X_q), \ X_1, \ldots, X_q \in T_p N.$$

A vector field Z on a smooth map $\phi : P \to M$ between two manifolds is a mapping $Z : P \to TM$ such that $\pi \circ Z = \phi$, where π is the projection $TM \to M$. The simplest case of a vector field on a mapping is a vector field Z along a curve $\gamma : I \to M$ defined on an open interval I, where Z smoothly assigns to each $t \in I$ a tangent vector to M at $\gamma(t)$. For instance, the velocity vector field γ' on γ is a vector field on the curve γ. Let $\mathcal{V}(\gamma)$ denote the set consisting of smooth vector field of M along γ.

For a pseudo-Riemannian manifold M, there is a natural way to define the vector rate of change Z' of a vector field $Z \in \mathcal{V}(\gamma)$.

Proposition 2.2. *Let $\gamma : I \to M$ be a curve in a pseudo-Riemannian manifold M. Then there exists a unique function $Z \mapsto Z' = \frac{DZ}{dt}$ from $\mathcal{V}(\gamma) \to \mathcal{V}(\gamma)$ such that*

(1) $(aZ_1 + bZ_2)' = aZ_1' + bZ_2',$

(2) $(\lambda Z)' = \left(\dfrac{d\lambda}{dt}\right)Z + \lambda Z'$,

(3) $(V_\gamma)'(t) = \nabla_{\gamma'(t)} V,$

where $a, b \in \mathbb{R}, \lambda \in \mathcal{F}(I), V \in \mathfrak{X}(M)$ and $t \in I$. Furthermore, we have

(4) $\dfrac{d}{dt}\langle Z_1, Z_2 \rangle = \langle Z_1', Z_2 \rangle + \langle Z_1, Z_2' \rangle.$

Proof. For the uniqueness, let us assume that an induced connection exists which satisfy only the first three properties. We can assume that γ lies in the domain of a single coordinate system $\{u_1, \ldots, u_n\}$. For a vector field $Z \in \mathcal{V}(\gamma)$, we have

$$Z(t) = \sum_{j=1}^{n}(Z(t)u_j)\partial_j = \sum_{j=1}^{n}(Zu_j)(t)\partial_j.$$

Let us denote the component function $Zu_j : I \to \mathbb{R}$ by Z_j. Then, by properties (1), (2) and (3), we find

$$Z' = \sum_{j=1}^{n}\frac{dZ_j}{dt}\partial_j|_\gamma + \sum_{j}Z_j(\partial_j|_\gamma)'$$

$$= \sum_{j=1}^{n}\frac{dZ_j}{dt}\partial_j + \sum_{j=1}^{n}Z_j\nabla_{\gamma'}(\partial_j).$$

Thus Z' is completely determined by the Levi-Civita connection ∇. This shows uniqueness.

On any subinterval J of I such that $\gamma(J)$ lies in a coordinate neighborhood, let us define Z' by the formula above. Then straightforward computations show that all four properties hold. Now, it follows from the uniqueness that these local definitions of Z' gives rise to a single vector field $Z' \in \mathcal{V}(\gamma)$. $\qquad \square$

The $Z' = DZ/dt$ in Proposition 2.2 is called the *induced covariant derivative*. For a vector field Z along a curve γ, we simply write Z' for $\nabla_{\gamma'} Z$ and also γ'' for $\nabla_{\gamma'} \gamma'$. In terms of Christoffel symbols we have

$$Z' = \sum_{k=1}^{n} \left\{ \frac{dZ_k}{dt} + \sum_{i,j=1}^{n} \Gamma_{ij}^k \frac{d(u_i \circ \gamma)}{dt} Z_j \right\} \partial_k.$$

A vector field Z on γ is called *parallel* if $Z' = 0$ holds identically along γ. Hence $Z = \sum_k Z_k \partial_k$ is a parallel vector field if and only if $Z_1, \ldots, Z_n$ satisfy the system of ordinary differential equations:

$$\frac{dZ_k}{dt} + \sum_{i,j=1}^{n} \Gamma_{ij}^k \frac{d(u_i \circ \gamma)}{dt} Z_j = 0, \quad k = 1, \ldots, n.$$

Proposition 2.3. *For $\gamma : I \to M$, $a \in I$ and $z \in T_{\gamma(a)} M$, there exists a unique parallel vector field Z on γ such that $Z(a) = z$.*

Proof. Follows from the fundamental existence and uniqueness theorem of systems of first order linear equations. $\qquad \square$

Consider a curve $\gamma : I \to M$. Let $a, b \in I$ and $z \in T_{\gamma(a)} M$. The function

$$P = P_{\gamma(a)}^{\gamma(b)}(\gamma) : T_{\gamma(a)} M \to T_{\gamma(b)} M$$

sending each $z \in T_{\gamma(a)} M$ to $Z(\gamma(b))$ is called *parallel translation along γ from $\gamma(a)$ to $\gamma(b)$*, where Z is the unique parallel vector field along γ such that $Z(a) = z$.

Proposition 2.4. *Parallel translation is a linear isometry.*

Proof. Let $\gamma : I \to M$ be a curve and $p = \gamma(a)$, $q = \gamma(b)$. Let $u, v \in T_p M$ correspond to parallel vector fields U, V. Since $U + V$ is also parallel, we have

$$P(u + v) = (U + V)(b) = U(b) + V(b) = P(u) + P(v).$$

Similarly, we have $P(cu) = cP(u)$. Hence P is a linear map. For U, V as above, we get

$$\frac{d}{dt} \langle U, V \rangle = \langle U', V \rangle + \langle U, V' \rangle = 0.$$

Thus $\langle U, V \rangle$ is constant. Hence

$$\langle P(u), P(v) \rangle = \langle U(b), V(b) \rangle = \langle U(a), V(a) \rangle = \langle u, v \rangle,$$

which implies that P is an isometry. $\square$

Definition 2.4. A *geodesic* in a pseudo-Riemannian manifold M is a curve $\gamma : I \to M$ whose velocity vector field γ' is parallel, or equivalently, it satisfies

$$\frac{d^2(u_k \circ \gamma)}{dt^2} + \sum_{i,j=1}^n \Gamma_{ij}^k(\gamma) \frac{d(u_i \circ \gamma)}{dt} \frac{d(u_j \circ \gamma)}{dt} = 0 \qquad (2.5)$$

for $k = 1, \ldots, n$.

It follows from the existence and uniqueness theorem of linear system of ordinary differential equations that, for any given point $p \in M$ and any given tangent vector $v \in T_pM$, there exists a unit geodesic γ_v such that $\gamma(0) = p$ and $\gamma'(0) = v$. A geodesic with largest possible domain is called a *maximal geodesic*.

A pseudo-Riemannian manifold M for which every maximal geodesic is defined on the entire real line is said to be *geodesic complete* or simply *complete*.

It follows from (2.5) that the geodesic of a pseudo-Euclidean space $\mathbb{E}_s^m$ are straight lines. Thus every pseudo-Euclidean n-plane $\mathbb{E}_s^m$ is geodetically complete.

In general, parallel translation from a point p to another point q depends on the particular curve jointing two points p and q. However, on a pseudo-Euclidean space $\mathbb{E}_s^m$ the natural coordinate vector fields are parallel and hence so their restrictions to any curve. Consequently, parallel translation from a point p to another point q along any curve is just the canonical isomorphism $v_p \to v_q$. This phenomenon is called *distant parallelism*.

For a given $v \in T_pM$, there is a unique geodesic γ_v such that $\gamma_v(0) = p$ with initial tangent vector $\gamma_v'(0) = v$. Let U_p be the set of vectors $v \in T_pM$ such that the geodesic γ_v is defined at least on $[0, 1]$. For a vector $v \in U_p$ the *exponential map* is defined by $\exp_p(v) = \gamma_v(1)$.

Definition 2.5. A subset S of a vector space is called *star shaped about* o if $v \in S$ implies $tv \in S$ for all $t \in [0, 1]$. For each point $p \in M$ there exists a neighborhood $\mathcal{U}$ of o in T_pM on which the exponential map $\exp_o$ is a diffeomorphism onto a neighborhood U of p on M. If $\mathcal{U}$ is starshaped about o, then U is called a *normal neighborhood* of p.

Definition 2.6. Let $\{e_1, \ldots, e_n\}$ be an orthonormal basis of T_pM so that $\langle e_i, e_j \rangle = \epsilon_i \delta_{ij}$. The *normal coordinate system* $\{y_1, \ldots, y_n\}$ determined by $e_1, \ldots, e_n$ assigns to each point $q \in U$ the vector coordinates relative to $e_1, \ldots, e_n$ of the corresponding point $\exp_p^{-1}(q) \in \mathcal{U} \subset T_pM$. In other words,

$$\exp_p^{-1}(q) = \sum_{i=1}^{n} y_i(q)e_i, \quad q \in U.$$

The following proposition is well-known.

Proposition 2.5. *Let* $\{y_1, \ldots, y_n\}$ *be a normal coordinate system about a point* $p \in M$. *Then* $g_{ij}(0) = \delta_{ij}$ *and* $\Gamma^i_{jk}(p) = 0$.

Definition 2.7. Let p be a point in a Riemannian manifold M. Let S_r be the hypersphere of T_pM with radius r centered at the origin o. Suppose that r is a sufficiently small positive number such that $\exp_p(S_r)$ lies in a normal coordinate neighborhood of p, then $\exp_p(S_r)$ is called a *geodesic hypersphere*. The set $B_r(p) = \{u \in M : d(u, p) \leq r\}$ is called a *geodesic ball* of radius r and with center p (cf. [Besse (1978); Chen and Vanhecke (1981)]).

2.5 Riemann curvature tensor

Gauss' "theorema egregium" shows that the Gauss curvature, defined as product of two two principal curvatures, of a surface in a Euclidean 3-space $\mathbb{E}^3$ is an isometric invariant of the surface itself. This lead G. Riemann to his invention of Riemannian geometry, whose most important feature is the generalization of Gauss curvature to arbitrary Riemannian manifolds. No significant changes are required in extending from Riemannian to pseudo-Riemannian manifolds.

For a pseudo-Riemannian manifold M with Levi-Civita connection ∇, the function $R : \mathfrak{X}(M) \times \mathfrak{X}(M) \times \mathfrak{X}(M) \to \mathfrak{X}(M)$ defined by

$$R(X, Y)Z = \nabla_X \nabla_Y Z - \nabla_Y \nabla_X Z - \nabla_{[X,Y]} Z$$

is a $(1, 3)$ tensor field, called the *Riemann curvature tensor*. Sometimes, we put

$$R(X, Y; Z, W) = \langle R(X, Y)Z, W \rangle.$$

Proposition 2.6. *The curvature tensor R satisfies the following properties:*

$$R(u, v)w = -R(v, u)w, \tag{2.6}$$

$$\langle R(u, v)w, z \rangle = -\langle R(u, v)z, w \rangle, \tag{2.7}$$

$$R(u, v)w + R(v, w)u + R(w, u)v = 0, \tag{2.8}$$

$$\langle R(u, v)w, z \rangle = \langle R(w, z)u, v \rangle \tag{2.9}$$

for vectors $u, v, w, z \in T_p M$, $p \in M$.

Proof. Since both ∇ and the bracket operation on vector fields are local operations, it suffices to work on any neighborhood of p. Moreover, because the identities are tensor equations, u, v, w, z can be extended to local vector fields U, V, W, Z on some neighborhood of p in any convenient way. In particular, we may choose the extensions in such way that all of their brackets are zero.

Since $R(U, V)W = [\nabla_U, \nabla_V]W - \nabla_{[U,V]}W$ and the bracket operation is skew-symmetric, (2.6) follows immediately from the definition of the curvature tensor.

For (2.7) we only need to show that $\langle R(u, v)w, z \rangle = 0$ by polarization. By Theorem 2.1(5), we have

$$\langle R(U, V)W, W \rangle = \langle \nabla_U \nabla_V W, W \rangle - \langle \nabla_V \nabla_U W, W \rangle$$

$$= \langle \nabla_U W, \nabla_V W \rangle - V\langle \nabla_U W, W \rangle + \langle \nabla_V W, \nabla_U W \rangle - U\langle \nabla_V W, W \rangle$$

$$= \frac{1}{2} UV\langle W, W \rangle - \frac{1}{2} VU\langle W, W \rangle = 0.$$

This proves (2.7), since $[U, V] = 0$.

For (2.8) we consider $\mathcal{S}$, the sum of cyclic permutations of U, V, W, to find

$$R(U, V)W + R(V, W)U + R(W, U)V$$

$$= \mathcal{S}R(U, V)W$$

$$= \mathcal{S}\nabla_U \nabla_V W - \mathcal{S}\nabla_V \nabla_U W$$

$$= \mathcal{S}\nabla_U \nabla_V W - \mathcal{S}\nabla_U \nabla_W V$$

$$= \mathcal{S}\nabla_U [Y, W] = 0.$$

If we put

$$S(u, v, w, z) = \langle R(u, v)w, z \rangle + \langle R(v, w)u, z \rangle + \langle R(w, u)v), z \rangle,$$

then a direct computation shows that

$$0 = S(u, v, w, z) - S(v, w, z, u) - S(w, z, u, v) + S(z, u, v, w)$$

$$= \langle R(u, v)w, z \rangle - \langle R(v, u)w, z \rangle - \langle R(w, z)u, v \rangle + \langle R(z, w)u, v \rangle.$$

Thus, by applying (2.6), we obtain (2.9). $\qquad\square$

Equation (2.8) is called the *first Bianchi identity*.

Proposition 2.7. *The curvature tensor of a pseudo-Riemannian manifold M satisfies the second Bianchi identity:*

$$(\nabla_W R)(U, V) + (\nabla_U R)(V, W) + (\nabla_V R)(W, U) = 0, \tag{2.10}$$

where $(\nabla_W R)(U, V)$ *is defined by*

$$\begin{aligned}((\nabla_W R)(U, V))Z &= \nabla_W(R(U, V)Z) - R(\nabla_W U, V)Z \\ &- R(U, \nabla_W V)Z - R(U, V)(\nabla_W Z).\end{aligned} \tag{2.11}$$

Proof. Clearly, (2.10) is a tensor identity. Let p be a given point in M. We consider a normal coordinate system on a neighborhood of p.

We choose the extensions U, V, W of vectors $u, v, w \in T_p M$ in such way that not only all brackets vanishes identically, but also the extensions have constant components with respect to the normal coordinate system. Hence, by Proposition 2.6, we have

$$(\nabla_W R)(U, V)Z = \nabla_W(R(U, V)Z) - R(U, V)(\nabla_W Z)$$

at p, which gives $(\nabla_W R)(U, V) = [\nabla_W, R(U, V)] = [\nabla_W, [\nabla_U, \nabla_V]]$ at p. Thus summing the above formula over the cyclic permutations of U, V, W yields the required identity at p. □

2.6 Sectional, Ricci and scalar curvatures

Since the Riemann curvature tensor is rather complicated, we consider a simpler real-valued function, the sectional curvature, which completely determines the curvature tensor.

At a point $p \in M$, a 2-dimensional linear subspace π of the tangent space $T_p M$ is called a *plane section*. For a given basis $\{v, w\}$ of the plane section π, we define a real number by

$$Q(v, w) = \langle v, v \rangle \langle w, w \rangle - \langle v, w \rangle^2.$$

The plane section π is called *non-degenerate* if and only if $Q(u, v) \neq 0$. $Q(u, v)$ is positive if $g|_\pi$ is definite, and it is negative if $g|_\pi$ is indefinite. The absolute value $\|Q(u, v)\|$ is the square of the area of the parallelogram with sides u and v.

For a non-degenerate plane section π at p, the number

$$K(u, v) = \frac{\langle R(u, v)v, u \rangle}{Q(u, v)}$$

is independent of the choice of basis $\{u, v\}$ for π, which is called the *sectional curvature* $K(\pi)$ of π.

A pseudo-Riemannian manifold is said to be *flat* if its sectional curvature vanishes identically. It is well-known that a pseudo-Riemannian manifold M is flat if and only if its curvature tensor vanishes at every point.

For any index s, the pseudo-Euclidean m-space $\mathbb{E}_s^m$ is flat. In fact, all Christoffel symbols vanish for a natural coordinate system. Hence the curvature tensor of $\mathbb{E}_s^m$ vanishes identically.

Definition 2.8. A multilinear function $F : T_pM \times T_pM \times T_pM \times T_pM \to \mathbb{R}$ is called *curvature-like* if F satisfies properties (2.6)-(2.9) for the function $(u, v, w, z) \to \langle R(u, v)w, z\rangle$.

For a curvature-like function F we have the following.

Lemma 2.3. *Let M be a pseudo-Riemannian manifold and $p \in M$. If F is a curvature-like function on T_pM such that*

$$K(u, v) = \frac{F(u, v, v, u)}{Q(u, v)}$$

whenever u, v span a non-degenerate plane at p, then

$$\langle R(u, v)w, z\rangle = F(u, v, w, z)$$

for all $u, v, w, z \in T_pM$.

Proof. Put $\delta(u, v, w, z) = F(u, v, w, z) - \langle R(u, v)w, z\rangle$. Then δ is also curvature-like. Since $\delta(u, v, v, u) = 0$ whenever u, v span a non-degenerate plane section at p, we obtain $\delta = 0$. $\qquad\square$

For sectional curvature K of indefinite Riemannian manifolds, we have the following result [Kulkarni (1979)].

Theorem 2.2. *Let M be a pseudo-Riemannian manifold of dimension ≥ 3 and index $s > 0$. Then, at each point $p \in M$, the following four conditions are equivalent:*

(1) *K is constant;*
(2) *$a \leq K$ or $K \leq b$;*
(3) *$a \leq K \leq b$ on indefinite planes;*
(4) *$a \leq K \leq b$ on definite planes,*

where a and b are real numbers.

It follows from Theorem 2.2 that the sectional curvature of an indefinite Riemannian manifold at each point is unbounded from above and below unless M has constant sectional curvature.

Definition 2.9. The *Ricci tensor* of a pseudo-Riemannian n-manifold M, denoted by Ric, is a symmetric $(0,2)$ tensor defined by

$$Ric(X,Y) = \text{Tr}\{Z \mapsto R(Z,X)Y\},$$

or equivalently,

$$Ric(X,Y) = \sum_{\ell=1}^{n} \epsilon_\ell \langle R(e_\ell, X)Y, e_\ell \rangle, \tag{2.12}$$

where $e_1, \ldots, e_m$ is an orthonormal frame.

It is well-known that $Ric(X,Y)$ is independent of choice of orthonormal frame $e_1, \ldots, e_n$. If the Ricci tensor vanishes, M is called *Ricci flat*.

Definition 2.10. A pseudo-Riemannian manifold M is called an *Einstein manifold* if $Ric = cg$ for some constant c. For a unit vector $u \in TM$, the *Ricci curvature* $Ric(u)$ is defined by $Ric(u) = Ric(u,u)$.

If M is a pseudo-Riemannian manifold of dimension ≥ 3 which satisfies $Ric = fg$ for some function $f \in \mathcal{F}(M)$, then M is always Einsteinian.

Definition 2.11. The *scalar curvature* τ of M is defined by

$$\tau = \sum_{i<j} K(e_i, e_j), \tag{2.13}$$

where $e_1, \ldots, e_n$ is an orthonormal frame of M. The scalar curvature τ is independent of the choice of the orthonormal frame.

Remark 2.2. Every flat manifold is Ricci flat. But the converse does not hold. Thus the Ricci curvature of a Riemannian n-manifold provides one way of measuring the degree to which the geometry determined by the given Riemannian metric on M differ from that of the Euclidean n-space. The scalar curvature represents the amount by which the volume of a geodesic ball in a curved Riemannian manifold deviates from that of the standard ball in Euclidean space.

In relativity theory, the Ricci tensor is related to the matter content of the universe via Einstein's field equation. It is the part of the curvature of spacetime that determines the degree to which matter will tend to converge or diverge in time. The scalar curvature is the Lagrangian density for the Einstein-Hilbert action, first proposed in [Hilbert (1915)], that yields the Einstein field equations through the principle of least action.

2.7   Indefinite real space forms

A pseudo-Riemannian manifold M is said to have *constant curvature* if its sectional curvature is constant. For a constant c the function F defined by

$$F(u, v, w, z) = c\{\langle u, z \rangle \langle v, w \rangle - \langle u, w \rangle \langle v, z \rangle\}$$

is curvature-like. Thus Lemma 2.3 implies that $F(u, v, v, u) = cQ(u, v)$. Hence if u, v span a non-degenerate plane section, we have

$$K(u, v) = c = \frac{F(u, v, v, u)}{Q(u, v)}.$$

Consequently, if a pseudo-Riemannian manifold M is of constant curvature c, then its curvature tensor R satisfies

$$R(u, v)w = c\{\langle v, w \rangle u - \langle u, w \rangle v\}. \tag{2.14}$$

Let $\mathbb{E}_t^n$ be the pseudo-Euclidean n-space equipped with the canonical pseudo-Euclidean metric of index t given by

$$g_0 = -\sum_{i=1}^{t} dx_i^2 + \sum_{j=t+1}^{n} dx_j^2, \tag{2.15}$$

where $(x_1, \ldots, x_n)$ is a rectangular coordinate system of $\mathbb{E}_t^n$.

Let c be a nonzero real number. We put

$$S_s^k(\mathbf{x}_0, c) = \left\{ \mathbf{x} \in \mathbb{E}_s^{k+1} : \langle \mathbf{x} - \mathbf{x}_0, \mathbf{x} - \mathbf{x}_0 \rangle = \frac{1}{c} > 0 \right\}, \quad s > 0, \tag{2.16}$$

$$H_s^k(\mathbf{x}_0, c) = \left\{ \mathbf{x} \in \mathbb{E}_{s+1}^{k+1} : \langle \mathbf{x} - \mathbf{x}_0, \mathbf{x} - \mathbf{x}_0 \rangle = \frac{1}{c} < 0 \right\}, \quad s > 0, \tag{2.17}$$

$$H^k(c) = \left\{ \mathbf{x} \in \mathbb{E}_1^{k+1} : \langle \mathbf{x}, \mathbf{x} \rangle = \frac{1}{c} < 0 \text{ and } x_1 > 0 \right\}, \tag{2.18}$$

where $\langle \ , \ \rangle$ is the associated scalar product and $\mathbf{x} = (x_1, \ldots, x_n)$.

$S_s^k(\mathbf{x}_0, c)$ and $H_s^k(\mathbf{x}_0, c)$ are pseudo-Riemannian manifolds of curvature c with index s, known as a *pseudo sphere* and a *pseudo-hyperbolic space*, respectively. The point $\mathbf{x}_0$ is called the center of $S_s^m(\mathbf{x}_0, c)$ and $H_s^m(\mathbf{x}_0, c)$. If $\mathbf{x}_0$ is the origin o, we simply denote $S_s^k(o, c)$ and $H_s^k(o, c)$ by $S_s^k(c)$ and $H_s^k(c)$, respectively. The pseudo-Riemannian manifolds $\mathbb{E}_s^k$, $S_s^k(c)$, $H_s^k(c)$ are the standard models of the *indefinite real space forms*. In particular, $\mathbb{E}_1^k$, $S_1^k(c)$, $H_1^k(c)$ are the standard models of *Lorentzian space forms*. Topologically, a de Sitter spacetime S_1^k is $\mathbb{R} \times S^{k-1}$. Thus when $k \geq 3$ a de Sitter spacetime is simply-connected.

The S_1^4 and H_1^4 are known as the *de Sitter spacetime* and *anti-de Sitter spacetime*, respectively; named after Willem de Sitter (1872-1934), a Dutch mathematician, physicist and astronomer.

When $s = 0$, the manifolds $\mathbb{E}^k$, $S^k(c)$ and $H^k(c)$ are of constant curvature, called *real space forms*. The Euclidean k-space $\mathbb{E}^k$, the k-sphere $S^k(c)$ and the hyperbolic k-space $H^k(c)$ are simply-connected complete Riemannian manifolds of constant curvature $0, c > 0$ and $c < 0$, respectively. A complete simply-connected pseudo-Riemannian k-manifold, $k \geq 3$, of constant curvature c and index s is isometric to $\mathbb{E}_s^k$, or $S_s^k(c)$ or $H_s^k(c)$ according to $c = 0$, or $c > 0$ or $c < 0$, respectively. We denote a k-dimensional indefinite space form of curvature curvature c and index s simply by $R_s^k(c)$. We simply denote the indefinite space form $R_0^k(c)$ with index $s = 0$ by $R^k(c)$.

2.8 Gradient, Hessian and Laplacian

Definition 2.12. Let M be a pseudo-Riemannian n-manifold. For $f \in \mathcal{F}(M)$, the *gradient* of f, denote by ∇f (or by $\operatorname{grad} f$), is the vector field dual to the differential df. In other word, ∇f is defined by

$$\langle \nabla f, X \rangle = df(X) = Xf \quad \forall X \in \mathfrak{X}(M). \tag{2.19}$$

In terms of a coordinate system $\{u_1, \ldots, u_n\}$ of M, we have

$$df = \sum_{j=1}^{n} \frac{\partial f}{\partial u_j} du_j \quad \text{and} \quad \nabla f = \sum_{i,j} g^{ij} \frac{\partial f}{\partial u_i} \partial_j. \tag{2.20}$$

Definition 2.13. If $X \in \mathfrak{X}(M)$ and $\{e_1, \ldots, e_n\}$ is an orthonormal frame, the *divergence* of X, denoted by $\operatorname{div} X$, is defined by

$$\operatorname{div} X = \sum_{j=1}^{n} \epsilon_j \langle \nabla_{e_i} X, e_i \rangle. \tag{2.21}$$

If we put $X = \sum_{j=1}^{n} X^j \frac{\partial}{\partial u_j}$, $X_i = \sum_{j=1}^{n} g_{ij} X^j$, then

$$\operatorname{div} X = \sum_{j=1}^{n} \left\{ \frac{\partial X_i}{\partial u_i} + \sum_{k=1}^{n} \Gamma_{jk}^j X_k \right\}. \tag{2.22}$$

Definition 2.14. The *Hessian* of $f \in \mathcal{F}(M)$, denoted by H^f, is the second covariant differential $\nabla(\nabla f)$, so that

$$H^f(X, Y) = XYf - (\nabla_X Y)f = \langle \nabla_X(\nabla f), Y \rangle, \quad X, Y \in \mathfrak{X}(M). \tag{2.23}$$

Definition 2.15. The *Laplacian* of $f \in \mathcal{F}(M)$, denoted by Δf, is defined by $\Delta f = -\operatorname{div}(\nabla f)$. In terms of a coordinate system $\{u_1, \ldots, u_n\}$, we have

$$\Delta f = -\sum_{i,j=1}^{n} \left\{ \frac{\partial^2 f}{\partial u_i \partial u_j} - \sum_{k=1}^{n} \Gamma_{ij}^k \frac{\partial f}{\partial u_k} \right\}. \tag{2.24}$$

The Laplacian (or the Laplace operator) is named after Pierre-Simon de Laplace (1749-1827), who first applied the operator to the study of celestial mechanics in the 1770s.

In terms of a natural coordinate system $\{x_1, \ldots, x_n\}$ of $\mathbb{E}_s^n$, we have

$$\nabla f = \sum_{j=1}^n \epsilon_j \frac{\partial f}{\partial x_j} \partial_j, \quad \text{div} \, X = \sum_{j=1}^n \frac{\partial X_j}{\partial x_j}, \quad \Delta f = -\sum_{j=1}^n \epsilon_j \frac{\partial^2 f}{\partial x_j^2}. \tag{2.25}$$

The next theorem is well-known as *Omori's Maximum Principle*.

Theorem 2.3. [Omori (1967)] *Let M be a complete Riemannian manifold whose sectional curvature has a lower bound. If a function $f \in \mathcal{F}(M)$ has an upper bound, then for any $\epsilon > 0$, there is a point $p \in M$ such that $\|\nabla f(p)\| < \epsilon$ and $m(p) = \max\{H^f(X, X) : \|X\| = 1, X \in T_pM\} < \epsilon$.*

We also have the following generalized Omori-Yau maximum principle (cf. [Cheng and Yau (1976)]).

Theorem 2.4. *Let N be a complete Riemannian manifold whose Ricci curvature is bounded from below and let u is a smooth non-negative function on N. If there is a positive constant $k > 0$ such that $\Delta u \leq -ku^2$ on N, then $u = 0$ on N.*

Definition 2.16. A *volume element* on a pseudo-Riemannian n-manifold M is a smooth n-form ω such that $\omega(e_1, \ldots, e_n) = \pm 1$ for every orthonormal frame on M. Notice that volume elements always exist at least locally.

Lemma 2.4. *A pseudo-Riemannian manifold has a global volume element if and only if it is orientable.*

Proof. If a pseudo-Riemannian manifold M has a global volume element ω, then the bases $\{v_1, \ldots, v_n\}$ for each T_pM with $\omega(v_1, \ldots, v_n) > 0$ constitute an orientation of T_pM. Conversely, if M is oriented, then for all positively oriented coordinate systems the local volume elements agree on the overlaps, hence give a global volume element. $\square$

2.9   Lie derivative and Killing vector fields

The Lie derivative was named after Sophus Lie in [Slebodziński (1931)]. On a manifold M the *Lie derivative* $\mathcal{L}$ is a tensor derivation such that for any $V \in \mathfrak{X}(M)$ we have

$$\mathcal{L}_V f = V f, \quad \mathcal{L}_V X = [V, X], \quad \forall f \in \mathcal{F}(M), \forall X \in \mathfrak{X}(M).$$

Definition 2.17. A *Killing vector field* on a pseudo-Riemannian manifold is a vector field X for which the Lie derivative of the metric tensor vanishes, i.e., $\mathcal{L}_X g = 0$. A *conformal-Killing vector field* is a vector field X for which $\mathcal{L}_X g = \lambda g$ for some function $\lambda \in \mathcal{F}(M)$.

Under the flow of a Killing vector field X, the metric tensor does not change. Thus a Killing vector field is an infinitesimal isometry. Killing vector fields were named after Wilhelm Killing (1847-1923).

Proposition 2.8. *On a pseudo-Riemannian manifold M, the following three conditions on a vector field X of M are equivalent:*

(1) X *is a Killing vector field.*
(2) $X \langle V, W \rangle = \langle [X, V], W \rangle + \langle V, [X, W] \rangle, \quad \forall V, W \in \mathfrak{X}(M).$
(3) ∇X *is a skew-adjoint relative to the metric tensor g, i.e.,*

$$\langle \nabla_V X, W \rangle = - \langle \nabla_W X, V \rangle, \quad \forall V, W \in \mathfrak{X}(M).$$

Proof. For all $V, W \in \mathfrak{X}(M)$, the following are equivalent:

$$\langle \nabla_V X, W \rangle + \langle \nabla_W X, V \rangle = 0;$$
$$\langle \nabla_X V, W \rangle - \langle [X, V], W \rangle = \langle [X, W], V \rangle - \langle \nabla_X W, V \rangle;$$
$$X \langle V, W \rangle = \langle [X, V], W \rangle + \langle V, [X, W] \rangle.$$

In view of the product rule, the last one is equivalent to $\mathcal{L}_X g = 0$. $\square$

Definition 2.18. Let N and M be pseudo-Riemannian manifolds with metrics g_N and g_M. An *isometry* $\psi : N \to M$ is a diffeomorphism that preserves metric tensors, i.e., $\psi^*(g_M) = g_N$.

Definition 2.19. A map $\phi : N \to M$ between two pseudo-Riemannian manifolds is called a *local isometry* at $p \in N$ if there is a neighborhood $U \subset N$ of p such that $\phi : U \to \phi(U)$ is a diffeomorphism satisfying

$$\langle u, v \rangle_p = \left\langle \phi_{*p}(u), \phi_{*p}(v) \right\rangle_{\phi(p)}, \quad \forall v \in T_p N, \, \forall p \in N. \tag{2.26}$$

Definition 2.20. A pseudo-Riemannian manifold N is said to be *locally isometric* to a pseudo-Riemannian manifold M if for each $p \in N$ there exists a neighborhood U of p and a local isometry $\phi : U \to \phi(U) \subset M$.

The following result shows that a local isometry is uniquely determined by its differential map at a point.

Lemma 2.5. *Let $\phi_1, \phi_2 : N \to M$ be two local isometries. If there is a point $p \in N$ such that $\phi_{1*p} = \phi_{2*p}$, then $\phi_1 = \phi_2$.*

Proof. Put $U = \{q \in N : \phi_{1*q} = \phi_{2*q}\}$. Then U is a closed subset of N. Let $q \in U$ and V a normal neighborhood of q. Then for each point $z \in V$ there is a vector $v \in T_qN$ such that $\gamma_v(1) = \exp_q(v) = z$. Hence

$$\phi_1(z) = \phi_1(\gamma_v(1)) = \gamma_{\phi_{1*}v}(1) = \gamma_{\phi_{2*}v}(1) = \phi_2(\gamma_v(1)) = \phi_2(z),$$

which implies that $\phi_1 = \phi_2$ on V. Hence we have $\phi_{1*z} = \phi_{2*z}$ for all $z \in V$. Thus U is also a open subset of M. Therefore, we have $U = M$. Consequently, we get $\phi_1 = \phi_2$.   □

Definition 2.21. A map $\phi : N \to M$ between two pseudo-Riemannian manifolds is called *conformal* if $\phi^*(g_M) = fg_N$ for some function $f \in \mathcal{F}(N)$ such that $f > 0$ or $f < 0$. In particular, if the function f is a nonzero real number, then ϕ is called a *homothety*.

Lemma 2.6. *Homothety preserves Levi-Civita connection of pseudo-Riemannian manifolds.*

Proof. Follows immediately from Koszul's formula.   □

2.10   Weyl conformal curvature tensor

Let M be a pseudo-Riemannian m-manifold. Associated with the Ricci tensor Ric, define a $(1,1)$-tensor Q by $\langle Q(X), Y \rangle = Ric(X, Y)$.

The *Weyl conformal curvature tensor*, denoted by C, is a tensor field of type $(1,3)$ defined by

$$C(X,Y)Z = R(X,Y)Z + \frac{1}{m}\{Ric(X,Z)Y - Ric(Y,Z)X$$
$$+ \langle X, Z \rangle QY - \langle Y, Z \rangle QX\} - \frac{2\tau}{m(m+1)}\{\langle X, Z \rangle Y - \langle Y, Z \rangle X\}.$$

It is well-known that the Weyl conformal curvature tensor C vanishes identically when $\dim M = 3$ and it is invariant under conformal changes of the metric.

Definition 2.22. A pseudo-Riemannian metric g on a manifold M is called *conformally flat* if it is conformally related with a flat pseudo-Euclidean metric. A manifold with a conformal flat pseudo-Riemannian metric is called a *conformally flat manifold*.

The following important result of [Weyl (1918)] is well-known.

Theorem 2.5. *A pseudo-Riemannian manifold M of dimension ≥ 4 is conformally flat if and only if the conformal curvature tensor C vanishes identical.*

Chapter 3

Hodge Theory and Spectral Geometry

Hodge theory is a branch of algebraic geometry, algebraic topology and complex manifold theory that deals with the decomposition of the cohomology groups of a complex projective algebraic variety. The theory was developed in the 1930s by W. V. D. Hodge (1903–1975) as an extension of de Rham cohomology. Hodge theory has major applications to Riemannian manifolds, Kähler manifolds and algebraic geometry.

Spectral geometry is an area concerning relationship between geometric structures of manifolds and spectra of canonically defined differential operators. In particular, between geometric structures of manifolds and the spectra of the Laplacian (acting on differentiable functions) and of the Hodge-Laplace operator (on differentiable forms). This subject concerns with two kinds of questions: direct and inverse problems. Direct problems study properties of the spectra of canonically defined differential operators. Inverse problems seek to identify features of the geometry from information about the eigenvalues of the Laplacian or of Hodge-Laplace operator. One of the earliest results on inverse problems was due to H. Weyl who used D. Hilbert's theory of integral equation in [Weyl (1912)] to show that the volume of a bounded domain in Euclidean space can be determined from the asymptotic behavior of the eigenvalues for the Dirichlet boundary value problem of the Laplacian. A refinement of Weyl's asymptotic formula obtained in [Minakshisundaram and Pleijel (1949)] produces a series of local spectral invariants involving covariant differentiations of the curvature tensor, which can be used to establish spectral rigidity for a special class of manifolds. However, it was shown in [Milnor (1964)] that there are 16-dimensional flat tori which are not isometric, but with the same spectrum. Milnor's examples tells us that the information of eigenvalues alone is not enough to determine the isometry class of a manifold.

3.1   Operators d, $*$ and δ

In differential geometry, the *exterior derivative* d extends the concept of the differential of a function on a smooth manifold M, which is a form of degree zero, to differential forms of higher degree.

The exterior derivative d has the property that $d^2 = 0$ and is the differential (coboundary) used to define de Rham cohomology on forms. Integration of forms gives a natural homomorphism from the de Rham cohomology to the singular cohomology of a smooth manifold. The exterior derivative of a differential form of degree k is a differential form of degree $k + 1$.

If f is in $\mathcal{F}(M)$, then the exterior derivative of f is its differential of f. That is, df is the unique 1-form such that, for every $X \in \mathfrak{X}(M)$,

$$df(X) = Xf,$$

where Xf is the directional derivative of f in the direction of X. Thus the exterior derivative of a 0-form is a 1-form.

The *exterior derivative* is defined to be the unique $\mathbb{R}$-linear mapping from k-forms to $(k + 1)$-forms satisfying the following properties:

(1) df is the differential of f for $f \in \mathcal{F}(M)$.
(2) $d(df) = 0$ for any $f \in \mathcal{F}(M)$.
(3) $d(\alpha \wedge \beta) = d\alpha \wedge \beta + (-1)^p(\alpha \wedge d\beta)$, where α is a p-form, i.e., d is a derivation of degree one on the exterior algebra of differential forms.

We denote the space of all k-forms on M by $\Omega^k(M)$. To deal with differential forms, one can work entirely in a local coordinate system $x_1, \ldots, x_n$ on a manifold M. First, the coordinate differentials $dx_1, \ldots, dx_n$ form a basic set of 1-forms within the coordinate chart. Given a multi-index $(i_1, \ldots, i_k)$ with $1 \leq i_j \leq n$ for $1 \leq j \leq k$, the exterior derivative of a k-form $\omega = \sum_{1 \leq i_1 < \cdots < i_k \leq n} f_{i_1 \cdots i_k} dx_{i_1} \wedge \cdots \wedge dx_{i_k}$ is

$$d\omega = \sum_{i=1}^{n} \sum_{1 \leq i_1 < \cdots < i_k \leq n} \frac{\partial f_{i_{i_1} \cdots i_1}}{\partial x_i} dx_i \wedge dx_{i_1} \wedge \cdots \wedge dx_{i_k}.$$

Alternatively, an explicit formula can be given for the exterior derivative of a k-form ω, when paired with $k+1$ arbitrary smooth vector fields $V_0, \ldots, V_k \in \mathfrak{X}(M)$. In this case, we have

$$d\omega(V_0, V_1, \ldots, V_k) = \sum_{i=0}^{k} (-1)^i V_i(\omega(V_0, \ldots, \hat{V}_i, \ldots, V_k))$$

$$+ \sum_{i<j} (-1)^{i+j} \omega([V_i, V_j], V_0, \ldots, \hat{V}_i, \ldots, \hat{V}_j, \ldots, V_k),$$

$$(3.1)$$

where $[\ ,\ ]$ is Lie bracket and $\hat{\ }$ denotes the omission of that element. In particular, for 1-forms we have

$$d\omega(X, Y) = X\omega(Y) - Y\omega(X) - \omega([X, Y]).$$

Definition 3.1. A differential form ω is called a *closed form* if $d\omega = 0$. The image of d are called *exact forms*.

Closed and exact forms are related, because of the identity $d^2\omega = 0$ for any k-form ω. This implies that every exact form is closed. The converse is true in contractible regions according to the *Poincaré lemma*.

The *de Rham complex* is the cochain complex of exterior differential forms on M, with the exterior derivative as the differential

$$0 \longrightarrow \Omega^0(M) \xrightarrow{d} \Omega^1(M) \xrightarrow{d} \Omega^2(M) \xrightarrow{d} \Omega^3(M) \longrightarrow \cdots, \qquad (3.2)$$

where $\Omega^0(M) = \mathcal{F}(M)$ is the space of smooth functions on M.

The idea of de Rham cohomology is to classify the different types of closed forms on a manifold M. One performs this classification by saying that two closed forms $\alpha, \beta \in \Omega^k(M)$ are *cohomologous* if they differ by an exact form, i.e., if $\alpha - \beta$ is exact. This classification induces an equivalence relation on the space of closed forms in $\Omega^k(M)$. One then defines the k-th *de Rham cohomology group* to be the set of equivalence classes, i.e., the set of closed forms in $\Omega^k(M)$ modulo the exact forms.

On an oriented Riemannian n-manifold M, if we choose an orthonormal local frame $e_1, \ldots, e_n$ of TM whose orientation is compatible with that of M, then $\omega^1 \wedge \cdots \wedge \omega^n$ is the volume element of M, where $\{\omega^1, \ldots, \omega^n\}$ is the dual frame of $\{e_1, \ldots, e_n\}$.

The *Hodge star operator* $*$, introduced by W. V. D. Hodge in the 1930s, is an isomorphism $* : \Omega^k(M) \to \Omega^{n-k}(M)$ defined as follows:

For a k-form $\omega = \sum_{i_1 < \cdots < i_k} f_{i_1 \cdots i_k} \omega^{i_1} \wedge \cdots \wedge \omega^{i_k}$, we define

$$*\omega = \sum_{j_1 < \cdots < j_{n-k}} \epsilon_{i_1 \cdots i_k j_1 \cdots j_{n-k}} f_{i_1 \cdots i_k} \omega^{j_1} \wedge \cdots \wedge \omega^{j_{n-k}},$$

where $\epsilon_{i_1 \cdots i_k j_1 \cdots j_{n-k}}$ is 0 if $i_1 \cdots i_k j_1 \cdots j_{n-k}$ does not form a permutation of $1, \ldots, n$; and 1 or -1, according to the permutation is even or odd.

The form $*\omega$ is called the *adjoint* (or the *Hodge dual*) of ω. The adjoint of 1 is just the volume form $\omega^1 \wedge \cdots \wedge \omega^n$.

Lemma 3.1. *The star operator* $*$ *satisfies*

(1) $*(\alpha + \beta) = *\alpha + *\beta$ *and* $*(f\alpha) = f(*\alpha)$;
(2) $*(*\alpha) = (-1)^{nk+k}\alpha$;

(3) $\alpha \wedge *\beta = \beta \wedge *\alpha$;
(4) $\alpha \wedge *\alpha = 0$ *if and only if* $\alpha = 0$,

where α *and* β *are* k-*forms and* f *is a* 0-*form.*

Proof. Follows from direct computation. □

Let α and β be k-forms given by

$$\alpha = \sum_{i_1 < \cdots < i_k} a_{i_1 \cdots i_k} \omega^{i_1} \wedge \cdots \wedge \omega^{i_k}, \quad \beta = \sum_{j_1 < \cdots < j_k} b_{j_1 \cdots j_k} \omega^{j_1} \wedge \cdots \wedge \omega^{j_k}.$$

Then

$$\alpha \wedge *\beta = \sum_{i_1 < \cdots < i_k} a_{i_1 \cdots i_k} b_{i_1 \cdots i_k} * 1.$$

For α and β we define a global inner product of α and β by

$$(\alpha, \beta) = \int_M \alpha \wedge *\beta, \tag{3.3}$$

whenever the integral converges.

Two k-forms α, β are called *orthogonal* if $(\alpha, \beta) = 0$.

Lemma 3.2. *The star operator* $*$ *satisfies*

(1) $(\alpha, \alpha) \geq 0$ *and is equal to zero if and only if* $\alpha = 0$;
(2) $(\alpha, \beta) = (\beta, \alpha)$;
(3) $(\alpha, \beta_1 + \beta_2) = (\alpha, \beta_1) + (\alpha, \beta_2)$;
(4) $(*\alpha, *\beta) = (\alpha, \beta)$,

where $\alpha, \beta, \beta_1, \beta_2$ *are* k-*forms.*

Proof. Follows from Lemma 3.1. □

The most important application of the Hodge star operator is to define the codifferential δ on Riemannian manifolds. The *codifferential* δ on a k-form on a Riemannian manifold is defined as

$$\delta \alpha = (-1)^{nk+n+1} * d * \alpha. \tag{3.4}$$

In contrast with the differential operator d, the codifferential operator δ involves the metric structure of M.

Lemma 3.3. *The codifferential operator* $\delta : \Omega^k(M) \to \Omega^{k-1}(M)$ *satisfies*

(1) $\delta(\alpha + \beta) = \delta\alpha + \delta\beta$;
(2) $\delta^2 \alpha = 0$;

(3) $*\delta\alpha = (-1)^k d * \alpha$ *and* $*d\alpha = (-1)^{k+1}\delta * \alpha$

where α, β *are* k-*forms.*

Proof. Follows from (3.4), Lemma 3.1 and the property $d^2 = 0$. □

Definition 3.2. A differential form α is called *coclosed* if $\delta\alpha = 0$. If $\alpha = \delta\beta$ for some form β, then α is called *coexact.*

3.2 Hodge-Laplace operator

The Hodge-Laplace operator Δ acting on differentiable forms is defined by $\Delta = d\delta + \delta d$.

Definition 3.3. A form α on M is called *harmonic* if $\Delta\alpha = 0$.

Proposition 3.1. *If M is an oriented closed Riemannian manifold and α and β are forms of degree k and $k + 1$, respectively, then*

$$(d\alpha, \beta) = (\alpha, \delta\beta), \tag{3.5}$$

i.e., δ is the adjoint of d. Consequently, the Hodge-Laplace operator Δ is self-adjoint. In particular, the Laplacian acting on $\mathcal{F}(M)$ is self-adjoint.

Proof. Since M is closed, the Stokes theorem implies $\int_M d(\alpha \wedge *\beta) = 0$. Thus, by using the properties of d, we find

$$\int_M d\alpha \wedge *\beta = (-1)^{k-1} \int_M \alpha \wedge d * \beta.$$

Hence we obtain (3.5) from (3.4). □

Proposition 3.1 implies immediately the following.

Corollary 3.1. *On an oriented closed Riemannian manifold M, we have*

(1) *A k-form is closed if and only if it is orthogonal to all coexact k-forms.*
(2) *A k-form is coclosed if and only if it is orthogonal to all exact k-forms.*

Another application of Proposition 3.1 is the following.

Proposition 3.2. *On an oriented closed Riemannian manifold, a form is harmonic if and only if $d\alpha = 0$ and $\delta\alpha = 0$.*

Proof. If α is a k-form on M, then Proposition 3.1 implies that

$$(\Delta\alpha, \alpha) = (d\delta\alpha, \alpha) + (\delta d\alpha, \alpha) = (d\alpha, d\alpha) + (\delta\alpha, \delta\alpha).$$

From this we conclude that α is harmonic if and only if α is closed and coclosed.   □

An important application of Proposition 3.2 is the following.

Corollary 3.2. *Every harmonic function on a closed Riemannian manifold is a constant.*

The following result is known as the *Divergence Theorem*.

Proposition 3.3. *If X is a vector field on an oriented closed Riemannian manifold M, then*

$$\int_M (\operatorname{div} X) * 1 = 0. \tag{3.6}$$

Proof. By Proposition 3.2 we have

$$\int_M (\operatorname{div} X) * 1 = -\int_M (\delta X^\#) * 1 = -(\alpha^\#, d1) = 0,$$

where $X^\#$ is dual 1-form of X given by $X^\#(Y) = g(X, Y)$, $Y \in \mathfrak{X}(M)$.   □

Corollary 3.3. *If f is differentiable function on a closed Riemannian manifold, then*

$$\int_M (\Delta f) * 1 = 0. \tag{3.7}$$

Proof. Since a differentiable function f on M is a 0-form, $\Delta f = \delta df$, the Divergence Theorem implies (3.7).   □

The following result is well-known as Hopf's lemma.

Corollary 3.4. *Let M be a closed Riemannian n-manifold. If f is a differentiable function on M such that $\Delta f \geq 0$ everywhere on M (or $\Delta f \leq 0$ everywhere on M), then f is a constant function.*

Proof. We may assume that M is orientable by taking the 2-fold covering of M if necessary. If $\Delta f \geq 0$ (or $\Delta f \leq 0$), then Corollary 3.3 implies $\Delta f = 0$. Hence f is a constant function according to Corollary 3.2.   □

3.3 Elliptic differential operators

Elliptic differential operators are differential operators that resemble the Laplacian. They can be defined by a positivity condition on the coefficients of the highest-order derivatives, which implies the key property that the principal symbol is invertible, or equivalently that, there are no real characteristic directions. Elliptic regularity implies that their solutions tend to be smooth functions if the coefficients in the operator are smooth. The eigenspaces and eigenvalues of elliptic differential operators have many nice properties.

Let U be an open set in $\mathbb{E}^n$ with natural coordinates $x_1, \ldots, x_n$. For each n-tuple $t = (t_1, \ldots, t_n)$ of non-negative integer we put

$$|t| = t_1 + \cdots + t_n, \quad D^t = \frac{\partial^{|t|}}{\partial_1^{t_1} \cdots \partial_n^{t_n}}. \tag{3.8}$$

A *linear differential operator* D of degree r over U takes the form:

$$D = \sum_{|t| \le r} a_t(p) D^t, \tag{3.9}$$

where $a_t(p) : L \to V$ is a homomorphism of vector spaces and depends differentiably on $p \in U$. For each $\mathbf{y} = (y_1, \ldots, y_n) \in \mathbb{E}^n$, we set

$$\sigma(D, \mathbf{y}) = \sum_{|t| = r} a_t(\mathbf{y}) \mathbf{y}^t, \tag{3.10}$$

where we use the multi-indices: $\mathbf{y}^t = y_1^{t_1} \cdots y_n^{t_n}$. $\sigma(D, \mathbf{y})$ is called the *characteristic polynomial* of D and

$$\sigma(D, \cdot) : \mathbb{E}^n \to \mathrm{Hom}(L, V)$$

is called the *symbol* of D.

A differential operator D is called *elliptic* if the characteristic polynomial $\sigma(D, \mathbf{y})$ of D has no real zeros except $\mathbf{y} = 0$ at $p \in U$. The notion of elliptic differential operators has a natural generation to manifolds defined as follows: Let E, F be two (complex) vector bundles over a manifold M. A differential operator is a linear map $D : \Gamma(E) \to \Gamma(F)$ between the spaces of cross sections, which when restricted to each coordinate neighborhood U of M (over which E and F are trivial) is expressible in the form (3.9). The differential operator D is again called *elliptic* if the characteristic polynomial $\sigma(D, \mathbf{y})$ of D has no real zeros except $\mathbf{y} = 0$ at each point $p \in M$.

Perhaps the most important elliptic differential operator is the Laplacian Δ. Take $M = \mathbb{E}^n$ and $E = F$ the trivial line bundle over M. Then

$$\Delta = -\sum_{i=1}^{n} \frac{\partial^2}{\partial x_i^2}.$$

Since $\sigma(\Delta, \mathbf{y}) = -|y|^2 = -\sum y_i^2$, Δ is an elliptic differential operator.

Let E, F be two complex vector bundles over a closed Riemannian manifold M. Let us equipped E and F with *Hermitian metrics*, i.e., metrics g which are compatible with the complex structure J; namely they satisfy $g(JX, JY) = g(X, Y)$. This allows us to define an inner product $(\ ,\ )$ of on $\Gamma(E)$, e.g. if $s, s' \in \Gamma(E)$,

$$(s, s') = \int_M \langle s, s' \rangle * 1.$$

Let $D : \Gamma(E) \to \Gamma(F)$ be an elliptic differential operator. It is possible to define the *adjoint* of D, $D^* : \Gamma(F) \to \Gamma(E)$, as a differential operator characterized by the property:

$$(Ds, u) = (s, D^* u), \quad s \in \Gamma(E), \ u \in \Gamma(F). \tag{3.11}$$

It can be verified that D^* is also elliptic.

It follows from (3.11) that the kernel, $\ker(D)$, of D and the image, $\mathrm{im}(D^*)$, of D^* are orthogonal with respect to $(\ ,\ )$. Moreover, $\ker(D)$ is precisely the complement of $\mathrm{im}(D^*)$. In fact, if s is orthogonal to $\mathrm{im}(D^*)$, then $0 = (s, D^* Ds) = (Ds, Ds)$, thus $Ds = 0$. Hence we have

$$\Gamma(E) = \ker(D) \oplus \mathrm{im}(D^*). \tag{3.12}$$

An elliptic differential operator $D : \Gamma(E) \to \Gamma(F)$ has many other important properties. For instance, an elliptic operator D is a *Fredholm operator*, i.e., D is a differential operator which has finite-dimensional kernel and cokernel and it has closed image. From (3.12) we may conclude that for any η in the orthogonal complement of $\ker(D)$, there exists a solution of the equation

$$D^* u = \eta. \tag{3.13}$$

Consider the special case: $E = F$ and $D = D^*$. For each $\lambda \in \mathbb{R}$, we put

$$\Gamma_\lambda = \{s \in \Gamma(E) : Ds = \lambda s\}. \tag{3.14}$$

Let $L^2(E)$ denote the completion of $\Gamma(E)$ with respect to $(\ ,\ )$. Then it is known that there are only countably many λ with $\Gamma_\lambda \neq 0$ and each $\Gamma_\lambda \neq 0$ is a finite-dimensional vector space. Moreover,

$$L^2(E) = \hat\oplus_\lambda \Gamma_\lambda, \tag{3.15}$$

where $\hat\oplus$ is the completion of orthogonal sum.

Because the Hodge-Laplace operator $\Delta : \Omega^k(M) \to \Omega^{k+1}(M)$ is elliptic and self-adjoint, we have

$$\Omega^k(M) = \hat\oplus_i V_{k,i}, \tag{3.16}$$

where $V_{k,i}$ is the i-th eigenspace of Δ acting on k-forms. In fact, the eigenvalues of $\Delta : \Omega^k(M) \to \Omega^k(M)$ satisfies

$$0 \leq \lambda_{k,1} < \lambda_{k,2} < \cdots < \lambda_{k,i} < \cdots \nearrow \infty.$$

Since the kernel of $\Delta : \Omega^k(M) \to \Omega^k(M)$ is finite-dimensional, we have the following well-known results.

Theorem 3.1. *If M is a closed Riemannian manifold, then for each $k \in \{0, 1, \ldots, n\}$ the space of harmonic k-forms is finite-dimensional.*

We simplify denote $V_{0,i}$ by V_i and $\lambda_{0,i}$ by λ_i.

Theorem 3.2. *If M is a closed Riemannian manifold, then*

$$\mathcal{F}(M) = \hat{\oplus}_i V_i. \tag{3.17}$$

Denote by $Spec^k(M)$ the set of all eigenvalues of $\Delta : \Omega^k(M) \to \Omega^k(M)$ enumerated with multiplicity. $Spec^k(M)$ is called the *spectrum of k-forms*. We simply denote $Spec^0(M)$ by $Spec(M)$.

3.4   Hodge-de Rham decomposition and its applications

Let M be an oriented closed Riemannian n-manifold. For each integer $k \in \{0, 1, 2, \ldots, n\}$, denote by $\Omega_d^k(M)$, $\Omega_\delta^k(M)$ and $\Omega_H^k(M)$ the subspaces of $\Omega^k(M)$ consisting of k-forms which are exact, coexact and harmonic, respectively.

Lemma 3.4. *The three subspaces $\Omega_d^k(M), \Omega_\delta^k(M)$ and $\Omega_H^k(M)$ are mutually orthogonal.*

Proof. If $\omega \in \Omega_d^k(M)$, ω is closed. Thus, Corollary 3.1 implies that ω is orthogonal to $\Omega_\delta^k(M)$.

If $\omega = d\alpha$ and $\beta \in \Omega_H^k(M)$, Propositions 3.1 and 3.2 imply that

$$(\omega, \beta) = (d\alpha, \beta) = (\alpha, \delta\beta) = 0.$$

This shows that $\Omega_d^k(M)$ is orthogonal to $\Omega_H^k(M)$. Similar argument applies to the remaining case. $\qquad\square$

The following is the *Hodge-de Rham Decomposition Theorem*.

Theorem 3.3. *A k-form α on an oriented closed Riemannian manifold M is uniquely decomposed into the orthogonal sum*

$$\alpha = \alpha_d + \alpha_\delta + \alpha_H, \tag{3.18}$$

where $\alpha_d \in \Omega_d^k(M)$, $\alpha_\delta \in \Omega_\delta^k(M)$ and $\alpha_H \in \Omega_H^k(M)$.

Proof. Theorem 3.1 implies that $\Omega_H^k(M)$ is finite-dimensional. Thus we may choose an orthonormal basis $\omega^1, \ldots, \omega^h$ of $\Omega_H^k(M)$, $h = \dim \Omega_H^k(M)$. Let α be any k-form on M, we put $\alpha_H = \sum_{i=1}^h (\alpha, \omega^i)\omega^i$. Then $\Delta \alpha_H = 0$. Moreover, $\alpha - \alpha_H$ is orthogonal to $\Omega_H^k(M)$. Since $\Delta^* = \Delta$, there exists a solution of the equation $\Delta u = \alpha - \alpha_H$. Hence there is a k-form β such that $\Delta \beta = \alpha - \alpha_H$. If we put $\alpha_d = d\delta\beta$ and $\alpha_\delta = \delta d\beta$, we obtain (3.18).

Suppose that $\alpha = \alpha_d' + \alpha_\delta' + \alpha_H'$ is another decomposition of α, then

$$(\alpha_d - \alpha_d') + (\alpha_\delta + \alpha_\delta') + (\alpha_H - \alpha_H') = 0.$$

Thus Lemma 3.4 implies that $\alpha_d = \alpha_d', \alpha_\delta = \alpha_\delta'$ and $\alpha_H = \alpha_H'$. This proves the uniqueness. $\qquad\square$

The next result of Hodge-de Rham is an application of Theorem 3.3.

Theorem 3.4. *Every cohomology class in the k-th cohomology group $H^k(M)$ is represented uniquely by a harmonic k-form.*

Proof. If $\Phi \in H^k(M)$, then $\Phi = \{\alpha\}$ for some closed k-form α. Thus, by Corollary 3.1, we have $\alpha = \alpha_d + \alpha_H$ with $\alpha_d \in \Omega_d^k(M)$ and $\alpha_H \in \Omega_H^k(M)$. Since α_d is exact, Φ is thus represented by α_H.

If β is another k-form which represents Φ, then $\alpha - \beta$ is exact. Then, by Theorem 3.3, we have $\alpha_H = \beta_H$. $\qquad\square$

The following result is known as *Poincaré's Duality Theorem.*

Theorem 3.5. *If M is an oriented closed Riemannian n-manifold, then there is a natural isomorphism:*

$$\mu : H^k(M) \cong H^{n-k}(M).$$

Proof. If $\Phi \in H^k(M)$, there exists a hormonic k-form α_H representing Φ. Since $\Delta(*\alpha_H) - *(\Delta\alpha) = 0$, $*\alpha_H$ is also harmonic. Hence if we put $\mu(\Phi) = [*\alpha_H] \in H^{n-k}(M)$, then μ defines an isomorphism from $H^k(M)$ onto $H^{n-k}(M)$. $\qquad\square$

A fundamental result in algebraic topology asserts that there exists a natural isomorphism between the k-th cohomology group $H^k(M)$ and the k-th homology group $H_k(M)$. Thus the k-th *Betti number* $b_k(M) = \dim H_k(M)$ is equal to $\dim H^k(M)$. Consequently, Theorem 3.5 implies

$$b_k(M) = b_{n-k}(M) \tag{3.19}$$

for any oriented closed n-manifold M.

Another consequence of Hodge-de Rham's decomposition theorem is the following known result.

Corollary 3.5. *If $\hat{M}$ is a compact covering manifold of an oriented closed n-manifold M, then $b_k(M) \leq b_k(\hat{M})$ for $k = 1, \ldots, n - 1$.*

Proof. We may equip M with a Riemannian metric. For each nonzero harmonic k-form α on M, there exists a periodic extension $\hat{\alpha}$ on $\hat{M}$ given by $\hat{\alpha} = \pi^*\alpha$, where π is the projection of the covering map and π^* is the induced map of π. Clearly, $\hat{\alpha}$ is also a nonzero harmonic k-form. Since linearly independent harmonic forms on M lift to linearly independent harmonic forms on $\hat{M}$, we obtain the desired result from Theorem 3.4. $\quad\square$

3.5 Heat equation and its fundamental solution

The heat equation was formulated at the beginning of the 19th century by Joseph Fourier (1768-1830). It is a parabolic partial differential equation that describes the distribution of heat in a given region over time; proved to be a powerful tool for analyzing the dynamic motion of heat as well as for solving an enormous array of diffusion-type problems in diverse scientific fields. For instance, the heat equation is used in probability theory and describes *random walks*. In financial mathematics it is used in the Black-Scholes model to determine the value of (finance) derivatives of a financial market [Black-Scholes (1973)]. The heat equation is also important in Riemannian geometry, topology, applied mathematics and engineering.

A *heat operator* on a compact Riemannian manifold M is the operator $L = \Delta + \frac{\partial}{\partial t}$ acting on functions defined on $M \times \mathbb{R}_+$ which is of class C^2 on the first variable and of class C^1 on the second. The *heat equation* on M works on functions $F : M \times \mathbb{R}_+ \to \mathbb{R}$ which satisfy

$$L(F) = 0, \quad F(p, 0) = f(0), \quad p \in M, \tag{3.20}$$

where $f : M \to \mathbb{R}$ is a given initial condition.

Definition 3.4. A *fundamental solution of the heat equation* on M is a function $h : M \times M \times \mathbb{R}_+ \to \mathbb{R}$ which satisfies the following conditions:

(h_1) h is continuous on the three variables of class C^2 on the first two variables and of class C^1 on the third;

(h_2) $L_2 h = 0$, where $L_2 = \Delta_2 + \frac{\partial}{\partial t}$ and Δ_2 is the Laplacian on the second variable;

(h_3) for each $p \in M$, $\lim_{t \to 0^+} h(p, \cdot, t) = \delta_p$, where δ_p is the Dirac distribution at p, i.e., for each function f on M with $p \in \text{supp}(f)$, we have

$$\lim_{t \to 0^+} \int_M h(p, x, t) f(x) dx = f(p),$$

where dx denotes the volume element of the second M.

The following result is well-known.

Theorem 3.6. *A fundamental solution of heat equation on M exists and is unique.*

For an eigenspace V_i of $\Delta : \mathcal{F}(M) \to \mathcal{F}(M)$, we choose an orthonormal basis $\varphi_1^i, \ldots, \varphi_{m_i}^i$ of V_i, $m_i = \dim V_i$. The set of $\{\varphi_\alpha^i\}_{i,\alpha}$ is called an *orthonormal set of eigenfunctions* of Δ. According to Theorem 3.2, associated with a function $f \in \mathcal{F}(M)$ we have

$$f = \sum_{\alpha, i} (\varphi_\alpha^i, f) \varphi_\alpha^i \quad \text{(in the } L^2\text{-sense).} \tag{3.21}$$

Proposition 3.4. *If $\{\varphi_\alpha^i\}$ is an orthonormal set of eigenfunctions, then for each $(p, x, t) \in M \times M \times \mathbb{R}_+$, the series*

$$\sum_{i, \alpha} e^{-\lambda_i t} \varphi_\alpha^i(p) \varphi_\alpha^i(x) \tag{3.22}$$

converges and the fundamental solution of heat equation is given by

$$h(p, x, t) = \sum_{i, \alpha} e^{-\lambda_i t} \varphi_\alpha^i(p) \varphi_\alpha^i(x). \tag{3.23}$$

The proof of Proposition 3.4 bases on the uniqueness of the fundamental solution of the heat equation. From the definition of L_2 we have

$$L_2 \left(\sum_{i, \alpha} e^{-\lambda_i t} \varphi_\alpha^i(p) \varphi_\alpha^i(x) \right)$$

$$= \sum_{i, \alpha} e^{-\lambda_i t} \varphi_\alpha^i(p) (\Delta \varphi_\alpha^i)(x) + \sum_{i, \alpha} \frac{\partial}{\partial t} \left(e^{-\lambda_i t} \right) \varphi_\alpha^i(p) \varphi_\alpha^i(x)$$

$$= \sum_{i, \alpha} \lambda_i e^{-\lambda_i t} \varphi_\alpha^i(p) \varphi_\alpha^i(x) - \sum_{i, \alpha} \lambda_i \left(e^{-\lambda_i t} \right) \varphi_\alpha^i(p) \varphi_\alpha^i(x)$$

$$= 0.$$

Moreover, for each $f \in \mathcal{F}(M)$, we also have

$$\lim_{t \to 0^+} \int \sum e^{-\lambda_i t} \varphi_\alpha^i(p) \varphi_\alpha^i(x) f(x) dx = \lim_{t \to 0^+} \sum e^{-\lambda_i t} (\varphi_\alpha^i, f) \varphi_\alpha^i$$

$$= \sum (\varphi_\alpha^i, f) \varphi_\alpha^i = f.$$

These show that (3.23) satisfies conditions (h_2) and (h_3). In fact (3.23) also satisfies (h_1). Thus (3.23) is a fundamental solution of the heat equation. Consequently, by the uniqueness, it is $h(p, x, t)$.

By integrating $h(x, x, t)$ over M and using (3.23) we obtain

Proposition 3.5. *For each $t > 0$, the series $\sum_i m(\lambda_i)e^{-\lambda_i t}$ converges and*

$$\int_N h(x, x, t)dt = \sum_i m(\lambda_i)e^{-\lambda_i t}, \tag{3.24}$$

where $m(\lambda_i)$ is the multiplicity of λ_i.

To construct the fundamental solution $h(p, s, t)$ of heat equation, we use a successive approximation method. Although $\mathbb{E}^n$ is not compact, there exists a unique fundamental solution of heat equation on $\mathbb{E}^n$ under the condition to be decreasing at infinity. The solution is given by

$$h_0(p, x, t) = (4\pi t)^{-\frac{n}{2}} e^{-\frac{d(p,x)^2}{4t}},$$

where $d(p, x)$ denotes the Euclidean distance between p and x.

Using the idea that on a Riemannian manifold M, the fundamental solution of heat equation differs little from the pull-back of h_0 by $\exp^{-1}$, one arrives at the following *Asymptotic Expansion of Minakshisundaram-Pleijel* after long computation.

Theorem 3.7. *For each closed Riemannian n-manifold M, there exist constants $a_i's$ $(i = 0, 1, 2, \dots)$ with*

$$\sum_j m(\lambda_j)e^{-\lambda_j t} \underset{t \to 0}{\sim} (4\pi t)^{-\frac{n}{2}} \sum_{i=0}^{\infty} a_i t^i. \tag{3.25}$$

The first three coefficients a_0, a_1, a_2 are given by

$$a_0 = \int_M *1 = \text{vol}(M); \tag{3.26}$$

$$a_1 = \frac{1}{6} \int_M \tau * 1 \quad \text{(already folklore in 1965)}; \tag{3.27}$$

$$a_2 = \frac{1}{360} \int_M (2||R||^2 - 2||Ric||^2 + 5n^2(n-1)^2\tau^2) * 1, \tag{3.28}$$

where a_2 was determined in [McKean and Singer (1967)]. The fourth coefficient a_3 was determined in [Sakai (1971)].

In particular, formulas (3.26) and (3.27) imply the following.

Corollary 3.6. *If M is a closed Riemannian manifold, then the volume and the total scalar curvature $\int_M \tau * 1$ of M are spectral invariants.*

By a *spectral invariant* we mean a Riemannian invariant which depends only on $Spec(M)$.

3.6 Spectra of some important Riemannian manifolds

If we consider the Laplacian Δ acting on $\mathcal{F}(M)$ of a Riemannian manifold M, then $\Delta = \delta d$. For a function $f \in \mathcal{F}(M)$, df is a 1-form. Denote by $(df)_\#$ the associated vector field of df, i.e., $df(v) = \langle (df)_\#, v \rangle$ for $v \in TM$. Then $\Delta f = -\mathrm{div}(df)_\#$.

Various expressions of Δ.

(a) Since df is a 1-form, the covariant derivative $\nabla(df)$ is a 2-form which is the Hessian of f. The trace of $\nabla(df)$ is $-\Delta f$.

(b) Let $p \in M$ and $u_1, \ldots, u_n$ a normal coordinate system about p. Then

$$(\Delta f)(p) = -\sum_{i=1}^{n} \frac{\partial^2 f}{\partial u_i^2}(p). \tag{3.29}$$

This is equivalent to say that, for each $p \in M$, pick an orthonormal set of geodesics $\{\gamma_i\}$ parametrized by arc length and passing through p at $s = 0$, then

$$(\Delta f)(p) = -\sum_{i=1}^{n} \frac{d^2(f \circ \gamma_i)}{ds^2}(0). \tag{3.30}$$

(c) In terms of a local coordinate system $\{x_1, \ldots, x_n\}$ of M, we have

$$\Delta f = -\frac{1}{\sqrt{\mathfrak{g}}} \sum_{i,j=1}^{n} \frac{\partial \left(\sqrt{\mathfrak{g}} \, g^{ij} (\partial f / \partial x_j) \right)}{\partial x_i}, \tag{3.31}$$

where $\mathfrak{g} = \det(g_{ij})$, g_{ij} the components of metric tensor g with respect to $x_1, \ldots, x_n$ and (g^{ij}) is the inverse matrix of (g_{ij}).

Definition 3.5. Let M and B be two pseudo-Riemannian manifolds. A *pseudo-Riemannian submersion* is a smooth map $\pi : M \to B$ which is onto and satisfies the following three axioms:

(S1) $\pi_*|_p$ is onto for all $p \in M$;
(S2) the fibers $\pi^{-1}(b)$, $b \in B$, are pseudo-Riemannian submanifolds of M;
(S3) π_* preserves scalar products of vectors normal to fibers.

Condition (S2) holds automatically when M is Riemannian. Just like warped and twisted products, vectors tangent to fibers are called *vertical* and those normal to fibers are called *horizontal*.

Proposition 3.6. *Let $\pi : (M, g_M) \to (B, g_B)$ be a Riemannian submersion with totally geodesic fibers. Then for $f \in \mathcal{F}(B)$ we have*

$$\Delta_M (f \circ \pi) = (\Delta_B f) \circ \pi. \tag{3.32}$$

Proof. Let $p \in M$ and let $e_1, \ldots, e_b, e_{b+1}, \ldots, e_m$ be an orthonormal basis of $T_p M$ such that $e_1, \ldots, e_b$ are horizontal and $e_{b+1}, \ldots, e_m$ are vertical with $b = \dim B$ and $m = \dim M$. Let $\{\gamma_i\}_{i=1,\ldots,m}$ be the corresponding orthonormal set of geodesics through p. Then we find from (3.30) that

$$\Delta_M(f \circ \pi) = -\sum_{i=1}^{b} \frac{d^2}{ds^2}(f \circ \pi \circ \gamma_i) - \sum_{j=b+1}^{m} \frac{d^2}{ds^2}(f \circ \pi \circ \gamma_j).$$

Since $\pi : (M, g_M) \to (B, g_B)$ is a Riemannian submersion with totally geodesic fibers, $\{\pi \circ \gamma_i\}$ form an orthonormal set of geodesic in B through p and $\gamma_{b+1}, \ldots, \gamma_m$ are geodesics of the fiber $\pi^{-1}(\pi(p))$. Thus we obtain $\Delta_M(f \circ \pi)(p) = (\Delta_B f)(\pi(p))$. Because the Laplacian is well-defined, it is independent of the choice of local coordinates. Thus we have (3.32). $\square$

An eigenfunction of $\Delta : \mathcal{F}(M) \to \mathcal{F}(M)$ is called a *proper function*.

Proposition 3.7. *Let $\pi : (M, g_M) \to (B, g_B)$ be a Riemannian submersion with totally geodesic fibers. Then proper functions of (B, g_B) are those functions $f \in \mathcal{F}(B)$ such that $\pi^*(f) = f \circ \pi$ are proper functions of (M, g_M).*

Proof. If f is a proper function of (B, g_B) with eigenvalue λ, then (3.32) yields $\Delta_M(f \circ \pi) = \lambda f \circ \pi$. Thus $f \circ \pi$ is a proper function of (M, g_M) with eigenvalue λ.

Conversely, if $f \circ \pi$ is a proper function of (M, g_M) with eigenvalue λ, then $\Delta_M(f \circ \pi) = \lambda(f \circ \pi) = (\Delta_B f) \circ \pi$. Since $f \circ \pi$ is constant along fibers, we have $\Delta_B f = \lambda f$. $\square$

A point $p \in S^n$ determines a unit vector $e_{n+1} \in \mathbb{E}^{n+1}$. Let $e_1, \ldots, e_n$ be an orthonormal basis of $T_p S^n$. Then $e_1, \ldots, e_n, e_{n+1}$ form an orthonormal basis of $T_p \mathbb{E}^{n+1}$. Let $\gamma_1, \ldots, \gamma_n$ be the associated orthonormal set of geodesics of S^n through p. If we regard $e_1, \ldots, e_{n+1}$ as points in $\mathbb{E}^{n+1}$, then the geodesic γ_i through p with velocity vector e_i at p is given by

$$\gamma_i(s) = (\cos s)e_{n+1} + (\sin s)e_i, \quad i = 1, \ldots, n.$$

Let $f \in \mathcal{F}(\mathbb{E}^{n+1})$ and $x_1, \ldots, x_{n+1}$ the Euclidean coordinates associated with $e_1, \ldots, e_{n+1}$. Consider the functions $(f \circ \gamma_i)(s) = f(\gamma_i(s))$. By using the chain rule, we find

$$\frac{d(f \circ \gamma_i)}{ds} = -(\sin s)\frac{\partial f}{\partial x_{n+1}} + (\cos s)\frac{\partial f}{\partial x_i},$$

$$\frac{d^2(f \circ \gamma_i)}{ds^2}(0) = \frac{\partial^2 f}{\partial x_i^2}(p) - \frac{\partial f}{\partial x_{n+1}}(p).$$

Thus, by applying (3.30), we get

$$\Delta(f|_{S^n})(p) = -\sum_{i=1}^{n} \frac{\partial^2 f}{\partial x_i^2}(p) + n\frac{\partial f}{\partial x_{n+1}}(p). \qquad (3.33)$$

On the other hand, the Laplacian $\tilde{\Delta}$ of $\mathbb{E}^{n+1}$ satisfies

$$(\tilde{\Delta}f)(p) = -\sum_{j=1}^{n+1} \frac{\partial^2 f}{\partial x_j^2}(p). \qquad (3.34)$$

By combining (3.33) and (3.34) we find

$$\Delta(f|_{S^n})(p) = (\tilde{\Delta}f)\big|_{S^n}(p) + \frac{\partial^2 f}{\partial x_{n+1}^2}(p) + n\frac{\partial f}{\partial x_{n+1}}(p).$$

Consequently, if we denote by r the distance function from a point in $\mathbb{E}^{n+1}$ to the origin, then

$$(\tilde{\Delta}f)\big|_{S^n} = \Delta(f|_{S^n}) - \frac{\partial^2 f}{\partial r^2}\Big|_{S^n} - n\frac{\partial f}{\partial r}\Big|_{S^n}. \qquad (3.35)$$

Consider a homogeneous polynomial $\tilde{P}$ of degree $k \geq 0$ on $\mathbb{E}^{n+1}$. Let $P = \tilde{P}|_{S^n}$. Then $\tilde{P} = r^k P$. Thus we have

$$\frac{\partial \tilde{P}}{\partial r} = kr^{k-1}P, \quad \frac{\partial^2 \tilde{P}}{\partial r^2} = k(k-1)r^{k-2}P. \qquad (3.36)$$

By substituting (3.36) into (3.35), we get

$$\tilde{\Delta}\tilde{P}\big|_{S^n} = \Delta P - k(n+k-1)P. \qquad (3.37)$$

In particular, if $\tilde{P}$ is harmonic, we obtain

$$\Delta P = k(n+k-1)P. \qquad (3.38)$$

Let $\tilde{\mathcal{H}}_k$ denote the vector space of harmonic homogeneous polynomials of degree k on $\mathbb{E}^{n+1}$ and $\mathcal{H}_k$ the restriction of $\tilde{\mathcal{H}}_k$ on S^n. Then

$$\dim \tilde{\mathcal{H}}_k = \binom{n+k}{k} - \binom{n+k-2}{k-2},$$

where the last term is assumed to be zero for $k = 0, 1$. Because $\dim \tilde{\mathcal{H}}_k > 0$ for $k \geq 0$, we obtain from (3.38) the following well-known result [Berger et al. (1971)].

Proposition 3.8. *The spectrum* $Spec(S^n(1))$ *of the unit n-sphere $S^n(1)$ is given by*

$$\lambda_k = k(n+k-1), \quad k \geq 0.$$

The multiplicity $m(\lambda_k)$ of λ_k are given by

$$m(\lambda_0) = 1, \quad m(\lambda_1) = n+1,$$

$$m(\lambda_k) = \frac{(n+k-2)(n+k-3)\cdots(n+1)n}{k!}(n+2k-1), \quad k \geq 2.$$

Moreover, the eigenspace V_k with eigenvalue λ_k is $\mathcal{H}_k$.

Considering the Riemannian covering map $\pi : (S^n, g_0) \to (RP^n, g_0)$. Proposition 3.7 implies that proper functions of RP^n are induced from those proper functions of S^n invariant under the antipodal map. Thus proper functions of RP^n are obtained from harmonic homogeneous polynomials of even degree. Combining this and Proposition 3.8 yields

Proposition 3.9. *The spectrum* $Spec(RP^n(1)), n > 1$, *of the real projective n-space of constant curvature one is given by*

$$\lambda_k = 2k(n + 2k - 1), \quad k \geq 0.$$

The multiplicity $m(\lambda_k)$ *of* λ_k *are given by*

$$m(0) = 1,$$
$$m(\lambda_k) = \frac{(n + 2k - 2)(n + 2k - 3) \cdots (n + 1)n}{(2k)!}(n + 4k - 1), \quad k \geq 1.$$

The spectra of the complex projective spaces and quaternion projective spaces are also well-known.

Proposition 3.10. *The spectrum* $Spec(CP^n(4)), n > 1$, *of the complex projective n-space of constant holomorphic sectional curvature 4 is*

$$\lambda_k = 4k(n + k), \quad k \geq 0.$$

Proposition 3.11. *The spectrum* $Spec(QP^n(4)), n > 1$, *of the quaternion projective n-space of constant quaternionic sectional curvature 4 is*

$$\lambda_k = 4k(2n + k + 1), \quad k \geq 0.$$

3.7 Spectra of flat tori

Consider the flat torus $\mathbb{E}^n/\Lambda$, where Λ is a *lattice* of $\mathbb{E}^n$. Put

$$\Lambda^* = \{u \in \mathbb{E}^n : \langle u, v \rangle \in \mathbb{Z} \ \forall v \in \Lambda\}. \tag{3.39}$$

Then Λ^* is also a lattice, called the *dual lattice* of Λ. We have $(\Lambda^*)^* = \Lambda$. If $v_1, \ldots, v_n$ is a basis of Λ, its dual basis $v_1^*, \ldots, v_n^*$ is a basis for Λ^*.

For each $x \in \Lambda^*$, define a function f_x on $\mathbb{E}^n$ by

$$f_x(y) = e^{2\pi i x(y)}, \tag{3.40}$$

where $i = \sqrt{-1}$ and $y \in \mathbb{E}^n$ on the right is regarded a vector. Then f_x defines a function on $\mathbb{E}^n/\Lambda$ also denoted by f_x. If we denote by x_i and

y_i the components of x and y with respect to the bases $v_1^*, \ldots, v_n^*$ and $v_1, \ldots, v_n$, respectively, then

$$f_x(y) = e^{2\pi i \sum_{i=1}^n x_i y_i}. \tag{3.41}$$

By taking differentiation of (3.41) with respect to y_j we find

$$\frac{\partial f_x(y)}{\partial y_j} = 2\pi i x_j f_x(y), \quad \frac{\partial^2 f_x(y)}{\partial y_j^2} = -4\pi ||x||^2 f_x(y). \tag{3.42}$$

Thus

$$\Delta f_x = 4\pi^2 ||x||^2 f_x, \tag{3.43}$$

which shows that $\lambda = 4\pi^2 ||x||^2$ is an eigenvalue of Δ with proper function f_x for each $x \in \Lambda^*$. To each eigenvalue λ, the corresponding eigenspace V_λ is generated by the f_x's with $||x||^2 = \frac{\lambda}{4\pi^2}$. The multiplicity $m(\lambda)$ of λ is equal to the number of $x \in \Lambda^*$ such that $||x||^2 = \lambda/4\pi^2$. We summarize this as the following result of [Milnor (1964)].

Proposition 3.12. *Let* $(\mathbb{E}^n/\Lambda, g_0)$ *be a flat n-torus and* Λ^* *the dual lattice of* Λ. *Then the spectrum of* $\mathbb{E}^n/\Lambda$ *is given by*

$$\{4\pi^2 ||x||^2 : x \in \Lambda^*\}. \tag{3.44}$$

The multiplicity of $\lambda = 4\pi^2 ||x||^2$ *is equal to the number of* $u \in \Lambda^*$ *with* $||u|| = ||x||$.

It was discovered in [Witt (1941)] that there exist two lattices in $\mathbb{E}^{16}$ not isometric but with the same number of elements of any given norm. By applying this result, J. Milnor showed that there are two 16-dimensional flat tori which are not isometric, but with the same spectrum. Consequently, the information of eigenvalues alone is not enough to determine the isometry class of a Riemannian manifold in general.

3.8 Heat equation and Jacobi's elliptic functions

One of the greatest accomplishments of Carl Jacobi (1804-1851) was his theory of elliptic functions and their relation to theta functions. Elliptic functions were studied by many of the great mathematicians of the 19th century, including N. H. Abel, A.-L. Cauchy and K. Weierstrass. Such functions are very useful in mathematics, physics and engineering, e.g. the equations of motion are integrable in terms of elliptic functions in the well known cases of the pendulum, Kepler's problem, the Euler top and the symmetric Lagrange top in a gravitational field. Theta functions are very importance in mathematical physics, in particular in fluid mechanics, because of their role in the inverse problem for periodic and quasi-periodic flows.

Let θ be the temperature at time t at any point in a solid material whose conducting properties are uniform and isotropic. If ρ is the material's density, s its specific heat and k its thermal conductivity, then θ satisfies the *heat conduction equation*:

$$\kappa \nabla^2 \theta = \frac{\partial \theta}{\partial t}, \tag{3.45}$$

where $\kappa = k/s\rho$ is the diffusivity. In the special case where there is no variation of temperature in the x- and y-directions the heat flow is everywhere parallel to the z-axis and the heat equation reduced to the form:

$$\kappa \frac{\partial^2 \theta}{\partial z^2} = \frac{\partial \theta}{\partial t}, \quad \theta = \theta(z,t). \tag{3.46}$$

Consider the boundary conditions: $\theta(0,t) = \theta(\pi,t) = 0$ and the initial condition: $\theta(z,0) = \pi\delta(z - \frac{\pi}{2})$ for $0 < z < \pi$, where $\delta(z)$ is Dirac's unit impulse function. Then the solution of this boundary value problem is

$$\theta(z,t) = 2 \sum_{n=0}^{\infty} (-1)^n e^{-(2n+1)^2 \kappa t} \sin(2n+1)z. \tag{3.47}$$

By writing $e^{-4\kappa t} = q$, the solution (3.47) assume the form

$$\theta_1(z,q) = 2 \sum_{n=0}^{\infty} (-1)^n q^{(n+\frac{1}{2})^2} \sin(2n+1)z, \tag{3.48}$$

which is the first of the four *theta functions*. When the precise value of q is not important we shall suppress the dependence upon q. If one changes the boundary conditions to $\frac{\partial \theta}{\partial z} = 0$ on $z = 0$ and $z = \pi$ with $\theta(z,0) = \pi\delta(z - \frac{\pi}{2})$ for $0 < z < \pi$, then the corresponding solution of the boundary value problem of the heat equation (3.46) is given by

$$\theta_4(z) = \theta_4(z,q) = 1 + 2 \sum_{n=1}^{\infty} (-1)^n q^{n^2} \cos 2nz. \tag{3.49}$$

The theta function $\theta_1(z)$ of (3.48) is periodic with period 2π. Incrementing z by $\frac{1}{2}\pi$ yields the second theta function:

$$\theta_2(z) = \theta_2(z,q) = \theta_1(z + \tfrac{\pi}{2}, q) = 2 \sum_{n=1}^{\infty} q^{(n+\frac{1}{2})^2} \cos(2n+1)z. \tag{3.50}$$

Similarly, incrementing z by $\frac{1}{2}\pi$ for θ_4 yields the third theta function:

$$\theta_3(z) = \theta_3(z,q) = \theta_4(z + \tfrac{\pi}{2}, q) = 1 + 2 \sum_{n=1}^{\infty} q^{n^2} \cos 2nz. \qquad (3.51)$$

The four theta functions $\theta_1, \theta_2, \theta_3, \theta_4$ can be extended to complex values for z and q such that $|q| < 1$. The *Jacobi's elliptic functions* $\mathrm{sn}(u)$, $\mathrm{cn}(u)$ and $\mathrm{dn}(u)$ are defined as ratios of theta functions:

$$\mathrm{sn}(u) = \frac{\theta_3(0)\theta_1(z)}{\theta_2(0)\theta_4(z)}, \quad \mathrm{cn}(u) = \frac{\theta_4(0)\theta_2(z)}{\theta_2(0)\theta_4(z)}, \quad \mathrm{dn}(u) = \frac{\theta_4(0)\theta_3(z)}{\theta_3(0)\theta_4(z)}, \qquad (3.52)$$

where $z = u/\theta_3^2(0)$. Define parameters k and k' by

$$k = \frac{\theta_2^2(0)}{\theta_3^2(0)}, \quad k' = \frac{\theta_4^2(0)}{\theta_3^2(0)},$$

which are called the *modulus* and *complementary modulus*; k and k' satisfy $k^2 + k'^2 = 1$. Using cn, dn and sn, one may define minor elliptic functions

$$\mathrm{nc}(u) = \frac{1}{\mathrm{cn}(u)}, \quad \mathrm{nd}(u) = \frac{1}{\mathrm{dn}(u)}, \quad \mathrm{ns}(u) = \frac{1}{\mathrm{sn}(u)},$$
$$\mathrm{cd}(u) = \frac{\mathrm{cn}(u)}{\mathrm{dn}(u)}, \quad \mathrm{cs}(u) = \frac{\mathrm{cn}(u)}{\mathrm{sn}(u)}, \quad \mathrm{dc}(u) = \frac{\mathrm{dn}(u)}{\mathrm{cn}(u)}, \quad etc. \qquad (3.53)$$

When it is required to state the modulus explicitly the elliptic functions of Jacobi will be written as $\mathrm{sn}(u,k)$, $\mathrm{cn}(u,k)$, $\mathrm{dn}(u,k)$, etc. The elliptic functions $\mathrm{sn}\,(u)$, $\mathrm{cn}\,(u)$ and $\mathrm{dn}\,(u)$ satisfy the following relations:

$$\mathrm{sn}^2(u) + \mathrm{cn}^2(u) = 1,$$
$$\mathrm{dn}^2(u) + k^2 \mathrm{sn}^2(u) = 1,$$
$$k^2 \mathrm{cn}^2(u) + k'^2 = \mathrm{dn}^2(u),$$
$$\frac{d}{du} \mathrm{sn}(u) = \mathrm{cn}(u)\mathrm{dn}(u), \qquad (3.54)$$
$$\frac{d}{du} \mathrm{cn}(u) = -\mathrm{sn}(u)\mathrm{dn}(u),$$
$$\frac{d}{du} \mathrm{dn}(u) = -k^2 \mathrm{sn}(u)\mathrm{cn}(u).$$

The Jacobi's *Theta function* $\Theta(u)$ and *zeta function* $Z(u)$ are defined by

$$\Theta(u) = \theta_4\!\left(\frac{\pi u}{2K}\right), \quad Z(u) = \frac{d}{du}(\ln \theta_4), \quad K = \frac{\pi}{2}\theta_3^2(0). \qquad (3.55)$$

The zeta and Theta functions satisfy $Z(u) = \frac{\Theta'(u)}{\Theta(u)}$. Jacobi also defined a function $\mathrm{am}(x)$ by means of the equation:

$$\mathrm{am}(x) = \int_0^x \mathrm{dn}(u)\,du, \qquad (3.56)$$

which is known as Jacobi's *amplitude*.

Chapter 4

Submanifolds

Differential geometry started during the seventeenth century as the study of curves in the Euclidean plane and of curves and surfaces in the Euclidean 3-space $\mathbb{E}^3$ by means of the techniques of differential calculus.

Gauss and Riemann made the emergence of differential geometry as a distinct discipline in the nineteenth century. It was Gauss who proved in 1827 that the intrinsic geometry of a surface S in $\mathbb{E}^3$ can be derived solely from the Euclidean inner product as applied to tangent vectors of S. In 1854, Riemann discussed in his famous inaugural lecture at Göttingen the foundations of geometry, introduced n-dimensional manifolds, formulated the concept of Riemannian manifolds and defined their curvature.

Under the impetus of Einstein's Theory of General Relativity (1915) a further generalization appeared; the positiveness of the inner product was weakened to nondegeneracy. Consequently, one has the notion of pseudo-Riemannian manifolds. Inspired by Kaluza-Klein's theory and string theory in particle physics, mathematicians and physicists study not only submanifolds of Riemannian manifolds but also submanifolds of pseudo-Riemannian manifolds in recent years.

The influence of the differential geometry of submanifolds upon branches of mathematics, sciences and engineering has been profound. For instance, the study of geodesics and minimal surfaces is intimately related to dynamics, the theory of functions of a complex variable, calculus of variations, and topology. The application of double helix to the investigation of DNA structure in molecular biology resulting in important advancing in genetic engineering and medicine. In recent times, submanifold theory also plays an important part in computer design, image processing, economic modeling, arts and vision as well as in mathematical physics (including Kaluza-Klein and string theories), and mathematical biology.

4.1   Cartan-Janet's and Nash's embedding theorems

Let $\phi : M \to N$ be a map between two manifolds. Then ϕ is called an *immersion* if $\phi_{*_p} : T_pM \to T_{\phi(p)}N$ is injective for all $p \in M$. If, in addition, ϕ is a homeomorphism onto $\phi(M)$, where $\phi(M)$ has the subspace topology induced from N, we say that ϕ is an *embedding*.

For most local questions of geometry, it is the same to work with either immersions or imbeddings. In fact, if $\phi : M \to N$ is an immersion of a manifold M into another manifold N, then for each point $p \in M$, there exists a neighborhood $U \subset M$ of p such that the restriction $\phi : U \to N$ is an embedding.

For an immersion $\phi : M \to N$, we have $\dim M \leq \dim N$. The difference $\dim N - \dim M$ is called the *codimension* of the immersion. Throughout this book we consider only immersions of codimension ≥ 1 unless mentioned otherwise.

Definition 4.1. An immersion $\phi : M \to N$ of a pseudo-Riemannian manifold into another pseudo-Riemannian manifold is called *isometric* if

$$\langle u, v \rangle_p = \langle \phi_{*_p} u, \phi_{*_p} v \rangle_{\phi(p)} \tag{4.1}$$

holds for all $u, v \in T_pM$, $p \in M$.

Let $\phi : M \to N$ be an isometric immersion. Then, for each point $p \in M$, there exists a neighborhood U of p such that $\phi : U \to N$ is an embedding. Thus, each vector $u \in T_pU$ gives rise to a vector $\phi_*u \in T_{\phi(p)}N$. We may identify $u \in T_pM$ with $\phi_*u \in T_{\phi(p)}N$. In this way, each tangent space T_pM is a nondegenerate subspace of $T_{\phi(p)}N$. Hence there is a direct sum decomposition

$$T_{\phi(p)}N = T_pM \oplus T_p^{\perp}M, \tag{4.2}$$

where $T_p^{\perp}M$ is a nondegenerate subspace of $T_{\phi(p)}N$, which is called the *normal subspace* of M at p. Vectors in $T_p^{\perp}M$ are said to be normal to M and those in T_pM are tangent to M. Thus, each vector $v \in T_{\phi(p)}N$ has a unit expression

$$v = \tan v + \operatorname{nor} v, \tag{4.3}$$

where $\tan v \in T_pM$ and $\operatorname{nor} v \in T_p^{\perp}M$. The orthogonal projections

$$\tan : T_{\phi(p)}N \to T_pM; \text{ and } \operatorname{nor} : T_{\phi(x)}N \to T_p^{\perp}M$$

are **R**-linear.

One of the most fundamental problems in submanifold theory is the problem of isometric immersibility. The embedding problem had been around since Riemann. The earliest publication by L. Schläfli (1814–1895) on isometric embedding appeared in [Schläfli (1873)].

The problem of isometric immersion (or embedding) admits an obvious analytic interpretation; namely, if $g_{ij}(u)$, $u = (u_1, \ldots, u_n)$, are the components of the metric tensor g in local coordinates $u_1, \ldots, u_n$ on a Riemannian n-manifold M, and $x_1, \ldots, x_m$ are the standard Euclidean coordinates of $\mathbb{E}^m$, then the condition for an isometric immersion in $\mathbb{E}^m$ is

$$\sum_{i=1}^{n} \frac{\partial x_j}{\partial u_i} \frac{\partial x_k}{\partial u_i} = g_{jk}(u),$$

i.e., we have a system of $\frac{1}{2}n(n+1)$ nonlinear partial differential equations in m unknown functions. If $m = \frac{1}{2}n(n+1)$, then this system is definite and so we would like to have a solution. Schläfli asserted that any Riemannian n-manifold can be isometrically imbedded in Euclidean space of dimension $\frac{1}{2}n(n+1)$. Apparently, it is appropriate to assume that he had in mind of analytic metrics and local analytic imbeddings. This was later called *Schläfli's conjecture*. M. Janet (1888–1984) published in [Janet (1926)] a proof of Schläfli's conjecture which states that a real analytic Riemannian n-manifold can be locally isometrically embedded into any real analytic Riemannian manifold of dimension $\frac{1}{2}n(n+1)$. É. Cartan (1869–1951) revised Janet's paper with the same title in [Cartan (1927)]; while Janet wrote the problem in the form of a system of partial differential equations which he investigated using rather complicated methods, Cartan applied his own theory of Pfaffian systems in involution. Both Janet's and Cartan's proofs contained obscurities. C. Burstin get rid of them in [Burstin (1931)]. This result of Cartan-Janet implies that every Einstein n-manifold ($n \geq 3$) can be locally isometrically embedded in $\mathbb{E}^{n(n+1)/2}$.

The Cartan-Janet theorem is dimensionwise the best possible, i.e., there exist real analytic Riemannian n-manifolds which do not possess smooth local isometric imbeddings into any Euclidean space of dimension strictly less than $\frac{1}{2}n(n+1)$. Not every Riemannian n-manifold can be isometrically immersed in $\mathbb{E}^m$ with $m \leq \frac{1}{2}n(n+1)$. For instance, not every Riemannian 2-manifold can be isometrically immersed in $\mathbb{E}^3$.

A global isometric embedding theorem was proved by John Forbes Nash (1928-) which states as follows [Nash (1956)].

Theorem 4.1. *Every closed Riemannian n-manifold can be isometrically imbedded in a Euclidean m-space $\mathbb{E}^m$ with $m = \frac{1}{2}n(3n + 11)$. Every non-*

closed Riemannian n-manifold can be isometrically imbedded in $\mathbb{E}^m$ *with* $m = \frac{1}{2}n(n+1)(3n+11)$.

R. E. Greene improved Nash's result in [Greene (1970)] and proved that every non-compact Riemannian n-manifold can be isometrically embedded in the Euclidean m-space $\mathbb{E}^m$ with $m = 2(2n+1)(3n+7)$. Also, it was proved independently in [Greene (1970)] and [Gromov and Rokhlin (1970)] that a local isometric embedding from a Riemannian n-manifold into $\mathbb{E}^{\frac{1}{2}n(n+1)+n}$ always exist.

Concerning the isometric embedding of pseudo-Riemannian manifolds, the following existence theorem was proved independently in [Clarke (1970); Greene (1970)].

Theorem 4.2. *Any pseudo-Riemannian n-manifold M_t^n with index t can be isometrically embedded in a pseudo-Euclidean m-space $\mathbb{E}_s^m$, for m and s large enough. Moreover, this embedding may be taken inside any given open set in $\mathbb{E}_s^m$.*

4.2   Formulas of Gauss and Weingarten

Let $\phi : M \to N$ be an isometric immersion. The Levi-Civita connection of the ambient manifold N will be denoted by $\tilde{\nabla}$. Since the discussion is local, we may assume, if we want, that M is embedded in N.

Let X be a vector field tangent to M. A vector field $\tilde{X}$ on N is called an *extension* of X if its restriction to $\phi(M)$ is X. If $\tilde{X}$ and $\tilde{Y}$ are extensions of vector fields X and Y on M, respectively, then $[\tilde{X}, \tilde{Y}]|_M$ is independent of the extensions. Moreover, we have

$$[\tilde{X}, \tilde{Y}]|_M = [X, Y]. \tag{4.4}$$

If X and Y are local vector fields of M and $\tilde{X}$ and $\tilde{Y}$ are local extensions of X and Y to N. Then the restriction of $\tilde{\nabla}_{\tilde{X}}\tilde{Y}$ on M is independent of the extensions $\tilde{X}, \tilde{Y}$ of X, Y. This can be seen as follows: Let $\tilde{X}_1$ be another extension of X, then we have $\tilde{\nabla}_{\tilde{X}-\tilde{X}_1}\tilde{Y} = 0$ on M since $\tilde{X} - \tilde{X}_1 = 0$ on M. Thus $\tilde{\nabla}_{\tilde{X}}\tilde{Y} = \tilde{\nabla}_{\tilde{X}_1}\tilde{Y}$. Moreover, it follows from (4.4) and Theorem 2.1(4) that the restriction of $\tilde{\nabla}_{\tilde{X}}\tilde{Y}$ on M is also independent of the extension $\tilde{Y}$.

We define $\nabla_X Y$ and $h(X, Y)$ by

$$\nabla_X Y = \tan \tilde{\nabla}_{\tilde{X}}\tilde{Y} \quad \text{and} \quad h(X, Y) = \text{nor} \, \tilde{\nabla}_{\tilde{X}}\tilde{Y}, \tag{4.5}$$

so that we have the following *formula of Gauss*

$$\tilde{\nabla}_{\tilde{X}}\tilde{Y} = \nabla_X Y + h(X, Y). \tag{4.6}$$

Proposition 4.1. *Let* $\phi : M \to N$ *be an isometric immersion of a pseudo-Riemannian manifold M into a pseudo-Riemannian manifold N. Then*

(i) ∇ *defined in (4.5) is the Levi-Civita connection of M and*
(ii) $h(X, Y)$ *is $\mathcal{F}(M)$-bilinear and symmetric.*

Proof. To prove statement (i), we verify properties (1)-(3) of Definition 2.3 and properties (4) and (5) of Theorem 2.1.

Properties (1) and (2) of Definition 2.3 follow from the corresponding properties of $\tilde{\nabla}$ on N and linearity of the projection $\tan : T_{\phi(p)}N \to T_pM$.

To verify property (3), let f be a function in $\mathcal{F}(M)$. Then

$$\nabla_X(fY) = (Xf)Y + f\nabla_X Y.$$

Thus, after taking the tangential components of both sides, we get property (3) of Definition 2.3.

Next, we prove property (4) of Theorem 2.1. Let us write

$$\tilde{\nabla}_{\tilde{X}}\tilde{Y} = \nabla_X Y + h(X, Y), \tag{4.7}$$

$$\tilde{\nabla}_{\tilde{Y}}\tilde{X} = \nabla_Y X + h(Y, X). \tag{4.8}$$

Since $\tilde{\nabla}$ is the Levi-Civita connection of N, it follows form (4.4), (4.7) and (4.8) that

$$[X, Y] = \nabla_X Y - \nabla_Y X, \tag{4.9}$$

$$h(X, Y) = h(Y, X), \tag{4.10}$$

which imply property (4) and that $h(X, Y)$ is symmetric.

To prove property (5) we start with

$$X\langle \tilde{Y}, \tilde{Z} \rangle = \langle \tilde{\nabla}_X \tilde{Y}, \tilde{Z} \rangle + \langle \tilde{Y}, \tilde{\nabla}_X \tilde{Z} \rangle. \tag{4.11}$$

From (4.6) we have

$$\langle \tilde{\nabla}_X \tilde{Y}, \tilde{Z} \rangle = \langle \nabla_X Y, Z \rangle + \langle h(X, Y), Z \rangle = \langle \nabla_X Y, Z \rangle. \tag{4.12}$$

Similarly, we have

$$\langle \tilde{Y}, \tilde{\nabla}_X \tilde{Z} \rangle = \langle Y, \nabla_X Z \rangle. \tag{4.13}$$

Hence we obtain from (4.11)-(4.13) that

$$X\langle Y, Z \rangle = \langle \nabla_X Y, Z \rangle + \langle Y, \nabla_X Z \rangle,$$

which is property (5).

Finally, we show that $h(X, Y)$ is $\mathcal{F}(M)$-bilinear. The additivity in X or Y is obvious. Now, for any $f \in \mathcal{F}(M)$, we have

$$\tilde{\nabla}_{fX} Y + h(fX, Y) = \tilde{\nabla}_{fX} \tilde{Y} = f \tilde{\nabla}_X \tilde{Y}$$
$$= f(\nabla_X Y + h(X, Y)),$$

which implies that $h(fX, Y) = fh(X, Y)$. By symmetry, we also obtain $h(X, fY) = fh(X, Y)$. □

We define $h : T(M) \times T(M) \to T^\perp M$ as the *second fundamental form* of N for the given immersion.

Let $e_1, \ldots, e_n$ and $e_{n+1}, \ldots, e_m$ be orthonormal bases of the tangent space $T_p M$ and of the normal space $T_p^\perp M$ at a point $p \in M$. If we put $h_{ij}^r = \langle h(e_i, e_j), e_r \rangle$; $i, j = 1, \ldots, n$; $r = n+1, \ldots, m$, then we have

$$h(e_i, e_j) = \sum_{r=n+1}^{n} \epsilon_r h_{ij}^r e_r, \quad \epsilon_r = \langle e_r, e_r \rangle. \tag{4.14}$$

We call h_{ij}^r the *coefficients of the second fundamental form*.

Let $\phi : M \to N$ be an isometric immersion. If ξ is a normal vector field of M in N and X is a tangent vector field of M, then we may decompose $\tilde{\nabla}_X \xi$ as

$$\tilde{\nabla}_X \xi = -A_\xi(X) + D_X \xi, \tag{4.15}$$

where $-A_\xi(X)$ and $D_X \xi$ are the tangential and normal components of $\tilde{\nabla}_X \xi$. It is easy to verify that $A_\xi(X)$ and $D_X \xi$ are smooth vector fields on N whenever X and ξ are smooth.

Remark 4.1. Formula (4.15) is known as the *formula of Weingarten*. This formula for surfaces in $\mathbb{E}^3$ was established by Julius Weingarten (1836-1910) in [Weingarten (1861)]. The operator A in (4.15) is called the *Weingarten map* or *shape operator*.

Proposition 4.2. *Let $\phi : M \to N$ be an isometric immersion of a pseudo-Riemannian manifold M into a pseudo-Riemannian manifold N. Then*

(a) *$A_\xi(X)$ is $\mathcal{F}(M)$-bilinear in ξ and X; hence, at each point $p \in M$, $A_\xi(X)$ depends only on ξ_p and X_p;*

(b) *For a normal vector field ξ and tangent vectors X, Y of M, we have*

$$\langle h(X, Y), \xi \rangle = \langle A_\xi(X), Y \rangle; \tag{4.16}$$

(c) *D is a metric connection on the normal bundle $T^\perp M$ with respect to the induced metric on $T^\perp M$, i.e., $D_X \langle \xi, \eta \rangle = \langle D_X \xi, \eta \rangle + \langle \xi, D_X \eta \rangle$ holds for any tangent vector field X and normal vector fields ξ, η.*

Proof. (a) Let f, k be two functions in $\mathcal{F}(M)$. We have

$$\tilde{\nabla}_{fX}(k\xi) = f\tilde{\nabla}_X(k\xi) = f\{(Xk)\xi + k\tilde{\nabla}_X\xi\}$$
$$= f(Xk)\xi - fkA_\xi(X) + fkD_X\xi,$$

which implies that

$$A_{k\xi}(fX) = fkA_\xi(X), \tag{4.17}$$
$$D_{fX}(k\xi) = f(Xk)\xi + fkD_X\xi. \tag{4.18}$$

Thus $A_x(X)$ is $\mathcal{F}(M)$-bilinear in ξ and X, since additivity is trivial.

(b) For arbitrary $X, Y \in \mathfrak{X}(M)$, we have

$$0 = \langle \tilde{\nabla}_X Y, \xi \rangle + \langle Y, \tilde{\nabla}_X \xi \rangle$$
$$= \langle \nabla_X Y, \xi \rangle + \langle h(X, Y), \xi \rangle - \langle Y, A_\xi(X) \rangle + \langle Y, D_X \xi \rangle$$
$$= \langle h(X, Y), \xi \rangle - \langle Y, A_\xi(X) \rangle$$

which gives (4.16).

(c) It follows from (4.18) that D defines an affine connection on $T^\perp M$. Moreover, for any normal vector fields ξ and η, we have

$$\tilde{\nabla}_X \xi = -A_\xi(X) + D_X\xi \text{ and } \tilde{\nabla}_X \eta = -A_\eta(X) + D_X\eta.$$

Hence

$$\langle D_X\xi, \eta \rangle + \langle \xi, D_X\eta \rangle = \langle \tilde{\nabla}_X\xi, \eta \rangle + \langle \xi, \tilde{\nabla}_X\eta \rangle = X\langle \xi, \eta \rangle.$$

Thus D is a metric connection on the normal bundle with respect to the induced metric on $T^\perp M$. $\qquad \square$

Definition 4.2. An isometric immersion $\phi : M \to N$ is called *totally geodesic* if the second fundamental form vanishes identically, i.e., $h \equiv 0$.

For a normal vector field ξ on M, if $A_\xi = \rho I$ for some $\rho \in \mathcal{F}(M)$, then ξ is called an *umbilical section*, or M is said to be umbilical with respect to ξ. If the submanifold M is umbilical with respect to every local normal vector field, then M is called a *totally umbilical submanifold*.

The *mean curvature vector H* of M in N is defined by

$$H = \left(\frac{1}{n}\right) \operatorname{Tr} h, \tag{4.19}$$

where Tr stands for trace and $n = \dim M$. If $\{e_1, \ldots, e_n\}$ is an orthonormal frame of M, then the mean curvature vector is given by

$$H = \left(\frac{1}{n}\right) \sum_{j=1}^{n} \epsilon_j h(e_j, e_j). \tag{4.20}$$

The length of the mean curvature vector is called the *mean curvature*.

Notation 4.1. Let M be a submanifold of a Riemannian manifold N. For each nonzero real number k, we simply denote by H^k the k-th power of the length of mean curvature vector of M in N, i.e., $H^k = ||H||^k$. In particular, we denote the *squared mean curvature* by H^2, i.e., $H^2 = \langle H, H \rangle$.

Definition 4.3. The metric connection D defined by (4.15) is called the *normal connection*. A normal vector field ξ on M is said to be *parallel in the normal bundle*, or simply *parallel* if $D\xi = 0$ holds identically. In particular, M is said to have *parallel mean curvature vector* if $DH = 0$ holds identically.

The second fundamental form of a totally umbilical submanifold satisfies

$$h(X, Y) = \langle X, Y \rangle H, \quad X, Y \in TM. \tag{4.21}$$

Definition 4.4. A pseudo-Riemannian submanifold M is called *minimal* if the mean curvature vector H vanishes identically, i.e., $H \equiv 0$. And M is called *quasi-minimal* if $H \neq 0$ and $\langle H, H \rangle = 0$ at each point of M [Rosca (1972)]. A spacelike submanifold in a spacetime is called *marginally trapped* if it is quasi-umbilical. (The concept of *trapped surfaces* was introduced in [Penrose (1965)], see section 3.8 of [Chen (2011b)] for more details.)

4.3   Shape operator of submanifolds

A shape operator of a Riemannian submanifold is always diagonalizable, but this is not the case for a shape operator of a pseudo-Riemannian submanifolds, in particular, for a Lorentzian submanifold. In order to express shape operator of a pseudo-Riemannian submanifolds, we need the following algebraic lemma from [O'Neill (1983)].

Lemma 4.1. *A linear operator S on $V \approx \mathbb{E}_s^n$ is self-adjoint if and only if V can be expressed as a direct sum of subspaces V_k that are mutually orthogonal (hence nondegenerate) and S-invariant and each $S|_{V_k}$ has matrix of form* either

$$\begin{pmatrix} \lambda & & & 0 \\ 1 & \lambda & & \\ & \ddots & \ddots & \\ & & 1 & \lambda \\ 0 & & & 1 & \lambda \end{pmatrix}$$

relative to a basis $v_1, \ldots, v_r$ $(r \geq 1)$ with all scalar products zero except $\langle v_i, v_j \rangle = \epsilon = \pm 1$ if $i + j = r + 1$, or

$$\begin{pmatrix} a & b & & & & & & & \\ -b & a & & & & & & & \\ 1 & 0 & a & b & & & & & \\ 0 & 1 & -b & a & & & 0 & & \\ & & 1 & 0 & a & b & & & \\ & & 0 & 1 & -b & a & & & \\ & & & & & & \ddots & & \\ & & & & & & & 1 & 0 & a & b \\ & 0 & & & & & & 0 & 1 & -b & a \end{pmatrix}, \quad b \neq 0,$$

relative to a basis $v_1, w_1, \ldots, v_m, w_m$ with all scalar products zero except $\langle v_i, v_j \rangle = 1 = -\langle w_i, w_j \rangle$ if $i + j = m + 1$. (Here r, ϵ and m depend on k.)

In particular, if V is Lorentzian, we have the following.

Lemma 4.2. *A self-adjoint linear operator S of an n-dimensional vector space V with a Lorentzian scalar product $\langle \ , \ \rangle$ can be put into one of the following four forms :*

$$\text{I. } S \sim \begin{pmatrix} a_1 & & & 0 \\ & a_2 & & \\ & & \ddots & \\ 0 & & & a_n \end{pmatrix}, \quad \text{II. } S \sim \begin{pmatrix} a_0 & 0 & & & 0 \\ 1 & a_0 & & & \\ & & a_3 & & \\ & & & \ddots & \\ 0 & & & & a_n \end{pmatrix},$$

$$\text{III. } S \sim \begin{pmatrix} a_0 & 0 & 0 & & & 0 \\ 0 & a_0 & 1 & & & \\ -1 & 0 & a_0 & & & \\ & & & a_4 & & \\ & & & & \ddots & \\ 0 & & & & & a_n \end{pmatrix}, \quad \text{IV. } S \sim \begin{pmatrix} a_0 & b_0 & & & 0 \\ -b_0 & a_0 & & & \\ & & a_3 & & \\ & & & \ddots & \\ 0 & & & & a_n \end{pmatrix},$$

where b_0 is assumed to be nonzero.

In both cases I and IV, S is represented with respect to an orthonormal basis $\{v_1, \ldots, v_n\}$ satisfying $\langle v_1, v_1 \rangle = -1$, $\langle v_i, v_j \rangle = \delta_{ij}$, $\langle v_1, v_i \rangle = 0$ for $2 \leq i, j \leq n$, while in cases II and III the basis $\{v_1, \ldots, v_n\}$ is pseudo-orthonormal satisfying $\langle v_1, v_1 \rangle = 0 = \langle v_2, v_2 \rangle = \langle v_1, v_i \rangle = \langle v_2, v_i \rangle$, for $3 \leq i \leq n$, $\langle v_1, v_2 \rangle = -1$, and $\langle v_i, v_j \rangle = \delta_{ij}$ otherwise.

4.4   Equations of Gauss, Codazzi and Ricci

Let $\phi : M \to N$ be an isometric immersion of a pseudo-Riemannian manifold M into a pseudo-Riemannian manifold N. Denote by R and ∇ the Riemann curvature tensor and the Levi-Civita connection of M, respectively. And denote by $\tilde{R}$ and $\tilde{\nabla}$ the corresponding notions of N. Then we have

$$\tilde{R}(X,Y)Z = \tilde{\nabla}_X \tilde{\nabla}_Y Z - \tilde{\nabla}_Y \tilde{\nabla}_X Z - \tilde{\nabla}_{[X,Y]}Z \tag{4.22}$$

for $X,Y,Z \in \mathfrak{X}(M)$. By applying Gauss' formula, we find

$$\begin{aligned}
\tilde{R}(X,Y)Z &= \tilde{\nabla}_X(\nabla_Y Z + h(Y,Z)) - \tilde{\nabla}_Y(\nabla_X Z + h(X,Z)) \\
&\quad - \nabla_{[X,Y]}Z - h([X,Y],Z) \\
&= R(X,Y)Z + h(X,\nabla_Y Z) - h(Y,\nabla_X Z) \\
&\quad - h([X,Y],Z) + \nabla_X h(Y,Z) - \nabla_Y h(X,Z).
\end{aligned}$$

On the other hand, using Weingarten's formula we find

$$\begin{aligned}
\tilde{R}(X,Y)Z &= R(X,Y)Z - A_{h(Y,Z)}X + A_{h(X,Z)}Y + h(X,\nabla_Y Z) \\
&\quad - h(Y,\nabla_X Z) - h([X,Y],Z) + D_X h(Y,Z) - D_Y h(X,Z).
\end{aligned} \tag{4.23}$$

Hence we obtain

Theorem 4.3. *Let $\phi : M \to N$ be an isometric immersion of a pseudo-Riemannian manifold M into a pseudo-Riemannian manifold N. Then for vector fields X,Y,Z,W tangent to M, we have*

$$\begin{aligned}
R(X,Y;Z,W) &= \tilde{R}(X,Y;Z,W) + \langle h(X,W), h(Y,Z)\rangle \\
&\quad - \langle h(X,Z), h(Y,W)\rangle,
\end{aligned} \tag{4.24}$$

$$(\tilde{R}(X,Y)Z)^{\perp} = (\bar{\nabla}_X h)(Y,Z) - (\bar{\nabla}_Y h)(X,Z), \tag{4.25}$$

where $R(X,Y;Z,W) = \langle R(X,Y)Z,W\rangle$, $(\tilde{R}(X,Y)Z)^{\perp}$ the normal component of $\tilde{R}(X,Y)Z$, and $\bar{\nabla}h$ the covariant derivative of h with respect to the van der Waerden-Bortololli connection $\bar{\nabla} = \nabla \oplus D$, i.e.,

$$(\bar{\nabla}_X h)(Y,Z) = D_X h(Y,Z) - h(\nabla_X Y, Z) - h(Y, \nabla_X Z). \tag{4.26}$$

If ξ and η are normal vector fields of M, we have

$$\begin{aligned}
\tilde{R}(X,Y;\xi,\eta) &= \langle \tilde{\nabla}_X \tilde{\nabla}_Y \xi, \eta\rangle - \langle \tilde{\nabla}_Y \tilde{\nabla}_X \xi, \eta\rangle - \langle \tilde{\nabla}_{[X,Y]}\xi, \eta\rangle \\
&= \langle \tilde{\nabla}_Y(A_\xi X), \eta\rangle - \langle \tilde{\nabla}_X(A_\xi Y), \eta\rangle + \langle \tilde{\nabla}_X D_Y \xi, \eta\rangle \\
&\quad - \langle \tilde{\nabla}_Y D_X \xi, \eta\rangle - \langle D_{[X,Y]}\xi, \eta\rangle \\
&= \langle h(Y, A_\xi X), \eta\rangle - \langle h(X, A_\xi Y), \eta\rangle + \langle D_X D_Y \xi, \eta\rangle \\
&\quad - \langle D_Y D_X \xi, \eta\rangle - \langle D_{[X,Y]}\xi, \eta\rangle.
\end{aligned}$$

If R^D denotes the curvature tensor of the normal bundle $T^\perp M$, i.e.,

$$R^D(X,Y)\xi = D_X D_Y \xi - D_Y D_X \xi - D_{[X,Y]}\xi, \qquad (4.27)$$

then

$$R^D(X,Y;\xi,\eta) = \tilde{R}(X,Y;\xi,\eta) + \langle [A_\xi, A_\eta](X), Y \rangle, \qquad (4.28)$$

where $[A_\xi, A_\eta] = A_\xi A_\eta - A_\eta A_\xi$.

Equations (4.24), (4.25) and (4.28) are called the *equations of Gauss, Codazzi*, and *Ricci*, respectively; known as the *fundamental equations*.

Remark 4.2. For surfaces in $\mathbb{E}^3$ the equation of Gauss was found in 1827 in principal, though not explicitly, by C. F. Gauss; the equation of Codazzi was given by Delfino Codazzi (1824-1875) in 1860, independently by Gaspare Mainardi (1800-1879) in 1856 and also by Karl M. Peterson (1828-1881) in his doctoral thesis [Peterson (1853)]. Aurel E. Voss (1845-1931) extended in 1880 both equations of Gauss and Codazzi to Riemannian submanifolds. The equation of Ricci was discovered in 1899 by Gregorio Ricci (1853-1925).

Proposition 4.3. *Let M_s^n be a pseudo-Riemannian n-manifold with index s isometrically immersed in an indefinite real space form $R_s^m(c)$ of constant curvature c. Then the Ricci tensor of M_s^n satisfies*

$$Ric(Y,Z) = (n-1)\langle Y,Z \rangle c + n\langle H, h(Y,Z)\rangle - \sum_i \epsilon_i \langle h(Y,e_i), h(Z,e_i)\rangle,$$

where $\{e_1,\ldots,e_n\}$ is an orthonormal frame of M_s^n.

Proof. The equation of Gauss yields

$$Ric(Y,Z) = \sum_{i=1}^n \epsilon_i \tilde{R}(e_i, Y; Z, e_i) + n\langle H, h(Y,Z)\rangle - \sum_i \epsilon_i \langle h(Y,e_i), h(Z,e_i)\rangle.$$

Combining this with (1.14) gives the Proposition. $\qquad \square$

An immediate consequence of Proposition 4.3 is the following.

Corollary 4.1. *If M_s^n is a minimal submanifold of $\mathbb{E}_s^m$, then $Ric \leq 0$, with the equality holding identically if and only if M_s^n is totally geodesic.*

Proposition 4.4. *Let M_t^n be an n-dimensional pseudo-Riemannian submanifold of an indefinite real space form $R_s^m(c)$. Then the scalar curvature, the mean curvature vector, and the second fundamental form of M_t^n satisfy*

$$\tau = \frac{n^2}{2}\langle H, H \rangle - \frac{1}{2}S_h + \frac{n(n-1)}{2}c, \qquad (4.29)$$

where S_h is defined as

$$S_h = \sum_{i,j=1}^{n} \epsilon_i \epsilon_j \left\langle h(e_i, e_j), h(e_i, e_j) \right\rangle, \tag{4.30}$$

with $\epsilon_i = \langle e_i, e_i \rangle$ and $e_1, \ldots, e_n$ being an orthonormal frame of M_t^n.

Proof. Let $e_1, \ldots, e_n$ be an orthonormal frame of M_t^n. Then the equation of Gauss gives

$$\sum_{i,j=1}^{n} \epsilon_i \epsilon_j \left\langle R(e_i, e_j)e_j, e_i \right\rangle = \sum_{i,j=1}^{n} \epsilon_i \epsilon_j \left\langle \tilde{R}(e_i, e_j)e_j, e_i \right\rangle$$

$$+ \sum_{i,j=1}^{n} \left\langle \epsilon_i h(e_i, e_i), \epsilon_j h(e_j, e_j) \right\rangle - \sum_{i,j=1}^{n} \epsilon_i \epsilon_j \left\langle h(e_i, e_j), h(e_i, e_j) \right\rangle. \tag{4.31}$$

Since the sectional curvature K of M satisfies

$$K(e_i \wedge e_j) = \epsilon_i \epsilon_j \left\langle R(e_i, e_j)e_j, e_i \right\rangle, \tag{4.32}$$

we find from (4.31) that

$$2\tau = \sum_{i,j=1}^{n} K(e_i \wedge e_j) = n(n-1)c + n^2 \left\langle H, H \right\rangle - S_h,$$

which gives (4.29). $\qquad\qquad\qquad\qquad\qquad\qquad\qquad\qquad\qquad\qquad\square$

An immediate consequence of Proposition 4.4 is the following.

Corollary 4.2. *If M_s^n is an n-dimensional minimal submanifold of an indefinite real space form $R_s^{n+r}(c)$, then*

$$2\tau \leq n(n-1)c. \tag{4.33}$$

Similarly, if M_s^n is a minimal submanifold in $R_{s+r}^{n+r}(c)$, then

$$2\tau \geq n(n-1)c. \tag{4.34}$$

Either equality holds identically if and only if M_s^n is totally geodesic.

Another application of Proposition 4.4 is the following.

Proposition 4.5. *Let M be a submanifold of a real space form $R^{n+r}(c)$ of constant curvature c. Then the scalar curvature of M satisfies*

$$\tau \leq \frac{n(n-1)}{2}(H^2 + c), \quad n = \dim M, \tag{4.35}$$

with the equality holding at p if and only if p is a totally umbilical point.

Proof. Choose an orthonormal basis $e_1, \ldots, e_n, e_{n+1}, \ldots, e_{n+r}$ at p such that e_{n+1} is parallel to the mean curvature vector and $e_1, \ldots, e_n$ diagonalize the shape operator $A_{n+1} = A_{e_{n+1}}$. It follows from Proposition 4.4 that

$$n^2 H^2 = 2\tau + \sum_{i=1}^{n} a_i^2 + \sum_{\alpha=n+2}^{r} \sum_{i,j=1}^{n} (h_{ij}^\alpha)^2 - n(n-1)c, \qquad (4.36)$$

where $a_1, \ldots, a_n$ are eigenvalues of A_{n+1}.

On the other hand, it follows from the Cauchy-Schwarz inequality that

$$\sum_{i=1}^{n} a_i^2 \geq nH^2, \qquad (4.37)$$

with the equality holding if and only if $a_1 = a_2 = \cdots = a_n$. Combining (4.36) and (4.37) gives

$$n(n-1)H^2 \geq 2\tau - n(n-1)c + \sum_{\alpha=n+2}^{r} \sum_{i,j=1}^{n} (h_{ij}^\alpha)^2, \qquad (4.38)$$

which implies inequality (4.35).

If the equality sign of (4.35) holds at a point $p \in M$, then it follows from (4.37) and (4.38) that $A_{n+2} = \cdots = A_{n+r} = 0$ and $a_1 = \cdots = a_n$. Therefore, p is a totally umbilical point. The converse is trivial. $\qquad \square$

Proposition 4.5 has some nice applications, e.g., it implies the following.

Corollary 4.3. *If the scalar curvature of an n-dimensional Riemannian submanifold M of $\mathbb{E}^m$ satisfies $\tau \geq n(n-1)/2$ at a point p, then every isometric immersion of M into any Euclidean space satisfies $H^2 \geq 1$ at p regardless of codimension.*

In particular, if $\tau = n(n-1)/2$ on M, then $H^2 \geq 1$ holds identically on M, with the equality holding identically if and only if M is an open part of a standard unit hypersphere in a totally geodesic $\mathbb{E}^{n+1} \subset \mathbb{E}^m$.

Proof. Inequality $H^2 \geq 1$ follows immediately from Proposition 4.5. If the scalar curvature of M is $n(n-1)/2$ at each point and if $\|H\| = 1$ holds identically on M, then M is totally umbilical. Thus M is an open portion of an ordinary unit n-sphere. $\qquad \square$

Similar to Proposition 4.5, we also have the following.

Corollary 4.4. *Let M be a spacelike submanifold of an indefinite real space form $R_r^{n+r}(c)$ of constant curvature c. Then*

$$\|H\|^2 \leq \frac{2\tau}{n(n-1)} - c, \quad n = \dim M, \qquad (4.39)$$

with equality holding at $p \in M$ if and only if p is a totally umbilical point.

For further applications of Proposition 4.5, see [Chen (1996a)].

4.5 Fundamental theorems of submanifolds

Now, we can state the fundamental theorems of submanifolds as follows. For the proofs see, for instance, [Eschenburg and Tribuzy (1993); Wettstein (1978)].

Theorem 4.4. (Existence) *Let* (M_t^n, g) *be a simply-connected pseudo-Riemannian n-manifold with index* t. *Suppose that there exists an* $(m - n)$-*dimensional pseudo-Riemannian vector bundle* $\nu(M_t^n)$ *with index* $s - t$ *over* M_t^n *and with curvature tensor* R^D *and also exists a* $\nu(M_t^n)$-*valued symmetric* $(0, 2)$ *tensor* h *on* M_t^n. *For a cross section* ξ *of* $\nu(M_t^n)$, *define* A_ξ *by* $g(A_\xi X, Y) = \langle h(X, Y), \xi \rangle$, *where* $\langle \ , \ \rangle$ *is the fiber metric of* $\nu(M_t^n)$. *If they satisfy* (4.24), (4.25) *and* (4.28), *then* M_t^n *can be isometrically immersed in an m-dimensional indefinite real space form* $R_s^m(c)$ *of constant curvature* c *in such way that* $\nu(M_t^n)$ *is the normal bundle and* h *is the second fundamental form.*

Theorem 4.5. (Uniqueness) *Let* $\phi, \phi' : M_t^n \to R_s^m(c)$ *be two isometric immersions of a pseudo-Riemannian n-manifold* M_t^n *into an indefinite space for* $R_s^m(c)$ *of constant curvature* c *with normal bundles* ν *and* ν' *equipped with their canonical bundle metrics, connections and second fundamental forms, respectively. Suppose there is an isometry* $\phi : M_t^n \to M_t^n$ *such that* ϕ *can be covered by a bundle map* $\bar{\phi} : \nu \to \nu'$ *which preserves the bundle metrics, the connections and the second fundamental forms. Then there is an isometry* Φ *of* $R_s^m(c)$ *such that* $\Phi \circ \phi = \phi'$.

Two submanifolds M_1 and M_2 of a pseudo-Riemannian manifold N are said to be *congruent* if there exists an isometry of N which carries one to the other. Congruent submanifolds have the same intrinsic and extrinsic geometry.

4.6 A universal inequality for submanifolds

Let M be an n-dimensional Riemannian manifold. For an integer $k \geq 0$, denote by $\mathcal{S}(n, k)$ the finite set consisting of k-tuples $(n_1, \ldots, n_k)$ of integers ≥ 2 satisfying $n_1 < n$ and $n_1 + \cdots + n_k \leq n$. Denote by $\mathcal{S}(n)$ the set of (unordered) k-tuples with $k \geq 0$ for a fixed positive integer n.

The cardinal number $\#\mathcal{S}(n)$ of $\mathcal{S}(n)$ is equal to $p(n) - 1$, where $p(n)$ denotes the number of partition of n which increases quite rapidly with n. For instance, for

$$n = 2, 3, 4, 5, 6, 7, 8, 9, 10, \ldots, 20, \ldots, 50, \ldots,$$

$$100, \ldots, 200,$$

the cardinal number $\#\mathcal{S}(n)$ are given respectively by

$$1, 2, 4, 6, 10, 14, 21, 29, 41, \ldots, 626, \ldots, 204\,225, \ldots,$$

$$190\,569\,291, \ldots, 3\,972\,999\,029\,387.$$

For each $(n_1, \ldots, n_k) \in \mathcal{S}(n)$ the author introduced in 1990s a Riemannian invariant $\delta(n_1, \ldots, n_k)$ by

$$\delta(n_1, \ldots, n_k)(x) = \frac{1}{2}(\rho(x) - \inf\{\rho(L_1) + \cdots + \rho(L_k)\}),$$

where $L_1, \ldots, L_k$ run over all k mutually orthogonal subspaces of $T_x M$ such that $\dim L_j = n_j$, $j = 1, \ldots, k$. For each $(n_1, \ldots, n_k) \in \mathcal{S}(n)$, put

$$a(n_1, \ldots, n_k) = \frac{1}{2}n(n-1) - \frac{1}{2}\sum_{j=1}^{k} n_j(n_j - 1) \qquad (4.40)$$

$$b(n_1, \ldots, n_k) = \frac{n^2(n + k - 1 - \sum_j n_j)}{2(n + k - \sum_j n_j)}. \qquad (4.41)$$

We have the following universal inequality for submanifolds.

Theorem 4.6. [Chen (1998, 2000a)] *For any n-dimensional submanifold M of a Riemannian m-manifold $R^m(c)$ of constant curvature c and for any k-tuple $(n_1, \ldots, n_k) \in \mathcal{S}(n)$, we have*

$$\delta(n_1, \ldots, n_k) \leq b(n_1, \ldots, n_k)H^2 + a(n_1, \ldots, n_k)c, \qquad (4.42)$$

where H^2 is the squared mean curvature of M in $R^m(c)$.

The equality case of inequality (4.42) holds at a point $x \in M$ if and only if, there exists an orthonormal basis $e_1, \ldots, e_m$ at x, such that the shape operators of M in $R^m(c)$ at x take the following form:

$$A_r = \begin{pmatrix} A_1^r & \cdots & 0 & 0 & \cdots & 0 \\ \vdots & \ddots & \vdots & \vdots & & \vdots \\ 0 & \cdots & A_k^r & 0 & \cdots & 0 \\ 0 & \cdots & 0 & \mu_r & \cdots & 0 \\ \vdots & & \vdots & \vdots & \ddots & \vdots \\ 0 & \cdots & 0 & 0 & \cdots & \mu_r \end{pmatrix}, \quad r = n+1, \ldots, m, \qquad (4.43)$$

where A_j^r are symmetric $n_j \times n_j$ submatrices which satisfy

$$\text{trace } A_1^r = \cdots = \text{trace } A_k^r = \mu_r.$$

For the extension of Theorem 4.6 to submanifolds in arbitrary Riemannian manifolds, see, e.g., [Chen (2011b), Theorem 13.3].

Definition 4.5. An isometric immersion $\phi : M \to R^m(c)$ of a Riemannian n-manifold into a real space form $R^m(c)$ is called $\delta(n_1, \ldots, n_k)$-*ideal* if the equality sign of (4.42) holds identically.

Remark 4.3. Roughly speaking, ideal immersions are the immersions which receive the least possible amount of tension at each point form its ambient space. In this sense, ideal immersions of a Riemannian manifold are nice immersions among all possible isometric immersions of the manifold.

Remark 4.4. The δ-invariants and the inequality (4.42) have many applications, see e.g., [Chen (2011b, 2013d)].

4.7   Reduction theorem of Erbacher-Magid

Let $\mathbf{R}_{i,j}^n$ denote the affine n-space equipped with the metric whose canonical form is

$$\begin{pmatrix} O_j & & \\ & -I_i & \\ & & I_{n-i-j} \end{pmatrix},$$

where I_k is the $k \times k$ identity matrix and O_j is the $j \times j$ zero matrix.

The metric is non-degenerate if and only if $j = 0$. The j in $\mathbf{R}_{i,j}^n$ measures the degenerate part. The metric of $\mathbf{R}_{i,1}^n = \mathbf{R}_0 \times \mathbb{E}_i^{n-1}$ vanishes on the first factor $\mathbf{R}_0$ and it is the standard pseudo-Euclidean metric with index i on the second factor $\mathbb{E}_i^{n-1}$.

Denote the natural embedding $\iota : \mathbf{R}_{i,1}^n \to \mathbb{E}_{i+1}^{n+1}$ of $\mathbf{R}_{i,1}^n$ into $\mathbb{E}_{i+1}^{n+1}$ given by

$$\iota((x_1, x_2, \ldots, x_n)) = (x_1, x_2, \ldots, x_n, x_1) \in \mathbb{E}_{i+1}^{n+1}$$

for $(x_1, \ldots, x_n) \in \mathbf{R}_{i,1}^n$. Then the light-like vector $\zeta_0 = (1, 0, \ldots, 0, 1)$ is a normal vector of $\mathbf{R}_{i,1}^n$ in $\mathbb{E}_{i+1}^{n+1}$.

Let $\phi : M \to N$ be an isometric immersion of a pseudo-Riemannian manifold into another pseudo-Riemannian manifold. At each point $p \in M$, the *first normal space* $\mathcal{N}^1(p)$ is defined to be the orthogonal complement of $\mathcal{N}^0(p) = \{\xi \in T_p^\perp M : A_\xi = 0\}$.

Definition 4.6. Let $\phi : M \to N$ be an isometric immersion of a pseudo-Riemannian manifold into another pseudo-Riemannian manifold. The first normal spaces are called *parallel* if, for any curve σ joining any two points $p, q \in M$, the parallel displacement of normal vectors along σ with respect to the normal connection maps $\mathcal{N}^1(p)$ onto $\mathcal{N}^1(q)$.

The following result is the reduction theorem of Erbacher-Magid (see [Erbacher (1971); Magid (1984)]).

Theorem 4.7. *Let $\phi : M_i^n \to \mathbb{E}_s^m$ be an isometric immersion of a pseudo-Riemannian n-manifold M_i^n with index i into $\mathbb{E}_s^m$. If the first normal spaces are parallel, then there exists a complete $(n+k)$-dimensional totally geodesic submanifold E^* such that $\psi(M_i^n) \subset E^*$, where k is the dimension of the first normal spaces.*

Proof. Under the hypothesis, the dimension of $\mathcal{N}^1$ is a constant, say k. If ξ is a normal vector field such that $\xi \in \mathcal{N}^1(p)$ for each $p \in M_i^n$, then $D_X \xi \in \mathcal{N}^1(p)$ for all $X \in T_p M_i^n$. Thus the first normal spaces $\mathcal{N}^1(p)$ form a parallel normal subbundle. Since D is a metric connection, the subspaces $\mathcal{N}^0(p)$ are also parallel with respect to the normal connection.

Let p_0 be a point of M_i^n. Consider the $(n + k)$-dimensional subspace E of $\mathbb{E}_s^m$ through $\phi(p_0)$ which is perpendicular to $\mathcal{N}^0(p_0)$, i.e.,

$$E = T_{p_0}(M_i^n) \oplus \mathcal{N}^1(p_0).$$

Then the degenerate part of E is $\mathcal{N}^0(p_0) \cap \mathcal{N}^1(p_0)$. Now, we claim that $\phi(M_i^n) \subset E$. This can be proved as follows: Let $\beta(t)$ be any curve in M_i^n starting at p_0. For any $\xi_0 \in \mathcal{N}^0(p_0)$, let ξ_t be the parallel displacement of ξ_0 along $\beta(t)$, so that $\xi_t \in \mathcal{N}^0(\beta(t))$. For the pseudo-Euclidean connection ∇, we have

$$\nabla_{\beta'(t)} \xi_t = -d\phi(A_{\xi_t}(\beta'(t))) + D_{\beta'(t)} \xi_t = 0,$$

which means that ξ_t is parallel in $\mathbb{E}_s^m$. Thus it is a constant vector. Now, we have

$$\frac{d}{dt} \langle \phi(\beta(t)) - \phi(p_0), \xi_0 \rangle = \langle d\phi(\beta'(t)), \xi_0 \rangle = \langle d\phi(\beta'(t)), \xi_t \rangle = 0.$$

Thus $\phi(\beta(t))$ lies in E. Since this is true for arbitrary curve $\beta(t)$ in M_i^n, $\phi(M_i^n) \subset E$. $\square$

In Erbacher-Magid's reduction theorem, $E^* = \mathbf{R}_{s,t}^{n+k}$ for some s, t and t need not be zero.

Definition 4.7. A pseudo-Riemannian submanifold M of a pseudo-Riemannian manifold is called *parallel* if $\bar{\nabla} h = 0$ identically.

The following is an easy consequence of the Reduction Theorem.

Corollary 4.5. *Let $\phi : M \to \mathbb{E}_s^m$ be an isometric immersion of a pseudo-Riemannian n-manifold M into pseudo-Euclidean m-space $\mathbb{E}_s^m$. If ϕ is a parallel immersion, then there exists a complete $(n + k)$-dimensional totally geodesic submanifold $E^* \subset \mathbb{E}_s^m$ such that $\phi(M) \subset E^*$, where k is the dimension of the first normal spaces.*

Proof. If ϕ is parallel, then $\bar{\nabla}h = 0$. Thus it follows from (4.26) that

$$D_X h(Y, Z) = h(\nabla'_X Y, Z) + h(Y, \nabla'_X Z)$$

for $X, Y, Z \in \mathfrak{X}(M)$. Hence the first normal spaces are parallel. Therefore, the corollary follows from the Reduction Theorem.   □

4.8   Two basic formulas for submanifolds

The following *formula of Beltrami* [Beltrami (1864)] (see also [Tazzioli (1997)]), named after Eugenio Beltrami (1835-1900), is a fundamental formula in submanifold theory.

Proposition 4.6. [Beltrami (1864)] *Let $\phi : M \to \mathbb{E}_s^m$ be an isometric immersion of a pseudo-Riemannian n-manifold M into a pseudo-Euclidean space. Then we have*

$$\Delta\phi = -nH, \tag{4.44}$$

where H is the mean curvature vector of the immersion.

Proof. Let c be any given vector in $\mathbb{E}_s^m$ and $p \in M$. If $\{e_1, \ldots, e_n\}$ is an orthonormal basis of T_pM, then we may extend $e_1, \ldots, e_n$ to an orthonormal frame $E_1, \ldots, E_n$ such that $\nabla_{E_i} E_j = 0$ at p for $i, j = 1, \ldots, n$, where ∇ is the Levi-Civita connection of M. Then we have

$$
\begin{aligned}
(\Delta \langle \phi, c \rangle)_p &= -\sum_{i=1}^n \epsilon_i e_i \langle E_i, c \rangle = -\sum_{i=1}^n \epsilon_i \langle \nabla_{e_i} E_i, c \rangle \\
&= -\sum_{i=1}^n \epsilon_i \langle h(e_i, e_i), c \rangle = -n \langle H, c \rangle (p)
\end{aligned}
\tag{4.45}
$$

for any $c \in \mathbb{E}_s^m$. Since both $\Delta\phi$ and H are independent of the choice of the local basis, we have $\langle \Delta\phi, c \rangle = -n \langle H, c \rangle$. Because the scalar product $\langle \; , \; \rangle$ is nondegenerate, (4.45) implies (4.44).   □

An immediate consequence of Proposition 4.6 is the following.

Corollary 4.6. *Every spacelike minimal submanifold M in a pseudo-Euclidean space $\mathbb{E}_s^m$ is non-closed.*

Proof. If M is a minimal submanifold, it follows from Proposition 4.6 that each natural coordinate function of $\mathbb{E}_s^m$ restricted to M is a harmonic function. Thus when M is spacelike and closed, each such function is constant. Thus M must be a point which is a contradiction since each submanifold is assumed to be of dimension at least one. $\qquad\square$

Another easy application of Proposition 4.6 is the following.

Corollary 4.7. *Every spacelike minimal submanifold M in a pseudo-hyperbolic space $H_s^m(-1)$ is non-closed.*

Proof. Let $\phi : M \to H_s^m(-1)$ be a minimal isometric immersion of a closed Riemannian n-manifold N into $H_s^m(-1)$. Without loss of generality we may regard $H_s^m(-1)$ as a hypersurface of $\mathbb{E}_{s+1}^{m+1}$ via the canonical imbedding (1.17). Then the mean curvature vector $\tilde{H}$ of M in $\mathbb{E}_{s+1}^{m+1}$ is $\mathbf{x} = \iota \circ \phi$. Therefore Proposition 4.6 yields $\Delta\mathbf{x} = -n\mathbf{x}$. Hence the Laplacian Δ on M has an eigenfunction with eigenvalue equal to $-n$. But this is impossible since $n > 0$ and M is assumed to be a closed manifold. $\qquad\square$

The next formula is basic and quite useful in finite type theory.

Proposition 4.7. [Chen (1979b, 1986b)] *Let $\phi : M \to \mathbb{E}_s^m$ be an isometric immersion of a pseudo-Riemannian n-manifold M into a pseudo-Euclidean space. Then we have the following formula:*

$$\Delta H = \Delta^D H + \sum_{i=1}^{n} \epsilon_i h(A_H e_i, e_i) + (\Delta H)^T, \tag{4.46}$$

where

$$(\Delta H)^T = \frac{n}{2}\nabla\langle H, H\rangle + 2\operatorname{Tr} A_{DH} \tag{4.47}$$

is the tangential component of ΔH, Δ^D is the Laplacian associated with the normal connection D and $e_1, \ldots, e_n$ is an orthonormal tangent frame of M.

Proof. Let c be a given constant vector field in $\mathbb{E}_s^m$ and let $e_1, \ldots, e_n$ be an orthonormal basis of T_pM, $p \in M$. We may extend $e_1, \ldots, e_n$ to an orthonormal frame $E_1, \ldots, E_n$ such that $\nabla_{E_i} E_j = 0$ at p for $i, j = 1, \ldots, n$.

For any vector fields $X, Y \in \mathfrak{X}(M)$, it follows from the formula of Weingarten that

$$X \langle H, c \rangle = \langle D_X H, c \rangle - \langle A_H X, c \rangle. \tag{4.48}$$

Thus

$$\begin{aligned} YX \langle H, c \rangle = \langle D_Y D_X H, c \rangle - \langle \nabla_Y (A_H X), c \rangle \\ - \langle A_{D_X H} Y, c \rangle - \langle h(A_H X, Y), c \rangle. \end{aligned} \tag{4.49}$$

Therefore, it follows from the definition of Δ that

$$\Delta H = \Delta^D H + \sum_{i=1}^{n} \epsilon_i \{ h(A_H e_i, e_i) + (\nabla_{e_i} A_H) e_i + A_{D_{e_i} H} e_i \}. \tag{4.50}$$

Put

$$\begin{aligned} \mathrm{Tr} \, A_{DH} = \sum_i \epsilon_i A_{D_{e_i} H} e_i, \\ \mathrm{Tr} \, (\nabla A_H) = \sum_i \epsilon_i (\nabla_{e_i} A_H) e_i. \end{aligned} \tag{4.51}$$

Then

$$\begin{aligned} \langle \mathrm{Tr}(\nabla A_H), X \rangle &= \sum_i \epsilon_i \langle (\nabla_{e_i} A_H) e_i, X \rangle \\ &= \sum_i \epsilon_i \{ \langle (\nabla_X A) e_i, e_i \rangle - \langle A_{D_X H} e_i, e_i \rangle + \langle A_{D_{e_i} H} e_i, X \rangle \} \\ &= \sum_i \{ \langle (\bar{\nabla}_X h)(e_i, e_i), H \rangle + \langle A_{D_{e_i} H} e_i, H \rangle \} \\ &= \sum_i \{ \langle D_X h(e_i, e_i), H \rangle + \langle A_{D_{e_i} H} e_i, H \rangle \} \\ &= n \langle D_X H, H \rangle + \langle \mathrm{Tr} \, A_{DH}, X \rangle \} \\ &= \frac{n}{2} X \langle H, H \rangle + \langle \mathrm{Tr} \, A_{DH}, X \rangle, \end{aligned}$$

which implies that

$$\mathrm{Tr} \, (\nabla A_H) = \frac{n}{2} \nabla \langle H, H \rangle + \mathrm{Tr} \, A_{DH}. \tag{4.52}$$

From (4.50), (4.51) and (4.52) we obtain (4.46). $\qquad \square$

An immediate consequence of Proposition 4.7 is the following.

Corollary 4.8. *Let M be a pseudo-Riemannian n-manifold isometrically immersed in a pseudo-Euclidean space. If M has parallel mean curvature vector, then we have*

$$(\Delta H)^T = 0 \tag{4.53}$$

$$\Delta H = \sum_{i=1}^{n} \epsilon_i h(A_H e_i, e_i), \tag{4.54}$$

where $e_1, \ldots, e_n$ is an orthonormal frame of TM.

4.9 Totally geodesic submanifolds

The simplest submanifolds are totally geodesic submanifolds.

Proposition 4.8. *Let $\phi : M \to N$ be an isometric immersion of a pseudo-Riemannian manifold M into a pseudo-Riemannian manifold N. Then the following three statements are equivalent.*

(1) *M is a totally geodesic submanifold of N;*
(2) *Geodesics of M are geodesics of N;*
(3) *For any $p \in M$ and any $v \in T_pM$, the geodesic γ_v with $\gamma_v(0) = p$ and $\gamma_v'(0) = v$ lies locally in M.*

Proof. Since M is totally geodesic in N, the second fundamental form vanishes. Thus if γ is a curve of M, then $\nabla_{\gamma'}\gamma' = \tilde{\nabla}_{\gamma'}\gamma'$, where ∇ and $\tilde{\nabla}$ are the Levi-Civita connections of M and N, respectively. Hence γ is a geodesic of N if and only if it is geodesic of M. This proves the equivalence of (1) and (2).

Now, suppose that $\gamma : I \to M$ is a geodesic of M with initial velocity vector v, then γ is also a geodesic of N. Hence, by the uniqueness of geodesics, we obtain (3).

If (3) holds, Gauss' formula implies that $h(v, v) = 0$. Thus, for vectors $v, w \in T_pM$,

$$0 = h(v + w, v + w)$$
$$= h(v, v) + 2h(v, w) + h(w, w)$$
$$= 2h(v, w).$$

Since this is true for any $p \in N$ and any $v, w \in T_pM$, we find $h = 0$ by polarization. Thus M is totally geodesic in N. $\square$

Proposition 4.9. *Up to rigid motions, an n-dimensional totally geodesic pseudo-Riemannian submanifold of a pseudo-Euclidean space $\mathbb{E}_s^m$ is an open portion of a pseudo-Euclidean linear subspace $\mathbb{E}_t^n$ of $\mathbb{E}_s^m$.*

Proof. Obviously, every pseudo-Euclidean linear subspace $\mathbb{E}_t^n$ of $\mathbb{E}_s^m$ is a totally geodesic submanifold of $\mathbb{E}_s^m$.

Now, assume that M is an n-dimensional totally geodesic pseudo-Riemannian submanifold of $\mathbb{E}_s^m$ such that the origin o of $\mathbb{E}_s^m$ lies in M. Then T_oM is a nondegenerate subspace of $T_o\mathbb{E}_s^m$. Because geodesics of $\mathbb{E}_s^m$ are lines, it follows from Proposition 4.8 that M is an open portion of the pseudo-Euclidean subspace whose tangent space at o is T_oM. $\square$

Recall that the pseudo m-sphere $S_s^m(c)$ is defined by

$$S_s^m(c) = \left\{ \mathbf{x} = (x_1, \ldots, x_{m+1}) \in \mathbb{E}_s^{m+1} : \langle \mathbf{x}, \mathbf{x} \rangle = \frac{1}{c} > 0 \right\}.$$

For $n < m$ and $0 \le t \le s$,

$$\left\{ (x_1, \ldots, x_t, 0, \ldots, 0, x_{s+1}, \ldots, x_{n+1}, 0, \ldots, 0) \in S_s^m(c) \right\}$$

defines a pseudo-Riemannian manifold of constant curvature c with index t, which is totally geodesic in $S_s^m(c)$. We call this totally geodesic submanifold of $S_s^m(c)$ a *pseudo n-sphere* of $S_s^m(c)$. Analogously, we call

$$\left\{ (x_1, \ldots, x_{t+1}, 0, \ldots, 0, x_{s+1}, \ldots, x_{n+1}, 0, \ldots, 0) \in H_s^m(c) \right\}$$

a *pseudo-hyperbolic n-subspace* of $H_s^m(c)$.

Similar to Proposition 4.9, we have the following.

Proposition 4.10. *Up to rigid motions, an n-dimensional totally geodesic pseudo-Riemannian submanifold of a pseudo m-sphere $S_s^m(c)$ is an open portion of a pseudo n-sphere of $S_s^m(c)$.*

Proposition 4.11. *Up to rigid motions, an n-dimensional totally geodesic pseudo-Riemannian submanifold of a pseudo-hyperbolic m-space $H_s^m(c)$ is an open portion of a pseudo-hyperbolic n-subspace of $H_s^m(c)$.*

Remark 4.5. The classification of totally geodesic submanifolds of symmetric spaces are rather complicated. For this reason, (M_+, M_-)-method was introduced in [Chen and Nagano (1978)] to investigate compact Lie groups and totally geodesic submanifolds of symmetric spaces (see also, [Chen (1987c, 2013e); Chen and Nagano (1988)]).

4.10 Parallel submanifolds

The class of *symmetric R-spaces* includes:

(a) all Hermitian symmetric spaces of compact type;
(b) Grassmann manifolds $O(p+q)/O(p) \times O(q)$, $Sp(p+q)/Sp(p) \times Sp(q)$;
(c) the classical groups $SO(m)$, $U(m)$, $Sp(m)$;
(d) $U(2m)/Sp(m)$, $U(m)/O(m)$;
(e) $(SO(p+1) \times SO(q+1))/S(O(p) \times O(q))$, where $S(O(p) \times O(q))$ is the subgroup of $SO(p+1) \times SO(q+1)$ consisting of matrices of the form

$$\begin{pmatrix} \epsilon & 0 & & \\ 0 & A & & \\ & & \epsilon & 0 \\ & & 0 & B \end{pmatrix}, \quad \epsilon = \pm 1, \quad A \in O(p), \quad B \in O(q);$$

(f) the Cayley projective plane $\mathcal{O}P^2$;

(g) the three exceptional spaces $E_6/Spin(10) \times T, E_7/E_6 \times T$, and E_6/F_4.

Clearly, totally geodesic submanifolds are parallel submanifolds. Affine subspaces of $\mathbb{E}^m$ and symmetric R-spaces minimally embedded in a hypersphere of $\mathbb{E}^m$ as described in [Takeuchi and Kobayashi (1968)] are parallel submanifolds of $\mathbb{E}^m$.

Parallel submanifolds of Euclidean spaces were classified in D. Ferus in [Ferus (1974)]. He proved that essentially these submanifolds exhaust all parallel submanifolds of $\mathbb{E}^m$.

Theorem 4.8. [Ferus (1974)] *A parallel submanifold M of the Euclidean m-space $\mathbb{E}^m$ is congruent to*

(1) $M = \mathbb{E}^{m_0} \times M_1 \times \cdots \times M_k \subset \mathbb{E}^{m_0} \times \mathbb{E}^{m_1} \times \cdots \times \mathbb{E}^{m_k} = \mathbb{E}^m, \ k \geq 0$, *or to*

(2) $M = M_1 \times \cdots \times M_k \subset \mathbb{E}^{m_1} \times \cdots \times \mathbb{E}^{m_k} = \mathbb{E}^m, \ k \geq 1$,

where each $M_i \subset \mathbb{E}^{m_i}$ is an irreducible symmetric R-space.

Notice that in case (1) M is not contained in any hypersphere of $\mathbb{E}^m$, but in case (2) M is contained in a hypersphere of $\mathbb{E}^m$.

If the unit $(m-1)$-sphere $S^{m-1}(1)$ is regarded as an ordinary hypersphere of $\mathbb{E}^m$, then a submanifold $M \subset S^{m-1}(1)$ is parallel if and only if $M \subset S^{m-1}(1) \subset \mathbb{E}^m$ is a parallel submanifold of $\mathbb{E}^m$. Hence Ferus' result implies the following.

Theorem 4.9. *A submanifold M of $S^{m-1}(1)$ is a parallel submanifold if and only if M is obtained by a submanifold of type (2) in Theorem 4.8.*

An immersion of a manifold M into a real space form $R^m(c)$ is called *full* if it is not contained in any totally geodesic hypersurface of $R^m(c)$.

Parallel submanifolds in hyperbolic spaces have been determined in [Takeuchi (1981)].

Theorem 4.10. *Let $H^m(c)$, $c < 0$, be the hyperbolic m-space defined by*

$$H^m(c) = \left\{ (x_0, \ldots, x_m) \in \mathbb{E}_1^{m+1} : x_1^2 + \cdots + x_m^2 - x_0^2 = \frac{1}{c}, \ x_0 > 0 \right\}.$$

If M is a complete, full parallel submanifold of $H^m(c)$, then

(i) *if M is not contained in any complete totally umbilical hypersurface of $H^m(c)$, then M is congruent to the product*

$$H^{m_0}(c_0) \times M_1 \times \cdots \times M_s \subset H^{m_0}(c_0) \times S^{m-m_0-1}(c') \subset H^{m_0}(c)$$

with $c_0 < 0$, $c' > 0$, $c_0^{-1} + c'^{-1} = c^{-1}$, $s \geq 0$, where $M_1 \times \cdots \times M_s \subset$ $S^{m-m_0-1}(c')$ *is a parallel submanifold as described in Ferus' result; and*

(ii) *if M lies in a complete totally umbilical hypersurface M of $H^m(c)$, then M is isometric to a complete simply-connected real space form $R^{m-1}(\bar{c})$, where $\bar{c} = c + H^2$, and $M \subset R^{m-1}(\bar{c})$ is of the type described in (i) when $\bar{c} < 0$, or described by Theorem 4.8 when $\bar{c} = 0$, or by Theorem 4.9 when $\bar{c} > 0$.*

Remark 4.6. The explicit classifications of parallel pseudo-Riemannian submanifolds in indefinite real space forms are much more complicated than parallel submanifolds in real space forms. Nevertheless, the classification of parallel pseudo-Riemannian surfaces in 4D indefinite space forms were done in a series of papers [Chen and Van der Veken (2009); Chen et al. (2010); Chen (2010d,e,g)] (see also [Graves (1979a,b); Magid (1984)]).

The complete explicit classification of spacelike and Lorentz parallel surfaces in indefinite real space forms with arbitrary codimension and arbitrary index were achieved in [Chen (2010a,b)].

4.11 Totally umbilical submanifolds

The following result is analogous to Proposition 4.8.

Proposition 4.12. [Ahn et. al. (1996)] *Let $\phi : M_t^n \to N_s^m$ be an isometric immersion of a pseudo-Riemannian n-manifold M_t^n with index $t \in [1, n-1]$ into another pseudo-Riemannian manifold N_s^m. Then M_t^n is totally umbilical if and only if null geodesics of M_t^n are geodesics of N_s^m.*

Proof. Under the hypothesis, assume that M_t^n is totally umbilical in N_s^m. If $\gamma : I \to M_t^n$ is a null geodesic of M_t^n, then $\gamma'(t)$ is a null vector for each $t \in I$. Then it follows from (4.21) that $h(\gamma'(t), \gamma'(t)) = 0$. Thus $\tilde{\nabla}_{\gamma'} \gamma' = 0$, which shows that γ is also a geodesic of N_s^m.

Conversely, if null geodesics of M_t^n are geodesics of N_s^m, then $h(v,v) = 0$ for null vectors v of M_t^n. At a point $p \in M_t^n$, let us choose an orthonormal basis $\{e_1, \ldots, e_n\}$ of $T_p M_t^n$ such that $\langle e_i, e_i \rangle = -1$ for $i = 1, \ldots, t$ and $\langle e_j, e_j \rangle = 1$ for $j = t+1, \ldots, n$. Then $e_i \pm e_j$ are null vectors for $i \in \{1, \ldots, t\}$ and $j \in \{t+1, \ldots, n\}$. Thus $h(e_i \pm e_j, e_i \pm e_j) = 0$, which implies that

$$h(e_i, e_j) = 0, \quad h(e_i, e_i) + h(e_j, e_j) = 0. \tag{4.55}$$

If $t \geq 2$, then $e_{i_1} + e_{i_2} + \sqrt{2} e_n$ is a null vector for $1 \leq i_1 \neq i_2 \leq s$. Thus we find $h(e_{i_1}, e_{i_2}) = 0$ by applying (4.55).

Similarly, if $n - t \geq 2$, we have $h(e_{j_1}, e_{j_2}) = 0$ for $t + 1 \leq j_1 \neq j_2 \leq n$. Consequently, we obtain (4.21). Thus M_t^n is totally umbilical in N_s^m. $\quad\square$

Lemma 4.3. *Let $\phi : M \to R_s^m(c)$ be an isometric immersion of a pseudo-Riemannian n-manifold M into an indefinite real space form $R_s^m(c)$. If M is totally umbilical, then*

(1) *H is a parallel normal vector field, i.e., $DH = 0$;*
(2) *$\langle H, H \rangle$ is constant;*
(3) *ϕ is a parallel immersion, i.e., $\bar{\nabla}h = 0$ identically on M;*
(4) *N is of constant curvature $c + \langle H, H \rangle$;*
(5) *$A_H = \langle H, H \rangle I$;*
(6) *M is a parallel submanifold.*

Proof. Under the hypothesis, we have (4.21). Thus

$$(\bar{\nabla}_Z h)(X, Y) = \langle X, Y \rangle D_Z H. \tag{4.56}$$

Hence, by the equation of Codazzi, we find $\langle X, Y \rangle D_Z H = \langle Z, Y \rangle D_X H$ for $X, Y, Z \in TM$. Since $\dim M > 1$, this shows that $DH = 0$. Therefore we get (1). Statement (2) follows immediately from (1) and the fact that D is a metric connection. (3) follows from (1) and (4.56). And (4) is an easy consequence of (2) and the equation of Gauss. Statement (5) follows immediately from (2.14) and (4.21). Finally, statement (6) follows from (4.21), (4.26) and statement (1). $\quad\square$

Lemma 4.4. *If $\phi : M \to \mathbb{E}_s^m$ is a totally umbilical immersion of a pseudo-Riemannian n-manifold M into the pseudo-Euclidean space $\mathbb{E}_s^m$, then M lies in an $(n + 1)$-dimensional totally geodesic submanifolds of $\mathbb{E}_s^m$ as a hypersurface.*

Proof. Follows from Lemma 4.3 and the Reduction Theorem of Erbacher-Magid. $\quad\square$

Proposition 4.13. *Let $\phi : M \to \mathbb{E}_s^m$ be an isometric immersion of a pseudo-Riemannian n-manifold M with index t into a pseudo-Euclidean m-space $\mathbb{E}_s^m$. If $n > 1$ and M is totally umbilical in $\mathbb{E}_s^m$, then M is congruent to an open portion of one of the following submanifolds:*

(1) *A totally geodesic pseudo-Euclidean subspace $\mathbb{E}_t^n \subset \mathbb{E}_s^m$;*
(2) *A pseudo n-sphere $S_t^n(c)$ lying in a totally geodesic pseudo-Euclidean $(n + 1)$-subspace $\mathbb{E}_t^{n+1} \subset \mathbb{E}_s^m$;*

(3) *A pseudo-hyperbolic n-space $H_t^n(c)$ lying in a totally geodesic pseudo-Euclidean $(n+1)$-subspace $\mathbb{E}_{t+1}^{n+1} \subset \mathbb{E}_s^m$;*
(4) *A flat quasi-minimal submanifold defined by*

$$\left(\sum_{i=1}^{t} x_i^2 - \sum_{j=t+1}^{n} x_j^2, x_1, \ldots, x_t, 0, \ldots, 0, x_{t+1}, \ldots, x_n, \sum_{i=1}^{t} x_i^2 - \sum_{j=t+1}^{n} x_j^2 \right).$$

The last case occurs only when $s > t$.

Proof. Let M be a totally umbilical pseudo-Riemannian submanifold of $\mathbb{E}_s^m$. Assume that the index of M is t. Then Lemma 4.3 implies that $\langle H, H \rangle$ is a constant, $DH = 0$ and ϕ is a parallel immersion.

Case (a): $H = 0$. In this case M is totally geodesic. Thus we obtain (1) by Proposition 4.8.

Case (b): $H \neq 0$. Corollary 4.1 implies that $\phi(M)$ is contained in a $(n+1)$-dimensional totally geodesic submanifold $E^* \subset \mathbb{E}_s^m$. From Lemma 4.3 we have $\tilde{\nabla}_X H = - \langle H, H \rangle X$ for $X \in T(M)$.

Case (b.1): $\langle H, H \rangle = \epsilon r^2, r > 0, \epsilon = \pm 1$. In this case, $\phi + (\epsilon/r^2)H$ is a constant vector, say ϕ_0. By applying a suitable translation we have $\phi_0 = 0$. Thus $\langle \phi, \phi \rangle = \epsilon/r^2$, which gives case (2) or case (3) according to H is spacelike or timelike, respectively.

Case (b.2): H *is lightlike.* In this case, E^* is a totally geodesic $\mathbf{R}_{1,t}^{n+1}$. Since H is a constant lightlike vector and M is totally umbilical, (4.21) and Gauss' equation imply that M is flat. Thus locally there exists a natural coordinate system $\{x_1, \ldots, x_n\}$ such that the metric tensor g_0 of M is given by

$$g_0 = -\sum_{i=1}^{t} dx_i^2 + \sum_{j=t+1}^{n} dx_j^2. \tag{4.57}$$

Hence it follows from Proposition 1.1, (4.6), (4.21) and (4.57) that

$$\phi_{x_i x_i} = H, \quad i = 1, \ldots, t;$$
$$\phi_{x_k, x_\ell} = 0, \quad 1 \leq k \neq \ell \leq n, \tag{4.58}$$
$$\phi_{x_j x_j} = -H, \quad j = t+1, \ldots, n.$$

After solving system (4.58) we find

$$\phi = \mathbf{c}_0 + \sum_{k=1}^{n} \mathbf{c}_k x_k + \frac{H}{2} \sum_{i=1}^{t} x_i^2 - \frac{H}{2} \sum_{j=t+1}^{n} x_j^2$$

for some vectors $\mathbf{c}_0, \mathbf{c}_1, \ldots, \mathbf{c}_n \in \mathbb{E}_s^m$. Since H is a constant lightlike vector, without loss of generality we may put $H = (2, 0, \ldots, 0, 2) \in \mathbb{E}_s^m$. Hence, after choosing suitable initial conditions, we obtain (4). $\qquad \square$

Remark 4.7. A different statement is given in [Ahn et. al. (1996)].

Lemma 4.5. *Let $\phi : M \to \mathbb{E}_s^m$ be an isometric immersion of a pseudo-Riemannian n-manifold M into $\mathbb{E}_s^m$ and $f_1, \ldots, f_\ell \in \mathcal{F}(M)$. Then*

$$\mathbf{y} = (f_1, \ldots, f_\ell, \phi, f_\ell, \ldots, f_1) : M \to \mathbb{E}_{s+\ell}^{m+2\ell} \qquad (4.59)$$

is a parallel isometric immersion if and only if the Hessian $\mathcal{H}_{f_1}, \ldots, \mathcal{H}_{f_\ell}$ of $f_1, \ldots, f_\ell$ are covariant constant and ϕ is a parallel immersion.

Proof. Since the metric of $\mathbb{E}_{s+\ell}^{m+2\ell}$ is

$$g_0 = - \sum_{i=s+\ell}^{s+\ell} dx_i^2 + \sum_{j=s+\ell+1}^{m+2\ell} dx_j^2$$

and ϕ is an isometric immersion, (4.59) is an isometric immersion. It follows from Gauss' formula and (4.59) that the second fundamental form $h_{\mathbf{y}}$ of $\mathbf{y}$ is

$$h_{\mathbf{y}} = \left(\mathcal{H}_{f_1}, \ldots, \mathcal{H}_{f_\ell}, h_\phi, \mathcal{H}_{f_\ell}, \ldots, \mathcal{H}_{f_1} \right),$$

where $h_{\mathbf{x}}$ is the second fundamental form of ϕ. Consequently, $\mathbf{y}$ is a parallel immersion if and only if the Hessians $\mathcal{H}_{f_1}, \ldots, \mathcal{H}_{f_\ell}$ are covariant constant and ϕ is a parallel immersion. $\qquad \square$

Let $\phi : M \to S_s^m(1/r^2)$ (resp. $\phi : M \to H_s^m(-1/r^2)$) be an isometric immersion from a pseudo-Riemannian n-manifold M into $S_s^m(1/r^2)$ (resp. $H_s^m(-1/r^2)$). Denote by

$$\mathbf{x} = \iota \circ \phi : M \to \mathbb{E}_s^{m+1}, \qquad \left(\text{resp. } \mathbf{x} = \iota \circ \phi : M \to \mathbb{E}_{s+1}^{m+1} \right)$$

the composition of ϕ with the inclusion map $\iota : S_s^m(1/r^2) \subset \mathbb{E}_s^{m+1}$ via (1.16) (resp. $\iota : H_s^m(-1/r^2) \subset \mathbb{E}_{s+1}^{m+1}$ via (1.17)). Let $\nabla, \tilde{\nabla}$ and ∇' be the Levi-Civita connections of $M, \mathbb{E}_i^{m+t}$ and $S_s^m(1/r^2)$ (resp. $H_s^m(-1/r^2)$). Denote by D' the normal connection of M in $S_s^m(1/r^2)$ (resp. in $H_s^m(-1/r^2)$) and by D the corresponding quantities of M in $\mathbb{E}_s^{m+1}$ (resp. in $\mathbb{E}_{s+1}^{m+1}$).

Lemma 4.6. *Let $\phi : M \to S_s^m(1/r^2)$ (resp. $\phi : M \to H_s^m(-1/r^2)$) be an isometric immersion of a pseudo-Riemannian n-manifold M into $S_s^m(1/r^2)$ (resp. M into $H_s^m(-1/r^2)$). Then the second fundamental form h and the mean curvature vector H of M in $\mathbb{E}_s^{m+1}$ (resp. M in $\mathbb{E}_{s+1}^{m+1}$) via $\mathbf{x} = \iota \circ \phi$ are related with the second fundamental form h' and the mean curvature vector H' of M in $S^m(1/r^2)$ (resp. M in $H_s^m(-1/r^2)$) by*

$$h(X, Y) = h'(X, Y) - \frac{\epsilon}{r^2} \langle X, Y \rangle \mathbf{x}, \qquad (4.60)$$

$$H = H' - \frac{\epsilon}{r^2} \mathbf{x}, \qquad (4.61)$$

where $\epsilon = 1$ or -1, depending on ϕ is given by $\phi : M \to S_s^m(1/r^2)$ or by $\phi : M \to H_s^m(-1/r^2)$. Moreover, we have $DH = D'H'$.

Proof. Under the hypothesis, the positive vector field is a normal vector field of M which is normal to $S_s^m(1/r^2)$ or to $H_s^m(-1/r^2)$. Since $\tilde{\nabla}_X \mathbf{x} = X$ for $X \in TM$, the Weingarten formula yields

$$A_{\mathbf{x}} = -I, \quad DH = D'H', \tag{4.62}$$

where A is the Weingarten map in $\mathbb{E}_s^{m+1}$ or in $\mathbb{E}_{s+1}^{m+1}$. Hence it follows from the formula of Gauss and (4.62) that

$$\tilde{\nabla}_X Y = \nabla'_X Y - \frac{\epsilon}{r^2}\mathbf{x} = \nabla_X Y + h'(X,Y) - \frac{\epsilon}{r^2}\mathbf{x}, \tag{4.63}$$

which gives (4.60). By taking the trace of (4.60) we get (4.61). $\qquad\square$

Corollary 4.9. *Let $\phi : M \to S_s^m(r^{-2})$ (resp. $\phi : M \to H_s^m(-r^{-2})$) be an isometric immersion of a pseudo-Riemannian manifold M in $S_s^m(r^{-2})$ (resp. $M \to H_s^m(-r^{-2})$) and $\iota : S_s^m(r^{-2}) \subset \mathbb{E}_s^{m+1}$ (resp. $\iota : H_s^m(-r^{-2}) \subset \mathbb{E}_{s+1}^{m+1}$) is the inclusion map defined in Section 1.6. Then we have:*

(1) *ϕ has parallel mean curvature vector if and only if $\mathbf{x} = \iota \circ \phi$ has parallel mean curvature vector.*

(2) *ϕ is a parallel immersion if and only if $\mathbf{x} = \iota \circ \phi$ is a parallel immersion.*

(3) *ϕ is totally umbilical if and only if $\mathbf{x} = \iota \circ \phi$ is totally umbilical.*

Proof. Follows immediately from Lemma 4.6 and (4.63). $\qquad\square$

Proposition 4.14. *Let $\phi : M_t^n \to S_s^m(1) \subset \mathbb{E}_s^{m+1}$ be an isometric immersion of a pseudo-Riemannian n-manifold M_t^n into a pseudo-Riemannian m-space $S_s^m(1)$ with $n > 1$. If M_t^n is totally umbilical in $S_s^m(1)$, then it is congruent to an open portion of one of the following submanifolds:*

(1) *A totally geodesic pseudo sphere $S_t^n(1) \subset S_s^m(1)$;*

(2) *A pseudo sphere $S_t^n(1/r^2)$, $r \in (0,1)$, lying in a totally geodesic pseudo-Euclidean $(n+2)$-subspace $\mathbb{E}_t^{n+2} \subset \mathbb{E}_s^{m+1}$ as*

$$\left\{ \left(\mathbf{y}, \sqrt{1-r^2} \right) \in \mathbb{E}_t^{n+2} : \langle \mathbf{y}, \mathbf{y} \rangle = r^2, \; \mathbf{y} \in \mathbb{E}_t^{n+1} \right\};$$

(3) *A pseudo sphere $S_t^n(1/r^2)$, $r > 1$, lying in a totally geodesic pseudo-Euclidean $(n+2)$-subspace $\mathbb{E}_{t+1}^{n+2} \subset \mathbb{E}_s^{m+1}$ as*

$$\left\{ \left(\sqrt{r^2-1}, \mathbf{y} \right) \in \mathbb{E}_{t+1}^{n+2} : \langle \mathbf{y}, \mathbf{y} \rangle = r^2, \; \mathbf{y} \in \mathbb{E}_t^{n+1} \right\};$$

(4) *A pseudo-hyperbolic space* $H_t^n(-1/r^2)$, $r > 0$, *lying in a totally geodesic pseudo-Euclidean* $(n+2)$-*subspace* $\mathbb{E}_{t+1}^{n+2} \subset \mathbb{E}_s^{m+1}$ *as*

$$\left\{ \left(\mathbf{y}, \sqrt{1+r^2} \right) \in \mathbb{E}_{t+1}^{n+2} : \langle \mathbf{y}, \mathbf{y} \rangle = -r^2, \mathbf{y} \in \mathbb{E}_{t+1}^{n+1} \right\};$$

(5) *A flat totally umbilical submanifold lying in a totally geodesic pseudo-Euclidean* $(n+3)$-*subspace* $\mathbb{E}_{t+1}^{n+3} \subset \mathbb{E}_s^{m+1}$ *as*

$$\left\{ \left(r\langle \mathbf{y}, \mathbf{y} \rangle - rb - \frac{r}{4}, r\mathbf{y}, \sqrt{1+br^2}, r\langle \mathbf{y}, \mathbf{y} \rangle - rb + \frac{r}{4} \right) \in \mathbb{E}_{t+1}^{n+2} : \mathbf{y} \in \mathbb{E}_t^n \right\},$$

where b, r *are real numbers satisfying* $br^2 \geq -1$ *and* $r > 0$;

(6) *A flat totally umbilical submanifold lying in a totally geodesic pseudo-Euclidean* $(n+3)$-*subspace* $\mathbb{E}_{t+2}^{n+3} \subset \mathbb{E}_s^{m+1}$ *as*

$$\left\{ \left(r\langle \mathbf{y}, \mathbf{y} \rangle + rb - \frac{r}{4}, \sqrt{br^2-1}, r\mathbf{y}, r\langle \mathbf{y}, \mathbf{y} \rangle + rb + \frac{r}{4} \right) \in \mathbb{E}_{t+2}^{n+2} : \mathbf{y} \in \mathbb{E}_t^n \right\},$$

where b, r *are real numbers satisfying* $br^2 \geq 1$ *and* $r > 0$.

Proof. In view of Corollary 4.9, it suffices to look for totally umbilical submanifolds of $\mathbb{E}_s^{m+1}$ from Proposition 4.13 such that, up to dilations and rigid motions, the totally umbilical submanifolds lie in $S_s^m(1) \subset \mathbb{E}_s^{m+1}$.

Obviously, a totally geodesic $\mathbb{E}_t^n$ in $\mathbb{E}_s^{m+1}$ does not lie in any pseudo hypersphere of $\mathbb{E}_s^{m+1}$. From Proposition 4.13(2) we get (2) and (3) of the Proposition. Similarly, we obtain (4) form Proposition 4.13(3). Finally, (5) and (6) are obtained from Proposition 4.13(4) after applying a suitable dilation and a rigid motion. □

Proposition 4.15. *Let* $\phi : M_t^n \to H_s^m(-1)$ *be an isometric immersion of a pseudo-Riemannian* n-*manifold* M_t^n *into a pseudo-hyperbolic* m-*space* $H_s^m(-1)$. *If* $n > 1$ *and* M_t^n *is totally umbilical in* $H_s^m(-1) \subset \mathbb{E}_{s+1}^{m+1}$, *then it is congruent to an open portion of one of the following submanifolds:*

(1) *A totally geodesic pseudo-hyperbolic* n-*space* $H_t^n(-1) \subset H_s^m(-1)$;

(2) *A pseudo* n-*sphere* $S_t^n(1/r^2)$, $r > 0$, *lying in a totally geodesic pseudo-Euclidean* $(n+2)$-*subspace* $\mathbb{E}_{t+1}^{n+2} \subset \mathbb{E}_{s+1}^{m+1}$ *as*

$$\left\{ \left(\sqrt{1+r^2}, \mathbf{y} \right) \in \mathbb{E}_{t+1}^{n+2} : \langle \mathbf{y}, \mathbf{y} \rangle = r^2, \mathbf{y} \in \mathbb{E}_{t+1}^{n+1} \right\};$$

(3) *A pseudo-hyperbolic* n-*space* $H_t^n(-1/r^2)$, $r > 1$, *lying in a totally geodesic pseudo-Euclidean* $(n+2)$-*subspace* $\mathbb{E}_{t+1}^{n+2} \subset \mathbb{E}_{s+1}^{m+1}$ *as*

$$\left\{ \left(\mathbf{y}, \sqrt{r^2-1} \right) \in \mathbb{E}_{t+1}^{n+2} : \langle \mathbf{y}, \mathbf{y} \rangle = -r^2, \mathbf{y} \in \mathbb{E}_{t+1}^{n+1} \right\};$$

(4) *A pseudo-hyperbolic n-space $H_t^n(-1/r^2)$, $r > 0$, lying in a totally geodesic pseudo-Euclidean $(n+2)$-subspace $\mathbb{E}_{t+2}^{n+2} \subset \mathbb{E}_{s+1}^{m+1}$ as*

$$\left\{ \left(\sqrt{1-r^2}, \mathbf{y} \right) \in \mathbb{E}_{t+2}^{n+2} : \langle \mathbf{y}, \mathbf{y} \rangle = -r^2, \mathbf{y} \in \mathbb{E}_{t+1}^{n+1} \right\};$$

(5) *A flat totally umbilical submanifold lying in a totally geodesic pseudo-Euclidean $(n+3)$-subspace $\mathbb{E}_{t+1}^{n+3} \subset \mathbb{E}_{s+1}^{m+1}$ as*

$$\left\{ \left(r\langle \mathbf{y}, \mathbf{y} \rangle - rb - \frac{r}{4}, r\mathbf{y}, \sqrt{br^2-1}, r\langle \mathbf{y}, \mathbf{y} \rangle - rb + \frac{r}{4} \right) \in \mathbb{E}_{t+2}^{n+3} : \mathbf{y} \in \mathbb{E}_t^n \right\},$$

where b, r are real numbers satisfying $br^2 \geq 1$ and $r > 0$;

(6) *A flat totally umbilical submanifold lying in a totally geodesic pseudo-Euclidean $(n+3)$-subspace $\mathbb{E}_{t+2}^{n+3} \subset \mathbb{E}_{s+1}^{m+1}$ as*

$$\left\{ \left(r\langle \mathbf{y}, \mathbf{y} \rangle - rb - \frac{r}{4}, \sqrt{1-br^2}, r\mathbf{y}, r\langle \mathbf{y}, \mathbf{y} \rangle - rb + \frac{r}{4} \right) \in \mathbb{E}_{t+3}^{n+3} : \mathbf{y} \in \mathbb{E}_t^n \right\},$$

where b, r are real numbers satisfying $br^2 \leq 1$ and $r > 0$.

Proof. Analogous to the proof of Proposition 4.14 it suffices to look for totally umbilical submanifolds of $\mathbb{E}_{s+1}^{m+1}$ from the list of Proposition 4.13 such that, up to dilations and rigid motions, the totally umbilical submanifolds lie in $H_s^m(-1) \subset \mathbb{E}_{s+1}^{m+1}$. Clearly, totally geodesic $\mathbb{E}_t^n$ in $\mathbb{E}_s^{m+1}$ does not lie in any pseudo-hyperbolic hypersurface of $\mathbb{E}_{s+1}^{m+1}$. From Proposition 4.13(2) we obtain (2) of the Proposition. Similarly, we obtain (3) and (4) form Proposition 4.13(3). (5) and (6) are obtained from Proposition 4.13(4) after applying suitable dilation and rigid motion. $\qquad\square$

4.12 Pseudo-umbilical submanifolds

Definition 4.8. A non-minimal pseudo-Riemannian submanifold M of a pseudo-Riemannian manifold M is called a *pseudo-umbilical submanifold* if there exists a function λ in $\mathcal{F}(M)$ such that

$$\langle h(X,Y), H \rangle = \lambda \langle X, Y \rangle, \quad X, Y \in \mathfrak{X}(M). \tag{4.64}$$

The following three results classify pseudo-umbilical submanifolds with parallel mean curvature vector in indefinite real space forms.

Proposition 4.16. *Let $\phi : M \to \mathbb{E}_s^m$ be an isometric immersion of a pseudo-Riemannian submanifold M into a pseudo-Euclidean m-space $\mathbb{E}_s^m$. Then ϕ is a pseudo-umbilical immersion with parallel mean curvature vector if and only if one of the following three cases occurs:*

(1) M is a minimal submanifold of a pseudo hypersphere $S_s^{m-1}(\mathbf{x}_0, 1/r^2)$
    for some $\mathbf{x}_0 \in \mathbb{E}_s^m$ and $r > 0$;
(2) M is a minimal submanifold of a pseudo-hyperbolic hyperspace
    $H_{s-1}^{m-1}(\mathbf{x}_0, -1/r^2)$ for some $\mathbf{x}_0 \in \mathbb{E}_s^m$ and $r > 0$;
(3) ϕ is congruent to $(f, \mathbf{z}, f)$, where $f \in \mathcal{F}(M)$, Δf is a nonzero real
    number and $\mathbf{z} : M \to \mathbb{E}_{s-1}^{m-2}$ is a minimal isometric immersion.

Case (2) occurs only when $s \geq 1$ and case (3) occurs only when $s \geq 1$ and
$m \geq \dim M + 2$.

Proof. Assume that M is a pseudo-umbilical submanifold with parallel
mean curvature vector in $\mathbb{E}_s^m$. Then $X\langle H, H \rangle = 2\langle H, D_X H \rangle = 0$ for any
$X \in TM$. Thus $\langle H, H \rangle$ is constant.

Case (a): $\langle H, H \rangle \neq 0$. We put

$$\langle H, H \rangle = \frac{\epsilon}{r^2}, \tag{4.65}$$

where $\epsilon = 1$ or -1 depending on H is spacelike or timelike. On the other
hand, it follows from (4.64) that $A_H = \lambda I$ for some function $\lambda \in \mathcal{F}(M)$.
Thus we find from (4.65) that $\epsilon = \lambda r^2$. Let us put

$$\hat{\phi} = \phi + \epsilon r^2 H. \tag{4.66}$$

Then $\tilde{\nabla}_X \hat{\phi} = X - \epsilon r^2 A_H X = 0$. Thus $\hat{\phi}$ is a constant vector, say $\mathbf{x}_0 \in \mathbb{E}_s^m$.
Hence

$$\langle \phi - \mathbf{x}_0, \phi - \mathbf{x}_0 \rangle = r^4 \langle H, H \rangle = \epsilon r^2,$$

which implies that M lies either in $S_s^{m-1}(\mathbf{x}_0, 1/r^2)$ or $H_s^{m-1}(\mathbf{x}_0, -1/r^2)$,
according to H is spacelike or timelike, respectively. Since H is either
normal to $S_r^{m-1}(\mathbf{x}_0, 1/r^2)$ or to $H_s^{m-1}(\mathbf{x}_0, -1/r^2)$, it follows from Lemma
4.6 that N is minimal in $S_r^{m-1}(\mathbf{x}_0, 1/r^2)$ or in $H_s^{m-1}(\mathbf{x}_0, -1/r^2)$.

Case (b): H is lightlike. It follows from (4.64) that $A_H = \langle H, H \rangle I = 0$.
Combining this with $DH = 0$ implies that H is a lightlike constant vector,
say $\zeta_0 \in \mathbb{E}_s^{m+1}$. Thus $X\langle \phi, \zeta_0 \rangle = 0$ for any $X \in TM$. So, $\langle \phi, \zeta_0 \rangle = c$ for
some real number c.

If we put $\zeta_0 = (1, 0, \ldots, 0, 1) \in \mathbb{E}_s^m$ and $\phi = (x_1, \ldots, x_m)$, then we
obtain $x_m = x_1 + c$. By applying a suitable translation, we find $c = 0$.
Thus the immersion $\mathbf{x} : N \to \mathbb{E}_s^m$ takes the form

$$\phi = (f, \mathbf{z}, f), \tag{4.67}$$

where f is function on N and $\mathbf{z} : N \to \mathbb{E}_{s-1}^{m-2}$ is an isometric immersion.
Now, by applying the Laplace operator Δ to (4.67) we find from Beltrami's

formula that $nH = (-\Delta f, nH_{\mathbf{z}}, -\Delta f)$, where $H_{\mathbf{z}}$ is the mean curvature vector of $\mathbf{z}$. Since $H = (1, 0, \ldots, 0, 1)$, we find $H_{\mathbf{z}} = 0$ and $\Delta f = -nr$. Thus $\mathbf{z}$ is a minimal immersion and Δf is a nonzero constant.

The converse can be verified easily. □

Proposition 4.17. *Let $\phi : M \to S_s^m(1)$ be an isometric immersion of a pseudo-Riemannian submanifold M into the pseudo m-sphere $S_s^m(1)$. Then ϕ is a pseudo-umbilical immersion with parallel mean curvature vector if and only if M lies in a non-totally geodesic, totally umbilical hypersurface of $S_s^m(1)$ as a minimal submanifold.*

Proof. Let $\phi : M \to S_s^m(1)$ be an isometric immersion of a pseudo-Riemannian submanifold M into $S_s^m(1)$. Then it follows from Lemma 4.6 that M is a pseudo-umbilical submanifold with parallel mean curvature vector in $S_s^m(1)$ if and only if M is a pseudo-umbilical submanifold of $\mathbb{E}_s^{m+1}$ via the composition $\mathbf{x} = \iota \circ \phi$, where ι is the inclusion $S_s^m(1) \subset \mathbb{E}_s^{m+1}$.

Let M be a pseudo-umbilical submanifold with parallel mean curvature vector in $S_s^m(1)$. Then the mean curvature vector H of M in $\mathbb{E}_s^{m+1}$ satisfies

$$A_H = \lambda I, \quad DH = 0, \quad \lambda \in \mathbf{R}. \tag{4.68}$$

Case (1): *H is spacelike.* If we put $\langle H, H \rangle = r^{-2}$, then as in the proof of Proposition 4.16, we have

$$\mathbf{x} - \mathbf{x}_0 = -r^2 H \tag{4.69}$$

for some vector $\mathbf{x}_0$. Hence M is a minimal submanifold of the pseudo hypersphere $S_s^m(\mathbf{x}_0, 1/r^2)$. Since M lies in $S_s^m(1) \cap S_s^m(\mathbf{x}_0, 1/r^2)$, we find

$$\langle \mathbf{x}, \mathbf{x} \rangle = 1 \quad \text{and} \quad \langle \mathbf{x} - \mathbf{x}_0, \mathbf{x} - \mathbf{x}_0 \rangle = \frac{1}{r^2}, \tag{4.70}$$

which imply that

$$\langle \mathbf{x}, \mathbf{x}_0 \rangle = c, \quad c = \frac{1 + \langle \mathbf{x}_0, \mathbf{x}_0 \rangle}{2} - \frac{1}{2r^2}. \tag{4.71}$$

Since M is non-minimal in $S_s^m(1)$, $\mathbf{x}_0 \neq 0$.

Case (1.1): $\langle \mathbf{x}_0, \mathbf{x}_0 \rangle = k^2 > 0$. If we put $\mathbf{x}_0 = (0, \ldots, 0, k^{-1}) \in \mathbb{E}_s^{m+1}$, then the first equation in (4.71) implies that the last canonical coordinate x_{m+1} of M satisfies $x_{m+1} = ck$. Hence M is contained in $\mathcal{H} = S_s^m(1) \cap E$, where E is the hyperplane defined by $x_{m+1} = ck$. An easy computation shows that $\xi = c\mathbf{x} - \mathbf{x}_0$ is a normal vector field of $\mathcal{H}$ in $S_s^m(1)$. Moreover, it follows from Gauss' formula that $A_\xi = -cI$. Hence $\mathcal{H}$ is a totally umbilical hypersurface of $S_s^m(1)$. Since M is non-minimal in $S_s^m(1)$, $\mathcal{H}$ must be a non-totally geodesic, totally umbilical hypersurface. Also, since H lies in Span $\{\mathbf{x}, \xi\}$, M is minimal in $\mathcal{H}$. Therefore M lies in a non-totally geodesic, totally umbilical hypersurface of $S^m_s(1)$ as a minimal submanifold.

Case (1.2): $\langle \mathbf{x}_0, \mathbf{x}_0 \rangle = -k^2 < 0$. By putting $\mathbf{x}_0 = (k^{-1}, 0, \ldots, 0)$ and applying the same arguments as case (1.1), we get the same conclusion as case (1.1).

Case (1.3): $\mathbf{x}_0$ *is lightlike*. We find from (4.71) that $\langle \mathbf{x}, \mathbf{x}_0 \rangle = c \in (-\infty, \frac{1}{2})$, which defines a hyperplane $\mathcal{E}$ of $\mathbb{E}^{m+1}_s$. Hence M lies in $\mathcal{H}_1 = S^m_s(1) \cap \mathcal{E}$. It is easy to verify that $\eta = \mathbf{x}_0 - c\mathbf{x}$ is a normal vector field of $\mathcal{H}_1$ in $S^m_s(1)$ such that $\langle \eta, \eta \rangle = -c^2$. Since $\tilde{\nabla}_X \eta = -cX$ for $X \in TM$, we get $A_\eta = cI$. This shows that $\mathcal{H}_1$ is totally umbilical in $S^m_s(1)$. Because M is non-minimal in $S^m_s(1)$, we have $c \neq 0$. Moreover, it follows from (4.69) and Lemma 4.6 that the mean curvature vector H' of M in $S^m_s(1)$ is η. Hence we have the same conclusion as case (1.1).

Case (2): H *is timelike*. A similar arguments as case (1) gives the same conclusion as case (1).

Case (3): H *is lightlike*. Just like in case (b) of the proof of Proposition 4.16, H is a constant lightlike vector in $\mathbb{E}^{m+1}_s$, say ζ_0, and that we have $\langle \mathbf{x}, \zeta_0 \rangle = c$ for some real number c. Let $\mathcal{E}$ denote the hyperplane defined by $\langle \mathbf{x}, \zeta_0 \rangle = c$. Then M is contained in $\mathcal{H}_1 = S^m_s(1) \cap \mathcal{E}$. It is easy to verify that $\eta = \zeta_0 - c\mathbf{x}$ is a normal vector field of $\mathcal{H}_1$ in $S^m_s(1)$ with $\langle \eta, \eta \rangle = -c^2$. Since $\tilde{\nabla}_X \eta = -cX$, we get $A_\eta = cI$. Thus $\mathcal{H}_1$ is totally umbilical in $S^m_s(1)$. Because M is non-minimal in $S^m_s(1)$, we find $c \neq 0$. Hence $\mathcal{H}_1$ is a non-totally geodesic, totally umbilical hypersurface. Moreover, it follows from (4.69) and Lemma 4.6 that the mean curvature vector of M in $S^m_s(1)$ is exactly η. Therefore M is minimal in $\mathcal{H}_1$.

The converse is easy to verify. $\square$

Proposition 4.18. *Let* $\phi : M \to H^m_s(-1)$ *be an isometric immersion of a pseudo-Riemannian manifold* M *into the pseudo-hyperbolic m-space* $H^m_s(-1)$. *Then* ϕ *is a pseudo-umbilical immersion with parallel mean curvature vector if and only if* M *is contained in a non-totally geodesic, totally umbilical hypersurface of* $H^m_s(-1)$ *as a minimal submanifold.*

Proof. This can be done in the same way as Proposition 4.17. $\square$

By applying Proposition 4.17 we have the following.

Corollary 4.10. *Let* $\phi : M \to S^m_s(1)$ *be an isometric immersion of a pseudo-Riemannian manifold* M *into* $S^m_s(1)$. *Then* ϕ *is a pseudo-umbilical* *immersion with unit timelike parallel mean curvature vector if and only if M lies in a flat totally umbilical hypersurface of $S_s^m(1)$ as a minimal submanifold.*

Proof. We follow the same notations as in the proof of Proposition 4.17. Assume $\phi : M \to S_s^m(1)$ is a pseudo-umbilical immersion with unit timelike parallel mean curvature vector H'. Then the mean curvature vector H of M in $\mathbb{E}_s^{m+1}$ is $H' - \mathbf{x}$, where $\mathbf{x} = \iota \circ \phi$ as before. Since H' is a unit timelike vector field, as in the proof of Proposition 4.16, we see that H is a lightlike constant vector, say $\zeta_0 \in \mathbb{E}_s^{m+1}$. Clearly, we have $\langle \mathbf{x}, \zeta_0 \rangle = -1$. So M lies in $\mathcal{H}_1 = S_s^m(1) \cap \mathcal{E}$, where $\mathcal{E}$ is defined by $\langle \mathbf{x}, \zeta_0 \rangle = -1$.

Since $\eta = \zeta_0 + \mathbf{x}$ is a unit timelike normal vector field of $\mathcal{H}_1$ in $S_s^m(1)$ and $A_\eta = -I$, the second fundamental form $\hat{h}$ of $\mathcal{H}_1$ in $S_s^m(1)$ satisfies $\hat{h}(X,Y) = -\langle X, Y \rangle \eta$, $X, Y \in T\mathcal{H}_1$. Hence the equation of Gauss implies that $\mathcal{H}_1$ is a flat totally umbilical hypersurface. Thus by using the same argument as given in the proof of Proposition 4.17 we conclude that M is minimal in $\mathcal{H}_1$.

The converse is easy to verify. $\qquad\square$

Corollary 4.11. *Let $\phi : M \to H_s^m(-1)$ be an isometric immersion of a pseudo-Riemannian submanifold M into $H_s^m(-1)$. Then ϕ is a pseudo-umbilical immersion with unit spacelike parallel mean curvature vector if and only if M lies in a flat totally umbilical hypersurface of $H_s^m(-1)$ as a minimal submanifold.*

Proof. This can be done in the same way as Corollary 4.10. $\qquad\square$

4.13   Minimal Lorentzian surfaces

The following lemma is an easy consequence of a result of [Larsen (1996)].

Lemma 4.7. *Locally there exists a coordinate system $\{x, y\}$ on a Lorentzian surface M such that the metric tensor is given by*

$$g = -E^2(x, y)(dx \otimes dy + dy \otimes dx) \tag{4.72}$$

for some positive function $E(x, y)$.

Proof. It is known that locally there exist isothermal coordinates (u, v) on a Lorentzian surface M such that the metric tensor takes the form:

$$g = F(u, v)^2(-du \otimes du + dv \otimes dv) \tag{4.73}$$

for some positive function F [Larsen (1996)]. By putting $x = u+v, y = u-v$, we obtain (4.72) from (4.73) with $E(x, y) = F(x, y)/\sqrt{2}$. $\square$

The Levi-Civita connection of metric tensor (4.72) satisfies

$$\nabla_{\frac{\partial}{\partial x}}\frac{\partial}{\partial x} = \frac{2E_x}{E}\frac{\partial}{\partial x}, \quad \nabla_{\frac{\partial}{\partial x}}\frac{\partial}{\partial y} = 0, \quad \nabla_{\frac{\partial}{\partial y}}\frac{\partial}{\partial y} = \frac{2E_y}{E}\frac{\partial}{\partial y} \tag{4.74}$$

and the Gauss curvature K is given by

$$K = \frac{2EE_{xy} - 2E_xE_y}{E^4}. \tag{4.75}$$

If we put

$$e_1 = \frac{1}{E}\frac{\partial}{\partial x}, \quad e_2 = \frac{1}{E}\frac{\partial}{\partial y}, \tag{4.76}$$

then $\{e_1, e_2\}$ forms a pseudo-orthonormal frame such that

$$\langle e_1, e_1 \rangle = \langle e_2, e_2 \rangle = 0, \ \langle e_1, e_2 \rangle = -1. \tag{4.77}$$

We define the *connection 1-form* ω by the following equations:

$$\nabla_X e_1 = \omega(X)e_1, \quad \nabla_X e_2 = -\omega(X)e_2. \tag{4.78}$$

From (4.74) and (4.76) we find

$$\nabla_{e_1} e_1 = \frac{E_x}{E^2}e_1, \ \nabla_{e_2} e_1 = -\frac{E_y}{E^2}e_1,$$
$$\nabla_{e_1} e_2 = -\frac{E_x}{E^2}e_2, \ \nabla_{e_2} e_2 = \frac{E_y}{E^2}e_2. \tag{4.79}$$

By comparing (4.78) and (4.79), we get

$$\omega(e_1) = \frac{E_x}{E^2}, \ \omega(e_2) = -\frac{E_y}{E^2}. \tag{4.80}$$

Let $\psi : M \to M_s^m$ be an isometric immersion of M into a pseudo-Riemannian m-manifold M_s^m with index s. Then it follows from (2.17) and (4.77) that the mean curvature vector is given by

$$H = -h(e_1, e_2). \tag{4.81}$$

Thus M is a minimal surface of M_s^m if and only if $h(e_1, e_2) = 0$ identically.

The following result from [Chen (2011a)] completely classifies minimal Lorentzian surfaces in an arbitrary indefinite pseudo-Euclidean m-space.

Theorem 4.11. *A Lorentzian surface in a pseudo-Euclidean m-space $\mathbb{E}_s^m$ is minimal if and only if, locally, the immersion takes the form*

$$L(x, y) = z(x) + w(y),$$

where z and w are null curves satisfying $\langle z'(x), w'(y) \rangle \neq 0$.

Proof. Let $L : M \to \mathbb{E}_s^m$ be an isometric immersion of a Lorentzian surface M into a pseudo-Euclidean m-space $\mathbb{E}_s^m$. Then, according to Lemma 4.7, we may choose a local coordinate system $\{x, y\}$ on M such that

$$g = -E^2(x, y)(dx \otimes dy + dy \otimes dx). \tag{4.82}$$

Then we have (4.74)-(4.81). If M is a minimal surface, then it follows from (4.81) that $h(e_1, e_2) = 0$ holds. Hence we may put

$$h(e_1, e_1) = \xi, \quad h(e_1, e_2) = 0, \quad h(e_2, e_2) = \eta \tag{4.83}$$

for some normal vector fields ξ, η. After applying Gauss' formula, (4.74), (4.76) and (4.83), we obtain

$$L_{xx} = \frac{2E_x}{E} L_x + E^2 \xi, \quad L_{xy} = 0, \quad L_{yy} = \frac{2E_y}{E} L_y + E^2 \eta. \tag{4.84}$$

After solving the second equation in (4.84), we find

$$L(x, y) = z(x) + w(y) \tag{4.85}$$

for some vector-valued functions $z(x), w(y)$. Thus, by applying (4.73) and (4.85), we obtain $\langle z', z' \rangle = \langle w', w' \rangle = 0$, $\langle z', w' \rangle = -E^2$. Therefore, z and w are null curves satisfying $\langle z', w' \rangle \neq 0$.

Conversely, if $L : M \to \mathbb{E}_s^m$ is an immersion of a Lorentzian surface M into $\mathbb{E}_s^m$ such that $L = z(x) + w(y)$ for some null curves z, w satisfying $\langle z', w' \rangle \neq 0$, then $\langle L_x, L_x \rangle = \langle L_y, L_y \rangle = 0, \langle L_x, L_y \rangle \neq 0$, $L_{xy} = 0$. Thus M is surface with induced metric given by

$$g = F(x, y)(dx \otimes dy + dy \otimes dx)$$

for some nonzero function F. Hence, after applying (4.81) and $L_{xy} = 0$, we conclude that L is a minimal immersion of a Lorentzian surface. $\quad\square$

Remark 4.8. Theorem 4.11 also obtained independent in [Anciaux (2011)].

For flat minimal Lorentzian surfaces in $\mathbb{E}_s^m$, we have [Chen (2011a)].

Corollary 4.12. *A flat Lorentzian surface in a pseudo-Euclidean m-space $\mathbb{E}_s^m$ is minimal if and only if, locally, the immersion takes the form*

$$L(x, y) = z(x) + w(y), \tag{4.86}$$

where z and w are null curves satisfying $\langle z', w' \rangle = const. \neq 0$.

Proof. Let $L : M \to \mathbb{E}_s^m$ be an isometric immersion of a flat Lorentzian surface M into a pseudo-Euclidean m-space $\mathbb{E}_s^m$. Then we may choose a local coordinate system $\{x, y\}$ on M satisfying

$$g = -(dx \otimes dy + dy \otimes dx). \tag{4.87}$$

Then we find from (4.84) that the immersion L satisfies

$$L_{xx} = \xi, \quad L_{xy} = 0, \quad L_{yy} = \eta \tag{4.88}$$

for some normal vector fields ξ, η. After solving the second equation in (4.88) we find

$$L(x, y) = z(x) + w(y) \tag{4.89}$$

for some vector functions z, w. Thus, by applying (4.87), we find $\langle z', z' \rangle = \langle w', w' \rangle = 0$ and $\langle z', w' \rangle = -1$. Consequently, z and w are null curves satisfying $\langle z'(x), w'(y) \rangle = -1$.

Conversely, consider a map L defined by (4.86) such that z and w are null curves satisfying $\langle z'(x), w'(y) \rangle = const. \neq 0$. Then we have

$$\langle L_x, L_x \rangle = \langle L_y, L_y \rangle = 0, \quad \langle L_x, L_y \rangle = const. \neq 0.$$

Thus, with respect to the induced metric, (4.86) defines an isometric immersion of a flat Lorentzian surface M into $\mathbb{E}_s^m$. The remaining part follows from Theorem 4.11. $\qquad \square$

Since every totally geodesic Lorentzian surface in an indefinite space form $R_s^m(c)$ is of constant curvature c, a natural question is the following.

Question 4.1. *Besides totally geodesic ones how many minimal Lorentzian surfaces of constant curvature c in $R_s^m(c)$ are there?*

Corollary 4.12 provides the answer to this basic question for $c = 0$. For $c \neq 0$, we have the following two results from [Chen (2011a)].

Theorem 4.12. *Let M be a Lorentzian surface of constant curvature one. Then an isometric immersion $\psi : M \to S_s^m(1)$ is minimal if and only if one of the following three cases occurs:*

(a) *M is an open portion of a totally geodesic $S_1^2(1) \subset S_s^m(1)$;*
(b) *the immersion $L = \iota \circ \psi : M \to S_s^m(1) \subset \mathbb{E}_s^{m+1}$ is locally given by*

$$L(x, y) = \frac{z(x)}{x + y} - \frac{z'(x)}{2}, \tag{4.90}$$

where $z(x)$ is a spacelike curve with constant speed 2 lying in the light cone $\mathcal{LC}$ satisfying $\langle z'', z'' \rangle = 0$ and $z''' \neq 0$;

(c) *the immersion* $L = \iota \circ \psi : M \rightarrow S_s^m(1) \subset \mathbb{E}_s^{m+1}$ *is locally given by*

$$L(x, y) = \frac{z(x) + w(y)}{x + y} - \frac{z'(x) + w'(y)}{2},$$

where z *and* w *are curves in* $\mathbb{E}_s^{m+1}$ *satisfying*

$$\left\langle \frac{z(x)+w(y)}{x+y} - \frac{z'(x)+w'(y)}{2}, \frac{z(x)+w(y)}{x+y} - \frac{z'(x)+w'(y)}{2} \right\rangle = 1,$$

$$2 \langle z + w, z''' \rangle = (x + y) \langle z' + w', z''' \rangle,$$

$$2 \langle z + w, w''' \rangle = (x + y) \langle z' + w', w''' \rangle.$$

Proof. Assume $\psi : M \rightarrow S_s^m(1)$ is an isometric immersion of a Lorentzian surface M of constant curvature one into $S_s^m(1)$. If M is totally geodesic, we get case (a). Thus we may assume that M is non-totally geodesic in $S_s^m(1)$.

Since M is of constant curvature one, we may choose local coordinates $\{x, y\}$ such that the metric tensor is given by

$$g = \frac{-2}{(x + y)^2} (dx \otimes dy + dy \otimes dx). \tag{4.91}$$

Hence the Levi-Civita connection satisfies

$$\nabla_{\frac{\partial}{\partial x}} \frac{\partial}{\partial x} = \frac{-2}{x + y} \frac{\partial}{\partial x}, \quad \nabla_{\frac{\partial}{\partial x}} \frac{\partial}{\partial y} = 0, \quad \nabla_{\frac{\partial}{\partial y}} \frac{\partial}{\partial y} = \frac{-2}{x + y} \frac{\partial}{\partial y}. \tag{4.92}$$

Let us put

$$\frac{\partial}{\partial x} = \frac{\sqrt{2} e_1}{x + y}, \quad \frac{\partial}{\partial y} = \frac{\sqrt{2} e_2}{x + y}. \tag{4.93}$$

Then we get

$$\langle e_1, e_1 \rangle = \langle e_2, e_2 \rangle = 0, \quad \langle e_1, e_2 \rangle = -1. \tag{4.94}$$

Because M is minimal in $S_s^m(1)$, it follows from (3.9) and (4.91) that $h(e_1, e_2) = 0$. Hence we may put

$$h(e_1, e_1) = \xi, \quad h(e_1, e_2) = 0, \quad h(e_2, e_2) = \eta \tag{4.95}$$

for some normal vector fields ξ, η. Without loss of generality, we may assume $\xi \neq 0$. Since $K = 1$, it follows from the equation of Gauss and (4.91) that $\langle \xi, \eta \rangle = 0$.

Case (i): $\eta = 0$. From formula of Gauss, (4.91)-(4.93), and (4.95), we get

$$L_{xx} = \frac{2\xi}{(x + y)^2} - \frac{2L_x}{x + y}, \quad L_{xy} = \frac{2L}{(x + y)^2}, \quad L_{yy} = -\frac{2L_y}{x + y}. \tag{4.96}$$

After solving the last two equations in (4.96) we obtain

$$L(x, y) = \frac{z(x)}{x + y} - \frac{z'(x)}{2} \tag{4.97}$$

for some $\mathbb{E}_s^{m+1}$-valued function $z(x)$. Since the metric tensor is given by (4.91), one finds

$$\langle L_x, L_x \rangle = \langle L_y, L_y \rangle = 0 \quad \text{and} \quad \langle L_x, L_y \rangle = -\frac{2}{(x + y)^2}.$$

Thus it follows from (4.96), (4.97) and $\langle L, L \rangle = 1$ that $z(x)$ satisfies

$$\langle z, z \rangle = \langle z'', z'' \rangle = 0 \quad \text{and} \quad \langle z', z' \rangle = 4.$$

Moreover, by substituting (4.97) into the first equation (4.96) we find

$$\xi = -\frac{(x + y)^2 z'''(x)}{4}. \tag{4.98}$$

Combining this with $\xi \neq 0$ gives $z'''(x) \neq 0$. Consequently, we get case (b).

Conversely, suppose that L is given by (4.90), where $z(x)$ is a spacelike curve with constant speed 2 lying in the light cone $\mathcal{LC} \subset \mathbb{E}_s^{m+1}$ satisfying $\langle z'', z'' \rangle = 0$ and $z''' \neq 0$. Then L satisfies (4.97) with ξ given by (4.98). From the assumption, we have

$$\langle z, z \rangle = \langle z, z' \rangle = \langle z'', z'' \rangle = 0, \quad \langle z', z' \rangle = -\langle z, z'' \rangle = 4. \tag{4.99}$$

By using (4.97) and (4.99) we know that $\langle L, L \rangle = 1$ and the induced metric tensor is given by (4.91). Moreover, the second equation in (4.96) shows that the second fundamental form of ψ satisfies $h(\partial/\partial x, \partial/\partial y) = 0$. Consequently, the immersion ψ is a minimal immersion.

Case (ii): $\eta \neq 0$. After applying formula of Gauss, (4.91)-(4.93), and (4.95), we obtain

$$L_{xx} = \frac{2\xi}{(x + y)^2} - \frac{2L_x}{x + y},$$

$$L_{xy} = \frac{2L}{(x + y)^2}, \tag{4.100}$$

$$L_{yy} = \frac{2\eta}{(x + y)^2} - \frac{2L_y}{x + y}.$$

The compatibility conditions of (4.100) are given by

$$\tilde{\nabla}_{\frac{\partial}{\partial y}} \xi = \frac{2\xi}{x + y}, \quad \tilde{\nabla}_{\frac{\partial}{\partial x}} \eta = \frac{2\eta}{x + y}. \tag{4.101}$$

Solving (4.101) gives

$$\xi = (x + y)^2 A(x), \quad \eta = (x + y)^2 B(y) \tag{4.102}$$

for some $\mathbb{E}_s^{m+1}$-valued functions $A(x), B(y)$. Substituting (4.102) into (4.100) yields

$$L_{xx} = A(x) - \frac{2L_x}{x+y}, \quad L_{xy} = \frac{2L}{(x+y)^2}, \quad L_{yy} = B(y) - \frac{2L_y}{x+y}. \quad (4.103)$$

After solving system (4.103), we obtain

$$L(x,y) = \frac{z(x) + w(y)}{x+y} - \frac{z'(x) + w'(y)}{2}, \quad (4.104)$$

where $z(x), w(y)$ are $\mathbb{E}_s^{m+1}$-valued functions satisfying

$$z'''(x) = -4A(x), \quad w'''(y) = -4B(y). \quad (4.105)$$

From $\langle L, L \rangle = 1$ and (4.104), we obtain condition (c.1) in Theorem 5.7.

By combining (4.103) and (4.105), we obtain

$$L_{xx} = -\frac{z'''}{4} - \frac{2L_x}{x+y}, \quad L_{xy} = \frac{2L}{(x+y)^2}, \quad L_{yy} = -\frac{w'''(y)}{4} - \frac{2L_y}{x+y}. \quad (4.106)$$

Since the metric tensor of M is given by (4.91), we find

$$\langle L_x, L_x \rangle = \langle L_y, L_y \rangle = 0, \quad \langle L_x, L_y \rangle = -\frac{2}{(x+y)^2}. \quad (4.107)$$

Because $\langle L, L \rangle = 1$, we have $\langle L_{xx}, L \rangle = -\langle L_x, L_x \rangle = 0$. Thus we obtain condition (c.2) from (4.104), (4.106) and (4.107).

Similarly, since $\langle L_{yy}, L \rangle = -\langle L_y, L_y \rangle = 0$, we have condition (c.3) from (4.104) and (4.106).

Conversely, assume that L is defined by

$$L(x,y) = \frac{z(x) + w(y)}{x+y} - \frac{z'(x) + w'(y)}{2}, \quad (4.108)$$

where $z(x), w(y)$ are curves satisfying conditions (c1), (c.2) and (c.3). Then it follows from (4.108) that L satisfies system (4.106). Also, it follows from (4.108) and condition (c.1) that $\langle L, L \rangle = 1$. Thus we have

$$\langle L, L_x \rangle = \langle L, L_y \rangle = 0, \quad (4.109)$$

which implies that

$$\begin{aligned}\langle L_x, L_x \rangle &= -\langle L, L_{xx} \rangle, \\ \langle L_x, L_y \rangle &= -\langle L, L_{xy} \rangle, \\ \langle L_y, L_y \rangle &= -\langle L, L_{yy} \rangle.\end{aligned} \quad (4.110)$$

By applying (4.110), (c.1) and the first equation in (4.110), we obtain

$$\langle L_x, L_x \rangle = -\langle L, L_{xx} \rangle = \frac{2}{(x+y)^2} \langle L_x, L_x \rangle, \quad (4.111)$$

which shows that $\langle L_x, L_x \rangle = 0$. Similarly, from (4.110) and (c.3) we find $\langle L_y, L_y \rangle = 0$. Also, after applying (c.1), (4.110) and the second equation in (4.106), we find $\langle L_x, L_y \rangle = -2/(x+y)^2$. Consequently, the induced metric tensor via L is given by (4.91). Finally, it follows from (3.9) and the second equation in (4.106) that $\psi : M \to S_s^m(1)$ is a minimal immersion. $\square$

Theorem 4.13. [Chen (2011a)] *Let M be a Lorentzian surface of constant Gauss curvature -1. Then an isometric immersion $\psi : M \to H_s^m(-1)$ is a minimal immersion if and only if one of the following three cases occurs:*

(a) *M is an open portion of a totally geodesic $H_1^2(-1) \subset H_s^m(-1)$;*
(b) *the immersion $L = \iota \circ \psi : M \to H_s^m(-1) \subset \mathbb{E}_{s+1}^{m+1}$ is locally given by*

$$L(x,y) = z(x) \tanh\left(\frac{x+y}{\sqrt{2}}\right) - \frac{z'(x)}{\sqrt{2}},$$

where $z(x)$ is a timelike curve with constant speed $\sqrt{2}$ lying in the light cone $\mathcal{LC} \subset \mathbb{E}_{s+1}^{m+1}$ satisfying $\langle z'', z'' \rangle = 4$ and $z''' \neq 2z'$;
(c) *the immersion $L = \iota \circ \psi : M \to H_s^m(-1) \subset \mathbb{E}_{s+1}^{m+1}$ is locally given by*

$$L(x,y) = (z(x) + w(y)) \tanh\left(\frac{x+y}{\sqrt{2}}\right) - \frac{z'(x) + w'(y)}{\sqrt{2}},$$

where z and w are curves satisfying

$$\left\langle (z+w) \tanh\left(\frac{x+y}{\sqrt{2}}\right) - \frac{z'+w'}{\sqrt{2}}, (z+w) \tanh\left(\frac{x+y}{\sqrt{2}}\right) - \frac{z'+w'}{\sqrt{2}} \right\rangle = -1,$$

$$\sqrt{2} \langle z+w, 2z' - z''' \rangle \tanh\left(\frac{x+y}{\sqrt{2}}\right) = \langle z'+w', 2z' - z''' \rangle,$$

$$\sqrt{2} \langle z+w, 2w' - w''' \rangle \tanh\left(\frac{x+y}{\sqrt{2}}\right) = \langle z'+w', 2w' - w''' \rangle.$$

Proof. This can be proved in a similar way as Theorem 4.12. $\square$

Remark 4.9. A natural extension of minimal surfaces are surfaces with parallel mean curvature vector. The classification of such surfaces is much more complicated than minimal ones. Nevertheless, the complete classifications of spacelike and Lorentzian surfaces with parallel mean curvature vector in indefinite real space forms were done in [Chen (2009b,c, 2010c)] (see also [Fu and Hou (2010)]). For a recent survey on these, see [Chen (2010g)].

4.14   Cartan's structure equations

Let M be a pseudo-Riemannian n-submanifold of a pseudo-Riemannian m-manifold N. Denote by $\tilde{\nabla}$ the Levi-Civita connection of N. Choose a local orthonormal frame $e_1, \ldots, e_n, e_{n+1}, \ldots, e_m$ of M such that $e_1, \ldots, e_n$ are tangent to M and $e_{n+1}, \ldots, e_m$ are normal to M. Let $\omega^1, \ldots, \omega^n$ be the dual frame of $e_1, \ldots, e_n$, i.e., $\omega_i(e_j) = \delta_{ij}$.

We shall make use of the following convention on the ranges of indices unless mentioned otherwise:

$$1 \leq \alpha, \beta, \gamma, \ldots \leq m; \ 1 \leq i, j, k, \ell \leq n; \ n+1 \leq r, s, t, \ldots \leq m.$$

Put $\langle e_\alpha, e_\beta \rangle = \epsilon_\alpha \delta_{\alpha\beta}; \ \alpha, \beta = 1, \ldots, m$, and

$$\tilde{\nabla} e_j = \sum_j \epsilon_k \omega_j^k e_k + \sum_r \epsilon_r \omega_j^r e_r,$$

$$\tilde{\nabla} e_r = \sum_k \epsilon_k \omega_r^k e_k + \sum_s \epsilon_s \omega_r^s e_s. \tag{4.112}$$

The 1-forms ω_α^β, $1 \leq \alpha, \beta \leq m$, are called the *connection forms*. From (4.112) we find

$$\omega_j^k = -\omega_k^j, \ \ \omega_j^r = -\omega_r^j, \ \ \omega_r^s = -\omega_s^r. \tag{4.113}$$

For the second fundamental form h of M, if we put

$$h(e_i, e_j) = \sum_r \epsilon_r h_{ij}^r e_r, \tag{4.114}$$

then it follows from (4.15) and (4.16) that $\omega_j^r(e_i) = -\langle \nabla_{e_i} e_r, e_j \rangle = h_{ij}^r$. Combining this with (4.113) gives $\omega_j^r = \sum_i h_{ij}^r \omega^i$.

The *Cartan structure equations* are given by

$$d\omega^i = -\sum_j \epsilon_j \omega_j^i \wedge \omega^j,$$

$$d\omega_j^i = \sum_{k,\ell,r} \epsilon_r (h_{ik}^r h_{j\ell}^r - h_{i\ell}^r h_{jk}^r) \omega^k \wedge \omega^\ell - \sum_k \epsilon_k \omega_k^i \wedge \omega_j^k + \tilde{\Omega}_j^i,$$

$$d\omega_i^r = -\sum_{j,k} \epsilon_j h_{jk}^r \omega^k \wedge \omega_i^j + \sum_{j,s} \epsilon_s h_{ij}^s \omega^j \wedge \omega_s^r + \tilde{\Omega}_i^r,$$

$$d\omega_s^r = \sum_{i,k,\ell} \epsilon_i (h_{ik}^r h_{i\ell}^s - h_{i\ell}^r h_{ik}^s) \omega^k \wedge \omega^\ell - \sum_t \epsilon_t \omega_t^r \wedge \omega_s^t + \tilde{\Omega}_s^r,$$

where

$$\tilde{\Omega}_\beta^\alpha = \frac{1}{2} \sum_{j,k} \tilde{K}_{\beta jk}^\alpha \omega^j \wedge \omega^k, \ \ \tilde{R}(e_\alpha, e_\beta) e_\gamma = \sum_\delta \epsilon_\delta \tilde{K}_{\beta\gamma\delta}^\alpha e_\delta.$$

Those $\tilde{\Omega}_\beta^\alpha$ are well-known as the *curvature 2-forms* of N, restricted to M.

Chapter 5

Total Mean Curvature

5.1 Introduction

The two most important invariants of a surface in a Euclidean 3-space $\mathbb{E}^3$ are the Gauss curvature G and the squared mean curvature H^2.

According to Theorema Egregium of Gauss, Gauss curvature is an intrinsic invariant and the integral of the Gauss curvature over a closed surface M gives the well-known Gauss-Bonnet formula

$$\int_M G dV = 2\pi\chi(M), \tag{5.1}$$

where $\chi(M)$ denotes the Euler number of M. For the Gauss curvature of a closed surface M in $\mathbb{E}^3$, there also exists an inequality due to [Chern and Lashof (1957)]; namely,

$$\int_M |G| dV \geq 4\pi(1+g), \tag{5.2}$$

where g is the genus of M.

The study of the total mean curvature

$$w(\phi) = \int_M H^2 dV \tag{5.3}$$

of a surface M in $\mathbb{E}^3$ goes back at least to Blaschke's school in the 1920's. The energy $w(\phi)$ is sometime called *bending energy*, which appears naturally in some physical contexts. For instance, the total mean curvature was proposed by S. D. Poisson (1781-1840) in [Poisson (1812)] and also by S. Germain (1776-1831) in [Germain (1921)] to describe elastic shells. The functional (5.3) also appears in the Helfrich model in [Helfrich (1973)] as one of the terms that contribute to the energy of cell membranes in mathematical biology.

Among others, G. Thomsen in [Thomsen (1923)] investigated the first variations of (5.3) and proved that the Euler-Lagrange equation[1] of the functional $w(\phi)$ is

$$\Delta H + 2H(H^2 - K) = 0. \tag{5.4}$$

W. Blaschke proved in his book [Blaschke (1929)] that $\int_M H^2 dV$ of a closed surface M in $\mathbb{E}^3$ is a conformal invariant, i.e, it keeps the same value under conformal mappings of $\mathbb{E}^3$. More than 40 years after Thomsen's work, T. J. Willmore reintroduced in [Willmore (1968)] the problem. He observed that combining the inequality $H^2 \geq G$, Gauss-Bonnet's formula and Chern-Lashof's inequality gives

$$w(\phi) = \int_M H^2 dV \geq 4\pi, \tag{5.5}$$

with equality holding if and only if M is a round sphere.

5.2   Total absolute curvature of Chern and Lashof

Let C be a closed oriented curve lying in a plane $\mathbb{E}^2$. As a point moves along C, the line through a fixed point o and parallel to the tangent line of C rotates through an angle $2n\pi$ or rotate n times about o. This integer n is called the *rotation index* of C. If C is a simple curve, $n = \pm 1$.

Two closed curves in a Euclidean 2-pane are called *regularly homotopic* if one can be deformed to the other through a family of closed smooth curves. Because the rotation index is an integer and it varies continuously through the deformation, it must keep constant. Hence two closed smooth curves have the same rotation index if they are regularly homotopic. A theorem of Graustein and Whitney says that the converse of this is also true (cf. [Whitney (1937)]). Thus the only invariant of a regular homotopy class is the rotation index.

Let $(x_1(s), x_2(s))$ be the Euclidean coordinate of a closed smooth curve C in $\mathbb{E}^2$ which is parametrized by its arc length s. Then we have

$$x_1''(s) = -\kappa(s)x_2'(s), \quad x_2''(s) = \kappa(s)x_1'(s), \tag{5.6}$$

where $\kappa(s)$ denotes the curvature of C. Let $\theta(s)$ be the angle between the tangent line and the x_1-axis. We have

$$d\theta = \frac{x_1'x_2'' - x_1''x_2'}{x_1'^2 + x_2'^2} ds = \kappa ds.$$

---

[1]Thomsen attributes the Euler-Lagrange equation (5.4) to W. Schadow.

From this we obtain the following formula:

$$\int_C \kappa(s)ds = 2n\pi, \quad n = \text{the rotation index.} \tag{5.7}$$

By applying (5.7) we conclude that the total absolute curvature of C satisfies the following inequality:

$$\int_C |\kappa(s)|ds \geq 2\pi, \tag{5.8}$$

with the equality holding if and only if C is a convex planar curve. This result was generalized in [Fenchel (1929)] to closed curves in $\mathbb{E}^3$ and in [Borsuk (1947)] to closed curves in $\mathbb{E}^m$ with $m > 3$. [Fary (1949)] and [Milnor (1950)] improved Fenchel-Borsuk's result to knotted curves as follows:

Theorem 5.1. *If C is a knotted closed curve in $\mathbb{E}^m$, then $\int_C |\kappa(s)|ds \geq 4\pi$.*

The Fenchel-Borsuk result was extended by S. S. Chern and R. K. Lashof in [Chern and Lashof (1957)] to arbitrary closed submanifolds in $\mathbb{E}^m$ which we will discuss as follows: Let $\phi : M \to \mathbb{E}^m$ be an isometric immersion of an n-dimensional closed manifold M into a Euclidean m-space. The normal bundle $T^\perp M$ of M in $\mathbb{E}^m$ is an $(m - n)$-dimensional vector bundle over M whose bundle space is the subspace $M \times \mathbb{E}^m$, consisting of all points (p, ξ) so that $p \in M$ and ξ is a normal vector of M at p. With respect to the induced metric from $\mathbb{E}^m$ the normal bundle $T^\perp(M)$ is a Riemannian $(m - n)$-plane bundle over M. Let B_1 be the subbundle of the normal bundle whose bundle space consists of all points (p, ξ) such that $p \in M$ and ξ is a unit normal vector at p. Then B_1 is a bundle of $(m - n - 1)$-spheres over M and is a Riemannian manifold of dimension $m - 1$ with the induced metric. Let dV denote the volume element of M. Then there is a differential form $d\sigma$ of degree $m - n - 1$ on B_1 such that its restriction to a fibre S_p is the volume element of the sphere S_p of unit normal vectors at p. Then $dV \wedge d\sigma$ is the volume element of S_1. We denote it by dV_{B_1}. In fact, this can be seen as follows: Suppose that $\mathbb{E}^m$ is oriented. By a frame $p, e_1, \ldots, e_m$ in $\mathbb{E}^m$ we mean a point $p \in \mathbb{E}^m$ and an ordered set of orthonormal vectors $e_1, \ldots, e_m$ whose orientation consistent with the of $\mathbb{E}^m$. Denote by $\mathcal{F}(\mathbb{E}^m)$ the space of all frames in $\mathbb{E}^m$. Then $\mathcal{F}(\mathbb{E}^m)$ is a fibre bundle over $\mathbb{E}^m$ with the structure group $SO(m)$.

In what follows it is convenient to agree to the following range of indices:

$$1 \leq i, j, k \leq n; \quad n + 1 \leq r, s, t \leq m; \quad 1 \leq A, B, C \leq m.$$

Let $(\tilde{\omega}^A)$ denote the dual 1-forms of (e_A) and $\tilde{\omega}_B^A$ the connection 1-forms defined by

$$\tilde{\nabla} e_B = \sum_A \tilde{\omega}_B^A e_A. \tag{5.9}$$

Then $\tilde{\omega}^A, \tilde{\omega}_B^A$ satisfy the following structure equations of Cartan:

$$d\tilde{\omega}^A = -\sum_B \tilde{\omega}_B^A \wedge \tilde{\omega}^B, \tag{5.10}$$

$$d\tilde{\omega}_B^A = -\sum_C \tilde{\omega}_C^A \wedge \tilde{\omega}_B^C, \quad \tilde{\omega}_B^A + \tilde{\omega}_A^B = 0. \tag{5.11}$$

We shall consider $\tilde{\omega}^A, \tilde{\omega}_B^A$ as forms on $\mathcal{F}(\mathbb{E}^m)$ in a natural way.

For an isometric immersion $\phi : M \to \mathbb{E}^m$, we identify a tangent vector with its image under ϕ_*. Let B denote the bundle whose bundle space is the set of $M \times \mathcal{F}(\mathbb{E}^m)$ consisting of $(p, \phi(p), e_1, \ldots, e_n, e_{n+1}, \ldots, e_m)$ such that $e_1, \ldots, e_n$ are tangent to M and $e_{n+1}, \ldots, e_m$ are normal to M. The projection $B \to M$ is denoted by ψ. We define the map $\psi_1 : B \to B_1$ by

$$\psi_1(p, \phi(p), e_1, \ldots, e_m) = (p, e_m). \tag{5.12}$$

Consider the maps

$$B \xrightarrow{\iota} M \times \mathcal{F}(\mathbb{E}^m) \xrightarrow{\lambda} \mathcal{F}(\mathbb{E}^m), \tag{5.13}$$

where ι is the inclusion and λ is the projection onto the second factor. Put

$$\omega^A = (\lambda \circ \iota)^* \tilde{\omega}^A, \quad \omega_B^A = (\lambda \circ \iota)^* \tilde{\omega}_B^A. \tag{5.14}$$

Since d and $\wedge$ commute with $(\lambda \circ \iota)^*$, (5.10) and (5.11) imply

$$d\omega^A = -\sum_B \omega_B^A \wedge \omega^B, \tag{5.15}$$

$$d\omega_B^A = -\sum_C \omega_C^A \wedge \omega_B^C, \quad \omega_B^A + \omega_A^B = 0. \tag{5.16}$$

From the definition of B it follows that $\omega^r = 0$ and $\omega^1, \ldots, \omega^n$ are linearly independent. If we restrict these 1-forms to M, then the volume element dV of M is given by

$$dV = \omega^1 \wedge \cdots \wedge \omega^n. \tag{5.17}$$

Moreover, the volume element of B_1 is

$$dV \wedge d\sigma = \omega^1 \wedge \cdots \wedge \omega^n \wedge \omega_{n+1}^m \wedge \cdots \wedge \omega_{m-1}^m, \tag{5.18}$$

where $d\sigma$ is $\omega_{n+1}^m \wedge \cdots \wedge \omega_{m-1}^m$, which is an $(m - n - 1)$-form on B_1.

From $\omega^r = 0$ we get $0 = d\omega^r = -\sum_i \omega_u^r \wedge w^i$. So, Cartan's lemma gives

$$\omega_i^r = \sum_j h_{ij}^r \omega^j. \tag{5.19}$$

In fact, if we denote by h the second fundamental form of M in $\mathbb{E}^m$, we have $h_{ij}^r = \langle h(e_i, e_j), e_r \rangle$. Let us consider the map $\nu : B_1 \to S^{m-1}$ of B_1 into the unit sphere S^{m-1} of $\mathbb{E}^m$ defined by $\nu(p, e) = e$. Denote by $d\Sigma$ the volume element of S^{m-1} in $\mathbb{E}^m$. Then (5.9) implies

$$d\Sigma = \tilde{\omega}_1^m \wedge \cdots \wedge \tilde{\omega}_{m-1}^m. \tag{5.20}$$

Therefore, by (5.19) and (5.20), we find

$$\nu^* d\Sigma = K(p, e_m)\omega^1 \wedge \cdots \wedge \omega^n \wedge \omega_{n+1}^m \wedge \cdots \wedge \omega_{m-1}^m, \tag{5.21}$$

where $K(p, e_m) = \det(h_{ij}^m)$ is called the *Lipschitz-Killingz-Killing curvature* at (p, e_m). The *total absolute curvature* $TA(\phi)$ is then defined by

$$TA(\phi) = \frac{1}{c_{m-1}} \int_{B_1} |\nu^* d\Sigma| = \frac{1}{c_{m-1}} \int_M G^*(p) dV, \tag{5.22}$$

where c_{m-1} is the volume of the unit $(m-1)$-sphere and

$$G^*(p) = \int_{S_p} |K(p, e_m)| d\sigma. \tag{5.23}$$

The Chern-Lashof inequality is the following.

Theorem 5.2. [Chern and Lashof (1957)] *Let $\phi : M \to \mathbb{E}^m$ be an immersion of an n-dimensional closed manifold M into $\mathbb{E}^m$. Then the total absolute curvature of ϕ satisfies the following inequality:*

$$TA(\phi) \geq b(M), \tag{5.24}$$

where $b(M)$ is the total Betti number.

Proof. For each unit vector $u \in S^{m-1}$ we define the height function h_u in the direction u by $h_u(p) = \langle u, \phi(p) \rangle$, $p \in M$. If ξ is a unit normal vector at p, i.e., $(p, \xi) \in B_1$, then we have $dh_\xi(p) = \langle \xi, d\phi(p) \rangle = 0$. Hence p is a critical point of the height function h_ξ. Therefore, if p is a critical point of the height function h_u, then $dh_u(p) = \langle u, d\phi(p) \rangle = 0$. Thus u is a unit vector normal to M at p, i.e., $(p, u) \in B_1$. Consequently, the number $\beta(h_u)$ of all critical points of h_u is equal to the number of points in M with u as its normal vector. Hence we obtain

$$\int_{B_1} |\nu^* d\Sigma| = \int_{u \in S^{m-1}} \beta(h_u) d\Sigma.$$

Since, for each $u \in S^{m-1}$, h_u has degenerate critical points if and only if u is a critical value of the map $\nu : B_1 \to S^{m-1}$. According to the theorem of Sard, the image of the set of critical points of ν has measure zero in S^{m-1}. Thus, for almost all $u \in S^{m-1}$, h_u is a non-degenerate function. Therefore $\beta(h_u)$ is well-defined and it is finite for almost all $u \in S^{m-1}$. Consequently, after applying Morse's theorem, we obtain inequality (5.24). □

Theorem 5.3. [Chern and Lashof (1957)] *Let $\phi : M \to \mathbb{E}^m$ be an immersion of an n-dimensional closed manifold M into $\mathbb{E}^m$. If $TA(\phi) < 3$, then M is homeomorphic to an n-sphere.*

Proof. If $TA(\phi) < 3$ holds, then there exists a set of positive measure on S^{m-1} such that if u is a unit vector in this set, the height function h_u has exactly two critical points.

On the other hand, by Sard's theorem, the image of the set of critical points under ν is of measure zero. Thus there is a unit vector u such that h_u has two non-degenerate critical points. Hence, after applying the theorem of Reeb, we conclude that M is homeomorphic to an n-sphere. □

For a hypersurface M in $\mathbb{E}^{n+1}$, if the tangent plane $T_x M$ at each $x \in M$ does not separate M into two parts, the M called a *convex hypersurface*.

Theorem 5.4. [Chern and Lashof (1957)] *Let $\phi : M \to \mathbb{E}^m$ be an immersion of an n-dimensional closed manifold M into $\mathbb{E}^m$. If $TA(\phi) = 2$, then $\phi(M)$ lies in a linear subspace $\mathbb{E}^{n+1}$ of $\mathbb{E}^m$. Moreover, M is imbedded as a convex hypersurface. The converse is also true.*

For the proof of this theorem, see [Chern and Lashof (1957)].

If $\dim M = 1$, then Chern-Lashof's results reduce to Fenchel-Borsuk's theorem. Chern-Lashof's result gave birth to the notion of *tight immersion* which serves as a natural generalization of convexity.

If $\phi : M \to \mathbb{E}^3$ is an immersion of a closed surface M, the Lipschitz-Killing curvature of M in $\mathbb{E}^3$ reduces to the Gauss curvature G, i.e.,

$$K(p, e_3) = \det(h_{ij}^3) = G(p). \tag{5.25}$$

Then, by (5.24), we obtain the following inequality of Chern-Lashof:

$$\int_M |G| dV \geq 2\pi b(M) \geq 2\pi(4 - \chi(M)), \tag{5.26}$$

where $\chi(M)$ is the Euler characteristic of M.

Analogous to Fary-Milnor's result on knotted curves, R. Langevin and H. Rosenberg obtained the following result on knotted tori.

Theorem 5.5. [Langevin and Rosenberg (1976)] *Let T be a knotted torus in $\mathbb{E}^3$. Then $\int_T |G| dV \geq 16\pi$.*

This result was improved by N. H. Kuiper and W. H. Meeks as follows.

Theorem 5.6. [Kuiper and Meeks (1984)] *Let T be a knotted torus in $\mathbb{E}^3$. Then $\int_T |G| dV > 16\pi$.*

5.3 Willmore's conjecture and Marques-Neves' theorem

For tubes in $\mathbb{E}^3$ we have the following.

Theorem 5.7. [Shiohama and Takagi (1970); Willmore (1971)] *Let M be a torus embedded in $\mathbb{E}^3$ such that the embedded surface is generated by carrying a small circle around a closed curve so that the center moves along the curve and plane of the circle is in the normal plane to the curve at each point. Then*

$$\int_M H^2 \, dV \geq 2\pi^2. \tag{5.27}$$

The equality sign holds if and only if the embedded surface is congruent to the anchor ring defined by

$$a\left(\left(\sqrt{2} + \cos u\right)\cos v, \left(\sqrt{2} + \cos u\right)\sin v, \sin u\right), \quad a > 0.$$

Proof. Let $\gamma(s)$ be the closed curve of length ℓ described in the theorem. Denote by κ and τ the curvature and torsion of γ. Then the position vector of M is given by

$$\mathbf{x}(s,v) = \gamma(s) + r(\cos v)\mathbf{N} + r(\sin v)\mathbf{B},$$

where $\mathbf{N}, \mathbf{B}$ are the principal and binormal vector fields of γ. A direct computation shows that the two principal curvatures of M are given by

$$\kappa_1 = \frac{1}{r}, \quad \kappa_2 = \frac{\kappa \cos v}{\kappa r \cos v - 1}.$$

Thus the squared mean curvature H^2 is given by

$$H^2 = \left(\frac{1 - 2\kappa r \cos v}{2r(1 - \kappa r \cos v)}\right)^2.$$

Hence

$$
\int_M H^2 dV = \int_0^\ell \int_0^{2\pi} \left(\frac{1 - 2\kappa r \cos v}{2r(1 - \kappa r \cos v)} \right)^2 dV
$$

$$
= \frac{\pi}{2r} \int_0^\ell \frac{ds}{\sqrt{1 - \kappa^2 r^2}} \tag{5.28}
$$

$$
\geq \int_0^\ell |\kappa| ds \geq 2\pi^2,
$$

where we use the fact that, for any real variable x, $x\sqrt{1-x^2}$ takes its maximum value $\frac{1}{2}$ at $x = 1/\sqrt{2}$; and also apply the well-known inequality of Fenchel-Borsuk.

If the equality sign of (5.28) holds, then the inequality in (5.28) is an equality. Thus, after applying Fenchel-Borsuk's theorem again, we conclude that γ is a convex planar curve. In this case, we have $\kappa = 1/\sqrt{2}r$. Hence γ is a circle of radius $\sqrt{2}r$. The converse is easy to verify. $\qquad\square$

T. J. Willmore (1919–2005) made the following.

Conjecture 5.1. [Willmore (1965)] $\int_M H^2 dV \geq 2\pi^2$ *holds for tori in* $\mathbb{E}^3$.

Willmore's conjecture has been known to be true in some particular cases. Among others, K. Shiohama and R. Tagaki in [Shiohama and Takagi (1970)], independent by Willmore in [Willmore (1971)], proved that the conjecture is true for tubes in $\mathbb{E}^3$. The author proved in [Chen (1983a)] that it is true for conformal images of all flat tori in S^3. J. Langer and D. Singer proved it for tori of revolution in [Langer and Singer (1984)]. R. Langevin and H. Rosenberg established $\int H^2 dV \geq 8\pi$ in [Langevin and Rosenberg (1976)] for any imbedded knotted torus in $\mathbb{E}^3$. A. Ros proved in [Ros (1999)] that the conjecture is true for stereographic projections of tori in S^3 that are invariant under the antipodal map. Ros also showed that the conjecture was true for tori in $\mathbb{E}^3$ that are symmetric with respect to a point [Ros (2001)].

After almost 50 years later, Willmore's conjecture was finally solved by F. C. Marques and A. Neves using Almgren-Pitts' min-max theory of minimal surfaces.

Marques and Neves' theorem states as follows.

Theorem 5.8. [Marques and Neves (2014)] *The integral of the square of the mean curvature of a torus immersed in* $\mathbb{E}^3$ *is at least* $2\pi^2$.

5.4 Total mean curvature and conformal invariants

Let (N, g) be a Riemannian m-manifold and ρ be a positive function on N. We put

$$\tilde{g} = \rho^2 g. \tag{5.29}$$

Then $\tilde{g}$ is called a *conformal change of the metric g*. Denote by $\tilde{\nabla}$ and ∇ the Levi-Civita connections of $\tilde{g}$ and g. Then $\tilde{\nabla}$ and ∇ are related by

$$\tilde{\nabla}_{\tilde{X}}\tilde{Y} - \nabla_{\tilde{X}}\tilde{Y} = (\tilde{X}\ln\rho)\tilde{Y} + (\tilde{Y}\ln\rho)\tilde{X} - g(\tilde{X}, \tilde{Y})U, \tag{5.30}$$

where $U = (d\rho)_\#$ is the vector field associated with $d\rho$.

Let M be an n-dimensional submanifold of N. Denote by g_M and $\tilde{g}_M$ the metrics on M induced from g and $\tilde{g}$, respectively. Then, for each normal vector field ξ of M in N, we have

$$\tilde{\nabla}_X\xi - \nabla_X\xi = (X\ln\rho)\xi - (\xi\ln\rho)X, \quad X \in TM. \tag{5.31}$$

Thus, after applying Weingarten's formula, we find

$$\tilde{D}_X\xi - D_X\xi = (X\ln\rho)\xi, \tag{5.32}$$

where $\tilde{D}$ and D are the normal connections of M with respect to $\tilde{\nabla}$ and ∇, respectively. Hence

$$\tilde{D}_X - D_X = (X\ln\rho)I. \tag{5.33}$$

Therefore, the normal curvature tensors $R^{\tilde{D}}$ and R^D are related by

$$R^{\tilde{D}}(X, Y) = R^D(X, Y) + D_X((Y\ln\rho)I) + (X\ln\rho)D_Y \\ - D_Y((X\ln\rho)I) - (Y\ln\rho)D_X - ([X, Y]\ln\rho)I. \tag{5.34}$$

Consequently, by the definition of Lie bracket, we obtain

$$R^{\tilde{D}}(X, Y) = R^D(X, Y), \quad \forall X, Y \in \mathfrak{X}(M). \tag{5.35}$$

This implies the following [Chen (1974a)].

Proposition 5.1. *Let M be a submanifold of a Riemannian manifold N. Then the normal curvature tensor R^D of M is a conformal invariant.*

Let h, A and $\tilde{h}, \tilde{A}$ be the second fundamental form and shape operator of M in (N, g) and $(N, \tilde{g})$, respectively. Then it follows from (5.30) that

$$g(\tilde{A}_\xi X, Y) = g(A_\xi X, Y) + g(X, Y)g(U, \xi). \tag{5.36}$$

Let $e_1, \ldots, e_n$ be the eigenvectors of A_ξ with respect to the metric on M induced from g. Then

$$\rho^{-1}e_1, \ldots, \rho^{-1}e_n \tag{5.37}$$

form an orthonormal frame of N with respect to the metric induced from $\tilde{g}$; and they are eigenvectors of $\tilde{A}_\xi$.

Denote by $\kappa_1(\xi), \ldots, \kappa_n(\xi)$ the eigenvalues of A_ξ and by $\tilde{\kappa}_1(\xi), \ldots, \tilde{\kappa}_n(\xi)$ that of $\tilde{A}_\xi$. Then (5.36) implies that

$$\tilde{\kappa}_i(\xi) = \kappa_i(\xi) + \lambda_\xi, \quad \lambda_\xi = g(U, \xi). \tag{5.38}$$

Since $\tilde{A}_\xi = \rho \tilde{A}_{\tilde{\xi}}$ and $\tilde{\xi} = \rho^{-1}\xi$ is a unit normal vector with respect to $\tilde{g}$, we find from (5.38) that

$$\rho(\tilde{\kappa}_i(\tilde{\xi}) - \tilde{\kappa}_j(\tilde{\xi})) = \kappa_i(\xi) - \kappa_j(\xi). \tag{5.39}$$

Now, let $\xi_{n+1}, \ldots, \xi_m$ be an orthonormal normal frame of M with respect to g. Then the mean curvature vector H of M in (N, g) is given by

$$H = \frac{1}{n} \sum_r \sum_i \kappa_i(\xi_r)\xi_r. \tag{5.40}$$

Let us put

$$\tau_{ext} = \frac{2}{n(n-1)} \sum_r \sum_{i<j} \kappa_i(\xi_r)\kappa_j(\xi_r). \tag{5.41}$$

Then τ_{ext} is well-defined. We call τ_{ext} the *extrinsic scalar curvature* with respect to g. In particular, if N is of constant curvature c, then Gauss' equation implies that

$$\tau_{ext} = \frac{2}{n(n-1)}\tau - c, \tag{5.42}$$

where τ is the scalar curvature.

By using (5.39)-(5.41), we find the following [Chen (1974a)].

Proposition 5.2. *If M is a submanifold of a Riemannian manifold, then*

$$(H^2 - \tau_{ext})g \tag{5.43}$$

is invariant under any conformal change of metric.

When M is closed, this implies immediately the following.

Proposition 5.3. *If M is an n-dimensional closed submanifold of a Riemannian manifold, then*

$$\int_M (H^2 - \tau_{ext})^{n/2} dV \tag{5.44}$$

is a conformal invariant.

When M is 2-dimensional, Proposition 5.3 becomes the following.

Proposition 5.4. *If M is a closed surface in a Riemannian m-manifold $\tilde{M}$, then $\int_M (H^2 + \tilde{K}(M)) dV$ is a conformal invariant, where $\tilde{K}(M)$ denotes the sectional curvature of $\tilde{M}$ restricted to the tangent planes of M.*

If M is $\mathbb{E}^m$, Proposition 5.4 implies the following [Chen (1973c)].

Corollary 5.1. *Let M be a closed surface in $\mathbb{E}^m$ and ψ is a diffeomorphism of $\mathbb{E}^m$ which induces a conformal change of metric on $\mathbb{E}^m$. Then*

$$\int_M H^2 dV_M = \int_{\psi(M)} H_\psi^2 dV_{\psi(M)}, \qquad (5.45)$$

where dV_M is the area element of the surface M.

When $m = 3$, this corollary is due to [Blaschke (1929)].

Definition 5.1. A surface M in $\mathbb{E}^m$ is said to be *conformally equivalent* to another surface $\bar{M}$ in $\mathbb{E}^m$ if M can be obtained from $\bar{M}$ via a conformal mapping of $\mathbb{E}^m$.

Corollary 5.1 shows that conformally equivalent surfaces in $\mathbb{E}^m$ have the same total mean curvature.

Definition 5.2. Let $\phi : M \to N$ be an isometric immersion of a Riemannian n-manifold M into another Riemannian manifold. Then ϕ is called *stationary* if ϕ is a critical point of the functional (5.44).

Remark 5.1. Since the functional $w(f) = \int_M H^2 dV$ is compatible with conformal geometry, the problem of finding the infimum of $\int_M H^2 dV$ in $\mathbb{E}^3$ can be stated as a problem in the unit 3-sphere S^3. In fact, if M is viewed, via the stereographic projection, as a surface S in S^3, then the functional above transforms into

$$\int_S (1 + H_S^2) dV_S, \qquad (5.46)$$

where H_S^2 is the squared mean curvature of S in S^3. The above relation shows that the area of a minimal surface $S \subset S^3$ equals the total mean curvature $\int_{\sigma(S)} H^2 dV_{\sigma(S)}$ of the stereographic projection of S in $\mathbb{E}^3$.

Many minimal closed surfaces in S^3 were discovered in [Lawson (1970)]. Lawson proved that every closed surface, but the real projective plane, can be minimally immersed into S^3. Also, every closed orientable surface can be embedded minimally in S^3. All *Lawson's minimal surfaces* are constructed by starting with classical solutions to the Plateau problem for specific piecewise geodesic curves in S^3 and extending these solutions to complete minimal surfaces of the desired type using a reflection principle.

A *Hopf torus* of S^3 is the inverse image of a closed spherical curve in S^2 via the Hopf fibration $\pi : S^3 \to S^2$. Pinkall used Hopf tori in S^3 to show that there exists surfaces in $\mathbb{E}^3$ satisfying (5.4) which cannot be obtained by stereographic projections of minimal surfaces in S^3 [Pinkall (1985)].

5.5   Total mean curvature for arbitrary submanifolds

According to Nash's embedding theorem, every closed Riemannian n-manifold can be isometrically embedded in $\mathbb{E}^n$ with $n = \frac{1}{2}n(3n + 11)$. On the other hand, most closed Riemannian n-manifold cannot be isometrically immersed in $\mathbb{E}^{n+1}$ as a hypersurface. For instance, every closed flat surface cannot be isometrically immersed in $\mathbb{E}^3$. Consequently, the theory of submanifolds with arbitrary codimension is far richer than the theory of hypersurfaces; in particular, much richer than surfaces in $\mathbb{E}^3$.

For the total mean curvature of arbitrary closed submanifolds in a Euclidean space with arbitrary codimension, we have the following general result from [Chen (1971)].

Theorem 5.9. *Let M be an n-dimensional closed submanifold of $\mathbb{E}^m$. Then we have*

$$\int_M H^n dV \geq c_n, \tag{5.47}$$

where c_n is the volume of unit n-sphere $S^n(1)$.

The equality sign of (5.47) holds if and only if M is embedded as an ordinary n-sphere in a totally geodesic $(n+1)$-subspace $\mathbb{E}^{n+1}$ when $n > 1$; and as a convex planar curve when $n = 1$.

Proof. Let $\phi : M \to \mathbb{E}^m$ be an isometric immersion of a closed Riemannian n-manifold M into $\mathbb{E}^m$. Let $p \in M$ and let B be the bundle space consisting of all frame $\{p, \phi(p), e_1, \ldots, e_n, e_{n+1}, \ldots, e_m\}$ such that $e_1, \ldots, e_n$ are orthonormal vectors in $T_p M$ and $e_{n+1}, \ldots, e_m$ are orthonormal vectors in $T_p^\perp M$. Let us choose the frame $\{p, \phi(p), e_1, \ldots, e_n, \bar{e}_{n+1}, \ldots, \bar{e}_m\}$ in B such that $\bar{e}_m$ is parallel to the mean curvature vector H at p. Then

$$||H|| = \frac{1}{n}(\bar{h}_{11}^m + \cdots + \bar{h}_{nn}^m), \tag{5.48}$$

$$\sum_{i=1}^{n} \bar{h}_{ii}^r = 0, \quad r = n+1, \ldots, m-1, \tag{5.49}$$

where $\bar{h}_{ij}^s = \langle h(e_i, e_j), \bar{e}_s \rangle$. On the other hand, for each (p, e_m) in the unit normal bundle B_1, we may put

$$e_m = \sum_{r=n+1}^{m} \cos \theta_r \bar{e}_r, \tag{5.50}$$

where θ_r is the angle between e_m and $\bar{e}_r$. For each $(p, e) \in B_1$, we put

$$m(p, e) = \frac{1}{n} \text{Tr}\,(A_e). \tag{5.51}$$

From (5.48)-(5.51) we find

$$m(p, e) = \sum_{r=n+1}^{m} (\cos \theta_r) m(p, \bar{e}_r) = \|H(p)\| \cos \theta_m. \tag{5.52}$$

Hence

$$\int_{B_1} |m(p, e_m)|^n \, dV \wedge d\sigma = \int_{B_1} H^n |\cos^n \theta_m| \, dV \wedge d\sigma$$
$$= \frac{2c_{m-1}}{c_n} \int_M H^n dV, \tag{5.53}$$

where dV denotes the volume element of M and $d\sigma$ is the volume element of fibers. Let e be a unit vector in $S^{m-1}(1)$. Consider the height function $h_e = \langle e, \phi(p) \rangle$ on M. Then $Xh_e = \langle e, X \rangle$ for $X \in \mathfrak{X}(M)$. Thus we find

$$XYh_e = \langle e, \nabla_X Y + h(X, Y) \rangle \tag{5.54}$$

for $X, Y \in \mathfrak{X}(M)$. Since h_e is continuous on M, h_e has at least one maximum and one minimum, say at q and q', respectively. Since e is normal to N at q and q', we obtain from (5.54) that

$$XYh_e = \langle A_e X, Y \rangle \tag{5.55}$$

at q, q'. This implies that the shape operator A is either non-negative or non-positive definite at (q, e) and at (q', e). Let U denote the set consisting of all elements in B_1 such that the eigenvalues of $\kappa_1(p, e), \ldots, \kappa_n(p, e)$ of A_e have the same sign. Then $S^{m-1}(1)$ is covered by U at least twice under the map $\nu : B_1 \to S^{m-1}(1)$ defined by $\nu(p, e) = e$. This shows that

$$\int_U \nu^* d\Sigma \geq 2c_{m-1}, \tag{5.56}$$

where $d\Sigma$ is the volume element of B_1.

Since $\kappa_1(p,e), \ldots, \kappa_n(p,e)$ have the same sign on U, we have

$$
\begin{aligned}
|m(p,e)|^n &= |\tfrac{1}{n}(\kappa_1(p,e) + \cdots + \kappa_n(p,e))|^n \\
&\geq |\kappa_1(p,e) \cdots \kappa_n(p,e)| \\
&= |K(p,e)|,
\end{aligned}
\tag{5.57}
$$

where $K(p,e) = \kappa_1(p,e) \cdots \kappa_n(p,e)$ is the *Lipschitz-Killing curvature* at (p,e). Hence, by using (5.53), (5.56), (5.57) and (5.21), we obtain

$$
\begin{aligned}
\int_N H^n dV &= \frac{c_n}{2c_{m-1}} \int_{B_1} |m(p,e)|^n dV \wedge d\sigma \\
&\geq \frac{c_n}{2c_{m-1}} \int_U \nu^* d\Sigma \geq c_n.
\end{aligned}
$$

This proves (5.47).

Now, assume that the equality case of (5.47) holds. Consider the map

$$
\mathbf{y} : B_1 \to \mathbb{E}^m : (p,e) \mapsto \phi(p) + re,
\tag{5.58}
$$

where r is a sufficiently small positive number such that (5.58) defines an immersion. In this way, we may regard B_1 as a hypersurface in $\mathbb{E}^m$. Because

$$
\langle e, d\mathbf{y} \rangle = \langle e, d\phi \rangle + r \langle e, de \rangle = 0,
$$

e is a unit normal vector field of B_1 in $\mathbb{E}^m$ at (p,e). Thus $e_1, \ldots, e_{m-1}$ form an orthonormal basis of $T_{(p,e)}B_1$. A direct computation shows that the principal curvatures of B_1 in $\mathbb{E}^m$ at (p,e) are given by

$$
\begin{aligned}
\bar{\kappa}_i(p,e) &= \frac{\kappa_i(p,e)}{1 + r\kappa_i(p,e)}, \quad i = 1, \ldots, n, \\
\bar{\kappa}_r(p,e) &= \frac{1}{r}, \quad r = n+1, \ldots, m-1.
\end{aligned}
\tag{5.59}
$$

Let $\bar{U} = \{(p,e) : \kappa_1(p,e) = \cdots = \kappa_n(p,e) \neq 0\}$ and $\bar{V} = B_1 - \bar{U}$. Under the hypothesis that the equality sign of (5.47) holds, we may prove that $m(p,e) = 0$ identically on $\bar{V}$.

Let $\pi : B_1 \to M$ be the projection defined by $\pi(p,e) = p$. Then $U = \pi(\bar{U})$ is totally umbilical in $\mathbb{E}^m$. Hence $m(p,e)$ is a nonzero constant on each connected component. Now, by applying the continuity of $m(p,e)$, we obtain $\pi(\bar{U}) = M$. Thus M is totally umbilical in $\mathbb{E}^m$. Consequently, for $n \geq 2$, M is embedded as an ordinary hypersphere in a totally geodesic $(n+1)$-subspace of $\mathbb{E}^m$. For $n = 1$, M is immersed as a convex planar curve according to Fenchel-Borsuk's theorem.

The converse is trivial. $\qquad\qquad\qquad\qquad\qquad\qquad\qquad\qquad\qquad \square$

An immediate consequence of Theorem 5.9 is the following.

Corollary 5.2. *If M is an n-dimensional minimal closed submanifold of a unit m-sphere $S^m(1)$, then*

$$\text{vol}(M) \geq c_n,$$

with the equality holding if and only if M is a great n-sphere of $S^m(1)$.

Proof. We may regard $S^m(1)$ as an ordinary hypersphere of $\mathbb{E}^{m+1}$. Then it follows from the minimality of M in $S^m(1)$ and Theorem 5.9 that

$$\text{vol}(M) = \int_M H^2 dV \geq c_n,$$

where H is the mean curvature vector M in $\mathbb{E}^{m+1}$.

The remaining part follows immediately from Theorem 5.9. $\qquad\square$

Similarly, we also have the following from Theorem 5.9.

Corollary 5.3. [Chen (1983c)] *If M is an n-dimensional minimal closed submanifold of the real projective m-space $RP^m(1)$ of constant curvature 1, then we have*

$$\text{vol}(M) \geq \frac{c_n}{2}, \tag{5.60}$$

with the equality holding if and only if M is a totally geodesic submanifold of RP^m.

Proof. Let M be an n-dimensional minimal closed submanifold of the real projective m-space $RP^m(1)$. Consider the canonical two-fold covering map $\pi : S^m(1) \to RP^m(1)$. Then $\pi^{-1}(M)$ is a minimal submanifold of $S^m(1)$ with $\text{vol}(\pi^{-1}(M)) \leq 2\,\text{vol}(M)$. Thus by applying Corollary 5.2 to $\pi^{-1}(M)$ we obtain inequality (5.60).

If the equality sign of (5.60) holds, then

$$\text{vol}(\pi^{-1}(M)) = 2\,\text{vol}(M) = c_n.$$

Thus, by Corollary 5.2, $\pi^{-1}(M)$ is a great n-sphere in $S^m(1)$. Hence M is a real projective n-space $RP^n(1)$ imbedded in $RP^m(1)$ as a totally geodesic submanifold.

The converse is trivial. $\qquad\square$

Corollary 5.4. [Chen (1983c)] *If M is an n-dimensional minimal closed submanifold of the complex projective m-space $CP^m(4)$ of constant holomorphic sectional curvature 4, then*

$$\text{vol}(M) \geq \frac{c_{n+1}}{2\pi}, \tag{5.61}$$

with the equality holding if and only if $M = CP^k, n = 2k$, imbedded as a totally geodesic complex submanifold of $CP^m(4)$.

Proof. Let M be an n-dimensional minimal closed submanifold of $CP^m(4)$. Consider the Hopf fibration $\pi : S^{2m+1}(1) \to CP^m(4)$. Denote by $\pi^{-1}(M)$ by $\tilde{M}$. Then $\pi : \tilde{M} \to M$ is a Riemannian submersion with totally geodesic fibres S^1. Let us consider the following commutative diagram:

$$\begin{array}{ccc} \tilde{M} & \xrightarrow{\tilde{\iota}} & S^{2m+1}(1) \\ \pi \downarrow & & \downarrow \pi \\ M & \xrightarrow{\iota} & \mathbb{C}P^m(4). \end{array}$$

Since M is minimal in $CP^m(4)$, $\tilde{M}$ is minimal in $S^{2m+1}(1)$. Thus by applying Corollary 5.2 to $\tilde{M}$, we find

$$\text{vol}(\tilde{M}) \geq c_{n+1}, \tag{5.62}$$

with equality holding if and only if $\tilde{M}$ is a great $(n+1)$-sphere in $S^{2m+1}(1)$.

On the other hand, because $\pi : \tilde{M} \to M$ is a Riemannian submersion with totally geodesic fibre S^1, we have

$$\text{vol}(\tilde{M}) = 2\pi \cdot \text{vol}(M). \tag{5.63}$$

By combining (5.62) and (5.63), we obtain (5.61).

If the equality sign of (5.61) holds, then $\tilde{M}$ is a great $(n + 1)$-sphere $S^{n+1}(1)$ of $S^{2m+1}(1)$. Since $\pi : S^{n+1}(1) \to M$ us a Riemannian submersion with fibre S^1, $n = 2k$ is even (cf. [Adem (1953)]). Therefore $\tilde{M}$ is a great $(2k + 1)$-sphere of $S^{2m+1}(1)$. Hence we conclude that M is a $CP^k(4)$ with was imbedded in $CP^m(4)$ as a totally geodesic complex submanifold.

The converse is trivial.   □

Similarly, we also have the following results.

Corollary 5.5. [Chen (1983c)] *If M is an n-dimensional minimal closed submanifold of the quaternionic projective m-space $QP^m(4)$ of constant quaternionic sectional curvature 4, then*

$$\text{vol}(M) \geq \frac{c_{n+3}}{2\pi^2},$$

with the equality holding if and only if $M = QP^k, n = 4k$, which is embedded as a totally geodesic quaternionic submanifold of $QP^m(4)$.

Corollary 5.6. [Chen (1983c)] *If M is an n-dimensional minimal closed submanifold of the Cayley plane $\mathcal{O}P^2(4)$ of maximal sectional curvature 4, then we have $\text{vol}(M) \geq c_n/2^n$.*

Theorem 5.9 implies that the total mean curvature of an n-dimensional closed submanifold of $\mathbb{E}^m$ is always bounded below by $c_n = \text{vol}(S^n)$. On the other hand, according to Theorem 5.2 of Chern-Lashof, the total absolute curvature is bounded below by a topological invariant; namely, the total Betti number $b(M)$.

It is natural to ask the following.

Problem 5.1. *Does the total mean curvature of a closed submanifold of $\mathbb{E}^m$ bound below by some topological invariants? In particular, if the total Betti number $b(M)$ of M is large, does the total mean curvature of M in $\mathbb{E}^m$ be "proportionally large"?*

In the following, we provide a partial answer to this question.

Let $||h||^2$ denote the squared norm of the second fundamental form of M in $\mathbb{E}^m$. Then by the equation of Gauss we have

$$2\tau = n^2 H^2 - ||h||^2 \tag{5.64}$$

and

$$(n-1)||h||^2 - 2\tau = \sum_{r,i,j} \left\{ n(h_{ij}^r)^2 - h_{ii}^r h_{jj}^r \right\}$$

$$= n \sum_r \sum_{i \neq j} (h_{ij}^r)^2 + \sum_r \sum_{i<j} (h_{ii}^r - h_{jj}^r)^2$$

$$\geq 0,$$

where τ is the scalar curvature of M defined by (2.13). Thus we have

$$-||h||^2 \leq 2\tau \leq (n-1)||h||^2. \tag{5.65}$$

Definition 5.3. An n-dimensional submanifold M of a Euclidean m-space $\mathbb{E}^m$ is called δ-*pinched in* $\mathbb{E}^m$ if we have

$$\delta||h||^2 \leq 2\tau \leq (n-1)||h||^2 \tag{5.66}$$

for some real number $\delta \geq -1$.

For δ-pinched submanifolds of $\mathbb{E}^m$ we have the following.

Proposition 5.5. [Chen (1976)] *Let M be an n-dimensional closed submanifold of $\mathbb{E}^m$. If M is δ-pinched, then*

$$\int_M H^n dV \geq \frac{1}{2} \left(\frac{1+\delta}{n} \right)^{n/2} c_n b(M), \tag{5.67}$$

where $b(M)$ denotes the total Betti number of M. The equality sign of (5.67) holds if and only if M is $(n-1)$-pinched in $\mathbb{E}^m$.

Proof. Let M be an n-dimensional closed manifold immersed in $\mathbb{E}^m$ and let $e_{n+1}, \ldots, e_m$ be a local orthonormal normal vector fields of M in $\mathbb{E}^m$. If e is a unit normal vector field on M, then $e = \sum_r \cos \theta_r e_r$ for some functions θ_r. Thus $A_e = \sum_r \cos \theta_r A_r$, $A_r = A_{e_r}$. Hence we find

$$||A_e||^2 = \sum_{r,s} \cos \theta_r \cos \theta_s \mathrm{Tr}(A_r A_s). \tag{5.68}$$

The right hand side of (5.68) is a quadratic form on $\cos \theta_{n+1}, \ldots, \cos \theta_m$. Hence we may choose local orthonormal normal vector fields $\bar{e}_{n+1}, \ldots, \bar{e}_m$ such that with respect to this frame field, we have

$$||A_e||^2 = \sum_r \zeta_r \cos^2 \theta_r, \quad \zeta_{n+1} \geq \cdots \geq \zeta_m, \tag{5.69}$$

$$\zeta_r = ||A_{\bar{e}_r}||^2. \tag{5.70}$$

Let B_1 denote the bundle of unit normal vectors of M in $\mathbb{E}^m$. We define a function f on B_1 by

$$f(p, e) = ||A_e||^2 \tag{5.71}$$

for $(p, e) \in B_1$. Since all of ζ_r are non-negative and $\sum_r \cos \theta_r = 1$, an inequality of Minkowski implies

$$\left(\int_{S_p} f^{\frac{n}{2}} d\sigma \right)^{2/n} = \left\{ \int_{S_p} \left(\sum_r \zeta_r \cos^2 \theta_r \right)^{2/n} d\sigma \right\}^{\frac{2}{n}}$$
$$\leq \sum_r \zeta_r \left(\int_{S_p} |\cos^n \theta_r| d\sigma \right)^{2/n}, \tag{5.72}$$

where S_p is the $(m - n - 1)$-sphere of unit normal vectors at p.

On the other hand, we have the following identity:

$$\int_{S_p} |\cos^n \theta| d\sigma = \frac{2c_{m-1}}{c_n}. \tag{5.73}$$

Thus, by (5.70), (5.72) and (5.73), we find

$$||h||^2 \geq \frac{2c_n}{c_{m-1}} \int_{S_p} f^{\frac{n}{2}} d\sigma. \tag{5.74}$$

Let $K(p, e)$ denote the determinant of A_e. Then by applying a relation between elementary symmetric functions we find $||A_e||^n \geq \sqrt{n^n} |K(p, e)|$. Thus, after applying (5.74), we obtain

$$\int_M ||h||^2 dV \geq \frac{\sqrt{n^n} c_n}{2c_{m-1}} \int_{B_1} |K(p, e)| dV \wedge d\sigma.$$

Now, by combining with with Chern-Lashof's inequality, we get

$$\int_M ||h||^2 dV \geq \frac{\sqrt{n^n} c_n}{2} b(M). \tag{5.75}$$

By the hypothesis of the proposition, M is δ-pinched in $\mathbb{E}^m$. Therefore we have

$$\tau \geq \frac{\delta}{2} ||h||^2. \tag{5.76}$$

On the other hand, by applying (5.64), we find from (5.76) that

$$H^2 \geq \frac{1+\delta}{n^2} ||h||^2. \tag{5.77}$$

Consequently, by combining (5.75) and (5.77) we obtain (5.67). If the equality sign of (5.67) holds, then the equality sign of (5.73) holds as well. From this we conclude that M is imbedded in a linear subspace $\mathbb{E}^{n+1}$ of $\mathbb{E}^m$ as a hypersurface satisfying $nH^2 = ||h||^2$. Therefore, after applying (5.64) once more, we conclude that M is $(n-1)$-pinched in $\mathbb{E}^m$. $\qquad\square$

When M is a minimal submanifold of a unit hypersphere $S^{m-1}(1)$ of $\mathbb{E}^m$, M is δ-pinched if and only if the scalar curvature of M satisfies

$$\tau \geq \frac{n^2}{2} \cdot \frac{\delta}{1+\delta}.$$

In particular, M is 0-pinched if and only if M has zero scalar curvature. In this case, Proposition 5.5 implies the following,

Theorem 5.10. [Chen (1976)] *Let M be an n-dimensional minimal closed submanifold of $S^m(1)$. If M has non-negative scalar curvature, then*

$$\mathrm{vol}(M) \geq \left(\frac{1}{n}\right)^{2/n} \frac{c_n}{2} b(M).$$

The following four results were proved in [Chen (1979b)].

Theorem 5.11. *For any isometric immersion of the real projective space $RP^{2n}(1)$ into $\mathbb{E}^m$, we have*

$$\int_{RP^{2n}(1)} H^{2n} dV \geq \left(\frac{2n+1}{n}\right)^n \frac{c_{2n}}{2}. \tag{5.78}$$

Theorem 5.12. *For any isometric immersion of the complex projective space $CP^n n(4)$ into $\mathbb{E}^m$, we have*

$$\int_{CP^n(4)} H^{2n} dV \geq \frac{1}{n!} \left(\frac{2(n+1)\pi}{n}\right)^n. \tag{5.79}$$

Theorem 5.13. *For any isometric immersion of the quaternion projective space $QP^n n(1)$ into $\mathbb{E}^m$, we have*

$$\int_{QP^n(4)} H^{4n} dV \geq \frac{2}{(2n+1)!} \left(\frac{(2n+3)\pi}{n} \right)^{2n}. \qquad (5.80)$$

Theorem 5.14. *Let T^{2n} be the flat $2n$-torus given by the direct product of $2n$ unit circles. Then for any isometric immersion of T^{2n} in $\mathbb{E}^m$, we have*

$$\int_{T^n} H^{2n} dV \geq \left(\frac{2\pi^2}{n} \right)^n, \qquad (5.81)$$

with the equality holding if and only if T^{2n} is immersed in a hypersphere with radius $r = \sqrt{2n}$ by imbedding of order $\{1\}$.

The following problem was posed by the author at the AMS Symposium on Differential Geometry held in Stanford University (cf. [Chen (1973d)]).

Problem 5.2. *Find relations between the total mean curvature and the topological invariants for closed submanifolds of $\mathbb{E}^m$?*

The following three conjectures posed in [Chen (1979b)] remain open.

Conjecture 5.2. *Inequality (5.78) holds for every immersion $RP^{2n} \to \mathbb{E}^m$.*

Conjecture 5.3. *Inequality (5.79) holds for every immersion $CP^n \to \mathbb{E}^m$.*

Conjecture 5.4. *Inequality (5.80) holds for every immersion $QP^n \to \mathbb{E}^m$.*

In views of Theorem 5.8 of Marques and Neves and Theorem 5.14, the author proposes the following conjecture.

Conjecture 5.5. *$\int H^n dV \geq \left(\frac{4}{n} \right)^{n/2} \pi^n$ holds for every immersed n-torus $(n \geq 3)$ in a Euclidean space with arbitrary codimension.*

5.6   A variational problem on total mean curvature

In order to gain more information about total mean curvature for Euclidean submanifolds, one may also apply standard techniques of calculus of variations. We will deal with this variational problem in this section.

Let M be an n-dimensional compact submanifold (with or without boundary ∂M) of a Euclidean m-space $\mathbb{E}^m$. Let x denote the position vector field of M in $\mathbb{E}^m$. Then

$$\mathbf{x} = \mathbf{x}(u_1, \ldots, u_n), \tag{5.82}$$

where $u_1, \ldots, u_n$ are the local coordinates of M. Let ξ be a unit normal vector fields of M in $\mathbb{E}^m$. We put

$$\bar{\mathbf{x}}(u_1, \ldots, u_n, t) = \mathbf{x}(u_1, \ldots, u_n) + t\varphi(u_1, \ldots, u_n)\xi(u_1, \ldots, u_n), \tag{5.83}$$

where φ is a differentiable function and t lies in small interval $(-\varepsilon, \varepsilon)$. If $\varphi \equiv 0$ on the boundary ∂M, (5.83) is called a *normal variation* of M in $\mathbb{E}^m$. We shall only consider normal variations which leave ∂M strongly fixed in the sense that both φ and its gradient vanish identically on ∂M. If M has no boundary, there is no restriction on the normal variations.

Throughout this section, we put $\mathbf{x}_i = \partial\mathbf{x}/\partial u_i$ and $g_{ij} = \langle \mathbf{x}_i, \mathbf{x}_j \rangle$. Then the induced metric tensor on M is given by $g = \sum_{i,j=1}^n g_{ij} du_i \otimes du_j$.

As before, let (g^{ij}) denote the inverse matrix of (g_{ij}). The volume element dV of M is given by

$$dV = *1 = W \, du_1 \wedge \cdots \wedge du_n, \tag{5.84}$$

where

$$W = \sqrt{\det(g_{ij})}. \tag{5.85}$$

Definition 5.4. Let ξ be the unit normal vector field in the direction of the mean curvature vector H of a compact submanifold M in $\mathbb{E}^m$. Then the variation (5.83) is called an *H-variation* of M in $\mathbb{E}^m$. Let δ denote the operator $\left(\frac{\partial}{\partial t}\right)|_{t=0}$. An n-dimensional submanifold M in $\mathbb{E}^m$ is called *H-stationary* if $\delta \int_M H^n dV = 0$ for all H-variations of M. And M is called *stationary* if $\delta \int_M H^n dV = 0$ for all normal variations.

It is clear that stationary submanifolds are H-stationary. When M is a hypersurface, an H-stationary hypersurface is always stationary.

Let $e_{n+1}, \ldots, e_m$ be a local frame of orthonormal normal vector fields on M such that $e_{n+1} = \xi$ and $\mathbf{x}_1, \ldots, \mathbf{x}_n, e_{n+1}, \ldots, e_m$ define the natural orientation of $\mathbb{E}^m$. Then we have

$$e_r = \frac{(-1)^{m+r}}{W}[\mathbf{x}_1, \ldots, \mathbf{x}_n, e_{n+1}, \ldots, \hat{e}_r, \ldots, e_m], \tag{5.86}$$

where $[v_1, \ldots, v_{m-1}]$ is the vector product of $m - 1$ vectors $v_1, \ldots, v_{m-1}$ in $\mathbb{E}^m$ and $\hat{\ }$ denotes the omitted term.

Let $\mathbf{x}_{ij} = \partial^2 \mathbf{x}/\partial u_i \partial u_j$ and $e_{ri} = \partial e_r/\partial u_i$. Then the formulas of Gauss and Weingarten are given by

$$\mathbf{x}_{ij} = \sum_k \Gamma_{ij}^k \mathbf{x}_k + \sum_r h_{ij}^r e_r, \tag{5.87}$$

$$e_{ri} = -\sum_j h_i^{rj} \mathbf{x}_j + \sum_s \ell_{ri}^s e_s, \tag{5.88}$$

where $h_i^{rt} = \sum g^{tj} h_{ij}^r$, $h_{ij}^r = \langle h(\mathbf{x}_i, \mathbf{x}_j), e_r \rangle$, $\ell_{ri}^s = \langle D_{\partial/\partial u_i} e_r, e_s \rangle$ and D is the normal connection. It follows from (5.83) and a direct computation the

$$\delta \mathbf{x} = \varphi e_{n+1}, \tag{5.89}$$

$$\delta \mathbf{x}_i = \varphi_i e_{n+1} - \varphi \sum_j h_i^{n+1j} \mathbf{x}_j + \varphi \sum_r \ell_{n+1i}^r e_r, \tag{5.90}$$

$$\delta g_{ij} = -2\varphi h_{ij}^{n+1}, \quad \delta g^{ij} = 2\varphi h^{n+1ij}, \tag{5.91}$$

$$\delta W = -\varphi \left(\operatorname{Tr} A_{e_{n+1}} \right) W, \tag{5.92}$$

where $\varphi_i = \partial \varphi/\partial u_i$ and $h^{rij} = \sum_t g^{it} h_t^{rj}$. Moreover, we also have

$$e_{rij} = \sum_s \frac{\partial}{\partial u_j}(\ell_{ri}^s) e_s + \sum_{t,s} \ell_{ri}^t \ell_{tj}^s e_s - \sum_k h_i^{rk} \mathbf{x}_{kj} \pmod{\mathbf{x}_k}, \tag{5.93}$$

$$\delta \mathbf{x}_{ij} = \varphi_{ij} e_{n+1} + \sum_s \left\{ \varphi_i \ell_{n+1j}^s + \varphi_j \ell_{n+1i}^s - \varphi \sum h_i^{n+1k} h_{kj}^s \right. \tag{5.94}$$

$$\left. + \varphi \frac{\partial}{\partial u_j}(\ell_{n+1i}^s) + \varphi \sum_r \ell_{n+1i}^r \ell_{rj}^s \right\} e_s \pmod{\mathbf{x}_k},$$

where

$$e_{rij} = \frac{\partial^2 e_r}{\partial u_i \partial u_j}, \quad \varphi_{ij} = \frac{\partial^2 \varphi}{\partial u_i \partial u_j}.$$

Hence we find

$$\langle e_{n+1}, \delta \mathbf{x}_{ij} \rangle = \varphi_{ij} - \varphi \sum_k h_i^{n+1k} h_{kj}^{n+1} - \varphi \sum_r \ell_{n+1i}^r \ell_{n+1j}^r, \tag{5.95}$$

and

$$\langle e_r, \delta x_{ij} \rangle = \varphi_i \ell_{n+1j}^r + \varphi_j \ell_{n+1i}^r - \varphi \sum_k h_i^{n+1k} h_{kj}^r + \varphi \frac{\partial \ell_{n+1i}^r}{\partial u_j}$$

$$+ \varphi \sum_s \ell_{n+1i}^s \ell_{sj}^r, \quad r = n+2, \ldots, m. \tag{5.96}$$

From (5.90) we find

$$\langle \mathbf{x}_i, \delta e_r \rangle = \begin{cases} -\varphi_i, & \text{if } r = n+1; \\ -\varphi \ell_{n+1i}^r, & \text{if } r = n+2, \ldots, m, \end{cases} \tag{5.97}$$

where we have used (5.86). From (5.86) we also have

$$\langle e_s, \delta(W e_r) \rangle = (-1)^{m+r} [\mathbf{x}_1, \ldots, \mathbf{x}_n, e_{n+1}, \ldots, \hat{e}_r, \ldots, e_m, \delta e_s] \quad (5.98)$$

for $r \neq s$, where we use the notation

$$[v_1, \ldots, v_m] = (-1)^{m-1} \langle v_1, [v_2, \ldots, v_m] \rangle \quad (5.99)$$

for m vectors $v_1, \ldots, v_m \in \mathbb{E}^m$. Since $\langle e_r, \delta e_r \rangle = 0$, (5.97) and (5.98) imply

$$\delta e_{n+1} = \sum_{r=n+2}^{m} \frac{(-1)^{m-n}}{W} [\mathbf{x}_1, \ldots, \mathbf{x}_n, e_{n+2}, \ldots, e_m, \delta e_r] e_r \quad (5.100)$$

$$- \sum_i \varphi^i x_i,$$

$$\delta e_r = \sum_{s \neq r} \frac{(-1)^{m+r-1}}{W} [\mathbf{x}_1, \ldots, \mathbf{x}_n, e_{n+1}, \ldots, \hat{e}_r, \ldots, e_m, \delta e_s] e_s \quad (5.101)$$

$$- \varphi \sum_i \ell_{n+1}^{ri} \mathbf{x}_i, \quad r = n+2, \ldots, m,$$

where $\varphi^i = \sum_t g^{ti} \varphi_t$ and $\ell_{n+1}^{ri} = \sum_t g^{ti} \ell_{n+1t}^r$.

From (5.87), (5.100) and (5.101) we find

$$\langle \mathbf{x}_{ij}, \delta e_{n+1} \rangle = \sum_{s=n+2}^{m} \frac{(-1)^{m-n} h_{ij}^s}{W} [\mathbf{x}_1, \ldots, \mathbf{x}_n, e_{n+2}, \ldots, e_m, \delta e_s] \quad (5.102)$$

$$- \sum_k \varphi_k \Gamma_{ij}^k,$$

$$\langle \mathbf{x}_{ij}, \delta e_r \rangle = \sum_{s \neq r} \frac{(-1)^{m+r-1} h_{ij}^s}{W} [\mathbf{x}_1, \ldots, \mathbf{x}_n, e_{n+1}, \ldots, \hat{e}_r, \ldots, e_m, \delta e_s]$$

$$(5.103)$$

$$- \sum_k \varphi \, \ell_{n+1k}^r \Gamma_{ij}^k.$$

Thus, by using (5.87), (5.94), (5.102) and (5.103), we find

$$\delta h_{ij}^{n+1} = \varphi_{i;j} - \varphi \sum_t h_i^{n+1t} h_{tj}^{n+1} - \varphi \sum_r \ell_{n+1i}^r \ell_{n+1j}^r \quad (5.104)$$

$$+ \sum_{s=n+2}^{m} \frac{(-1)^{m-n} h_{ij}^s}{W} [\mathbf{x}_1, \ldots, \mathbf{x}_n, e_{n+2}, \ldots, e_m, \delta e_s],$$

$$\delta h_{ij}^r = \phi_i \ell_{n+1j}^r + \varphi_j \ell_{n+1i}^r - \varphi \sum_t h_i^{n+1t} h_{tj}^r + \varphi \frac{\partial \ell_{n+1i}^r}{\partial u_j} \quad (5.105)$$

$$+ \varphi \sum_s \ell_{n+1i}^s \ell_{sj}^r - \varphi \sum_k \ell_{n+1k}^r \Gamma_{ij}^k$$

$$+ \sum_{s \neq r} \frac{(-1)^{m+r-1} h_{ij}^s}{W} [\mathbf{x}_1, \ldots, \mathbf{x}_n, e_{n+1}, \ldots, \hat{e}_r, \ldots, e_m, \delta e_s],$$

for $r = n + 2, \ldots, m$, where $\varphi_{i;j} = \varphi_{ij} - \sum_k \varphi_k \Gamma_{ij}^k$ are the components of the Hessian $\nabla d\varphi$ of φ on M.

From (5.91) and (5.104) we obtain

$$\delta(\operatorname{Tr} A_{n+1}) = -\Delta\varphi + \varphi\|A_{n+1}\|^2 - \varphi\ell^2$$
$$- \sum_{s=n+2}^{m} (\operatorname{Tr} A_s) \langle e_{n+1}, \delta e_s \rangle, \tag{5.106}$$

where $\ell^2 = \sum_{r,t} \ell_{n+1t}^r \ell_{n+1}^{rt}$. Similarly, we also have

$$\delta(\operatorname{Tr} A_r) = \varphi \operatorname{Tr}(A_r A_{n+1}) + 2 \sum_{i,j} g^{ij} \varphi_i \ell_{n+1j}^r$$
$$+ \varphi \sum_{i,j} g^{ji} \ell_{n+1i;j}^r + \varphi \sum_{i,j,s} g^{ji} \ell_{n+1i}^s \ell_{sj}^r - \sum_{s \neq r} (\operatorname{Tr} A_s) \langle e_r, \delta e_s \rangle, \tag{5.107}$$

where

$$\ell_{si;j}^r = \frac{\partial \ell_{si}^r}{\partial u_j} - \sum_k \ell_{sk}^r \Gamma_{ij}^k.$$

From (5.106) and (5.107) we get

$$\delta H^2 = \frac{2}{n^2} \Bigg\{ (\operatorname{Tr} A_{n+1}) \left(\varphi\|A_{n+1}\|^2 - \varphi\ell^2 - \Delta\varphi \right)$$
$$+ \sum_{r=n+2}^{m} (\operatorname{Tr} A_r) \Big[\varphi \operatorname{Tr}(A_r A_{n+1}) + 2 \langle d\varphi, \omega_{n+1}^r \rangle \tag{5.108}$$
$$+ \varphi \sum_{ij} g^{ji} \ell_{n+1i;j}^r + \varphi \sum_{ij} g^{ji} \ell_{n+1i}^s \ell_{sj}^r \Big] \Bigg\},$$

where $De_s = \sum_r \omega_s^r e_r$.

If the normal variation (5.83) is an H-variation, then $\operatorname{Tr} A_r = 0$ for $r = n + 2, \ldots, m$. Thus (5.108) reduces to

$$\delta\alpha = \frac{1}{n}\{-\Delta\varphi - \varphi\,\ell^2 + \varphi\|A_{n+1}\|^2\} \tag{5.109}$$

with $\alpha = \|H\|$. Hence, by applying (5.84), (5.92) and (5.109), we have

$$\delta \int_M \alpha^c dV$$
$$= \int_M \left\{ \frac{c}{n}\alpha^{c-1}\left(\varphi\|A_{n+1}\|^2 - \Delta\varphi - \varphi\ell^2\right) - n\varphi\alpha^{c-1} \right\} dV \tag{5.110}$$

for $c \geq 0$. If we integrate by parts to get rid of the derivative of φ, we obtain

$$\int_M (\alpha^{c-1}\Delta\varphi)dV = \int_M \varphi(\Delta\alpha^{c-1})dV, \tag{5.111}$$

where the boundary terms one would expect vanishes after integration by parts because of our hypothesis on φ on ∂M.

By combining (5.110) and (5.111) we find

$$\delta \int_M \alpha^c dV$$

$$= \frac{c}{n} \int_M \varphi \left\{ \Delta\alpha^{c-1} - \alpha^{c-1}\ell^2 - \frac{n^2}{c}\alpha^{c+1} + \alpha^{c-1}||A_{n+1}||^2 \right\} dV.$$

From this we conclude that $\delta \int_M \alpha^c dV = 0$ for all H-variations of M if and only if

$$c\Delta\alpha^{c-1} - c\alpha^{c-1}\ell^2 - n^2\alpha^{c+1} + c\alpha^{c-1}||A_{n+1}||^2 = 0. \qquad (5.112)$$

In particular, if $c = n$, this gives us the following.

Theorem 5.15. [Chen and Houh (1975)] *If M is an n-dimensional compact submanifold of $\mathbb{E}^m$, then M is H-stationary if and only if the mean curvature α of M satisfies*

$$\Delta\alpha^{n-1} + \alpha^{n-1}\left\{\ell^2 + n\alpha^2 - ||A_{n+1}||^2\right\} = 0, \qquad (5.113)$$

where A_{n+1} is the Weingarten map with respect to the unit vector in the direction of mean curvature vector.

From this theorem we obtain immediately the following.

Corollary 5.7. *Let M be an n-dimensional compact submanifold in $\mathbb{E}^m$. If M is stationary, then we have*

$$\Delta\alpha^{n-1} + \alpha^{n-1}\left\{\ell^2 + n\alpha^2 - ||A_{n+1}||^2\right\} = 0,$$

Corollary 5.8. *Let M be an n-dimensional compact hypersurface in $\mathbb{E}^{n+1}$. Then M is stationary if only if*

$$\Delta\alpha^{n-1} + \alpha^{n-1}\left\{n(n-1)\alpha^2 - 2\tau\right\} = 0,$$

where τ is the scalar curvature of M.

Remark 5.2. When M is a surface in $\mathbb{E}^3$, Corollary 5.8 is already known to Thomsen in [Thomsen (1923)].

By applying Theorem 5.15 we may obtain the following.

Theorem 5.16. [Chen and Houh (1975)] *Let M be a compact submanifold of $\mathbb{E}^m$ with parallel mean curvature vector. Then M is stationary (or H-stationary) if and only if M is either a minimal submanifold of $\mathbb{E}^m$ or a minimal submanifold of a hypersphere of $\mathbb{E}^m$.*

Proof. Let M be a submanifold of $\mathbb{E}^m$ with parallel mean curvature vector. Then M has constant mean curvature α. Suppose that M is stationary or H-stationary. If $\alpha = 0$, M is a minimal submanifold of $\mathbb{E}^m$.

Next, assume that $\alpha \neq 0$. Then by $DH = 0$, we get $\ell = 0$. Thus it follows from (5.113) that

$$0 = ||A_{n+1}||^2 - n\alpha^2 = \frac{1}{n}\sum_{i<j}(\kappa_i - \kappa_j)^2,$$

where $\kappa_1, \ldots, \kappa_n$ are the eigenvalues of A_{n+1}. From these we know that M is pseudo-umbilical in $\mathbb{E}^m$, Hence, by applying Proposition 4.16, we conclude that M is a minimal submanifold of a hypersphere of $\mathbb{E}^m$.

The converse is trivial. $\square$

Theorem 5.17. [Chen and Houh (1975)] *The only closed pseudo-umbilical stationary (or H-stationary) submanifolds of $\mathbb{E}^m$ are minimal submanifolds of a hypersphere of $\mathbb{E}^m$.*

Proof. Since M is assumed to be pseudo-umbilical in $\mathbb{E}^m$, we have either $\alpha = 0$ or $n\alpha^2 = ||A_{n+1}||^2$. Thus Theorem 5.15 implies that

$$\Delta\alpha^{n-1} - \alpha^{n-1}\ell^2 = 0. \tag{5.114}$$

Hence we find

$$-\Delta\alpha^{2(n-1)} = 2\alpha^{2(n-1)}\ell^2 + ||d\alpha^{n-1}||^2 = 0.$$

Therefore, by applying the Divergence Theorem, we conclude that α is a constant which is non-zero. Hence (5.114) implies that $\ell^2 = 0$. Therefore the mean curvature vector of M is parallel in the normal bundle. Consequently, by applying Proposition 4.16, we conclude that M is a minimal submanifold of a hypersphere of $\mathbb{E}^m$. $\square$

Theorem 5.18. [Chen (1972b)] *The only closed stationary hypersurfaces of odd-dimensional in $\mathbb{E}^{n+1}$ are the ordinary hyperspheres.*

Proof. Let M be an odd-dimensional closed stationary hypersurface of $\mathbb{E}^{n+1}$. Then Corollary 5.8 implies

$$\Delta\alpha^{n-1} + \alpha^{n-1}\{n(n-1)\alpha^2 - 2\tau\} = 0.$$

Let $\kappa_1, \ldots, \kappa_n$ be the principal curvatures of M in $\mathbb{E}^{n+1}$. Then we have

$$\alpha = \frac{1}{n}(\kappa_1 + \cdots + \kappa_n), \quad \tau = \sum_{i<j}\kappa_i\kappa_j.$$

Therefore $\alpha^2 \geq 2\tau/(n(n-1))$. Hence we find

$$\Delta\alpha^{n-1} = \alpha^{n-1}(2\tau - n(n-1)\alpha^2) \leq 0.$$

Consequently, by applying Hopf's lemma, we obtain $\Delta\alpha^{n-1} = 0$, which implies that $2\tau = \alpha^2$. Therefore M is totally umbilical in $\mathbb{E}^{n+1}$. So, M is a hypersphere of $\mathbb{E}^{n+1}$. □

Theorem 5.19. [Chen (1972b)] *If n is even, the only closed stationary hypersurfaces of $\mathbb{E}^{n+1}$ whose mean curvature does not change sign are the ordinary hyperspheres.*

Proof. This can be proved in a similar way as Theorem 5.18. □

Remark 5.3. It follows from Theorem 5.16 that minimal submanifolds of a hypersphere of $\mathbb{E}^m$ are H-stationary in $\mathbb{E}^m$. However, there exist many H-stationary or stationary compact submanifolds which are not of this type. For instance, consider the anchor ring in $\mathbb{E}^3$ defined by

$$x_1 = (a + b\cos u)\cos v,$$
$$x_2 = (a + b\cos u)\sin v,$$
$$x_3 = b\sin u.$$

By direct computation, we find

$$\Delta\alpha + 2\alpha(\alpha^2 - G) = \frac{a(a^2 - 2b^2)}{4b^3 r^3}.$$

This shows that the anchor ring is stationary if and only if $a = \sqrt{2}b$.

Remark 5.4. J. Weiner defined a surface M in a 3-dimensional Riemannian manifold $\tilde{M}$ to be stationary if

$$\delta \int_M (H^2 + R')dV = 0$$

for any variation of M in $\tilde{M}$ [Weiner (1978)]. Since $\int_M (H^2 + R')dV$ is a conformal invariant, the equation

$$\Delta\alpha^2 + \alpha(2\alpha^2 - ||A_3||^2) - 0$$

itself is invariant under conformal transformations. In particular, if σ denotes the stereographic projection from $S^3(1)$ onto $\mathbb{E}^3$, then an immersion $\phi : M \to S^3(1)$ is stationary if and only if the composition $\sigma \circ \phi : M \to \mathbb{E}^3$ is stationary. Thus, by using Lawson's examples of minimal closed surfaces in S^3, we obtain closed stationary surfaces in $\mathbb{E}^3$ of arbitrary genus (cf.

Corollary 1 of [Weiner (1978)]). In fact, the stationary anchor ring given in Remark 5.3 is one of the examples obtained from stereographic projection in which the minimal surface in $S^3(1)$ is the square torus (also known as the Clifford torus).

Remark 5.5. Surfaces in $\mathbb{E}^3$ which satisfy the Euler-Lagrange equation (5.4) are sometime called *Willmore surfaces*. The stereographic projections of minimal surfaces of S^3 are Willmore surfaces due to conformal property of $w(\phi)$.

There are many other Willmore surfaces of different type. For instance, R. L. Bryant found and classified in [Bryant (1984)] immersed Willmore spheres in $\mathbb{E}^3$ and U. Pinkall constructed in [Pinkall (1985)] infinitely many embedded Willmore tori in $\mathbb{E}^3$. The existence of surfaces that minimizes the Willmore energy was established in [Simon (1993)] and later extended in [Bauer and Kuwert (2003)].

Remark 5.6. Let M be a surface in a Riemannian 3-manifold $(\tilde{M}, \tilde{g})$. Denote by Δ the Laplacian on M with respect to the induced metric g. If $\tilde{g}^* = \rho^2 \tilde{g}$ is a conformal change of metric on $\tilde{M}$.

Denote by g^* the metric on M induced from $\tilde{g}^*$. Then the Laplacian Δ^* of (M, g^*) satisfies

$$\Delta^* = \rho^{-2}\Delta. \tag{5.115}$$

Define an operator $\Box$ on M by

$$\Box = \Delta + (2\alpha^2 - ||h||^2)I. \tag{5.116}$$

Then M is stationary in $\tilde{M}$ if and only if $\Box\alpha = 0$.

From (5.39) and (5.116) we find

$$\Box^* = \rho^{-2}\Box, \tag{5.117}$$

where $\Box^*$ denotes the corresponding operator on M with respect to g^*.

5.7 Surfaces in $\mathbb{E}^m$ which are conformally equivalent to flat surfaces

In this section, we shall improve inequality (5.47) for an important family of surfaces in a Euclidean m-space $\mathbb{E}^m$ with arbitrary m. According to Nash's imbedding theorem, every Riemannian surface can be isometrically imbedded in $\mathbb{E}^{51}$.

Definition 5.5. A surface M in $\mathbb{E}^m$ is called *conformally equivalent to a flat surface* if it is the image of a flat surface under a conformal mapping of $\mathbb{E}^m$, i.e., M is equivalent to a flat surface up to conformal mappings or diffeomorphisms of $\mathbb{E}^m$. Such a surface is simply called a *C-surface* in [Chen (1981e)].

Theorem 5.20. [Chen (1973a, 1981e)] *If M is a closed surface in $\mathbb{E}^m$ which is conformally equivalent to a flat closed surface in $\mathbb{E}^m$, then we have*

$$\int_M H^2 dV \geq 2\pi^2. \tag{5.118}$$

The equality sign of (5.118) holds if and only if M is conformal Clifford torus, i.e., it is conformally equivalent to a square torus in a totally geodesic $\mathbb{E}^4 \subset \mathbb{E}^m$.

Proof. Since the total mean curvature of a closed surface in $\mathbb{E}^m$ is a conformal invariant according to Corollary 5.1, it suffices to prove the theorem only for closed flat surfaces in $\mathbb{E}^m$. For each $p \in M$, we denote by A the shape operator:

$$A : T_p^\perp(M) \to \text{End}(T_p(M), T_p(M)) \tag{5.119}$$

by $A\xi = A_\xi$. Let O_p denote the kernel of A. Then we have $\dim O_p \geq m - 5$. Let N_p be the subspace of the normal space $T_p^\perp(M)$ at p given by

$$T_p^\perp(M) = N_p \oplus O_p, \quad N_p \perp O_p.$$

Then we have $A(e) = 0$ for any $e \in O_p$. We choose an orthonormal normal frame $e_3, \ldots, e_m$ at p in such a way that $e_6, \ldots, e_m \in O_p$. Then for each unit normal vector e at p we have

$$e = \sum_{r=3}^m \cos \theta_r e_r. \tag{5.120}$$

Thus the Lipschitz-Killing curvature at (p, e) is given by

$$K(p, e) = \left(\sum_{r=3}^5 \cos \theta_r h_{11}^r \right) \left(\sum_{s=3}^5 \cos \theta_s h_{22}^s \right) - \left(\sum_{t=3}^5 \cos \theta_t h_{12}^t \right)^2. \tag{5.121}$$

The right-hand side of (5.121) is a quadratic form on $\cos \theta_r$. Hence, after choosing suitable e_3, e_4, e_5 at p, we have

$$K(p, e) = \lambda_1(p) \cos^2 \theta_3 + \lambda_2(p) \cos^2 \theta_4 + \lambda_3(p) \cos^2 \theta_5 \tag{5.122}$$

with $\lambda_1 \geq \lambda_2 \geq \lambda_3$. Moreover, since M is a flat surface, we also have

$$\lambda_1 + \lambda_2 + \lambda_3 = 0, \quad \lambda_A = \det(A_{2+A}). \tag{5.123}$$

In particular, we have $\lambda_1 \geq 0$ and $\lambda_3 \leq 0$.

We consider the cases $\lambda_2 \geq 0$ and $\lambda_2 < 0$, separately.

Case 1: $\lambda_2 \geq 0$. From (5.123) we have

$$K(p, e) = \lambda_1(\cos^2 \theta_3 - \cos^2 \theta_5) + \lambda_2(\cos^2 \theta_4 - \cos^2 \theta_5). \qquad (5.124)$$

Hence

$$\int_{S_p} |K(p, e)| d\sigma$$

$$= \int_{S_p} |\lambda_1(\cos^2 \theta_3 - \cos^2 \theta_5) + \lambda_2(\cos^2 \theta_4 - \cos^2 \theta_5)| d\sigma \qquad (5.125)$$

$$\leq \lambda_1 \int_{S_p} |\cos^2 \theta_3 - \cos^2 \theta_5| d\sigma + \lambda_2 \int_{S_p} |\cos^2 \theta_4 - \cos^2 \theta_5| d\sigma.$$

On the other hand, by a formula on spherical integration, we have

$$\int_{S_p} |\cos^2 \theta_r - \cos^2 \theta_s| d\sigma = \frac{2c_{m-1}}{\pi^2}, \quad r \neq s. \qquad (5.126)$$

Hence, after applying (5.125) and (5.126), we obtain

$$\int_{S_p} |K(p, e)| d\sigma \leq \frac{2c_{m-1}}{\pi^2}(\lambda_1(p) + \lambda_2(p)). \qquad (5.127)$$

From the definition of the squared mean curvature we have

$$\begin{aligned}
4H^2 &= (h_{11}^3 + h_{22}^3)^2 + (h_{11}^4 + h_{22}^4)^2 + (h_{11}^5 + h_{22}^5)^2 \\
&\geq (h_{11}^3)^2 + (h_{22}^3)^2 + 2\lambda_1 + 2(h_{12}^3)^2 + (h_{11}^4)^2 \\
&\quad + (h_{22}^4)^2 + 2\lambda_2 + 2(h_{12}^4)^2 \\
&\geq 4(\lambda_1 + \lambda_2).
\end{aligned} \qquad (5.128)$$

By combining (5.127) and (5.128) we obtain

$$H^2 \geq \frac{\pi^2 G^*(p)}{2c_{m-1}}, \qquad (5.129)$$

where $G^*(p) = \int_{S_p} |K(p, e)| d\sigma$.

Case 2: $\lambda_2 < 0$. From (5.123) we find

$$K(p, e) = \lambda_2(\cos^2 \theta_4 - \cos^2 \theta_3) + \lambda_3(\cos^2 \theta_5 - \cos^2 \theta_3). \qquad (5.130)$$

Thus it follows from (5.123) and (5.126) that

$$\begin{aligned}
S^*(p) &\leq -\lambda_2 \int_{S_p} |\cos^2 \theta_4 - \cos^2 \theta_3| d\sigma \\
&\quad - \lambda_3 \int_{S_p} |\cos^2 \theta_5 - \cos^2 \theta_3| d\sigma \\
&= 2\lambda_1 \frac{c_{m-1}}{\pi^2}.
\end{aligned} \qquad (5.131)$$

On the other hand, we also have

$$4H^2 \geq (h_{11}^3)^2 + (h_{22}^3)^2 + 2\lambda_1 + 2(h_{12}^3)^2$$
$$\geq 4\lambda_1 + 4(h_{12}^3)^2 \tag{5.132}$$
$$\geq 4\lambda_1.$$

Therefore we also obtain

$$H^2 \geq \frac{\pi^2 G^*(p)}{2c_{m-1}}. \tag{5.133}$$

Consequently, we have inequality (5.129) in general. Therefore, by taking integration of both sides of (5.129), we find

$$\int_M H^2 dV \geq \frac{\pi^2}{2} b(M), \tag{5.134}$$

by virtue of Theorem 5.2. Now, since M is a closed flat surface, M is either diffeomorphic to a 2-torus or to a Klein bottle. In both cases we have $b(M) = 4$. Thus (5.134) gives

$$\int_M H^2 dV \geq 2\pi^2. \tag{5.135}$$

Now, let us suppose that the equality sign of (5.135) holds, then the inequalities in (5.125) and (5.131) become equalities. Hence at least one of λ_1, λ_2 is zero for the first case and at least one of λ_2, λ_3 is zero for the second case. However, this implies that the second case cannot occur. Thus we find $\lambda_2 = 0$ identically on M. Furthermore, since the equality sign of (5.128) holds, we have

$$h_{11}^3 = h_{22}^3, \quad h_{11}^4 = h_{22}^4,$$
$$h_{11}^5 + h_{22}^5 = 0, \quad h_{12}^3 = h_{12}^4 = 0.$$

Because $\lambda_2 = 0$, these imply

$$h_{11}^3 = h_{22}^3, \quad h_{11}^5 + h_{22}^5 = 0, \quad h_{12}^3 = h_{ij}^4 = 0.$$

Hence, by choosing some suitable orthonormal frame $e_1, e_2, e_3, \ldots, e_m$, we obtain

$$A_3 = \begin{pmatrix} a & 0 \\ 0 & a \end{pmatrix}, \quad A_4 = \begin{pmatrix} a & 0 \\ 0 & -a \end{pmatrix},$$
$$A_5 = \cdots = A_m = 0. \tag{5.136}$$

So, the normal connection is flat. Therefore there exist locally orthonormal normal frame $\bar{e}_3, \ldots, \bar{e}_m$ such that (cf. [Chen (1973b), page 99])

$$D\bar{e}_3 = \cdots = D\bar{e}_m = 0. \tag{5.137}$$

Let us put

$$\bar{e}_r = \sum_{s=3}^{m} a_{rs}e_s, \quad r = 3, \ldots, m. \tag{5.138}$$

Then (a_{rs}) is an orthonormal $(m-2) \times (m-2)$-matrix. Since M is assumed to be a flat surface and our study is local. we may assume that M is covered by a coordinate system (x, y) such that the metric tensor of M has the form $g = dx^2 + dy^2$. Let X_1 and X_2 denote the coordinate vector fields $\partial/\partial x$ and $\partial/\partial y$, respectively.

We define a function $\theta = \theta(x, y)$ by

$$\begin{aligned} X_1 &= \cos\theta e_1 + \sin\theta e_2, \\ X_2 &= -\sin\theta e_1 + \cos\theta e_2. \end{aligned} \tag{5.139}$$

With respect to the frame field $X_1, X_2, \bar{e}_3, \ldots, \bar{e}_m$, the Weingarten map satisfies

$$A_{\bar{e}_r} = \begin{pmatrix} a(a_{r}1 + a_{r2}\cos 2\theta) & -aa_{r2}\sin 2\theta \\ -aa_{r2}\sin 2\theta & a(a_{r1} - a_{r2}\cos 2\theta) \end{pmatrix} \tag{5.140}$$

for $r = 3, \ldots, m$.

The Codazzi equation is given by

$$\begin{aligned} \frac{\partial}{\partial y}(a(a_{r1} + a_{r2}\cos 2\theta)) &= -\frac{\partial}{\partial x}aa_{r2}\sin 2\theta, \\ \frac{\partial}{\partial y}(aa_{r2}\sin 2\theta) &= -\frac{\partial}{\partial x}(a(a_{r1} - a_{r2}\cos 2\theta)). \end{aligned} \tag{5.141}$$

By multiplying a_{r1} to the first equation in (5.141) and summing over r and by applying the fact that $(a_{rs}) \in O(m-2)$, we get

$$\frac{\partial \ln a}{\partial y} = \sum_r \left\{ \left(\frac{\partial a_{r1}}{\partial y}\right) a_{r2}\cos 2\theta + \left(\frac{\partial a_{r1}}{\partial x}\right) a_{r2}\sin 2\theta \right\}. \tag{5.142}$$

Similarly, by multiplying a_{r1} to the second equation in (5.141) and by summing over r and by applying the fact that $(a_{rs}) \in O(m-2)$, we get

$$\frac{\partial \ln a}{\partial x} = \sum_r \left\{ \left(\frac{\partial a_{r1}}{\partial y}\right) a_{r2}\sin 2\theta - \left(\frac{\partial a_{r1}}{\partial x}\right) a_{r2}\cos 2\theta \right\}. \tag{5.143}$$

By multiplying a_{r2} to the first equation in (5.141) and by summing over r we find

$$\begin{aligned} \sum_r a_{r2}\frac{\partial a_{r1}}{\partial y} + \frac{\partial \ln a}{\partial x}\sin 2\theta &+ \frac{\partial \ln a}{\partial y}\cos 2\theta \\ &= 2\frac{\partial \theta}{\partial y}\sin 2\theta - 2\frac{\partial \theta}{\partial x}\cos 2\theta. \end{aligned}$$

After substituting (5.141) and (5.142) into this equation, we obtain

$$\sum_r a_{r2} \frac{\partial a_{r1}}{\partial y} = \sin 2\theta \frac{\partial \theta}{\partial y} - \cos 2\theta \frac{\partial \theta}{\partial x}. \tag{5.144}$$

Similarly, by multiplying a_{r2} to the second equation in (5.141) and by summing over r and using (5.141) and (5.142), we get

$$\sum_r a_{r2} \frac{\partial a_{r1}}{\partial x} = -\cos 2\theta \frac{\partial \theta}{\partial y} - \sin 2\theta \frac{\partial \theta}{\partial x}. \tag{5.145}$$

Substituting (5.144) and (5.145) into (5.142) and (5.143), we find

$$\frac{\partial \ln a}{\partial x} = \frac{\partial \theta}{\partial y}, \quad \frac{\partial \ln a}{\partial y} = -\frac{\partial \theta}{\partial x}.$$

From this we get

$$\left(\frac{\partial^2}{\partial x^2} + \frac{\partial^2}{\partial y^2} \right) (\ln a) = 0. \tag{5.146}$$

Therefore we have $\Delta(\ln a) = \Delta(\ln H^2) = 0$. Because H^2 is a non-negative differentiable function on a closed surface, it follows that H^2 is a positive constant.

Let us put

$$e_3' = \cos \theta e_3 + \sin \theta e_4, \quad e_4' = \sin \theta e_3 - \cos \theta e_4,$$
$$e_5' = e_5, \dots, e_m' = e_m. \tag{5.147}$$

Then with respect to $e_1, e_2, e_3', \dots, e_m'$ we have

$$A_{e_3'} = \begin{pmatrix} \sqrt{2}a & 0 \\ 0 & 0 \end{pmatrix}, \quad A_{e_4'} = \begin{pmatrix} 0 & 0 \\ 0 & \sqrt{2}a \end{pmatrix}, \quad A_{e_5'} = \cdots = A_{e_m'} = 0. \tag{5.148}$$

It follows from (5.148) and Cartan's structure equations that distributions $T_i = \text{Span}\{\mathbb{R}e_i\}$, $i = 1, 2$, are parallel. Therefore, it follows from the de Rham decomposition theorem that $M = C_1 \times C_2$, where C_i is the maximal integral curve of T_i. Moreover, because $h(e_1, e_2) = 0$, a result of [Moore (1971)] implies that M is a product product submanifold where C_1 lies in a linear k-subspace and C_2 lies in a linear $(m - k)$-subspace of $\mathbb{E}^m$. Hence, by a result of [Kuiper (1958)], the total absolute curvature of M in $\mathbb{E}^m$ is the product of the total absolute curvatures of the curves C_1 and C_2. Because the total absolute curvature of M in $\mathbb{E}^m$ is equal to 4 by (5.133), the total absolute curvatures of C_1 and C_2 are both equal to 2. Consequently, Fenchel-Borsuk's result implies that C_1 and C_2 are planar curves with constant curvature $\sqrt{2}a$ by (5.148). Therefore C_1 and C_2 are circles of the same radius. Consequently, M is a square torus lying fully in a totally geodesic $\mathbb{E}^4 \subset \mathbb{E}^m$.

The converse is clear. $\qquad \square$

From Theorem 5.19 we obtain immediately the following.

Corollary 5.9. *If M is a closed surface in $\mathbb{E}^m$ which is conformally equivalent to a flat Klein bottle, then we have*

$$\int_M H^2 dV > 2\pi^2.$$

In views of Corollary (5.9), it is natural to ask the following.

Problem 5.3. *Let B be a Klein bottle. What is the $\inf_\phi \int_B H_\phi^2(B) dV$ for all immersions $\phi : B \to \mathbb{E}^m$?*

5.8   Total mean curvatures for surfaces in $\mathbb{E}^4$

Let M be a surface in $\mathbb{E}^m$. Choose an orthonormal normal frame $e_3, \ldots, e_m$ of M in $\mathbb{E}^m$. For a unit normal vector e at p, we put

$$e = \sum_{r=3}^{m} \cos \theta_r e_r.$$

Thus we have $A_e = \sum_{r=3}^m \cos \theta_r A_r$. Hence we find

$$K(p, e) = \det(A_e) = \det \left(\sum_r \cos \theta_r A_r \right). \tag{5.149}$$

Since the right-hand side of (5.149) is a quadratic form of $\cos \theta_3, \ldots, \cos \theta_m$, we may choose a suitable local orthonormal normal frame $\bar{e}_3, \ldots, \bar{e}_m$ such that with respect to this frame we have

$$K(p, e) = \sum_{r=3}^{m} \lambda_{r-2} \cos^2 \theta_r A_r, \quad \lambda_1 \geq \lambda_2 \geq \cdots \geq \lambda_{m-2}. \tag{5.150}$$

This special frame is known as *Otsuki's frame*. We call λ_A the A-th curvature of the surface M in $\mathbb{E}^m$ for $A = 1, \ldots, m - 2$. When $m = 4$, we simply denote λ_1 by λ_1 and denote λ_2 by μ.

For a surface in $\mathbb{E}^4$ we have

$$K(p, e) = \lambda(p) \cos^2 \theta + \mu(p) \sin^2 \theta, \tag{5.151}$$

with $\lambda(p) \geq \mu(p)$.

We need the following lemmas.

Lemma 5.1. *For s surface M in $\mathbb{E}^4$, we have $\mu \leq 0$ everywhere on M.*

Proof. Let e be a unit normal vector at $p \in M$ which is perpendicular to the mean curvature vector H. The we have $K(p, e) \leq 0$. Thus in view of (5.151) we find $\mu \leq 0$. $\square$

Lemma 5.2. *If M is a closed surface in $\mathbb{E}^4$ with $\mu = 0$, then M is homeomorphic to a 2-sphere.*

Proof. Under the hypothesis, it follows from (5.151) that

$$\int_0^{2\pi} K(p, e)d\theta = \int_0^{2\pi} (\lambda \cos^2 \theta + \mu \sin^2 \theta)d\theta$$

$$= (\lambda + \mu) \int_0^{2\pi} \sin^2 \theta = \pi G, \quad (5.152)$$

where we use the identity $G = \lambda + \mu$. Since $\mu = 0$, we see that the total absolute curvature $G^*(p)$ at p satisfies

$$G^*(p) = \int_0^{2\pi} |K(p, e)|d\theta = \int_0^{2\pi} K(p, e)d\theta = \pi G(p). \quad (5.153)$$

On the other hand, it follows from the inequality of Chern-Lashof that

$$\int G^*(p)dV \geq 2b(M)\pi^2, \quad (5.154)$$

where $b(M)$ is the total Betti number of M. Therefore, it follows from Gauss-Bonnet's formula, (5.153), and (5.154) that $b(M) = \chi(M)$, where $\chi(M)$ is the Euler characteristic of M. Hence we conclude that M has no torsion and $\chi(M) = 2$. Thus M is homeomorphic to a 2-sphere. $\square$

Lemma 5.3. *If M is a closed surface in $\mathbb{E}^4$, then $U = \{p \in M : \lambda(p) > 0\}$ is a nonempty open subset.*

Proof. If $\lambda \leq 0$ everywhere, we have

$$G^*(p) = -\int_0^{2\pi} K(p, e)d\theta = -\pi G. \quad (5.155)$$

Thus it follows from Gauss-Bonnet's formula, (5.154) and (5.155) that $b(M) \leq -\chi(M)$ which is impossible. $\square$

Lemma 5.4. *Let M be a surface in $\mathbb{E}^4$. If $\lambda(p) \geq 0$, then we have*

$$H^2(p) \geq \lambda(p), \quad (5.156)$$

with equality sign holding if and only if M is pseudo-umbilical at p.

Proof. We choose the local frame fields as Frenet frames. Then we have

$$\lambda = h_{11}^3 h_{22}^3 - (h_{12}^3)^2, \quad \mu = h_{11}^4 h_{22}^4 - (h_{12}^4)^2. \tag{5.157}$$

Thus we get

$$
\begin{aligned}
4H^2 &= (h_{11}^3 + h_{22}^3)^2 + (h_{11}^4 + h_{22}^4)^2 \\
&= \sum_{r=3}^{4} \{(h_{11}^r)^2 + (h_{22}^r)^2 + 2(h_{12}^r)^2\} + 2G \\
&\geq 2h_{11}^3 h_{22}^3 + 2|h_{11}^4 h_{22}^4| + 2(h_{12}^3)^2 + 2(h_{12}^4)^2 + 2G \\
&\geq 2\lambda - 2\mu + 2G = 4\lambda,
\end{aligned}
\tag{5.158}
$$

which proves (5.156).

If the equality sign of (5.156) holds, then the inequalities in (5.158) are actuarially equalities. Hence we obtain

$$h_{11}^3 = h_{22}^3, \quad h_{12}^3 = 0, \quad h_{11}^4 = -h_{22}^4$$

at p. Thus M is pseudo-umbilical at p.

The converse is easy to verify. □

Next, we improve inequality (5.47) for some family of closed surfaces in the Euclidean 4-space.

Theorem 5.21. [Chen (1973a)] *If M is a closed surface with non-positive Gauss curvature in $\mathbb{E}^4$, then we have*

$$\int_M H^2 dV \geq 2\pi^2. \tag{5.159}$$

If $H^2 > 0$, then the equality sign of (5.159) holds if and only if M is square torus in $\mathbb{E}^4$.

Proof. Let M be a closed surface in $\mathbb{E}^4$. Denote by B_1 the bundle of unit normal vectors of M in $\mathbb{E}^4$ and by $\pi : B_1 \to M$ the natural projection of B_1 onto M.

Let us put

$$W = \{p \in M : \lambda(p) \geq 0.$$

Then by the hypothesis we have

$$
\begin{aligned}
|K(p,e)| &= |\lambda \cos^2 \theta + \mu \sin^2 \theta| \\
&= |\lambda \cos 2\theta + G \sin^2 \theta| \\
&\leq \lambda |\cos 2\theta| - G \sin^2 \theta
\end{aligned}
\tag{5.160}
$$

on $\pi^{-1}(W)$, where $e = \cos\theta e_3 + \sin\theta e_4$ and e_3, e_4 is an Otsuki frame of M in $\mathbb{E}^4$. Thus we find

$$\int_{\pi^{-1}(W)} |K(p,e)|dV \wedge d\sigma \leq 4\int_W \lambda dV - \pi\int_W GdV. \tag{5.161}$$

On $B_1 \setminus \pi^{-1}(W)$ we have

$$\int_{B_1\setminus\pi^{-1}(W)} |K(p,e)|dV \wedge d\sigma = -\int_{B_1\setminus\pi^{-1}(W)} K(p,e)dV \wedge d\sigma$$
$$= -\pi\int_{M\setminus W} G(p)dV. \tag{5.162}$$

On the other hand, by definition we have

$$\begin{aligned}
4H^2 &= \sum_{r=3}^{m}(\operatorname{Tr} A_r)^2 = 2G + \sum_{r=3}^{4}||A_r||^2 \\
&\geq 2(\lambda+\mu) + 2(\lambda-\mu) \\
&= 4\lambda
\end{aligned} \tag{5.163}$$

on W. Thus we get

$$\begin{aligned}
\int_M H^2 dV &\geq \int_W \lambda dV \\
&\geq \frac{1}{4}\int_{\pi^{-1}(W)} |K(p,e)|dV \wedge d\sigma + \frac{\pi}{4}\int_W GdV \\
&\quad + \frac{1}{4}\int_{B_1\setminus\pi^{-1}(W)} |K(p,e)|dV \wedge d\sigma + \frac{\pi}{4}\int_{M\setminus W} GdV \\
&= \frac{1}{4}\int_{B_1} |K(p,e)|dV \wedge d\sigma + \frac{\pi}{4}\int_M GdV \\
&\geq \frac{\pi^2}{2}(b(M) + \chi(M)) = 2\pi^2.
\end{aligned} \tag{5.164}$$

This proves inequality (5.159). If the equality sign of (5.159) holds, then all the inequalities in (5.160)-(5.164) become equalities. If $H^2 > 0$, then (5.164) implies $W = M$. Therefore, it follows from (5.163) that M is pseudo–umbilical in $\mathbb{E}^4$. Also, from (5.160) we find

$$|\lambda\cos 2\theta + G\sin^2\theta| = \lambda|\cos 2\theta| - G\sin^2\theta$$

for all θ. Thus $G = 0$, i.e., M is a flat surface. Consequently, it follows from the proof of Theorem 5.20 that M is a square torus in $\mathbb{E}^4$.

The converse is obvious. $\qquad\square$

Theorem 5.22. [Chen (1973a)] *Let M be a closed surface in $\mathbb{E}^4$ with non-negative Gauss curvature. If*

$$\int_M H^2 dV \le (2+\pi)\pi, \tag{5.165}$$

then M is homeomorphic to a 2-sphere.

Proof. Let M be a closed surface with non-negative Gauss curvature in $\mathbb{E}^4$. Then it follows from (5.151) that

$$|K(p,e)| = |G(p)\cos^2\theta - \mu(p)\cos 2\theta|$$
$$\le G(p)\cos^2\theta - \mu(p)|\cos 2\theta|.$$

By integrating both sides of this from 0 to 2π, we get

$$\lambda(p) \ge \frac{1}{4}G^*(p) + \left(1 - \frac{\pi}{4}\right)G(p). \tag{5.166}$$

$$-\mu(p) \ge \frac{1}{4}G^*(p) - \frac{\pi}{4}G(p). \tag{5.167}$$

It is easy to see that the equality sign of (5.166) and (5.167) hold only when $\mu = 0$.

We find from (5.166), (5.167) and Lemma 5.4 that

$$\int_M H^2 dV \ge \frac{1}{4}\int_M G^* dV + \left(1 - \frac{\pi}{4}\right)\int_M G dV. \tag{5.168}$$

Now, let us assume that M is a real projective plane immersed in $\mathbb{E}^4$. Then we have $b(M) = 3$. Thus it follows from (5.168) and Chern-Lashof's inequality that

$$\int_M H^2 dV \ge (2+\pi)\pi. \tag{5.169}$$

By combining (5.165) and (5.169) we obtain the equality case of inequality (5.169). Hence both the quality signs of (5.166) and (5.167) hold. Therefore we have $\mu = 0$. Thus M must be homeomorphic to a 2-sphere according to Lemma 5.2. But this is a contradiction. Consequently, M cannot be a real projective plane.

On the other hand, if (5.165) holds, the Gauss curvature of M must be positive somewhere on M according to Theorem 5.21. Therefore M is not homeomorphic either to a torus or to a Klein bottle. Consequently, M is homeomorphic to a 2-sphere. $\qquad\square$

Definition 5.6. Let $\phi : M \to \mathbb{E}^{2n}$ be an immersion of an oriented closed n-manifold into $\mathbb{E}^{2n}$. By applying a regular deformation of ϕ if necessary, $\phi(M)$ intersects itself transversally. Thus $\phi(M)$ intersects itself at isolated points. At each point p of self-intersection, we assign $+1$ if the direct sum orientation of the two complementary tangent planes equals the given orientation on $\mathbb{E}^{2n}$, and we assign -1 otherwise. The *self-intersection number* I_ϕ of ϕ is defined to be the sum of local contributions from all the points of self-intersections.

It is well-known that the self-intersection number I_ϕ is an immersion invariant up to regular homotopy of M in $\mathbb{E}^{2n}$.

We mention the following result for later use.

Theorem 5.23. [Smale (1959)] *Two immersions from the 2-sphere S^2 into $\mathbb{E}^4$ are regular homotopic if and only if they have the same self-intersection number.*

This theorem says that the self-intersection number is the only regular homotopic invariant of S^2 in $\mathbb{E}^4$.

For surfaces in $\mathbb{E}^4$ we also have the following result.

Theorem 5.24. [Wintgen (1979)] *Let $\phi : M \to \mathbb{E}^4$ be an immersion of an oriented closed surface in $\mathbb{E}^4$. Then we have*

$$\int_M H^2 dV \geq 4\pi(1 + |I_\phi| - g), \tag{5.170}$$

where g denotes the genus of M

Proof. We choose an orthonormal local frame e_1, e_2, e_3, e_4 in $\mathbb{E}^4$ such that, restricted to M, e_1, e_2 are tangent to M and e_3, e_4 are normal to M. Then the Gauss curvature G and the normal curvature G^D are given respectively by

$$G = R(e_1, e_2; e_2, e_1) = \sum_{r=3}^{4} \{h_{11}^r h_{22}^r - (h_{12}^r)^2\},$$

$$G^D = R^D(e_1, e_2; e_4, e_3) = h_{12}^3(h_{22}^4 - h_{11}^4) - h_{12}^4(h_{22}^3 - h_{11}^3).$$

Thus the squared mean curvature H^2 satisfies

$$H^2 = \frac{1}{4}\left\{(h_{11}^3 + h_{22}^3)^2 + (h_{11}^4 + h_{22}^4)^2\right\}$$

$$= \frac{1}{4}(h_{11}^3 - h_{22}^3)^2 + \frac{1}{4}(h_{11}^4 - h_{22}^4)^2 + (h_{12}^3)^2 + (h_{12}^4)^2 + G$$

$$\geq |h_{12}^4||h_{11}^3 - h_{22}^3| + |h_{12}^3||h_{11}^4 - h_{22}^4| + G$$

$$\geq |G^D| + G.$$

Hence we find

$$\int_M H^2 dV \geq \int_M |G^D| dV + \int_M G dV. \tag{5.171}$$

It is well-known that Gauss-Bonnet's theorem implies the integral of Gauss curvature G gives $2\pi\chi(M)$. Moreover, we known from [Little (1969)] that the integral of the normal curvature G^D is equal to $2\pi\chi(T^\perp M)$, where $\chi(T^\perp M)$ denotes the Euler number of the normal bundle of M. Thus (5.171) implies

$$\int_M H^2 dV \geq 2\pi\{\chi(M) + |\chi(T^\perp M)|\}. \tag{5.172}$$

On the other hand, by a result of [Lashof and Smale (1958)], we have $\chi(T^\perp M) = 2I_\phi$. After combining these with (5.171), we find (5.170). □

By combining Theorem 5.23 and Theorem 5.24 we have the following.

Theorem 5.25. [Wintgen (1979)] *Let* $\phi : S^2 \to \mathbb{E}^4$ *be an immersion of 2-sphere into* $\mathbb{E}^4$. *If*

$$\int_M H^2 dV < 8\pi, \tag{5.173}$$

then ϕ *is regularly homotopic to the standard imbedding of* S^2 *into a hyperplane of* $\mathbb{E}^4$.

Definition 5.7. If $\phi : M \to \mathbb{E}^4$ is an imbedding of a closed surface M into $\mathbb{E}^4$, the fundamental group $\pi_1(\mathbb{E}^4 \setminus \phi(M))$ of $\mathbb{E}^4 \setminus \phi(M)$ is called the *knot group* of ϕ. The minimal number of generators of knot group of ϕ is called the *knot number* of ϕ.

P. Wintgen proved the following simple relation between total mean curvature and knot number.

Theorem 5.26. [Wintgen (1978)] *Let* $\phi : M \to \mathbb{E}^4$ *be an imbedding of a closed surface* M *into* $\mathbb{E}^4$. *Then we have*

$$\int_M H^2 dV \geq 4\pi\rho, \tag{5.174}$$

where ρ *is the knot number of* ϕ.

Proof. We need the following lemma.

Lemma 5.5. [Sunday (1976)] *Let* $h_u, u \in S^3(1)$, *be a height function of* M *in* $\mathbb{E}^4$ *which has only non-degenerate critical points on* M. *Then the number* $\beta_0(h_u)$ *of local minima satisfies*

$$\beta_0(h_u) \geq \rho. \tag{5.175}$$

Without loss of generality we can assume that h_u takes different values at the critical points p_i $(i = 0, 1, \ldots, t)$ written in the order induced from h_u. Let c_j be the real number with

$$c_0 < h_u(p_0) < c_1 < h_u(p_1) < \cdots < h_u(p_t) < c_{t+1}.$$

By a classical result of van Kampen for the fundamental group of the spaces

$$H_j = \{p \in \mathbb{E}^4 \setminus M : \langle p, u \rangle \leq c_j\},$$

we have

$$\pi_1(H_{j+1}) \approx \pi_1(H_j) + \text{one generator, if } p_j \text{ is a local minimum;}$$
$$\pi_1(H_{j+1}) \approx \pi_1(H_j) + \text{one relation, if } p_j \text{ is a saddle minimum;}$$
$$\pi_1(H_{j+1}) \approx \pi_1(H_j), \text{ if } p_j \text{ is a local maximum.}$$

The lemma follows from these relations.

We denote by $\beta_2(h_u)$ the number of local maxima of h_u. Since

$$\beta_2(h_u) = \beta_0(h_{-u}),$$

Lemma 5.5 implies $\beta_2(h_u) \geq \rho$. For each critical point p of h_u, u is normal to M at p. Moreover. A_u is semi-definite if h_u is either local maximum or local minimum at p. Let U denote the set of all elements (p, e) in B_1 such that A_e is semi-definite. Then according to the above observation we see that the unit sphere S^3 is covered by U at least 2ρ times under the map $\nu : B_1 \to S^3$. Thus by a similar argument as given in the proof of Theorem 5.9 we obtain (5.174). $\square$

Remark 5.7. For a surface in $\mathbb{E}^4$, Theorem 5.26 improves Theorem 5.9 if the knot number is at least 2.

5.9 Normal curvature and total mean curvature of surfaces

Let $\phi : M \to R^m(c)$ be an isometric immersion of an oriented closed surface M into a real space form of constant curvature c. By Ricci's equation the normal curvature tensor R^D satisfies

$$R^D(X, Y)\xi = h(X, A_\xi Y) - h(A_\xi X, Y) \tag{5.176}$$

for X, Y tangent to M and ξ normal to M.

Let e_1, e_2 be an orthonormal tangent frame. We put

$$h_{ij} = h(e_i, e_j), \quad i, j = 1, 2.$$

We define $a \wedge b$ to be the endomorphism

$$(a \wedge b)(c) = \langle b, c \rangle \, a - \langle a, c \rangle \, b. \qquad (5.177)$$

Then (5.176) becomes

$$R^D(e_1, e_2) = (h_{11} - h_{22}) \wedge h_{12}. \qquad (5.178)$$

The squared mean curvature H^2 and the Gauss curvature G are given by

$$4H^2 = |h_{11} + h_{22}|^2,$$
$$G = \langle h_{11}, h_{22} \rangle - |h_{12}|^2 + c. \qquad (5.179)$$

For each point $p \in M$ we put

$$E_p = \{h(X, X) : X \in T_p(M), |X| = 1\}. \qquad (5.180)$$

If $X = \cos\theta e_1 + \sin\theta e_2$, then we have

$$h(X, X) = H + \cos 2\theta \left(\frac{h_{11} - h_{22}}{2} \right) + (\sin 2\theta) h_{12}. \qquad (5.181)$$

This shows that E_p is an ellipse in the normal space $T_p^\perp(M)$ centered at H. Moreover, as X goes once around the unit tangent circle, $h(X, X)$ goes twice around the ellipse. This ellipse E_p is called *ellipse of curvature* at p. Notice that this ellipse could degenerate into a line segment or a point. The ellipse E_p is degenerate if and only if $R^D = 0$ at p.

If $R^D \neq 0$, then $h_{11} - h_{22}$ and h_{12} are linearly independent, which define a 2-plane subbundle $\mathcal{N}$ of the normal bundle $T^\perp(M)$. This plane bundle inherits a Riemannian connection from that of $T^\perp(M)$. Let e_3, e_4 be an orthonormal oriented frame of $\mathcal{N}$. We define the normal curvature G^D of M in $R^m(c)$ by

$$G^D = \langle R^D(e_1, e_2) e_4, e_3 \rangle. \qquad (5.182)$$

Since M and $\mathcal{N}$ are oriented, G^D is globally defined.

Let $\mathcal{N}^\perp$ be the oriented complementary subbundle of $\mathcal{N}$ in $T^\perp(M)$. Then we have the following splitting of the normal bundle:

$$T^\perp(M) = \mathcal{N} \oplus \mathcal{N}^\perp.$$

From the definition of $\mathcal{N}^\perp$ we have

$$R^D(e_1, e_2)\xi = 0 \quad \text{if } \xi \in \mathcal{N}^\perp. \qquad (5.183)$$

Let $\sigma_0 = \sigma_0(M)$ denote the bundle of symmetric endomorphism of the tangent bundle $T(M)$. Define a map $\varphi : \mathcal{N} \to \sigma_0$ by

$$\varphi(\xi) = A_\xi - \frac{1}{2}(\mathrm{Tr}\, A_\xi)I, \quad \xi \in \mathcal{N}. \qquad (5.184)$$

Because $R^D \neq 0$ by assumption, $[A_{e_3}, A_{e_4}] \neq 0$. Thus (5.184) implies that φ is an isomorphism. We denote by $\chi(\mathcal{N})$ the Euler characteristic of the oriented 2-plane bundle $\mathcal{N}$ over M.

The following result is an extension of a result of [Asperti (1980)], [Little (1969)] and [Dajczer (1980)].

Proposition 5.6. [Asperti et al. (1982)] *For an oriented closed surface in a real space form $R^m(c)$ with nowhere vanishing normal curvature tensor, we have $\chi(\mathcal{N}) = 2\chi(M)$.*

Proof. As before, let $\sigma_0 = \sigma_0(M)$ denote the bundle of symmetric endomorphism endowed with the orientation induced by that of $\mathcal{N}$ via φ. Since φ is orientation-preserving isomorphism, we have $\chi(\mathcal{N}) = \chi(\sigma_0(M))$.

For each $X \in T_p(M)$, let $B(X)$ denote the element in $\sigma_0(M)$ at p which is defined by

$$B(X)Y = 2\langle X, Y \rangle X - \langle X, X \rangle Y.$$

Then

$$B((\cos t)X + (\sin t)X^\perp) = (\cos 2t)B(X) + (\sin 2t)B(X)^\perp,$$

where $X^\perp$ is a vector in $T_p(M)$ such that $|X^\perp| = |X|$, $X \perp X^\perp$ and $\{X, X^\perp\}$ gives the orientation of M. Thus the index formula for the Euler characteristic applied to a generic vector field X and to $B(X)$, respectively, yields the proposition. □

The following is a generalization of Theorem 5.24.

Theorem 5.27. [Guadalupe and Rodriquez (1983)] *Let $\phi : M \to R^m(c)$ be an isometric immersion of an oriented closed surface M into an oriented real space form $R^m(c)$. Then we have*

$$\int_M H^2 dV \geq 2\pi\chi(M) + \left| \int_M G^D dV \right| - c\,\mathrm{vol}(M). \tag{5.185}$$

The equality sign of (5.185) holds if and only if G^D does not change sign and the ellipse of curvature is a circle at every point.

Proof. Under the hypothesis of the theorem, it follows from (5.176) and (5.182) that

$$G^D = |h_{11} - h_{22}||h_{12}|. \tag{5.186}$$

Thus (5.179) and (5.186) imply

$$
\begin{aligned}
0 \leq (|h_{11} - h_{22}| - 2|h_{12}|)^2 \\
= |h_{11} - h_{22}|^2 + 4|h_{12}|^2 - 4|h_{11} - h_{22}|\,|h_{12}| \\
= |h_{11}|^2 + |h_{22}|^2 + 2|h_{12}|^2 - 2G - 4|G^D| + 2c \\
= ||h||^2 - 2G - 4|G^D| + 2c.
\end{aligned}
$$

On the other hand, we also have

$$
\begin{aligned}
4H^2 = |h_{11} + h_{22}|^2 \\
= |h_{11}|^2 + |h_{22}|^2 + 2\langle h_{11}, h_{22} \rangle \\
= |h_{11}|^2 + |h_{22}|^2 + 2|h_{12}|^2 + 2G - 2c \\
= ||h||^2 + 2G - 2c.
\end{aligned}
$$

Hence we find

$$
H^2 + c \geq G + |G^D| \tag{5.187}
$$

with equality sign holding if and only if we have

$$
h_{11} - h_{22} = 2h_{12},
$$

i.e., the ellipse of curvature is a circle. After integrating (5.187) over M, we obtain inequality (5.185). Moreover, it is easy to verify that the equality sign of (5.185) holds if and only if G^D does not change sign and the ellipse of curvature is always a circle. □

Corollary 5.10. [Guadalupe and Rodriquez (1983)] *Let M be an oriented closed surface M in $\mathbb{E}^4$. If the normal curvature G^D is positive everywhere, then we have*

$$
\int_M H^2 dV \geq 12\pi. \tag{5.188}
$$

The equality sign of (5.188) holds if and only if the ellipse of curvature is a circle at every point.

Proof. Let M be an oriented closed surface M in $\mathbb{E}^4$. If G^D is positive at every point on M, we have

$$
\chi(T^\perp M) = \frac{1}{2\pi} \int_M G^D dV > 0.
$$

Thus M is homeomorphic to S^2. Hence we find $\chi(T^\perp M) = 2\chi(M) = 4$, which implies (5.188) by virtue of (5.185).

The remaining part is easy to verify. □

Remark 5.8. For a proof of an extension of the above given pointwise Wintgen inequality using conformal surface invariants and further results in this context concerning the conformal geometry of submanifolds, see [Rouxel (1982)].

Chapter 6

Submanifolds of Finite Type

6.1 Introduction

The theory of finite type submanifolds began in the late 1970s through author's attempts to find the best possible estimates of the total mean curvature of a closed submanifold of Euclidean space and to find a notion of "degree" for submanifolds of Euclidean space.

The main objects of studies in algebraic geometry are algebraic varieties. Because an algebraic variety is defined by using algebraic equations, one can define the degree of an algebraic variety by its algebraic structure. The concept of degree plays a fundamental role in algebraic geometry.

On the other hand, every Riemannian manifold can be realized as a Riemannian submanifold in some Euclidean space with sufficiently high codimension according to Nash's embedding theorem. However, one lacks the notion of "degree" for Euclidean submanifolds. Inspired by these simple observations, the author introduced in the late 1970s the notions of "order" and "type" for submanifolds of Euclidean spaces. By applying the notion of order, the author was able to establish some sharp estimates of the total mean curvature for closed Euclidean submanifolds in terms of their orders. Moreover, by utilizing the notions of type, the notions of finite type submanifolds and of finite type maps were naturally introduced.

The family of finite type submanifolds is huge, which contains many important families of submanifolds; including all minimal submanifolds of Euclidean space; all minimal submanifolds of hyperspheres; all parallel submanifolds as well as all equivariantly immersed compact homogeneous submanifolds. Just like minimal submanifolds, submanifolds of finite type can be described by a spectral variation principle, namely as critical points of directional deformations.

On one hand, the notion of finite type submanifolds provides a natural way to combine the spectral geometry with the theory of submanifolds; as well as with maps; e.g., with Gauss map. On the other hand, one can apply finite type theory to investigate spectral geometry of submanifolds.

The first results on finite type theory were collected in [Chen (1984)]. Since then many mathematicians contributed to this theory. A report on the progress in this subject up to 1996 was presented in [Chen (1996b)]. In this second edition we will present numerous important new results on finite type theory obtained after the publication of the first edition.

6.2   Order and type of submanifolds and maps

Let (M, g) be a closed Riemannian n-manifold with Levi-Civita connection ∇. Then the Laplacian Δ of (M, g) acting as an elliptic differential operator on $\mathcal{F}(M)$. The eigenvalues of Δ form a discrete infinite sequence:

$$0 = \lambda_0 < \lambda_1 < \lambda_2 < \ldots \nearrow \infty.$$

Let $V_t = \{f \in \mathcal{F}(M) : \Delta f = \lambda_t f\}$ be the eigenspace of Δ with eigenvalue λ_t. Then V_0 is 1-dimensional and V_t is finite-dimensional for $t \geq 1$. Define an inner product $(\ ,\ )$ on $\mathcal{F}(M)$ by

$$(f, h) = \int_M fh \, dV. \tag{6.1}$$

Then $\sum_{t=0}^{\infty} V_t$ is dense in $\mathcal{F}(M)$ (in L^2-sense). If we denote by $\hat{\oplus}_{t=0}^{\infty} V_t$ the completion of $\sum_{t=0}^{\infty} V_t$, we have $\mathcal{F}(M) = \hat{\oplus}_k V_k$.

For each $f \in \mathcal{F}(M)$ let f_t denote the projection of f onto the subspace V_t. Then we have the following spectral decomposition:

$$f = \sum_{t=0}^{\infty} f_t \quad \text{(in } L^2\text{-sense)}. \tag{6.2}$$

Because $\dim V_0 = 1$, for each non-constant $f \in \mathcal{F}(M)$ there is a positive integer $p \geq 1$ such that $f_p \neq 0$ and $f - f_0 = \sum_{t \geq p} f_t$, $f_0 \in V_0$.

If there are infinite many nonzero f_t's, we put $q = +\infty$. Otherwise, there is an integer q $(q \geq p)$, such that

$$f_q \neq 0 \quad \text{and} \quad f - f_0 = \sum_{t=p}^{q} f_t. \tag{6.3}$$

If we allow q to be $+\infty$, we have decomposition (6.3) in general. The set

$$T(f) = \{t > 0 : f_t \neq 0\} \tag{6.4}$$

is called the *order* of f. The smallest element p in $T(f)$ is the *lower order* of f, which is denoted by $l.o.(f)$. The supremum of $T(f)$ is called the *upper order* of f, denoted by $u.o.(f)$. A function f in $\mathcal{F}(M)$ is said to be of *finite type* if $T(f)$ is a finite set, equivalently, if its spectral decomposition (6.2) contains only finitely many nonzero terms. Otherwise f is said to be *of infinite type*. The function f is of k-*type* if $T(f)$ contains k elements.

For an isometric immersion $\phi : M \to \mathbb{E}_s^m$ of a closed Riemannian manifold M into a pseudo-Euclidean m-space $\mathbb{E}_s^m$, we put

$$\phi = (x_1, \ldots, x_m),$$

where x_A is the A-th Euclidean coordinate of M via ϕ. For each x_A,

$$x_A - (x_A)_0 = \sum_{t=p_A}^{q_A} (x_A)_t, \quad A = 1, \ldots, m.$$

For an isometric immersion $\phi : M \to \mathbb{E}^m$, put $p = \inf_A\{p_A\}$ and $q = \sup_A\{q_A\}$, where A ranges among all A with $x_A - (x_A)_0 \neq 0$. Then p and q are well-defined such that p is a positive integer and q is either $+\infty$ or an integer $\geq p$. Hence we have the spectral decomposition of ϕ in vector form:

$$\phi = \phi_0 + \sum_{t=p}^{q} \phi_t, \quad \Delta\phi_t = \lambda_t\phi_t, \quad \phi_t \neq 0. \tag{6.5}$$

Since M is closed, ϕ_0 in (6.5) is the *center of mass*, of M in $\mathbb{E}_s^m$, i.e.,

$$\phi_0 = \frac{\int_M \phi \, dV}{\mathrm{vol}(M)}.$$

By applying (6.5), we define the *order of the immersion* ϕ by

$$T(\phi) = \{0 \neq t \in \mathbb{N} : \phi_t \neq 0 \text{ in } (6.5)\}.$$

The immersion ϕ is said to be *of finite type* if $T(\phi)$ consists of only finite elements; and it is *of k-type* if $T(\phi)$ contains exactly k elements. It is *of infinite type* if it is not of finite type. Similarly we have the lower order and the upper order of the immersion or the map.

Given two $\mathbb{E}_s^m$-valued functions v, w on M, define the scalar product by

$$(v, w) = \int_M \langle v, w \rangle \, dV, \tag{6.6}$$

where $\langle v, w \rangle$ is the pseudo-Euclidean scalar product of v, w.

For $\phi : M \to \mathbb{E}_s^m$, the components of the spectral decomposition (6.5) are mutually orthogonal, i.e., $(\phi_t, \phi_s) = 0$, $t \neq s$.

In general, one cannot make the spectral decomposition of a function on a non-closed Riemannian manifold. However, it remains possible to define the notion of finite type immersions for non-closed manifolds as follows: An isometric immersion $\phi : M \to \mathbb{E}_s^m$ of a non-closed Riemannian manifold M

is said to be of *finite type* if the position vector can be decomposed into a finite sum of vector eigenfunctions of the Laplacian of M:

$$\phi = c_0 + \phi_1 + \cdots + \phi_k, \tag{6.7}$$

where c_0 is a constant vector and $\phi_1, \ldots, \phi_k$ are non-constant vector-valued eigenfunctions of Δ. If eigenvalues associated with $\phi_1, \ldots, \phi_k$ are mutual distinct, then the submanifold is said to be of k-type. In particular, if one of the eigenvalues of $\phi_1, \ldots, \phi_k$ is zero, then M is said to be of *null k-type*. For instance, a *null 2-type* submanifold admits the following spectral decomposition:

$$\phi = c_0 + \phi_1 + \phi_2, \quad \Delta\phi_1 = 0, \quad \Delta\phi_2 = \lambda\phi_2, \quad 0 \neq \lambda \in \mathbf{R}. \tag{6.8}$$

After choosing c_0 to be the origin of $\mathbb{E}_s^m$, (6.8) reduces to

$$\phi = \phi_1 + \phi_2, \quad \Delta\phi_1 = 0, \quad \Delta\phi_2 = \lambda\phi_2, \quad 0 \neq \lambda \in \mathbf{R}. \tag{6.9}$$

The notions of order and type of submanifolds defined above can be extended to smooth maps in a natural way; in particular to the Gauss map of a Riemannian submanifold of a pseudo-Euclidean space.

The next lemma is a generalization of Corollary 4.6.

Proposition 6.1. *For integers $k \geq 1$ and $s \geq 0$, every null k-type spacelike submanifold of the pseudo Euclidean m-space $\mathbb{E}_s^m$ is non-closed.*

Proof. Follows from the fact that every harmonic function of a closed Riemannian manifold is constant. □

Obviously, if two Riemannian submanifolds of a pseudo-Euclidean space are congruent, they have same type number.

Lemma 6.1. *Let $\phi : M \to \mathbb{E}_s^m$ be a k-type map of a closed Riemannian manifold M into $\mathbb{E}_s^m$. If $T(\phi) = \{t_1, \ldots, t_k\}$ and $L : \mathbb{E}_s^m \to \mathbb{E}_t^N$ is a linear map, then the composition $L \circ \phi : M \to \mathbb{E}_t^N$ is a finite type map. Moreover, we have*

(1) *the type number of $L \circ \phi$ is at most k;*
(2) *the order $T(L \circ \phi)$ of $L \circ \phi$ is a subset of the order $T(\phi)$ of ϕ.*

Proof. Follows from the fact that a linear combination of finite linear combinations of eigenfunctions of Δ is a linear combination of eigenfunctions of Δ. □

Lemma 6.2. *Let* $\phi : M \to \mathbb{E}_s^m$ *be a smooth map of a closed Riemannian manifold* M *into* $\mathbb{E}_s^m$ *and let* $\iota : \mathbb{E}_s^m \subset \mathbb{E}_{s+i}^{m+r}$ *be an inclusion map. Then* $T(\iota \circ \phi) = T(\phi)$.

Proof. Follows from Lemma 6.1 and that the inclusion map is linear. $\square$

6.3 Minimal polynomial criterion

The following *Minimal Polynomial Criterion* is very useful to determine whether a submanifold is of finite type.

Theorem 6.1. [Chen (1983a)] *Let* $\phi : M \to \mathbb{E}_s^m$ *be an isometric immersion of a closed Riemannian manifold* M *into a pseudo-Euclidean space* $\mathbb{E}_s^m$ *and let* H *be the mean curvature vector of* M *in* $\mathbb{E}_s^m$. *Then we have:*

(a) *M is of finite type if and only if there exists a nontrivial polynomial $Q(t)$ such that $Q(\Delta)H = 0$.*
(b) *If M is of finite type, then there exists a unique monic polynomial $P(t)$ of least degree with $P(\Delta)H = 0$.*
(c) *If M is of finite type, then the type number of M is equal to the degree of the unique monic polynomial P.*

The same results holds if H were replaced by $\phi - \phi_0$.

Proof. Under the hypothesis, let us consider the spectral decomposition given by (6.5). If ϕ is of finite type, then $q < \infty$. Thus (6.5) and Beltrami's formula give

$$n\Delta^i H = -\sum_{t=p}^q \lambda_t^{i+1} \phi_t, \quad i = 0, 1, 2, \dots. \tag{6.10}$$

Put

$$c_1 = -\sum_{t=p}^q \lambda_t,$$

$$c_2 = \sum_{t<r} \lambda_t \lambda_r,$$

$$\dots$$

$$c_{q-p+1} = (-1)^{q-p+1} \lambda_p \cdots \lambda_q.$$

Then we have

$$\Delta^{q-p+1} H + c_1 \Delta^{q-p} H + \cdots + c_{q-p+1} H = 0. \tag{6.11}$$

Thus if we put

$$Q(t) = t^{q-p+1} + c_1 t^{q-p} + \cdots + c_{q-p+1}, \tag{6.12}$$

we get $Q(\Delta)H = 0$. Conversely, if H satisfies $Q(\Delta)H = 0$ for some nontrivial polynomial $Q(t)$ of degree k, then Corollary 4.8 implies $k \geq 1$, since M is closed. Let us consider the spectral decomposition (6.5). From (6.10), (6.12) and $Q(\Delta)H = 0$ we find

$$\sum_{t=1}^{\infty} \lambda_t(\lambda_t^k + c_1\lambda_t^{k-1} + \cdots + c_{k-1}\lambda_t + c_k)\phi_t = 0. \tag{6.13}$$

For each integer $r > 0$, (6.13) implies that

$$\sum_{t=1}^{\infty} \lambda_t(\lambda_t^k + c_1\lambda_t^{k-1} + \cdots + c_{k-1}\lambda_t + c_k) \int_M \langle \phi_t, \phi_r \rangle \, dV = 0.$$

Since $(\phi_t, \phi_r) = \int_M \langle \phi_t, \phi_r \rangle \, dV = 0$ for $r \neq t$, we obtain

$$(\lambda_t^k + c_1\lambda_t^{k-1} + \cdots + c_{k-1}\lambda_t + c_k)||\phi_r||^2 = 0. \tag{6.14}$$

Thus, for each $\phi_t \neq 0$ in (6.5), the corresponding eigenvalue λ_t is a root of the polynomial (6.12). Therefore the decomposition (6.5) contains only finite nonzero terms. Consequently, ϕ is of finite type. This proves (a).

Now, assume that ϕ is of finite type. Then it follows from (a) that there exists a nontrivial polynomial $Q(t)$ such that $Q(\Delta)H = 0$. Let $P(t)$ is a monic polynomial of least possible degree satisfying $P(\Delta)H = 0$. Clearly, $P(t)$ is unique. Thus we have (b). Statement (c) follows from the proof of statement (a). The remaining part is clear. $\qquad\square$

Remark 6.1. Theorem 6.1 is false if M is non-closed. For instance, consider the surface $\phi : \mathbb{E}^2 \to \mathbb{E}_1^4$ defined by $\phi(u, v) = \frac{1}{6}(u^3, 3u^2, 6u - u^3, 6v)$. A direct computation shows that $\Delta^2\phi = 0$, but ϕ is of infinite type.

Definition 6.1. The unique monic polynomial P given in Theorem 6.1(b) is called the *minimal polynomial* of the finite type submanifold.

Remark 6.2. Clearly, congruent submanifolds of finite type share the same minimal polynomial. On the other hand, if two finite type submanifolds M_1 and M_2 of $\mathbb{E}_s^m$ have the same minimal polynomial, then we have:

(1) M_1 and M_2 have the same type number.
(2) The Laplacian of M_1 and M_2 have at least k common eigenvalues given by the roots of the minimal polynomial.
(3) The orders of M_1 and M_2 are the same whenever M_1 and M_2 are isospectral.

The isometries of a Riemannian manifold M form a Lie group $I(M)$, which is compact whenever M is closed. Let G be a closed subgroup of $I(M)$. The Riemannian manifold is called a *G-homogeneous* if G acts transitively on M. If the group G is not important, we simply say that M is a *homogeneous Riemannian manifold*.

Definition 6.2. Let $M = G/H$ be a homogeneous Riemannian manifold and M a Riemannian manifold with group $I(M)$ of isometries. An isometric immersion $\psi : M \to M$ is called *G-equivariant* if there is a homomorphism $\zeta : G \to I(M)$ such that $f(g(x)) = \zeta(g)f(x)$ for each $g \in G$ and $x \in M$.

An immediate application of Theorem 6.1 is the following.

Proposition 6.2. *Let M be a compact homogeneous Riemannian manifold. If M is equivariantly isometrically immersed in $\mathbb{E}_s^m$, then M is of finite type. Moreover, the type number of M is at most m.*

Proof. Let x be a given point of M. Then $H, \Delta H, \ldots, \Delta^m H$ at x are linearly dependent. Thus there is a polynomial $Q(t)$ of degree m such that $Q(\Delta)H = 0$ at x. Because M is equivariantly isometrically immersed in $\mathbb{E}_s^m$, $Q(\Delta)H = 0$ holds identically on M. Thus M is of finite type according to Theorem 6.1. Moreover, since the degree of the minimal polynomial is less than or equal to m, the type number is at most m. $\qquad\square$

Theorem 6.2. [Chen and Petrović (1991)] *Let $\phi : M \to \mathbb{E}_s^m$ be an isometric immersion of a Riemannian manifold M into $\mathbb{E}_s^m$. If there exists a nontrivial polynomial Q such that $Q(\Delta)H = 0$, then M is either of infinite type or of finite type with type number $k \leq \deg Q + 1$.*

Proof. Let $Q = t^d + c_1 t^{d-1} + \cdots + c_d$ be a nontrivial polynomial such that $Q(\Delta)H = 0$. Suppose that M is of k-type with finite k. Then we have the spectral decomposition

$$\phi = \mathbf{c}_0 + \phi_1 + \cdots + \phi_k, \quad \Delta\phi_i = \ell_i\phi_i, \tag{6.15}$$

where $\ell_1, \ldots, \ell_k$ are k mutually distinct eigenvalues of Δ. After applying Beltrami's formula, (6.15) implies that

$$-n\Delta^j H = \ell_1^{j+1}\phi_1 + \cdots + \ell_k^{j+1}\phi_k, \quad j = 0, 1, 2, \ldots. \tag{6.16}$$

Thus, by using $Q(\Delta)H = 0$, we find that

$$\ell_1 Q(\ell_1)\phi_1 + \cdots + \ell_k Q(\ell_k)\phi_k = 0. \tag{6.17}$$

Now, by applying Δ^j to (6.17), we find

$$\ell_1^{j+1}Q(\ell_1)\phi_1 + \cdots + \ell_k^{j+1}Q(\ell_k)\phi_k = 0, \quad j = 0, 1, 2, \ldots. \tag{6.18}$$

Since $\ell_1, \ldots, \ell_k$ are distinct, (6.18) yields $Q(\ell_1) = \cdots = Q(\ell_k) = 0$. Hence we have $k \leq d + 1$. □

For a non-closed M, the existence of a nontrivial polynomial Q satisfying $Q(\Delta)H = 0$ does not insure finite type of M (cf. Example 7.2). On the other hand, we have the next two results which guarantee submanifolds to be of finite type.

Theorem 6.3. [Chen and Petrović (1991)] *Let $\phi : M \to \mathbb{E}_s^m$ be an isometric immersion of a Riemannian manifold M into $\mathbb{E}_s^m$. If there exist a vector $c \in \mathbb{E}_s^m$ and a nontrivial polynomial P with simple roots such that $P(\Delta)(\phi - c) = 0$, then M is of finite type with type number $k \leq \deg P$.*

Proof. Assume that $P(t) = (t - \ell_1) \cdots (t - \ell_k)$ is the polynomial with simple roots satisfying $P(\Delta)(\phi - c) = 0$. Consider the following system

$$\phi - c = \phi_1 + \phi_2 + \cdots + \phi_k,$$
$$\Delta\phi = \ell_1\phi_1 + \ell_2\phi_2 + \cdots + \ell_k\phi_k,$$
$$\cdots \tag{6.19}$$
$$\Delta^{k-1}\phi = \ell_1^{k-1}\phi_1 + \ell_2^{k-1}\phi_2 + \cdots + \ell_k^{k-1}\phi_k$$

for some vector $c \in \mathbb{E}_s^m$. Since $\ell_1, \ldots, \ell_k$ are mutually distinct, we may solve (6.19) for $\phi_1, \ldots, \phi_k$ in terms of $\phi - c, \Delta\phi, \ldots, \Delta^{k-1}\phi$ to obtain

$$\prod_{j\neq i}(\ell_j - \ell_i)\phi_i = s_{i,k-1}(\phi - c) - s_{i,k-2}\Delta\phi + \cdots$$
$$+ (-1)^{k-2}s_{i,1}\Delta^{k-2}\phi + (-1)^{k-1}\Delta^{k-1}\phi, \tag{6.20}$$

where $s_{i,j}$ is the j-th elementary symmetric function of $\ell_1, \ldots, \ell_{i-1}, \ell_{i+1}, \ldots, \ell_k$. In other words, we have

$$s_{i,k-1} = \prod_{j\neq i}\ell_j, \quad s_{i,k-2} = \prod_{\substack{j\neq h \\ j,h\neq i}}\ell_j\ell_h, \quad \ldots, \quad s_{i,1} = \sum_{j\neq i}\ell_j.$$

From the assumption we have

$$\Delta^k\phi - s_1\Delta^{k-1}\phi + \cdots + (-1)^{k-1}s_{k-1}\Delta\phi + (-1)^k s_k(\phi - c) = 0, \tag{6.21}$$

where s_j is the j-th elementary symmetric function of $\ell_1, \ldots, \ell_k$. It follows from (6.19)-(6.21) that

$$\prod_{j\neq i}(\ell_j - \ell_i)\Delta\phi_i = \ell_i\prod_{j\neq i}(\ell_j - \ell_i)\phi_i \tag{6.22}$$

for $i = 1, \ldots, k$. Since $\ell_1, \ldots, \ell_k$ are mutually distinct, we find from (6.22) that $\Delta\phi_i = \ell_i\phi_i$. Thus, by applying (6.19), we find $\phi = c + \phi_1 + \cdots + \phi_k$, which shows that M is of finite type with type number $k \leq \deg P$. □

For finite type curves, we have the following simple characterization which shows that the existence of a non-trivial polynomial P satisfying $P(\Delta)H = 0$ is enough to guarantee finite type of a curve.

Proposition 6.3. [Chen and Petrović (1991)] *Let $\gamma(s)$ be a unit speed curve in $\mathbb{E}^m$. If there exists a non-trivial polynomial P such that $P(\Delta)H = 0$, then γ is of finite type.*

Remark 6.3. Using the same proof given in [Chen and Petrović (1991)], Proposition 6.3 can be extended to unit speed curves in $\mathbb{E}^m_s$.

6.4 A variational minimal principle

Let $\phi : M \to \mathbb{E}^m_s$ be an isometric immersion of a closed Riemannian manifold M into a pseudo-Euclidean space $\mathbb{E}^m_s$. Associated to each $\mathbb{E}^m_s$-valued vector field ξ on M, there is a deformation ϕ_t, defined by

$$\phi_t(u) := \phi(u) + t\xi(u), \quad u \in M, \quad t \in (-\epsilon, \epsilon), \qquad (6.23)$$

where ϵ is a sufficiently small positive number. For each t, ϕ_t gives rise to a submanifold $M_t = \phi_t(M)$. Let $\mathcal{V}(t)$ denote the volume of M_t.

Definition 6.3. Let $\mathfrak{D}$ be the class of all such deformations acting on the submanifold M and $\mathfrak{E}$ a nonempty subclass of $\mathfrak{D}$. Then M is called a *critical point of the volume functional* $\mathcal{V}(t)$ in the class $\mathfrak{E}$ if $\mathcal{V}'(0) = 0$ for each deformation in $\mathfrak{E}$; and M is *stable* in the class $\mathfrak{E}$ if $\mathcal{V}''(0) \geq 0$ for each deformation in $\mathfrak{E}$.

A closed submanifold M of $\mathbb{E}^m_s$ is said to satisfy the *variational minimal principle in the class* $\mathfrak{E}$ if M minimizes the volume functional $\mathcal{V}(t)$ for all deformations in $\mathfrak{E}$. In other words, M satisfies the variational minimal principle in the class $\mathfrak{E}$ if M is a critical point of the volume functional and is stable among all deformations of M in the class $\mathfrak{E}$.

Let c be a vector in $\mathbb{E}^m_s$ and $f \in \mathcal{F}(M)$. Then the deformation given by

$$\phi^{fc}_t(u) := \phi(u) + tf(u)c, \quad u \in M, \ t \in (-\epsilon, \epsilon) \qquad (6.24)$$

is called a *directional deformation* in the direction c (cf. [Voss (1956)]). Such deformations occur when an object moves along a fixed direction.

Let V_i denote the eigenspace of the Laplacian Δ on M associated with i-th eigenvalue λ_i. For each $q \in \mathbb{N}$, let $\mathcal{C}_q$ be the class of all directional deformations defined by smooth functions $f \in \sum_{i \geq q} V_i$ for (6.24). Then

$$\mathcal{C}_0 \supset \mathcal{C}_1 \supset \mathcal{C}_2 \supset \cdots \supset \mathcal{C}_k \supset \cdots,$$

where $\mathcal{C}_0 = \mathcal{F}(M)$. Clearly, if a closed submanifold M of $\mathbb{E}^m_s$ satisfies the variational minimal principle for one class $\mathcal{C}_k$, then it automatically satisfies the variational minimal principle in the class $\mathcal{C}_\ell$ for each $\ell \geq k$.

Theorem 6.4. (Variational minimal principle) [Chen et al. (1993a)]

(a) *There are no closed Riemannian submanifolds M in $\mathbb{E}^m_s$ which satisfy the variational minimal principle in the classes $\mathcal{C}_0$ and $\mathcal{C}_1$.*

(b) *A closed Riemannian submanifold M of $\mathbb{E}^m_s$ is of finite type if and only if it satisfies the variational minimal principle in the class $\mathcal{C}_r$ for some $r \geq 2$.*

(c) *Every closed Riemannian submanifold M of finite type in $\mathbb{E}^m_s$ is stable in the class of $\mathcal{C}_r$ for any $r \geq u.o.(M) + 1$, where $u.o.(M)$ is the upper order of N.*

Proof. Let $c \in \mathbb{E}^m_s$ and $f \in \mathcal{F}(M)$. Consider the directional deformation ϕ^{fc}_t defined by (6.24). Let $e_1, \ldots, e_n$ be an orthonormal local frame field on M. Extended $e_1, \ldots, e_n$ by $(\phi_t)_* e_1, \ldots, (\phi_t)_* e_n$. Put

$$g_{ij}(t) = \langle (\phi_t)_* e_i, (\phi_t)_* e_j \rangle. \tag{6.25}$$

Then

$$\mathcal{V}(t) = \int_M \sqrt{\det(g_{ij}(t))} dV. \tag{6.26}$$

On the other hand, it follows from (6.24) that

$$(\phi_t)_* e_i = e_i + t(f_i)c, \quad \frac{\partial}{\partial t} = fc, \tag{6.27}$$

where $f_i = e_i f$. From (6.25) and (6.27) we find

$$g_{ij}(t) = \delta_{ij} + t(f_i \langle c, e_j \rangle + f_j \langle c, e_i \rangle) + t^2 f_i f_j \langle c, c \rangle. \tag{6.28}$$

By using (6.26) and (6.28), we obtain

$$\mathcal{V}(t) = \int_M \{1 + 2t \langle c, \nabla f \rangle + t^2(\langle c^\perp, c^\perp \rangle \|\nabla f\|^2 + \langle c, \nabla f \rangle^2)\}^{\frac{1}{2}} dV, \tag{6.29}$$

where $\langle \, , \, \rangle$ is the scalar product of $\mathbb{E}^m_s$, $c^\perp$ the normal component of c, ∇f the gradient of f, and dV the volume element of M. From (6.29) we obtain

$$\mathcal{V}'(0) = \int_M \langle c, \nabla f \rangle dV, \quad \mathcal{V}''(0) = \int_M \langle c^\perp, c^\perp \rangle \|\nabla f\|^2 dV. \tag{6.30}$$

Let c^T be the tangential component of c and $(c^T)^\sharp$ the 1-form on M dual to c^T. Then we have

$$(c^T)^\sharp = dh_c, \tag{6.31}$$

where h_c is the height function of M in $\mathbb{E}^m_s$ with respect to c. By using (6.31) we find

$$\delta(c^T)^\sharp = \Delta h_c = \langle \Delta\phi, c \rangle, \tag{6.32}$$

where $\mathbf{x}$ is the position vector field of M in $\mathbb{E}^m_s$.

On the other hand, the Beltrami formula yields

$$\Delta\phi = -nH, \quad n = \dim M. \tag{6.33}$$

Hence, by using the first equation of (6.30), (6.32) and (6.33), we find

$$\mathcal{V}'(0) = \int_M H_c f dV, \tag{6.34}$$

where $H_c = \langle H, c \rangle$. If M is a closed submanifold in $\mathbb{E}^m_s$ which satisfies the variational minimal principle in the class $\mathcal{C}_1$, then (6.34) yields

$$(H_c, f) = \int_M H_c f dV = 0. \tag{6.35}$$

Since (6.35) holds for every $f \in \sum_{t=1}^\infty V_t$ and $c \in \mathbb{E}^m_s$, we get $H = 0$. But this is impossible since M is closed. Hence M cannot satisfy the variational minimal principle in the class $\mathcal{C}_1$. Thus it also does not satisfy the variational minimal principle in the class $\mathcal{C}_0$. This proves (a).

To prove (b) let us assume that M is of finite type. Then we have a finite spectral decomposition:

$$\phi = \phi_0 + \phi_{i_1} + \cdots + \phi_{i_k}, \tag{6.36}$$

where $\mathbf{x}_0$ is a constant vector and $\Delta\phi_{i_j} = \lambda_{i_j}\phi_{i_j}$ with $\lambda_{i_1} < \cdots < \lambda_{i_k}$. From (6.34) and (6.36) we find

$$-nH = \lambda_{i_1}\phi_{i_1} + \cdots + \lambda_{i_k}\phi_{i_k}, \tag{6.37}$$

which implies that $H_c \in \sum_{j\leq\ell} V_j$, where ℓ is the upper order of M. If we put $r = \ell + 1$, then (6.34) implies $\mathcal{V}'(0) = 0$ for any deformation in $\mathcal{C}_r$.

Conversely, if M satisfies variational minimal principal in $\mathcal{C}_r$, for some $r \geq 2$, then (6.35) yields $H_c \in \sum_{i<r} V_i$. Since this holds for any $c \in \mathbb{E}^m_s$, H has finite spectral decomposition

$$H = H_0 + H_1 + \cdots + H_{r-1}, \quad \Delta H_j = \lambda_j H_j. \tag{6.38}$$

If we put $P(t) = t \prod_{j=1}^{r-1}(t - \lambda_j)$, we obtain from (6.38) that $P(\Delta)H = 0$. Thus Theorem 6.1 implies that M is of finite type. This proves (b)

It follows from the proof of (b) that a finite type submanifold M satisfies the variational minimal principle in the class of $\mathcal{C}_r$ for $r \geq u.o.(M) + 1$. On the other hand, the second equation in (6.30) implies that $\mathcal{V}''(0) \geq 0$. Thus M is stable in the class $\mathcal{C}_r$. This proves (c). $\qquad\square$

6.5 Finite type immersions of homogeneous spaces

Let o be a point in a closed Riemannian manifold M. Denote by K the isotropy subgroup at o (i.e., K is the *stabilizer* of o). Then $M = G/K$. The linear isotropy representation is the orthogonal representation of K over the tangent space T_oM at o defined by $K \to O(T_oM) : \phi \mapsto (\phi_*)_o$, where $(\phi_*)_o$ is the differential of ϕ at o. A homogeneous Riemannian manifold is called *isotropy-irreducible* if the linear isotropy representation is irreducible. An isotropy-irreducible homogeneous Riemannian manifold is simply called an *irreducible homogeneous Riemannian manifold*.

Let $M = G/K$ be a compact irreducible homogeneous Riemannian manifold. For each eigenvalue λ of the Laplacian on M, we denote by m_λ the multiplicity of the eigenvalue λ. Let $\phi_1, \ldots, \phi_{m_\lambda}$ be an orthonormal basis of the eigenspace V_λ of the Laplacian Δ with eigenvalue λ. Define a map $\Psi_\lambda : M \to \mathbb{E}^{m_\lambda}$ by

$$\Psi_\lambda(u) = c_\lambda(\phi_1(u), \ldots, \phi_{m_\lambda}(u)) \qquad (6.39)$$

which is a 1-type isometric immersion for some suitable constant $c_\lambda > 0$. Corollary 6.3 implies that Ψ_λ is a minimal immersion into $S_0^{m_\lambda-1}(1)$.

If λ_i is the i-th nonzero eigenvalue of Laplacian of G/K, then $\psi_i = \Psi_{\lambda_i}$ is called the *i-th standard immersion* of $M = G/K$.

For the unit n-sphere $S^n(1)$, the first standard immersion of $S^n(1)$ is the standard imbedding of $S^n(1)$ as a Euclidean hypersphere. The second standard immersion of $S^n(1)$ is the *Veronese immersion* of $S^n(1)$, which is the first standard imbedding of the real projective n-space $RP^n(1)$.

Not every 1-type isometric immersion of order $\{k\}$ of an irreducible symmetric space of compact type into Euclidean space is the k-th standard isometric immersion. This is not true even for a Riemannian 3-sphere. In fact, N. Ejiri constructed in [Ejiri (1981)] a 1-type immersion of the 3-sphere $S^3(4)$ of radius 4 into $S^6(1) \subset \mathbb{E}^7$ which is not a standard immersion.

Definition 6.4. Let $\psi_{t_i} : M \to \mathbb{E}^{m_i}$, $i = 1, \ldots, k$, be standard immersions of a compact isotropy-irreducible homogeneous Riemannian manifold M into the Euclidean m_i-space $\mathbb{E}^{m_i}$ and let $c_{t_1}, \ldots, c_{t_k}$ be real numbers such that $c_{t_1}^2 + \cdots + c_{t_k}^2 = 1$. Then

$$D(c_{t_1}, \ldots, c_{t_k}) = (c_{t_1}\psi_{t_1}, \ldots, c_{t_k}\psi_{t_k})$$

defines an isometric immersion of M into $\mathbb{E}^{m_1+\cdots+m_k}$, which is called a *standard diagonal immersion* of M.

Notice that if the standard immersions $\psi_{t_1}, \ldots, \psi_{t_k}$ in Definition 6.4 arisen from k distinct eigenvalues of Δ, then $D(c_{t_1}, \ldots, c_{t_k})$ is of k-type. The following result shows that every finite type immersion of a compact irreducible homogeneous Riemannian manifold is indeed a "screw" diagonal immersion in general.

Theorem 6.5. [Deprez (1988); Chen (1996b)] *Let M be an irreducible compact homogeneous Riemannian manifold. Then we have:*

(1) *For every standard diagonal immersion $D(a_{t_1}, \ldots, a_{t_k}) : M \to \mathbb{E}^m$ and every linear map $L : \mathbb{E}^m \to \mathbb{E}^\ell$, the composition $L \circ D(a_{t_1}, \ldots, a_{t_k})$ is a finite type immersion whose type number is at most k and whose order is a subset of the order $T(D(a_{t_1}, \ldots, a_{t_k}))$ of $D(a_{t_1}, \ldots, a_{t_k})$.*

(2) *If $\phi : M \to \mathbb{E}^m$ is a finite type isometric immersion whose center of mass is chosen to be the origin of $\mathbb{E}^m$, then for each standard diagonal immersion $D(c_{t_1}, \ldots, c_{t_k}) : M \to \mathbb{E}^N$ with $T(\phi) \subset T(D(c_{t_1}, \ldots, c_{t_k}))$, there is a linear map $L : \mathbb{E}^N \to \mathbb{E}^m$ such that $\phi = L \circ D(c_{t_1}, \ldots, c_{t_k})$.*

Proof. Statement (1) follows immediately from Lemma 6.1. Now, let us assume that $\phi : M \to \mathbb{E}^m$ is a finite type isometric immersion of a compact irreducible homogeneous Riemannian manifold M whose center of mass is chosen to be the origin of $\mathbb{E}^m$. Let $T(\phi) = \{t_1, \ldots, t_\ell\}$.

For each t_i, let $\{\phi_1^{t_i}, \ldots, \phi_{m_{t_i}}^{t_i}\}$ be an orthonormal basis of the eigenspace V_{t_i} of Δ with eigenvalue λ_{t_i}. Consider the map

$$\phi = \sum_{i=1}^{\ell} \left(a_1^i \phi_1^{t_i} + \cdots + a_{m_{t_i}}^i \phi_{m_{t_i}}^{t_i} \right), \tag{6.40}$$

with $a_1^1, \ldots, a_{m_{t_1}}^1, \ldots, a_1^\ell, \ldots, a_{m_{t_\ell}}^\ell \in \mathbb{E}^m$. We consider a standard diagonal immersion $D(c_{t_1}, \ldots, c_{t_k}) : M \to \mathbb{E}^N$ of order $\{t_1, t_2, \ldots, t_k\}$ such that $k \geq \ell$. Then $D(c_{t_1}, \ldots, c_{t_k})$ is of the form:

$$\left(c_{t_1} b_{t_1} \phi_1^{t_1}, \ldots, c_{t_1} b_{t_1} \phi_{m_{t_1}}^{t_1}, \ldots, c_{t_k} b_{t_k} \phi_1^{t_k}, \ldots, c_{t_k} b_{t_k} \phi_{m_{t_k}}^{t_k} \right) \tag{6.41}$$

for some positive numbers $b_{t_1}, \ldots, b_{t_k}$. Let L be the linear map from $\mathbb{E}^N$ into $\mathbb{E}^m$ defined by

$$L(c_{t_1} b_{t_1} e_1) = a_1^1, \ldots, L(c_{t_1} b_{t_1} e_{m_{t_1}}) = a_{m_{t_1}}^1,$$

$$\vdots$$

$$L(c_{t_\ell} b_{t_\ell} e_{m_{t_1} + \cdots + m_{t_{\ell-1}} + 1}) = a_1^\ell, \ldots \tag{6.42}$$

$$L(c_{t_\ell} b_{t_\ell} e_{m_{t_1} + \cdots + m_{t_\ell}}) = a_{m_{t_\ell}}^\ell,$$

$$L(e_{m_{t_1} + \cdots + m_{t_\ell} + 1}) = \cdots = L(e_N) = 0,$$

where $\{e_1, \ldots, e_N\}$ is a standard basis of $\mathbb{E}^N$. Then $\phi = L \circ D(c_{t_1}, \ldots, c_{t_k})$, which implies statement (2). $\square$

Definition 6.5. A curve $\gamma : I \to \mathbb{E}^m$ defined on an open interval I is called a *W-curve of rank r*, if for $t \in I$ the derivatives $\gamma'(t), \gamma''(t), \ldots, \gamma^{(r)}(t)$ are linearly independent and $\gamma'(t), \gamma''(t), \ldots, \gamma^{(r+1)}(t)$ are linearly dependent, and if the Frenet curvatures $\kappa_1, \ldots, \kappa_{r-1}$ of γ are constant.

A unit speed W-curve of rank $2k$ (resp. $2k + 1$) in $\mathbb{E}^m$ is of the form:

$$\gamma(s) = a_0 + \sum_{i=1}^{k} (a_i \cos(\mu_i s) + b_i \sin(\mu_i s)), \qquad (6.43)$$

$$\left(\text{resp.,} \ \gamma(s) = a_0 + b_0 s + \sum_{i=1}^{k} (a_i \cos(\mu_i s) + b_i \sin(\mu_i s)), \right) \qquad (6.44)$$

where $a_i, b_i \in \mathbb{E}^m$ and $\mu_1 < \cdots < \mu_k$ are positive real numbers.

Remark 6.4. When dim $M = 1$, Theorem 6.5(2) means that every finite type closed curve in Euclidean space is the composition of a W-curve and a linear map between two Euclidean spaces.

6.6   Curves of finite type

For a periodic function $f = f(s)$ with period $2r\pi$, $f(s)$ has a *Fourier series* expansion given by

$$f(s) = \frac{a_0}{2} + a_1 \cos \frac{s}{r} + b_1 \sin \frac{s}{r} + a_2 \cos \frac{2s}{r} + b_2 \sin \frac{2s}{r} + \cdots$$

where a_k, b_k are the Fourier coefficients given by

$$a_k = \frac{1}{\pi r} \int_{-\pi}^{\pi} f(s) \cos \frac{ks}{r} \, ds, \quad k = 0, 1, 2, \ldots,$$

$$b_k = \frac{1}{\pi r} \int_{-\pi}^{\pi} f(s) \sin \frac{ks}{r}, \, ds, \quad k = 1, 2, \ldots.$$

Let $\gamma : S^1(r) \to \mathbb{E}^m$ be an isometric immersion of a circle with radius r into a Euclidean m-plane $\mathbb{E}^m$. Denote by s the arc length function of $S^1(r)$. In terms of Fourier series expansion, we have the following.

Proposition 6.4. [Chen (1983a)] *Let $\gamma : S^1 \to \mathbb{E}^m$ be a closed curve in $\mathbb{E}^m$. Then γ is of finite type if and only if the Fourier series expansion of each coordinate function of γ has only finite nonzero terms.*

Proof. Let γ be a closed smooth curve in $\mathbb{E}^m$ whose length is $2\pi r$. Denote by s the arc length of γ. We put $\gamma^{(j)} = d^j \gamma / ds^j$ for $j = 0, 1, 2, \ldots$. Because $\Delta = -d^2/ds^2$, we have $\Delta^j \gamma = (-1)^j \gamma^{2j}$, $j = 0, 1, 2, \ldots$.

If γ is of finite type, then Theorem 6.1 implies that each coordinate function x_A of γ in $\mathbb{E}^m$ satisfies the following differential equation:

$$x_A^{(2k+2)} + c_1 x_A^{(2k)} + \cdots + c_k x_A^{(2)} = 0 \tag{6.45}$$

for some integer $k \geq 1$ and constants $c_1, \ldots, c_k$. Since the solutions of (6.45) are periodic with period $2\pi r$, each coordinate function x_A is a finite linear combination of

$$1, \ \cos \frac{n_i s}{r}, \ \sin \frac{m_i s}{r}, \ \ n_i, m_i \in \mathbb{Z}. \tag{6.46}$$

Hence each x_A is of the following form:

$$x_A = c_A + \sum_{t=p_A}^{q_A} \left\{ a_A(t) \cos \frac{ts}{r} + b_A(t) \sin \frac{ts}{r} \right\} \tag{6.47}$$

for some constants $a_A(t), b_A(t), c_A \in \mathbb{E}^m$ and positive integers p_A, q_A. Thus each coordinate function x_A has a Fourier series expansion with only finite nonzero terms. Conversely, if each x_A has a Fourier series expansion with only finite nonzero terms, then γ takes the form:

$$\gamma = c + \sum_{t=p}^{q} \left\{ a_t \cos \frac{ts}{r} + b_t \sin \frac{ts}{r} \right\} \tag{6.48}$$

for $a_t, b_t, c \in \mathbb{E}^m$ and integers p, q with $q \geq p \geq 1$. From (6.48) we get

$$\Delta \gamma = \sum_{t=p}^{q} \left(\frac{t}{r} \right)^2 \left\{ a_t \cos \frac{ts}{r} + b_t \sin \frac{ts}{r} \right\}. \tag{6.49}$$

If we put $\gamma_t = a_t \cos \left(\frac{ts}{r} \right) + b_t \sin \left(\frac{ts}{r} \right)$, then (6.48) and (6.49) yield

$$\gamma = c + \sum_{t=p}^{q} \gamma_t, \ \ \Delta \gamma_t = \lambda_t \gamma_t,$$

which is the spectral decomposition of γ. Since q is a finite number, γ is of finite type. $\qquad \square$

It follows from Proposition 6.4 that every k-type closed curve γ in $\mathbb{E}^m$ can be written as

$$\gamma(s) = a_0 + \sum_{i=p}^{q} (a_i \cos(\lambda_{t_i} s) + b_i \sin(\lambda_{t_i} s)), \tag{6.50}$$

with $||a_q|| = ||b_q|| \neq 0$. Similarly, every null finite type curve in $\mathbb{E}^m$ can be expressed as

$$\gamma(s) = a_0 + b_0 s + \sum_{i=1}^{k}(a_i \cos(\lambda_{t_i} s) + b_i \sin(\lambda_{t_i} s)), \quad b_0 \neq 0, \qquad (6.51)$$

where a_i and b_i are not simultaneously zero for each $i \in \{1, 2, \ldots, k\}$. If we denote by p and q the lower and upper orders of γ, respectively, then we also have $||a_q|| = ||b_q|| \neq 0$. For simplicity we put

$$\begin{aligned} A_{ij} &= \langle a_i, a_j \rangle - \langle b_i, b_j \rangle, \quad D_{ij} = \langle a_i, a_j \rangle + \langle b_i, b_j \rangle, \\ \bar{A}_{ij} &= \langle a_i, b_j \rangle + \langle a_j, b_i \rangle, \quad \bar{D}_{ij} = \langle a_i, b_j \rangle - \langle a_j, b_i \rangle. \end{aligned} \qquad (6.52)$$

The following corollary follows immediately from (6.50) and (6.51).

Corollary 6.1.

(a) *Every k-type curve of $\mathbb{E}^m$ lies in an affine $2k$-subspace $\mathbb{E}^{2k}$ of $\mathbb{E}^m$.*

(b) *Every null k-type curve of $\mathbb{E}^m$ lies in an affine $(2k-1)$-subspace $\mathbb{E}^{2k-1}$ of $\mathbb{E}^m$. In particular, every null 2-type curve is a right circular helix in an affine 3-subspace.*

We need the following.

Lemma 6.3. *Let γ be a finite type closed curve of length $2\pi r$ in $\mathbb{E}^m$. Let p and q denote the lower and upper orders of γ. Then we have*

$$\sum_{i=p}^{q} i^2 \{||a_i||^2 + ||b_i||^2\} = 2r^2. \qquad (6.53)$$

$$\sum_{i+i'=k} ii' A_{ii'} - 2 \sum_{i-i'=k} ii' D_{ii'} = 0, \qquad (6.54)$$

$$\sum_{i+i'=k} ii' \langle a_i, b_{i'} \rangle + \sum_{i-i'=k} ii' \bar{D}_{ii'} = 0, \qquad (6.55)$$

for $1 \leq k \leq 2q$, where $a_i, b_i; p \leq i \leq q$ are vectors in $\mathbb{E}^m$ defined by (6.50).

Proof. From (6.50) and $\langle \gamma'(s), \gamma'(s) \rangle = 1$, we find

$$\begin{aligned} r^2 = \sum_{i'=p}^{q} \{ \langle a_i, a_{i'} \rangle \sin(\lambda_{t_i} s) \sin(\lambda_{t'_i} s) \\ + \langle b_i, b_{i'} \rangle \cos(\lambda_{t_i} s) \cos(\lambda_{t'_i} s) - 2 \langle a_i, b_{i'} \rangle \sin(\lambda_{t_i} s) \cos(\lambda_{t'_i} s) \}. \end{aligned} \qquad (6.56)$$

From this we find

$$2r^2 = \sum_{i,i'} \left\{ \langle a_i, a_{i'} \rangle \left[\cos(\lambda_{t_i} s - \lambda_{t_{i'}} s) - \cos(\lambda_{t_i} s + \lambda_{t_{i'}} s) \right] \right.$$

$$+ \langle b_i, b_{i'} \rangle \left[\cos(\lambda_{t_i} s - \lambda_{t_{i'}} s) + \cos(\lambda_{t_i} s + \lambda_{t_{i'}} s) \right] \cos(\lambda_{t'_i} s) \} \tag{6.57}$$

$$- 2 \langle a_i, b_{i'} \rangle \left[\sin(\lambda_{t_i} s - \lambda_{t_{i'}} s) + \sin(\lambda_{t_i} s + \lambda_{t_{i'}} s) \right] \cos(\lambda_{t'_i} s) \}.$$

Because $1, \cos(s/r), \sin(s/r), \ldots, \cos(2qs/r), \sin(2qs/r)$ are independent, we find (6.53), (6.54) and (6.55) from (6.57). $\qquad \square$

The most basic result on finite type curves is the following.

Theorem 6.6. [Chen (1983a)] *The only closed planar curve of finite type are circles.*

Proof. If γ is a finite type closed curve in $\mathbb{E}^2$, then Lemma 6.3 gives

$$||a_1|| = ||b_q|| \neq 0, \quad \langle a_q, b_q \rangle = 0. \tag{6.58}$$

If γ is of 1-type, then (6.50) and (6.58) imply that γ is a circle.

Now, let us assume that γ is of k-type with $k \geq 2$. From (6.58) we know that we may choose a suitable Euclidean coordinates of $\mathbb{E}^2$ such that

$$a_q = (\alpha, 0), \quad b_q = (0, \alpha). \tag{6.59}$$

Put $k = 2q - 1$. It follows from (6.54) and (6.55) of Lemma 6.3 that

$$\langle b_q, b_{q-1} \rangle = \langle a_q, a_{q-1} \rangle, \quad \langle a_q, b_{q-1} \rangle = -\langle a_{q-1}, b_q \rangle. \tag{6.60}$$

By using (6.59) and (6.60), we know that a_{q-1} and b_{q-1} take the form:

$$a_{q-1} = (u_{q-1}, v_{q-1}), \quad b_{q-1} = (-v_{q-1}, u_{q-1}). \tag{6.61}$$

From (6.59), (6.60) and (6.61) we find

$$||a_i|| = ||b_i||, \quad \langle a_i, b_i \rangle = 0, \quad i = q - 1, q,$$
$$\langle a_q, a_{q-1} \rangle = \langle b_q, b_{q-1} \rangle, \quad \langle a_{q-1}, b_q \rangle = -\langle a_q, b_{q-1} \rangle. \tag{6.62}$$

Now, we assume that

$$||a_i|| = ||b_i||, \quad \langle a_i, b_i \rangle = 0,$$
$$\langle a_i, a_j \rangle = \langle b_i, b_j \rangle, \quad \langle a_i, b_j \rangle = -\langle a_j, b_i \rangle, \quad i \neq j, \tag{6.63}$$

hold for $h \leq i, j \leq q$ with $h > p$. From (6.54), (6.55) and (6.62) we get

$$\langle a_{h-1}, b_q \rangle = \langle a_{h-1}, a_q \rangle, \quad \langle a_{h-1}, b_q \rangle + \langle a_q, b_{h-1} \rangle = 0. \tag{6.64}$$

From (6.59), (6.63) and (6.64) we obtain

$$a_i = (u_i, v_i), \quad b_i = (-v_i, u_i), \quad i = h - 1, \ldots, q. \tag{6.65}$$

Consequently, we find

$$||a_i|| = ||b_i||, \quad \langle a_i, b_i \rangle = 0, \quad \langle a_i, b_j \rangle + \langle a_j, b_i \rangle = 0, \quad \langle a_i, a_j \rangle = \langle b_i, b_j \rangle$$

for $i \neq j$ and $h - 1 \leq i, j \leq q$. Hence we obtain by induction that

$$||a_i|| = ||b_i||, \quad \langle a_i, b_i \rangle = 0, \quad \langle a_i, b_j \rangle + \langle a_j, b_i \rangle = 0, \quad \langle a_i, a_j \rangle = \langle b_i, b_j \rangle$$

for $i \neq j$ and $p \leq i, j \leq q$. Substituting these into (6.54) and (6.55) gives

$$\sum_{i-j=k} ij \langle a_i, a_j \rangle = \sum_{i-j=k} ij \langle a_i, b_j \rangle = 0, \qquad (6.66)$$

which gives $\langle a_q, a_p \rangle = \langle a_q, b_p \rangle = 0$. Since γ is of k-type with $k \geq 2$, a_p, a_q, b_p, b_q are nonzero orthogonal vectors in $\mathbb{E}^2$, which is impossible. $\square$

Similarly, we also have the following.

Theorem 6.7. *Curves of finite type in $\mathbb{E}^2$ are open portions of circles and lines.*

This result was mentioned in [Chen (1988a)]. Its detailed proof was presented in [Chen et al. (1990b)].

We have the following general result for closed curves of finite type.

Theorem 6.8. [Chen et al. (1990b)] *Every closed curve of finite type in a Euclidean space is a rational curve.*

Proof. A closed curve of k-type can be written as

$$\gamma(s) = a_0 + \sum_{j=1}^{k} \left\{ a_j \cos \frac{p_j s}{r} + b_j \sin \frac{p_j s}{r} \right\}, \qquad (6.67)$$

where $p_1 < \cdots < p_k$ are positive integers and $a_0, a_{p_1}, \ldots, a_{p_k}, b_{p_1}, \ldots, b_{p_k}$ are vectors, not simultaneously zero. The integers $p_1, \ldots, p_k$ are sometime called the *frequency numbers* of the finite type curve γ.

Let $d = \gcd(p_1, \ldots, p_k)$ be the greatest common divisor of $p_1, \ldots, p_k$. By using Legendre polynomials, we immediately find that the components of $\gamma(s)$ can be written as a polynomial of degree p_j/d in $\cos(ds/r)$ and $\sin(ds/r)$. If we make the following substitutions

$$\cos \frac{ds}{r} = \frac{1 - j^2}{1 + j^2}, \quad \sin \frac{ds}{r} = \frac{2j}{1 + j^2},$$

then it follows that γ is a rational curve of degree $2p_k/d$. $\square$

Remark 6.5. Theorem 6.8 is false if the curve is non-closed; e.g., a right circular helix in $\mathbb{E}^3$ is a null 2-type curve which is not rational.

Remark 6.6. It was proved in [Petrović-Torgašev (2008)] that every finite type closed curve in $\mathbb{E}^m$ lies in the zero set of a polynomial of m variables.

For k-type curves with $k \geq 2$, we have the following.

Theorem 6.9. [Chen et al. (1990c)] *For every integer $k \geq 2$ there exist infinitely many non-equivalent curves of k-type in $\mathbb{E}^3$.*

Proof. We treat 2-type curves in $\mathbb{E}^3$ first. For each $\varepsilon \in \mathbb{R}_0^+$,

$$\gamma(s) = \frac{12}{\varepsilon^2 + 36} \left(\varepsilon \sin s, -\frac{\varepsilon^2}{12} \cos s + \cos 3s, -\frac{\varepsilon^2}{12} \sin s + \sin 3s \right)$$

is a 2-type curve. Easy computation shows that

$$A_{22}(\varepsilon) = A_{21}(\varepsilon) = \bar{A}_{22}(\varepsilon) = \bar{A}_{21}(\varepsilon) = \bar{A}_{11}(\varepsilon) = \bar{D}_{21}(\varepsilon) = 0,$$

$$A_{11}(\varepsilon) = \frac{-144\varepsilon^2}{(\varepsilon^2 + 36)^2}, \quad D_{22}(\varepsilon) = \frac{288}{(\varepsilon^2 + 36)^2},$$

$$D_{21}(\varepsilon) = \frac{-24\varepsilon^2}{(\varepsilon^2 + 36)^2}, \quad D_{11}(\varepsilon) = \frac{2\varepsilon^4 + 144\varepsilon^2}{(\varepsilon^2 + 36)^2},$$

where $A_{ij}, D_{ij}, \bar{A}_{ij}, \bar{D}_{ij}$ are defined by (6.52). It is known that $\gamma(\varepsilon_1)$ and $\gamma(\varepsilon_2)$ are equivalent only if

$$\frac{D_{22}(\varepsilon_1)}{D_{22}(\varepsilon_2)} = \frac{D_{21}(\varepsilon_1)}{D_{21}(\varepsilon_2)}.$$

Hence only if $\varepsilon_1 = \varepsilon_2$. Next, let $k \geq 3$. First, choose $p_1, p_2, \ldots, p_k \in \mathbb{N}_0$ such that $2p_1 = p_k - p_2$ and, for a number $d \in \mathbb{N}_0$ such that $(k-2)d = 2p_1$, $p_{i+1} = p_i + d$ for all $i \in \{2, 3, \ldots, k-1\}$. We will determine suitable $c, u_2, \ldots, u_k, r \in \mathbb{R}_0$ such that the curve defined by

$$\gamma(s) = \sum_{i=1}^{k} \left(a_i \cos \frac{p_i s}{r} + b_i \sin \frac{p_i s}{r} \right),$$

$$a_1 = (0, 0, 0), \quad b_1 = \left(0, 0, \frac{c}{p_1} \right), \tag{6.68}$$

$$a_i = \left(\frac{u_i}{p_i}, 0, 0 \right), \quad b_i = \left(0, \frac{u_i}{p_i}, 0 \right), \quad i = 2, \ldots, k,$$

is a k-type curve parametrized by arc length. Easy computation shows that

$$A_{ij} = A_{i1} = D_{i1} = \bar{A}_{rs} = \bar{D}_{rs} = 0,$$

$$A_{11} = -\frac{c^2}{p_1^2}, \quad D_{ij} = \frac{2u_i u_j}{p_i p_j}, \quad D_{11} = \frac{c^2}{p_1^2}, \tag{6.69}$$

for all $i, j \in \{2, \ldots, k\}$; $r, s \in \{1, \ldots, k\}$. It is direct to verify that γ is a curve of finite type parametrized by arc length if and only if

$$\sum_{i=2}^{k} u_i^2 + \frac{c^2}{2} = r^2, \tag{6.70}$$

$$c^2 + 4u_2 u_k = 0, \tag{6.71}$$

$$\sum_{i=2}^{k-\ell} u_i u_{i+\ell} = 0 \quad \text{for all } \ell \in \{1, 2, \ldots, k-3\}. \tag{6.72}$$

For $u_2, \ldots, u_k \in \mathbb{R}_0$, let us put $p(X) = \sum_{i=0}^{k-2} u_{i+2} X^i$.

The following lemma provides a way to find solutions for the system (6.70)-(6.72) of equations.

Lemma 6.4. $c, u_2, \ldots, u_k$ *satisfy equations* (6.71) *and* (6.72) *when and only when* $c^2 = -4u_2 u_k$ *and there exist* $\lambda \in \mathbb{R}_0^-$ *and* $\mu \in \mathbb{R}_0^+$ *such that*

$$p(X)p\left(\frac{1}{X}\right) = \lambda(X^{k-2} + X^{2-k}) + \mu.$$

Proof. It is clear that

$$p(X)p\left(\frac{1}{X}\right) = u_2 u_k X^{k-2} + \sum_{\ell=1}^{k-3} \left(\sum_{i=2}^{k-\ell} u_i u_{i+\ell}\right) X^\ell + \sum_{i=2}^{k} u_i^2$$

$$+ \sum_{\ell=1}^{k-3} \left(\sum_{i=2}^{k-\ell} u_i u_{i+\ell}\right) X^{-\ell} + u_2 u_k X^{2-k}. \tag{6.73}$$

Thus $u_2, \ldots, u_k$ satisfy (6.72) if and only if

$$p(X)p\left(\frac{1}{X}\right) = \lambda(X^{2+k} + X^{2-k})\alpha + \mu$$

for $\lambda, \mu \in \mathbb{R}$. Moreover, it is clear that $\lambda \in \mathbb{R}_0^-$ and $\mu \in \mathbb{R}_0^+$. $\qquad \square$

For each $\mu \in (2, \infty)$, we determine a polynomial $p(X)$ which satisfies the equation $p(X)p\left(\frac{1}{X}\right) = -(X^{2+k} + X^{2-k}) + \mu$. To do so, let us put

$$\alpha = \sqrt[k-2]{\frac{\mu + \sqrt{\mu^2 - 4}}{2}}.$$

Then we have

$$p(X)p\left(\frac{1}{X}\right) = -\frac{(X^{k+2} - \alpha^{k-2})(X^{k-2} - \alpha^{2-k})}{X^{k-2}}$$

$$= -\frac{(X - \alpha)\left(X - \alpha^{-1}\right)\left(\sum_{i=0}^{k-3} X^i \alpha^{k-3-i}\right)\left(\sum_{i=0}^{k-3} X^i \alpha^{3+i-k}\right)}{X^{k-2}}. \tag{6.74}$$

It is now easy to see that $p(X) = \sqrt{\alpha^{4-k}}(X - \alpha^{-1}) \sum_{i=0}^{k-3} X^i \alpha^{k-3-i}$ is a possible choice for $p(X)$. This gives, after multiplication with a suitable constant, the following solution for the system (6.70)-(6.72) of equations:

$$u_k = 1, \quad u_2 = -\alpha^{k-4},$$
$$u_i = (\alpha^2 - 1)\alpha^{k-2-i} \text{ for all } i \in \{3, \ldots, k-1\}, \tag{6.75}$$
$$c = 2\sqrt{\alpha^{k-4}}, \quad r = \sqrt{\alpha^{2k-6} + \alpha^{-2} + 2\alpha^{k-4}}.$$

Call, for $\alpha \in (1, \infty)$, the curve

$$\gamma_\alpha(s) = \sum_{i=1}^{k} \left(a_i \cos\left(\frac{p_i s}{r}\right) + b_i \sin\left(\frac{p_i s}{r}\right) \right),$$

where the a_i, b_i, u_i, c and r are determined by (6.68) and (6.75). It is direct to verify that γ_{α_1} and γ_{α_2} are equivalent only if

$$\frac{D_{kk}(\alpha_1)}{D_{kk}(\alpha_2)} = \frac{D_{kk-1}(\alpha_1)}{D_{kk-1}(\alpha_2)}.$$

Hence, by (6.63) and (6.75), only if $\alpha_1 = \alpha_2$. $\quad\square$

All 2-type curves in Euclidean spaces were classified as follows.

Theorem 6.10. [Chen et al. (1986)] *Let γ be a 2-type closed curve in $\mathbb{E}^m$. Then either γ is a W-curve of rank 4 or it is congruent to*

$$\gamma(s) = \frac{12}{p\sqrt{(\varphi^2+\delta^2+\varepsilon^2+36)^2 - 4\varphi^2\varepsilon^2}} \left(\varphi \cos\frac{ps}{r} + \sin\frac{ps}{r}, \right.$$

$$\varepsilon \sin\frac{ps}{r}, \frac{1}{12}(\varphi^2-\delta^2-\varepsilon^2)\cos\frac{ps}{r} - \frac{\delta\varphi}{6}\sin\frac{ps}{r} + \cos\frac{3ps}{r},$$

$$\left. \frac{\delta\varphi}{6}\cos\frac{ps}{r} + \frac{1}{12}(\varphi^2-\delta^2-\varepsilon^2)\sin\frac{ps}{r} + \sin\frac{3ps}{r}, 0, \ldots, 0 \right),$$

where p is an integer ≥ 1 and $r, \varepsilon, \varphi, \delta$ are real numbers with $r, \varepsilon > 0$.

Remark 6.7. All 2-type curves in $\mathbb{E}^3$ are unknotted [Chen et al. (1986)]. On the other hand, some knotted 4-type curves in $\mathbb{E}^3$ were constructed in [Blair (1995)].

Problem 6.1. *It is not known whether there exist knotted 3-type curves in $\mathbb{E}^3$.*

Remark 6.8. The classification of non-closed 2-type curves were given in [Dillen et al. (1991)].

Remark 6.9. It was proved in [Chen et al. (1986)] that all closed 2-type spherical curves are W-curves. On the other hand, it was shown in [Deprez et al. (1990)] that every 2-type curve in $\mathbb{E}^3$ lies either on a hyperboloid of revolution of one sheet or on a cone of revolution.

Curves of finite type in $\mathbb{E}^3$ whose images lie in a quadratic surface, other than spheres, were investigated in [Deprez et al. (1990)].

Theorem 6.11. [Deprez et al. (1990)] *Let γ be a closed curve of finite type in $\mathbb{E}^3$ whose image lies in a quadratic surface Q. Then either*

(1) *γ is a circle, or*
(2) *the quadratic surface Q is one of the following surfaces:*

  (2.a) *an ellipsoid of revolution which is not a sphere,*
  (2.b) *a (one- or two-sheeted) hyperboloid of revolution, or*
  (2.c) *a circular cone.*

Theorem 6.11 implies the following.

Corollary 6.2. [Deprez et al. (1990)] *A finite type curve in $\mathbb{E}^3$ is a circle if it lies in one of the following surfaces: a general ellipsoid, a general hyperboloid, an elliptic cone, a hyperbolic cylinder or an elliptic cylinder.*

Examples of curves of finite type lying on certain ellipsoids of revolution, on circular cones or on one-sheeted hyperboloids of revolution were constructed in [Deprez et al. (1990)].

Problem 6.2. *It is not known whether there are curves of finite type on two-sheeted hyperboloids of revolution.*

3-type curves lying ellipsoids of revolution, on hyperboloids or on cones of revolutions have been determined in [Petrović et al. (1995, 1996)]. It was proved in [Deprez et al. (1990)] that there do not exist 3-type curves on paraboloids. In fact, the only finite type curves on paraboloids are circles.

We also have the following results on 3-type curves.

Theorem 6.12. [Blair (1995)] *A closed 3-type curve in $\mathbb{E}^3$ is either a curve which lies on a quadric of revolution or a curve whose frequency ratio is $1 : 3 : 7$ and the curve belongs to a 3-parameter family of such curves, or the frequency ratio is $1 : 3 : 5$ and the curve belongs to a 5-parameter family of such curves. Some curves with frequency ratio $1 : 3 : 5$ or $1 : 3 : 7$ also lie on quadrics of revolution.*

Curves of constant precession in a Euclidean space are defined by the property that, being transformed with unit speed, their *centrode* revolves about a fixed axis with constant angle and constant speed (see [Scofield (1995)]). Here by centrode for a curve $\gamma(s)$ we meant the vector

$$C(s) = \tau(s)T(s) + \kappa(s)B(s),$$

which at every point $\gamma(s)$ of the curve determines the axis of instantaneous rotation of the Frenet frame $\{T(s), N(s), B(s)\}$ moving along the curve.

P. D. Scofield obtained explicit parameterizations of curves of constant precession in $\mathbb{E}^3$. In particular, he proved the following.

Theorem 6.13. [Scofield (1995)] *All curves of constant precession in $\mathbb{E}^3$ are either of 2-type or of 3-type.*

The following results for finite type spherical curves were obtained in [Chen et al. (1990a)].

Theorem 6.14. *Every curve of finite type in a 2-sphere is a circle.*

Theorem 6.15. *Every curve of finite type in a 3-sphere is either a circle or a closed W-curve of rank 4.*

For further results on finite type curves in Euclidean spaces or spheres or hyperbolic 3-pace, see [Chen et al. (1990a); Chung et al. (1995); Dillen et al. (1995a); Ishikawa and Miyasato (1993); Petrović-Torgašev (2002, 2004); Petrović et al. (1995, 1996); Pyo and Lee (1998); Šućurović (2000, 2001); Verstraelen (1990)].

6.7 Classification of 1-type submanifolds

The following result provides the classification of 1-type submanifolds.

Theorem 6.16. [Chen (1986b)] *Let $\phi : M \to \mathbb{E}_s^m$ be an isometric immersion of a pseudo-Riemannian manifold into $\mathbb{E}_s^m$. Then M is of 1-type if and only if it is one of the following three types of submanifolds:*

(a) *A minimal submanifold of $\mathbb{E}_s^m$;*
(b) *A minimal submanifold of a pseudo-sphere $S_s^{m-1}(c) \subset \mathbb{E}_s^m$, $c > 0$;*
(c) *A minimal submanifold of a pseudo-hyperbolic space $H_{s-1}^{m-1}(c) \subset \mathbb{E}_s^m$, $c < 0$.*

Proof. If M is of 1-type in $\mathbb{E}_s^{m+1}$, we have the spectral decomposition:

$$\phi = \mathbf{c} + \phi_p, \tag{6.76}$$

where $\mathbf{c} \in \mathbb{E}_s^{m+1}$ and ϕ_p is a non-constant map such that $\Delta\phi_p = \lambda_p\phi_p$ for some eigenvalue λ_p. If $\lambda_p = 0$, then Beltrami's formula implies that M is minimal in $\mathbb{E}_s^{m+1}$. This gives (a). If $\lambda_p \neq 0$, Beltrami's formula gives

$$nH = \lambda_p(\mathbf{c} - \phi). \tag{6.77}$$

By differentiating (6.77), we obtain $n\tilde{\nabla}_X H = -\lambda_p X, \forall X \in TM$. Thus we find $nA_H = \lambda_p I$ and $DH = 0$, which imply that M is a pseudo-umbilical submanifold with parallel mean curvature vector. Hence, by Proposition 4.16, the immersion is given either by (b) or (c), or it is congruent to

$$\phi = (f, \psi, f), \tag{6.78}$$

where $\psi : M \to \mathbb{E}_{s-1}^{m-1}$ is a minimal immersion and $\Delta f = r$ for some nonzero real number r. If the last case occurs, (6.77), (6.78) and Beltrami's formula imply that $\lambda_p\phi = (r, 0, r) + \lambda_p\mathbf{c}$, which implies that ϕ is a constant vector. Hence the last case is impossible. The converse can be verified easily. $\square$

Theorem 6.16 implies the following result.

Corollary 6.3. [Takahasi (1966)] *If $\phi : M \to \mathbb{E}^m$ is an isometric immersion, then $\Delta\phi = \lambda\phi$ for some $\lambda \in \mathbf{R}$ if and only if either M is a minimal submanifold of $\mathbb{E}^m$ or a minimal submanifold of a hypersphere of $\mathbb{E}^m$.*

6.8 Submanifolds of finite type in Euclidean space

The class of submanifolds of finite type is very large. For instance, it follows from Theorem 6.9 that, for each integer $k \geq 2$, there exist k-type submanifolds in $\mathbb{E}^{n+2}$. On the other hand, very few hypersurfaces of finite type are known. For example, no surfaces of finite type in Euclidean 3-space are known, other than minimal surfaces, circular cylinders and spheres.

In views of these facts, the author made the following two conjectures in [Chen (1987a, 1991b, 1996b)].

Conjecture 6.1. *The only finite type closed hypersurfaces of a Euclidean space are the hyperspheres.*

Conjecture 6.2. *The only finite type surfaces in $\mathbb{E}^3$ are minimal surfaces, and open portions of spheres and circular cylinders.*

In this section we present many results obtained by various geometers which support these two conjectures.

Lemma 6.5. *Every anchor ring in* $\mathbb{E}^3$ *is of infinite type.*

Proof. Let A be an anchor ring in $\mathbb{E}^3$. Without loss of generality, we may assume that the position vector of A is given by

$$\mathbf{x}(t,\theta) = \left(\delta \cos \frac{t}{a}, \delta \sin \frac{t}{a}, r \sin \theta \right), \tag{6.79}$$

where $\delta = a + r \cos \theta$ for some real number a, r satisfying $a > r > 0$. It is easy to check that the Laplacian of A is given by

$$\Delta = \frac{\sin \theta}{r \delta} \frac{\partial}{\partial \theta} - \frac{1}{r^2} \frac{\partial^2}{\partial \theta^2} - \frac{a^2}{\delta^2} \frac{\partial^2}{\partial t^2}. \tag{6.80}$$

Put $z = r \sin \theta$ which is the third coordinate of A. We have

$$\Delta z = \frac{\sin \theta \cos \theta}{\delta} + \frac{\sin \theta}{r}. \tag{6.81}$$

By direct computation we find

$$\Delta^2 z = \frac{\sin^3 \theta \cos \theta}{\delta^3} + \frac{1}{\delta^2} P_2(\cos \theta, \sin \theta), \tag{6.82}$$

where $P_2(u,v)$ is a polynomial of degree 3. For each integer $m > 0$, we have

$$\Delta \left(\frac{\sin^m \theta \cos \theta}{\delta^m} \right) = -\frac{m^2 \sin^{m+2} \theta \cos \theta}{\delta^{m+2}}$$

$$- \frac{\sin^m \theta}{r \delta^{m+1}} \left\{ 2m^2 \cos^2 \theta - (2m-1) \sin^2 \theta \right\} \tag{6.83}$$

$$+ \frac{\sin^{m-2} \theta \cos \theta}{r^2 \delta^m} \left\{ (3m+1) \sin^2 \theta - m(m-1) \cos^2 \theta \right\}.$$

From (6.80) and (6.82) we find

$$\Delta^3 z = -\Delta \left(\frac{\sin^3 \theta \cos \theta}{\delta^3} \right) - \frac{4 \sin^2 \theta}{\delta^4} P_2(\cos \theta, \sin \theta)$$

$$+ \frac{1}{r \delta^3} \left\{ 2 \cos \theta \, P_2(\cos \theta, \sin \theta) - 3 \sin \theta \frac{d}{d\theta} P_2(\cos \theta, \sin \theta) \right\} \tag{6.84}$$

$$- \frac{1}{r^2 \delta^2} \frac{d^2}{d\theta^2} P_2(\cos \theta, \sin \theta).$$

It follows from (6.83) and (6.84) that

$$\Delta^3 z = 3^2 \cdot \frac{\sin^5 \theta \cos \theta}{\delta^5} + \frac{1}{\delta^4} P_3(\cos \theta, \sin \theta),$$

where $P_3(u, v)$ is a polynomial of degree 5. By induction we obtain

$$\Delta^{k+1} z = (-1)^k 1^2 \cdot 3^2 \cdots (2k-1)^2 \cdot \frac{\sin^{2k+1} \theta \cos \theta}{\delta^{2k+1}}$$
$$+ \frac{1}{\delta^{2k}} P_{k+1}(\cos \theta, \sin \theta) \tag{6.85}$$

for $k \geq 1$, where $P_{k+1}(u, v)$ is a polynomial of degree $2k + 1$.

If the anchor ring A is of finite type, then by the minimal polynomial criterion (Theorem 6.1), there exist constants $c_1, \ldots, c_k$ for some $k \geq 1$ such that $\Delta^{k+1} \mathbf{x} + c_1 \Delta^k \mathbf{x} + \cdots + c_k \Delta \mathbf{x} = 0$. In particular, we have

$$\Delta^{k+1} z + c_1 \Delta^k z + \cdots + c_k \Delta z = 0. \tag{6.86}$$

From (6.81), (6.85) and (6.86) we conclude that there exists a polynomial Q of degree $2k + 1$ such that

$$\frac{\sin^{2k+1} \theta \cos \theta}{a + r \cos \theta} = Q(\cos \theta, \sin \theta),$$

which implies that

$$\frac{\sin \theta \cos \theta (1 + \cos \theta)^k (1 - \cos \theta)^k}{a + r \cos \theta} = Q(\cos \theta, \sin \theta)$$

for some polynomial Q. But this is impossible for any $k \geq 1$ because $a > r > 0$. Consequently, A must be of infinite type. $\square$

The first classification result for surfaces of finite type in a Euclidean 3-space is the following.

Theorem 6.17. [Chen (1987a)] *A tube in $\mathbb{E}^3$ is of finite type if and only if it is an open portion of a circular cylinder.*

Proof. Let $\gamma(s)$ be a closed unit speed curve of length ℓ and $T = \gamma'(s)$. Denote by N, B the principal and binormal vector fields of γ and κ and τ the curvature and torsion of γ. Let $T_r(\gamma)$ be the tube with a sufficiently small radius r around γ. Then the position vector of $T_r(\gamma)$ is given by

$$\mathbf{x}(s, \theta) = \gamma(s) + r(\cos \theta) N + r(\sin \theta) B. \tag{6.87}$$

We put

$$\beta = \kappa' \cos \theta + \kappa \tau \sin \theta, \quad \mu = 1 - r\kappa \cos \theta. \tag{6.88}$$

If $\beta = 0$, then $\tau = 0$ and κ is a constant. Thus γ lies either in a circle or in a line. If γ lies in a circle, the tube is an anchor ring, which is of infinite type according to Lemma 6.5. If γ lies in a line, the tube is an open portion of a circular cylinder.

Now, let us assume that $\beta \neq 0$. In this case, a direct computation yields

$$\Delta \mathbf{x} = -\left(\frac{\kappa \cos \theta}{\mu}\right) N + \frac{1}{r} N. \tag{6.89}$$

It follows from (6.87) and (6.88) that

$$\Delta = -\frac{1}{\mu^3} \left\{ r\beta \frac{\partial}{\partial t} - \left[r\tau\beta + \tau'\mu - \frac{1}{r} \left(\kappa\mu^2 \sin\theta \right) \right] \frac{\partial}{\partial \theta} \right. \tag{6.90}$$
$$\left. + \mu \frac{\partial^2}{\partial t^2} - 2\tau\mu \frac{\partial^2}{\partial t \partial \theta} + \frac{1}{r^2} (\mu^3 + r^2 \mu \tau^2) \frac{\partial^2}{\partial \theta^2} \right\}.$$

For simplicity, we put

$$V = -\sin\theta \, N + \cos\theta \, B, \quad n = \cos\theta \, N + \sin\theta \, B.$$

Then V, T, n form an orthonormal frame such that V and T are tangent to the tube and n is normal to the tube. We have

$$\frac{\partial \mathbf{x}}{\partial s} = \mu T + r\tau V, \quad \frac{\partial \mathbf{x}}{\partial \theta} = rV, \tag{6.91}$$

$$\frac{\partial T}{\partial \theta} = 0, \quad \frac{\partial V}{\partial \theta} = -n, \quad \frac{\partial n}{\partial \theta} = V, \tag{6.92}$$

and

$$\frac{\partial T}{\partial s} = -\kappa \sin\theta \, V + \kappa \cos\theta \, n,$$

$$\frac{\partial V}{\partial s} = \kappa \sin\theta \, T - \tau n, \tag{6.93}$$

$$\frac{\partial n}{\partial s} = -\kappa \cos\theta \, T + \tau V.$$

By applying (6.89), (6.90), (6.92) and (6.93) and a long straightforward computation, we may obtain

$$\Delta^2 \mathbf{x} = \frac{1}{r\mu^4} \left\{ \beta(1 - 4r\kappa \cos\theta)T + r \left(\frac{\partial \beta}{\partial s} - \tau \frac{\partial \beta}{\partial \theta} \right) n \right\}$$
$$+ \frac{\kappa^2}{r\mu^3} (1 - 2r\kappa \cos^3\theta)n - \frac{\kappa}{r^2\mu^2} (3\sin\theta \, V + \cos\theta \, n) \tag{6.94}$$
$$- \frac{\kappa \cos\theta}{r^2\mu} n + \frac{n}{r^3} + \left(\frac{3r\beta^2}{\mu^5} \right) n.$$

In particular, this shows that $\Delta^2 \mathbf{x}$ can be written as

$$\Delta^2 \mathbf{x} = \left(\frac{3r\beta^2}{\mu^5} \right) n + \frac{1}{\mu^4} R_2(\cos\theta, \sin\theta), \tag{6.95}$$

where R_2 is a $\mathbb{E}^3$-valued polynomial of two variables with coefficients given by some functions of s.

We need the following.

Lemma 6.6. *For any integers $j, k \geq 1$, we have*

$$\Delta\left(\frac{\beta^k}{\mu^j}\right) = -\frac{j(j+2)r^2\beta^{k+2}}{\mu^{j+4}} + Q_{k,j}(\cos\theta, \sin\theta) \qquad (6.96)$$

for some polynomial $Q_{k,j}$ of two variables with with some functions of s as its coefficients.

This lemma follows from (6.90), (6.90) and a direct computation.

By applying (6.90), (6.95) and Lemma 6.6 and induction, we find

$$\begin{aligned}
\Delta^{k+1}\mathbf{x} &= \frac{(-1)^{k+1} \cdot (4k-1)!}{2^{2k-1} \cdot (2k-1)!} \cdot \frac{r^{2k-1}\beta^{2k}}{\mu^{4k+1}}\, n \\
&+ \frac{1}{\mu^{4k}} H_{k+1}(\cos\theta, \sin\theta), \quad k \geq 1,
\end{aligned} \qquad (6.97)$$

where H_{k+1} is a $\mathbb{E}^3$-valued polynomial of two variables with some functions of s as its coefficients. If the tube is of finite type, then by the minimal polynomial criterion, there are constants $c_1, \ldots, c_k$, $k \geq 1$, such that

$$\Delta^{k+1}\mathbf{x} + c_1\Delta^k\mathbf{x} + \cdots + c_k\Delta\mathbf{x} = 0.$$

Thus, by applying (6.89), (6.94) and (6.97), we see that there exists a polynomial Q of two variables with functions of s as coefficients such that

$$\frac{(\kappa'\cos\theta + \kappa\tau\sin\theta)^{2k}}{1 - r\kappa\cos\theta} = Q(\cos\theta, \sin\theta).$$

Since r is small, this is impossible unless $\kappa \equiv 0$ which implies that T is a circular cylinder. The converse is easy to verify. $\qquad \square$

A classical result of [Catalan (1842)] states that the only ruled minimal surfaces in $\mathbb{E}^3$ are the plane and the helicoid. The next theorem is a vast generalization of Catalan's result.

Theorem 6.18. [Chen et al. (1990b)] *A ruled surface M in $\mathbb{E}^m$ is of finite type if and only if it is either a cylinder over a curve of finite type or an open portion of a helicoid lying in a totally geodesic $\mathbb{E}^3 \subset \mathbb{E}^m$.*

Proof. Let M be a ruled surface in $\mathbb{E}^m$. We consider the following two cases separately.

Case 1. M is a cylinder. Assume M is a cylinder over a unit speed curve $\gamma(s)$ lying in an affine hyperplane $\mathbb{E}^{m-1}$ defined by the equation $x_m = 0$.

Then a parameterization of M is given by $\mathbf{x}(s,t) = \gamma(s) + te_n$, where e_n is a unit vector in x_m-direction. The Laplacian Δ of M is given by $\Delta = -\frac{\partial^2}{\partial s^2} - \frac{\partial^2}{\partial t^2}$, and the Laplacian Δ' of γ is $\Delta' = -\frac{\partial^2}{\partial s^2}$. Certainly we have $\Delta x_n = \Delta t = 0$. Thus M is of finite type if and only if each component of $\gamma(s)$ can be written as a finite sum of eigenfunctions of Δ, i.e.,

$$\gamma(s) = \Gamma_0 + \sum_{i=1}^{k} \Gamma_i(s,t) \tag{6.98}$$

with $\Delta\Gamma_i = \lambda_i\Gamma_i$. Assume that all the λ_i are mutually different. If we apply $\prod_{i=2}^{k}(\Delta - \lambda_i)$ to (6.98), we see that Γ_1 does not depend on t. Similarly we find that none of Γ_i depend on t. Moreover

$$\Delta'\Gamma_i(s) = -\frac{\partial^2}{\partial s^2}\Gamma_i(s) = -\frac{\partial^2}{\partial s^2}\Gamma_i(s) - \frac{\partial^2}{\partial t^2}\Gamma_i(s) = \Delta\Gamma_i(s) = \lambda_i\Gamma_i(s)$$

for all i. Hence every component of γ can be written as a finite sum of eigenfunctions of Δ'. This means that γ is of finite type. Thus M is of finite type if and only if γ is of finite type. Moreover, if γ is of k-type, then M is of $(k+1)$-type, unless one of the eigenfunctions which appear in the decomposition of γ has eigenvalue 0, in which case M is of k-type too.

Case 2. M is not cylindrical. In this case, we can decompose M into open pieces such that on each piece we can find a parameterization of the form $\mathbf{x}(s,t) = \alpha(s) + t\beta(s)$, where α and β are curves in $\mathbb{E}^m$ such that $\langle \alpha', \beta \rangle = 0$, $\langle \beta, \beta \rangle = 1$ and $\langle \beta', \beta' \rangle = 1$. If we define a function q by

$$q = t^2 + 2\langle \alpha', \beta' \rangle t + \langle \alpha, \alpha' \rangle = \|\alpha' + t\beta'\|^2,$$

it is easy to show that the Laplacian Δ of M can be expressed as follows

$$\Delta = -\frac{\partial^2}{\partial t^2} - \frac{1}{q}\frac{\partial^2}{\partial s^2} + \frac{1}{2}\frac{\partial q}{\partial s}\frac{1}{q^2}\frac{\partial}{\partial s} - \frac{1}{2}\frac{\partial q}{\partial t}\frac{1}{q}\frac{\partial}{\partial t}.$$

We need the following lemma which can be proved by a straightforward computation.

Lemma 6.7. *If P is a polynomial in t with functions in s as coefficients and $\deg P = d$, then*

$$\Delta\left(\frac{P(t)}{q^m}\right) = \frac{\tilde{P}(t)}{q^{m+3}},$$

where $\tilde{P}$ is a polynomial in t with functions in s as coefficients such that $\deg(\tilde{P}) \leq d + 4$.

From now on we suppose that M is of k-type. Hence there exist numbers $c_1, \ldots, c_k$ such that

$$\Delta^{k+1}\mathbf{x} + c_1\Delta^k\mathbf{x} + \cdots + c_k\Delta\mathbf{x} = 0. \tag{6.99}$$

We know that every component of $\mathbf{x}$ is a linear function in t with functions in s as coefficients. By applying Lemma 6.7, we easily obtain that

$$\Delta^r\mathbf{x} = \frac{P_r(t)}{q^{3r-1}},$$

where P_r is a vector whose components are polynomials in t with functions in s as coefficients and $\deg(P_r) \leq 1 + 4r$. Hence if r goes up by one, the degree of the numerator of any component of $\Delta^r\mathbf{x}$ goes up by at most 4, while the degree of the denominator goes up by 6. Hence the sum (6.99) can never be zero, unless of course $\Delta\mathbf{x} = 0$. But then M is minimal. $\square$

By combining Theorems 6.7 and 6.18 we obtain the following.

Theorem 6.19. [Chen et al. (1990b)] *A ruled surface in $\mathbb{E}^3$ is of finite type if and only if it is an open portion of a plane, a circular cylinder or a helicoid.*

The following is an immediate consequence of Theorem 6.19.

Corollary 6.4. *A flat surface in $\mathbb{E}^3$ is of finite type if and only if it is an open part of a plane or of a circular cylinder.*

Remark 6.10. Theorem 6.18 was extended to ruled submanifolds of finite type in a Euclidean space in [Dillen (1992)].

For algebraic surfaces of degree 2 we have the following.

Theorem 6.20. [Chen and Dillen (1990b)] *The only quadrics in $\mathbb{E}^3$ of finite type are open portions of spheres and of the circular cylinders.*

Let $S^p(r)$ be the hypersphere of $\mathbb{E}^{p+1}$ with radius r centered at the origin. Denote by $M_{p,q}$ the product of two spheres:

$$S^p\left(\sqrt{\frac{p}{p+q}}\right) \times S^q\left(\sqrt{\frac{q}{p+q}}\right) \subset S^{p+q+1}(1) \subset \mathbb{E}^{p+q+2}.$$

Let $C_{p,q}$ be the $(p+q+1)$-dimensional cone in $\mathbb{E}^{p+q+2}$ with vertex at the origin shaped on $M_{p,q}$. It is easy to verify that $C_{p,0}$ and $C_{0,q}$ are hyperplanes in $\mathbb{E}^{p+2}$ and $\mathbb{E}^{q+2}$, respectively, and $C_{p,q}$ is an algebraic hypersurface of degree two for each pair (p,q) of positive integers.

Theorem 6.20 was extended to the following.

Theorem 6.21. [Chen et al. (1992)] *A quadric hypersurface M of $\mathbb{E}^{n+1}$ is of finite type if and only if it is one of the following hypersurfaces:*

(1) *a hypersphere;*
(2) *one of the algebraic cones $C_{p,n-p-1}$, $0 < p < n - 1$;*
(3) *a spherical hypercylinder $\mathbb{E}^k \times S^{n-k}$, $0 < k < n$;*
(4) *the product of a linear subspace $\mathbb{E}^\ell$ and one of the algebraic cones $C_{p,n-\ell-p-1}$ with $0 < p < n - \ell - 1$.*

O. J. Garay considered finite type cones and proved the following.

Theorem 6.22. [Garay (1988b)] *Open portions of hyperplanes are the only cones in a Euclidean space of finite type.*

A *spiral surface* in $\mathbb{E}^3$ is a surface generated by rotating a plane curve γ about an axis L contained in the plane of the curve γ and simultaneously transforming γ homothetically relative to a point of L.

For spiral surfaces of finite type we have the following result.

Theorem 6.23. [Baikoussis and Verstraelen (1995)] *A spiral surface in $\mathbb{E}^3$ is of finite type if and only if it is a minimal surface.*

In 1822, C. Dupin defines a *cyclide* to be a surface of $\mathbb{E}^3$ which is the envelope of the family of spheres tangent to three fixed spheres [Dupin (1822)]. It was shown to be equivalent to requiring that both sheets of the focal set degenerate into curves. The cyclides are equivalently characterized by requiring that the lines of curvatures in both families be arcs of circles or straight lines. Thus one can obtain three obvious examples: a torus of revolution, a circular cylinder and a circular cone. It turns out that all cyclides can be obtained from these three by inversions in a sphere in E^3 (cf. e.g., [Cecil and Ryan (1985)]). In the next theorem, the term cyclide of Dupin is used in the narrow sense, i.e., exclusive the special limiting cases of spheres, cylinders, cones and tori of revolution.

Theorem 6.24. [Defever et al. (1993, 1994)] *All closed and non-closed cyclides of Dupin are of infinite type.*

A hypersurface M of $\mathbb{E}^{n+1}$ is called a *translation hypersurface* if it is a nonparametric hypersurface of the form:

$$x_{n+1} = P_1(x_1) + \cdots + P_n(x_n),$$

where each P_i is a function of one variable. If $P_1, \ldots, P_n$ are polynomials, the hypersurface is called a *polynomial translation hypersurface*.

For polynomial translation hypersurfaces, we have the following.

Theorem 6.25. [Dillen et al. (1995c)] *A polynomial translation hypersurface of a Euclidean space is of finite type if and only if it is a hyperplane.*

M. Barros investigated surfaces in $\mathbb{E}^3$ which are images under stereographic projection of minimal surfaces in S^3 and proved the following.

Theorem 6.26. [Barros (1992)] *There exist no 2-type surfaces in $\mathbb{E}^3$ which are images under stereographic projection of a minimal surface in S^3.*

A surface in $\mathbb{E}^3$ is called a *rotational surface* if it is generated by a curve γ on a plane π when π is rotated around a straight line L in π. By choosing π to be the xz-plane and line L to be the z-axis, a rotational surface can be parameterized by

$$\mathbf{x}(u, v) = (f(u) \cos v, f(u) \sin v, g(u)). \qquad (6.100)$$

A rotational surface of revolution defined by (6.100) is said to be *of polynomial kind* if both $f(u)$ and $g(u)$ are polynomial functions in u; it is said to be *of rational kind* if g is a rational function in f, i.e., g is the quotient of two polynomial functions in f (cf. [Chen and Ishikawa (1993)]).

For rotational surfaces, we have the following results.

Theorem 6.27. [Chen and Ishikawa (1993)] *Let M be a rotational surface of polynomial kind. Then M is of finite type if and only if it is either an open portion of a plane or an open portion of a circular cylinder.*

Theorem 6.28. [Chen and Ishikawa (1993)] *Let M be a rotational surface of rational kind. Then M is a surface of finite type if and only if it is an open portion of a plane.*

Theorem 6.29. [Hasanis-Vlachos (1993)] *Let M be a rotational surface with constant mean curvature in $\mathbb{E}^3$. If M is of finite type, then it is an open portion of a plane, of a sphere or of a circular cylinder.*

Theorem 6.30. [Arroyo et al. (1998)] *Let M be a finite type closed surface of revolution in $\mathbb{E}^3$. If there are at least two ellipses in M through each point of M, then it is an ordinary sphere.*

All of the results in this section support Conjectures 6.1 and 6.2.

For finite type graph hypersurfaces in isotropic geometry, see [Chen et al. (2014)].

6.9   2-type spherical hypersurfaces

A Riemannian manifold M immersed in a hypersphere $S^{m-1}(r) \subset \mathbb{E}^m$ is called *mass-symmetric* in $S^{m-1}(r)$ if the center of mass of M is the center of the hypersphere $S^{m-1}(r)$ in $\mathbb{E}^m$. Let $S_o^{m-1}(r)$ denote the hypersurface with radius r and centered at the origin $o \in \mathbb{E}^m$.

Lemma 6.8. *Every closed minimal submanifold M of $S^{m-1}(r) \subset \mathbb{E}^m$ is mass-symmetric in $S^{m-1}(r)$.*

Proof. Without loss of generality, we may assume that the center of $S^{m-1}(r)$ is the origin o of $\mathbb{E}^m$. By Beltrami's formula and Hopf's lemma, we find $\int_M H\,dV = 0$, where H is the mean curvature vector of M in $\mathbb{E}^m$. Since M is a minimal in $S^{m-1}(1)$, we get $H = -\mathbf{x}$, where $\mathbf{x}$ is the position vector field of M. Then $\int_M \mathbf{x}\,dV = -\int_M H\,dV = 0$, which implies that M is mass-symmetric in $S^{m-1}(r)$. $\qquad\square$

For an n-dimensional submanifold M in $S^{m-1}(r) \subset \mathbb{E}^m$, we denote by H and $\hat{H}$ the mean curvature vectors of M in $\mathbb{E}^m$ and in $S^{m-1}(r)$, respectively. Let H^2 and $\hat{H}^2$ denote the corresponding squared mean curvatures. If we choose an orthonormal normal frame $e_{n+1}, \ldots, e_m$ of M in $\mathbb{E}^m$ such that e_{n+1} is parallel to H, and define a normal vector field $\mathfrak{a}(H)$ by

$$\mathfrak{a}(H) = \sum_{r=n+2}^{m} \mathrm{Tr}(A_H A_r)e_r, \tag{6.101}$$

then $\mathfrak{a}(H)$ is called the *allied mean curvature vector*. Submanifolds with $\mathfrak{a}(H) = 0$ are called *Chen submanifolds* (cf. [Rouxel (1993); Gheysens et al. (1983)]). Same definitions apply to submanifolds in arbitrary Riemannian manifold; in particular, to submanifolds in $S^{m-1}(r)$.

In terms of $\mathfrak{a}(H)$, formula (4.46) can be simply expressed as

$$\Delta H = \Delta^D H + \|A_{n+1}\|^2 H + \mathfrak{a}(H) + (\Delta H)^T, \tag{6.102}$$

where

$$(\Delta H)^T = \frac{n}{2}\nabla\langle H, H\rangle + 2\,\mathrm{Tr}\,A_{DH} \tag{6.103}$$

is the tangential component of ΔH.

The next result is quite useful for studying 2-type spherical submanifolds.

Theorem 6.31. [Chen (1983b)] *Let M be a 2-type mass-symmetric submanifold of a hypersphere $S^{m-1}(r) \subset \mathbb{E}^m$. Then we have:*

(a) $\hat{H}^2 = \left(\dfrac{r}{n}\right)^2 \left(\dfrac{n}{r^2} - \lambda_p\right)\left(\lambda_q - \dfrac{n}{r^2}\right),$

(b) $(\Delta H)^T = 0,$

(c) $\Delta^{\hat{D}}\hat{H} + \mathfrak{a}(\hat{H}) + \left(||A_\xi||^2 + \dfrac{n}{r^2}\right)\hat{H} = (\lambda_p + \lambda_q)\hat{H},$

where $n = \dim M$, p and q are the lower and upper orders of M in $\mathbb{E}^m$ and ξ is a unit vector in the direction of $\hat{H}$.

Conversely, if (a), (b) and (c) hold, then M is of 2-type in $\mathbb{E}^m$.

Proof. We may assume that the center of $S^{m-1}(r)$ is the origin o. Then the mean curvature vectors H and $\hat{H}$ are related by

$$H = \hat{H} - \frac{\mathbf{x}}{r^2}, \tag{6.104}$$

where $\mathbf{x}$ is the position vector of M in $\mathbb{E}^m$. It follows from (6.104) that

$$H^2 = \hat{H}^2 + \frac{1}{r^2}. \tag{6.105}$$

Because M is of 2-type, M is non-minimal in $\mathbb{E}^m$ as well as in $S^{m-1}(r)$. Let ξ be the unit vector parallel to $\hat{H}$. We may choose an orthonormal normal frame $e_{n+1}, \ldots, e_m$ such that

$$H = \alpha e_{n+1}, \quad e_{n+2} = \frac{\xi + \hat{\alpha}\mathbf{x}}{r\alpha}, \tag{6.106}$$

with $\alpha = ||H||$ and $\hat{\alpha} = ||\hat{H}||$. Since $A_\mathbf{x} = -I$, we find

$$\mathrm{Tr}(A_H A_{n+2}) = \frac{\hat{\alpha}}{r\alpha}(||A_\xi||^2 - n\hat{H}^2),$$

$$\mathrm{Tr}(A_H A_r) = \mathrm{Tr}(A_{\hat{H}} A_r), \quad r = n+3, \ldots, m.$$

From these we obtain

$$\mathfrak{a}(H) = \mathfrak{a}(\hat{H}) + \frac{\hat{\alpha}}{r\alpha}(||A_\xi||^2 - n\hat{H}^2)e_{n+2}, \tag{6.107}$$

where $\mathfrak{a}(\hat{H})$ denotes the allied mean curvature vector of M in $S^{m-1}(r)$.

Since $D\mathbf{x} = 0$, we have

$$\Delta^D H = \Delta^{\hat{D}}\hat{H} \tag{6.108}$$

which is perpendicular to $\mathbf{x}$, where $\hat{D}$ is the normal connection of M in $S^{m-1}(r)$. By combining (6.102) and (6.104)-(6.108), we obtain

$$\Delta H = \Delta^{\hat{D}}\hat{H} + \mathfrak{a}(\hat{H}) + (\Delta H)^T + \left(||A_\xi||^2 + \frac{n}{r^2}\right)\hat{H} - \frac{nH^2}{r^2}\mathbf{x}. \tag{6.109}$$

Because M is of 2-type, there exists a polynomial P of degree 2 such that $P(\Delta)(\mathbf{x} - c) = 0$ for some vector c. Hence it follows from (6.109) that

$$\Delta^{\hat{D}}\hat{H} + \mathfrak{a}(\hat{H}) + (\Delta H)^T + \left(\|A_\xi\|^2 + \frac{n}{r^2} \right)\hat{H} - \frac{nH^2}{r^2}\mathbf{x} \tag{6.110}$$
$$+ c_1\hat{H} - \frac{c_1}{r^2}\mathbf{x} - \frac{c_2}{n}\mathbf{x} = 0$$

for some constants c_1, c_2. Since $(\Delta H)^T$ is tangent to M and other terms in (6.110) are normal to M, we find $(\Delta H)^T = 0$, which gives Statement (b).

Since $\mathbf{x}$ is normal to $S^{m-1}(r)$ and others are tangent to $S^{m-1}(r)$, we find from $\mathbf{x}$-components in (6.110) that

$$H^2 = \hat{H}^2 + \frac{1}{r^2} = -\frac{c_1}{n} - \frac{c_2 r^2}{n^2}. \tag{6.111}$$

On the other hand, (6.110) yields

$$P(\Delta)(\mathbf{x}) = 0, \tag{6.112}$$

where $P(t) = t^2 + c_1 t + c_2$. Because M is of 2-type and of order $\{p, q\}$, (6.112) implies that $c_1 = -\lambda_p + \lambda_q$ and $c_2 = \lambda_p \lambda_q$. Consequently, we obtain Statement (a) from (6.111). Statement (c) follows from Statements (a) and (b) and equation (6.110).

The converse follows easily from Lemma 6.8 and the minimal polynomial criterion. □

The following is an immediate consequence of Theorem 6.31.

Corollary 6.5. *If M is a 2-type mass-symmetric submanifold of the unit hypersphere $S^{m-1}(1) \subset \mathbb{E}^m$, then we have*

$$\lambda_p < n < \lambda_q, \tag{6.113}$$

where $n = \dim M$ and p and q are the lower and upper orders of M.

Proof. Statement (a) of Theorem 6.31 implies that $\lambda_p \leq n \leq \lambda_q$. Either equality sign holds only when $\hat{H} = 0$. In this case, M is minimal in $S^{m-1}(1)$ which implies that M is of 1-type, not of 2-type. □

For 2-type spherical hypersurfaces, we have the following.

Theorem 6.32. [Chen (1983b)] *Let M be a mass-symmetric 2-type hypersurface of $S^{n+1}(r) \subset \mathbb{E}^{n+2}$. Then we have:*

(1) *The squared mean curvature of M in $\mathbb{E}^{n+2}$ is a constant given by*

$$H^2 = \frac{1}{n}(\lambda_p + \lambda_q) - \left(\frac{r}{n} \right)^2 \lambda_p \lambda_q. \tag{6.114}$$

(2) *The scalar curvature of M is a constant given by*

$$\tau = \frac{n-1}{2}(\lambda_p + \lambda_q) - \frac{r^2}{2}\lambda_p\lambda_q. \tag{6.115}$$

(3) *The squared norm $||h||^2$ of the second fundamental form of M in $\mathbb{E}^{n+2}$ is a constant given by*

$$||h||^2 = \lambda_p + \lambda_q. \tag{6.116}$$

Proof. If M is a 2-type mass-symmetric hypersurface of $S^{n+1}(r)$, then we obtain (6.114) from (6.105) and Theorem 6.31(a). Because M has constant mean curvature and the codimension is one, we have $\hat{D}\hat{H} = \mathfrak{a}(\hat{H}) = 0$. Hence $\Delta^{\hat{D}}\hat{H} = 0$. Thus we find from Theorem 6.31(c) that

$$||h||^2 = ||A_\xi||^2 + \frac{n}{r^2} = \lambda_p + \lambda_q. \tag{6.117}$$

Equation (6.115) is an immediate consequence of (6.114), (6.116) and

$$2\tau = n^2 H^2 - ||h||^2. \tag{6.118}$$

The last equation follows immediately from (4.29) of Proposition 4.4. $\quad\square$

The next result shows that 2-type spherical hypersurfaces have constant scalar curvature if it has constant mean curvature.

Theorem 6.33. [Chen (1986a); Chen et al. (1987)] *Let M be a 2-type hypersurface of $S^{n+1}(r) \subset \mathbb{E}^{n+2}$. If M has constant mean curvature, then it has constant scalar curvature.*

Proof. Let M be a 2-type hypersurface of $S^{n+1}(r) \subset \mathbb{E}^{n+2}$. Without loss of generality, we may assume that $S^{n+1}(r)$ is centered at the origin with $r = 1$. Then, by Theorem 6.1, we have

$$\Delta H = bH + c(\mathbf{x} - \mathbf{x}_0), \tag{6.119}$$

where $\mathbf{x}_0$ is a constant vector in $\mathbb{E}^{n+2}$ and b, c are constants satisfying

$$b = \lambda_p + \lambda_q, \quad c = \frac{\lambda_p\lambda_q}{n}, \tag{6.120}$$

where p and q are the lower and upper order of M. If M has constant mean curvature, then it follows from (6.102) and (6.103) that

$$\Delta H = ||h||^2\hat{H} - nH^2\mathbf{x}, \tag{6.121}$$

where $\hat{H}$ is the mean curvature vector of M in $S^{n+1}(1)$. Because $H = \hat{H} - \mathbf{x}$, we find from (6.119) and (6.120) that

$$b\hat{H} - b\mathbf{x} + c(\mathbf{x} - \mathbf{x}_0) = ||h||^2\hat{H} - nH^2\mathbf{x}. \tag{6.122}$$

By taking the inner product of (6.122) with $\mathbf{x}$, we obtain

$$c \langle \mathbf{x}, \mathbf{x}_0 \rangle = nH^2 + c - b, \tag{6.123}$$

which is a constant. If $x_0 \neq 0$, then M lies in the intersection of a hyperplane of $\mathbb{E}^{n+2}$ and the hypersphere $S^{n+1}(1)$. Hence M is an open portion of a hypersurface of $S^{n+1}(1)$. Thus M is of 1-type which is a contradiction. So, we get $x_0 = 0$. Thus it follows from (6.122) that $||h||^2 = \lambda_p + \lambda_q$. Hence M has constant scalar curvature τ according to (6.118). $\qquad \square$

Theorem 6.34. [Chen (1986a)] *Let M be a hypersurface of $S^{n+1}(r) \subset \mathbb{E}^{n+2}$. If M has nonzero constant mean curvature and constant scalar curvature, then M is either of 1-type or of 2-type. Further, when M is closed, M is mass-symmetric in $S^{n+1}(r)$ unless M is totally umbilical.*

Proof. Under the hypothesis, we get $DH = 0$. Without loss of generality, we may assume that $S^{n+1}(r)$ is a unit hypersphere centered at the origin. From Corollary 4.8 we find

$$\Delta H = ||h||^2 \hat{H} - nH^2 \mathbf{x}. \tag{6.124}$$

Since M has constant mean curvature and constant scalar curvature, $||h||$ is a constant. Because $H = H' - \mathbf{x}$, (6.124) gives

$$\Delta H - ||h||^2 H + (nH^2 - ||h||^2) \mathbf{x} = 0. \tag{6.125}$$

Hence, by Theorem 6.3, M is either of 1-type or of 2-type.

Now, suppose that M is a closed hypersurface. Then it follows from (6.125), Hopf's lemma and Beltrami's formula that

$$(nH^2 - ||h||^2) \int_M \mathbf{x} \, dV = 0.$$

If M not totally umbilical in $S^{n+1}(1)$, then we have $nH^2 \neq ||h||^2$. Thus we get $\int_M \mathbf{x} \, dV = 0$, which shows that M is mass-symmetric in $S^{n+1}(r)$. $\qquad \square$

A hypersurface of an $(n+1)$-sphere S^{n+1} is called *isoparametric* if it has constant principal curvatures.

Corollary 6.6. [Chen (1983b)] *Every isoparametric hypersurface of $S^{n+1}(r) \subset \mathbb{E}^{n+2}$ is either of 1-type or of 2-type.*

Proof. Follows immediately from Theorem 6.34. $\qquad \square$

Remark 6.11. Since there exist many non-minimal isoparametric hypersurfaces in $S^{n+1}(r)$, Corollary 6.6 shows that there are many 2-type hypersurfaces in $S^{n+1}(r)$.

As a converse of Theorem 6.34 we have the following.

Theorem 6.35. [Hasanis-Vlachos (1991b)] *Every 2-type hypersurface of* $S^{n+1}(r) \subset \mathbb{E}^{n+2}$ *has nonzero constant mean curvature and constant scalar curvature.*

Proof. Let M be a 2-type hypersurface of $S^{n+1}(r) \subset \mathbb{E}^{n+2}$. Without loss of generality, we may assume that $S^{n+1}(r)$ is a unit hypersphere centered at the origin. Then the position vector-valued function $\mathbf{x}$ of M in $\mathbb{E}^{n+2}$ admits a spectral decomposition:

$$\mathbf{x} = \mathbf{a} + \mathbf{x}_p + \mathbf{x}_q \tag{6.126}$$

for some vector $\mathbf{a} \in \mathbb{E}^{n+2}$ and vector functions $\mathbf{x}_p, \mathbf{x}_q$ satisfying

$$\Delta \mathbf{x}_p = \lambda_p \mathbf{x}_p, \quad \Delta \mathbf{x}_q = \lambda_q \mathbf{x}_q, \quad \lambda_p < \lambda_q. \tag{6.127}$$

By applying Beltrami's formula, we find from (6.126) and (6.127) that

$$\Delta H = (\lambda_p + \lambda_q)H + \frac{\lambda_p \lambda_q}{n}(\mathbf{x} - \mathbf{a}). \tag{6.128}$$

We decompose the vector $\mathbf{a}$ at each point in M as

$$\mathbf{a} = \mathbf{a}_T + \langle \mathbf{a}, \mathbf{x} \rangle \mathbf{x} + \langle \mathbf{a}, \xi \rangle \xi, \tag{6.129}$$

where $\mathbf{a}_T$ is the tangential component and ξ is a unit normal vector field of M in $S^{n+1}(1)$. Clearly, we have

$$H = \hat{\alpha} e_{n+1} - \mathbf{x}, \quad \hat{\alpha} = \frac{1}{n} \mathrm{Tr}\, A_\xi. \tag{6.130}$$

From (6.128), (6.129) and (6.130) we obtain

$$\Delta H = -\frac{\lambda_p \lambda_q}{n}\mathbf{a}_T - \left(\lambda_p + \lambda_q + \frac{\lambda_p \lambda_q}{n}\langle \mathbf{a}, \mathbf{x}\rangle - \frac{\lambda_p \lambda_q}{n}\right)\mathbf{x}$$
$$+ \hat{\alpha}(\lambda_p + \lambda_q)\xi - \frac{\lambda_p \lambda_q}{n}\langle \mathbf{a}, \xi \rangle \xi. \tag{6.131}$$

Comparing (6.110) and (6.131) gives

$$2nA_\xi(\nabla \hat{\alpha}) + n^2 \hat{\alpha} \nabla \hat{\alpha} = -\lambda_p \lambda_q \mathbf{a}_T, \tag{6.132}$$

$$n + n\hat{H}^2 = \lambda_p + \lambda_q + \frac{\lambda_p \lambda_q}{n}\langle \mathbf{a}, \mathbf{x}\rangle - \frac{\lambda_p \lambda_q}{n}, \tag{6.133}$$

$$\Delta \hat{\alpha} + n\hat{\alpha} + \hat{\alpha}\, \mathrm{Tr}\, A_\xi^2 = \hat{\alpha}(\lambda_p + \lambda_q) - \frac{\lambda_p \lambda_q}{n}\langle \mathbf{a}, \xi \rangle. \tag{6.134}$$

By differentiating (6.130) with respect to a tangent vector X of M and applying formulas of Gauss and Weingarten, we obtain

$$0 = \nabla_X \mathbf{a}_T - \langle X, \mathbf{a}_T \rangle \mathbf{x} + \langle A_\xi \mathbf{a}_T, X \rangle \xi + (X\langle \mathbf{a}, \mathbf{x}\rangle)\mathbf{x}$$
$$+ \langle \mathbf{a}, \mathbf{x}\rangle X + (X\langle \mathbf{a}, \xi\rangle)\xi - \langle \mathbf{a}, \xi\rangle A_\xi X.$$

Taking the tangential and the normal components of both sides of this equation gives

$$\nabla_X \mathbf{a}_T = \langle \mathbf{a}, \xi \rangle A_\xi X - \langle \mathbf{a}, \mathbf{x} \rangle X,$$
$$X \langle \mathbf{a}, \mathbf{x} \rangle = \langle X, \mathbf{a}_T \rangle, \quad X \langle \mathbf{a}, \xi \rangle = - \langle A_\xi \mathbf{a}_T, X \rangle. \tag{6.135}$$

The last two equations in (6.135) can be expressed as

$$\nabla \langle \mathbf{a}, \mathbf{x} \rangle = \mathbf{a}_T, \tag{6.136}$$
$$\nabla \langle \mathbf{a}, \xi \rangle = -A_\xi \mathbf{a}_T. \tag{6.137}$$

By taking the gradient of both sides of (6.133) and using (6.136), we find

$$n^2 \nabla \hat{H}^2 = \lambda_p \lambda_q \mathbf{a}_T. \tag{6.138}$$

Combining this with (6.134) gives

$$A_\xi (\nabla \hat{\alpha}) = -\frac{3}{2} n \hat{\alpha} \nabla \hat{\alpha}. \tag{6.139}$$

Obviously, (6.133) yields

$$\lambda_p \lambda_q \langle \mathbf{a}, \mathbf{x} \rangle = n^2 \hat{H}^2 + c_1, \tag{6.140}$$

where $c_1 = n^2 - n(\lambda_p + \lambda_q) + \lambda_p \lambda_q$ is a constant. Now, by combining (6.137), (6.138) and (6.139), we find $\nabla(\lambda_p \lambda_q \langle \mathbf{a}, \xi \rangle) = n^3 \nabla \hat{\alpha}^3$, from which we obtain

$$\lambda_p \lambda_q \langle \mathbf{a}, \xi \rangle = n^3 \hat{\alpha}^3 + c_2 \tag{6.141}$$

for some constant c_2. Combining (6.130), (6.138), (6.140) and (6.141) gives

$$\lambda_p^2 \lambda_q^2 ||\mathbf{a}||^2 = n^4 ||\nabla \hat{H}^2||^2 + (n^2 \hat{H}^2 + c_1)^2 + (n^3 \hat{\alpha}^3 + c_2)^2. \tag{6.142}$$

Now, we want to claim that M has constant mean curvature in $S^{n+1}(1)$. So, let us assume $\nabla \hat{\alpha} \neq 0$ on some neighborhood V of a point $p_o \in M$. Let e_1 be the unit vector field in the direction of $\nabla \hat{\alpha}$ on V.

Lemma 6.9. *The integral curves of e_1 on V are geodesics of M, i.e.,* $\nabla_{e_1} e_1 = 0$ *on V.*

Proof. It follows from (6.139) that e_1 is an eigenvector field of A_ξ with eigenvalue $-\frac{3n}{2} \hat{\alpha}$. Differentiating (6.138) with respect $X \in T(V)$ we have

$$2n^2 (X \hat{\alpha}) \nabla \hat{\alpha} + 2n^2 \hat{\alpha} \nabla_X (\nabla \hat{\alpha}) = \lambda_p \lambda_q \nabla_X \mathbf{a}_T$$

for which by using (6.135) and $\nabla \hat{\alpha} = ||\nabla \hat{\alpha}|| e_1$, we obtain

$$2n^2 (X \hat{\alpha}) ||\nabla \hat{\alpha}|| e_1 + 2n^2 \hat{\alpha} X(||\nabla \hat{\alpha}||) e_1 + 2n^2 \hat{\alpha} ||\nabla \hat{\alpha}|| \nabla_X e_1$$
$$= \lambda_p \lambda_q (\langle \mathbf{a}, \xi \rangle A_\xi X - \langle \mathbf{a}, \mathbf{x} \rangle X). \tag{6.143}$$

Because $\nabla_{e_1} e_1$ is perpendicular to e_1, it follows from (6.143) with $X = e_1$ that $\hat{\alpha} ||\nabla \hat{\alpha}|| \nabla_{e_1} e_1 = 0$. Therefore, we have $\nabla_{e_1} e_1 = 0$. $\qquad \square$

Let $M_1 = \{p \in M : \nabla \hat{H}^2(p) = 0\}$ and $M_2 = M \setminus M_1$. It is clear that, on each connected component of $int(M_1)$, $\hat{H}^2$ is a nonzero constant. Otherwise, each connected component of $int(M_1)$ is of 1-type by Takahashi's theorem, which is a contradiction. Now, we take a connected component W of M_2. It follows from Lemma 6.9 that integral curves of e_1 on W are geodesics of M. Taking an integral curve $\gamma(s)$ of e_1 which starts from $p_o \in W$, we set $h(s) = \text{Tr}(A_\xi)|_{\gamma(s)}$, where s is the arc length of γ.

By restriction of (6.142) to $\gamma(s)$ and using $e_1\hat{\alpha} = ||\nabla\hat{\alpha}||$, we have

$$\lambda_p^2 \lambda_q^2 ||\mathbf{a}||^2 = 4n^2 h^2 ||\nabla\hat{\alpha}||^2 + (h^2 + c_1)^2 + (h^3 + c_2)^2.$$

Because $e_1 h = h'$, we obtain

$$(h')^2 = \frac{1}{4h^2}(\lambda_p^2 \lambda_q^2 ||\mathbf{a}||^2 - c_1^2 - c_2^2) - \frac{1}{4}(h^4 + h^2) - \frac{1}{2}(c_2 h + c_1). \quad (6.144)$$

After taking differentiation of (6.144) with respect to s, we obtain

$$hh'' + (h')^2 = -\frac{1}{4}(3h^4 + 2h^2 + 3c_2 h + 2c_1). \quad (6.145)$$

Let X be a tangent vector perpendicular to e_1. Then we have $X\hat{\alpha} = 0$. Since e_1 is an eigenvector of A_ξ, e_1 is perpendicular to $\langle \mathbf{a}, \xi \rangle A_\xi X - \langle \mathbf{a}, \mathbf{x} \rangle X$. By taking the inner product of (6.143) with e_1, we find $\hat{\alpha}X(||\nabla\hat{\alpha}||) = 0$ and hence (6.143) becomes

$$2n^2 \hat{\alpha} ||\nabla\hat{\alpha}|| \nabla_X e_1 = \lambda_p \lambda_q (\langle \mathbf{a}, \xi \rangle A_\xi X - \langle \mathbf{a}, \mathbf{x} \rangle X) \quad (6.146)$$

for any X perpendicular to e_1. We choose an orthonormal frame $e_1, \ldots, e_n$ on V such that $e_1, \ldots, e_n$ are eigenvectors of A_ξ with corresponding eigenvalues $\kappa_1 = -\frac{3}{2}n\hat{\alpha}, \kappa_2, \ldots, \kappa_n$. Then we have $\sum_{i=1}^n \kappa_i = \frac{5}{2}n\hat{\alpha}$. Therefore, by applying the equation of Codazzi $e_1\kappa_i = (\kappa_1 - \kappa_i)\langle \nabla_{e_i} e_1, e_i \rangle$ and (6.146) with $X = e_i$ $(i > 1)$, we derive that

$$2n^2 \hat{\alpha} ||\nabla\hat{\alpha}|| e_1 \kappa_i = (\kappa_1 - \kappa_i)\lambda_p \lambda_q (\langle \mathbf{a}, \xi \rangle \kappa_i - \langle \mathbf{a}, \mathbf{x} \rangle).$$

The last equation implies

$$2n^2 \hat{\alpha} ||\nabla\hat{\alpha}|| \sum_{i=2}^n e_1 \kappa_i = -\lambda_p \lambda_q \langle \mathbf{a}, \xi \rangle \sum_{i=2}^n \kappa_i$$

$$+ \lambda_p \lambda_q (\kappa_1 \langle \mathbf{a}, \xi \rangle + \langle \mathbf{a}, \mathbf{x} \rangle) \sum_{i=2}^n \kappa_i - (n-1)\kappa_1 \lambda_p \lambda_q \langle \mathbf{a}, \mathbf{x} \rangle,$$

from which we find

$$n\lambda_p \lambda_q \langle \mathbf{a}, \xi \rangle \hat{H}^2 = -5n^2 \hat{\alpha} ||\nabla\hat{\alpha}||^2 - \frac{3}{2}n\hat{H}^2 \lambda_p \lambda_q \langle \mathbf{a}, \xi \rangle$$

$$+ \left(\frac{3n+2}{2}\right) \hat{\alpha}\lambda_p \lambda_q \langle \mathbf{a}, \mathbf{x} \rangle, \quad (6.147)$$

where we have used $\sum_{i=2}^{n} \kappa_i^2 = n^2 \hat{H}^2 - \kappa_1^2 = n^2 \hat{H}^2 - \frac{9}{4} n^2 \hat{H}^2$. From (6.146) we obtain

$$2n^2 \hat{\alpha} ||\nabla \hat{\alpha}|| \langle \nabla_{e_i} e_1, e_i \rangle = \kappa_i \lambda_p \lambda_q \langle \mathbf{a}, \xi \rangle - \lambda_p \lambda_q \langle \mathbf{a}, \mathbf{x} \rangle. \tag{6.148}$$

Because $e_2 \hat{\alpha} = \cdots = e_n \hat{\alpha} = 0$ and $\nabla_{e_1} e_1 = 0$, we have

$$\Delta \hat{\alpha} = -e_1 e_1 \hat{\alpha} - (e_1 \alpha) \sum_{i=2}^{n} \langle \nabla_{e_i} e_1, e_i \rangle.$$

Since $e_1 \hat{\alpha} = ||\nabla \hat{\alpha}||$, we obtain from (6.148) that

$$n\hat{\alpha} \Delta \hat{\alpha} = -n\hat{\alpha} e_1 e_1 \hat{\alpha} - \frac{5}{4} \hat{\alpha} \lambda_p \lambda_q \langle \mathbf{a}, \xi \rangle + \frac{n-1}{2n} \lambda_p \lambda_q \langle \mathbf{a}, \mathbf{x} \rangle. \tag{6.149}$$

By restricting(6.134) to $\gamma(s)$ and taking into account of (6.140), (6.141), (6.147) and (6.149), we find

$$(\lambda_p + \lambda_q)h - (h^3 + c_2) = -h'' - \frac{5}{4}(h^3 + c_2) + \frac{n-1}{2h}(h^2 + c_1)$$
$$+ nh - \frac{h}{h^3 + c_2} \left\{ 5h(h')^2 + \frac{3}{2}h^2(h^3 + c_2) - \frac{3n+2}{2}h(h^2 + c_1) \right\},$$

or after some calculations

$$-hh'' - \frac{5h^3(h')^2}{h^3 + c_2} = \frac{7}{4}h^4 + \left(\lambda_p + \lambda_q - \frac{3n-1}{2} \right) h^2$$
$$+ \frac{c_2}{4}h - \frac{n-1}{2}c_1 - \frac{(3n+2)h^3}{2(h^3 + c_2)}(h^2 + c_1). \tag{6.150}$$

Summing (6.145) and (6.150) we find

$$\frac{c_2 - 4h^3}{h^3 + c_2}(h')^2 = h^4 + \left(\lambda_p + \lambda_q - \frac{3n}{2} \right) h^2 - \frac{c_2}{2}h - \frac{nc_1}{2}$$
$$- \frac{(3n+2)h^3}{2(h^3 + c_2)}(h^2 + c_1). \tag{6.151}$$

Finally, by substituting (6.144) into (6.151), we obtain

$$0 = (\lambda_p + \lambda_q - 3n - 2)h^7 - \frac{5c_2}{4}h^6 - (2n+3)c_1 h^5 - \frac{n-1}{2}c_1 c_2 h^2$$
$$+ \left(\lambda_p + \lambda_q - \frac{6n-1}{4} \right) c_2 h^4 + (\lambda_p^2 \lambda_q^2 ||\mathbf{a}||^2 - c_1^2 - c_2^2) \left(h^3 - \frac{c_2}{4} \right). \tag{6.152}$$

This equation is valid along $\gamma(s)$. Hence we have

$$c_1 = c_2 = 0, \quad \lambda_p + \lambda_q = 3n + 2, \quad \lambda_p \lambda_q ||\mathbf{a}|| = 0,$$

which contradicts to $M_2 \neq \emptyset$, since $\lambda_p \lambda_q ||\mathbf{a}|| = 0$. Consequently, M has constant mean curvature. Therefore, by Theorem 6.33, M also has constant scalar curvature. $\square$

For spherical 2-type hypersurfaces we also have the next two results from [Chen (1991c)].

Theorem 6.36. *Let M be a hypersurface of $S^{n+1}(1) \subset \mathbb{E}^{n+2}$ with at most two distinct principal curvatures. Then M is of 2-type if and only if M is an open portion of the direct product of two spheres $S^j(r_1) \times S^{n-j}(r_2)$ with $r_1^2 + r_2^2 = 1$ and $(r_1, r_2) \neq (\sqrt{j/n}, \sqrt{(n-j)/n})$.*

Theorem 6.37. *Let M be a hypersurface of $S^{n+1}(1) \subset \mathbb{E}^{n+2}$ with $n \geq 4$. Then M is conformally flat and of 2-type if and only if M is an open portion of $S^1(r_1) \times S^{n-1}(r_2)$ such that the with $r_1^2 + r_2^2 = 1$ and $(r_1, r_2) \neq (\sqrt{1/n}, \sqrt{(n-1)/n})$.*

When $n = 2$, Theorem 6.34 reduces to the following.

Theorem 6.38. *Let M be a surface of $S^3(1) \subset \mathbb{E}^4$. Then M is of 2-type if and only if it is open portion of the product of two plane circles of different radii, i.e., $M \subset S^1(a) \times S^1(b)$ with $a \neq b$ and $a^2 + b^2 = 1$.*

Remark 6.12. Theorem 6.13 was first proved in [Chen (1979c)] under the assumption that M is a mass-symmetric 2-type closed surface in $S^3(1)$. It was proved in [Barros and Garay (1987)] that the mass-symmetric condition can be removed. Further, it was shown in [Hasanis-Vlachos (1991a)] that this result is of local nature.

Let x, y, z be the natural coordinates of $\mathbb{E}^3$ and $u_1, \ldots, u_5$ that of $\mathbb{E}^5$. The mapping defined by

$$u_1 = \frac{yz}{\sqrt{3}}, \ u_2 = \frac{xz}{\sqrt{3}}, \ u_3 = \frac{xy}{\sqrt{3}}, \ u_4 = \frac{x^2 - y^2}{2\sqrt{3}},$$

$$u_5 = \frac{1}{6}(x^2 + y^2 - 2z^2)$$

gives rise to an isometric immersion of $S^2(\frac{1}{3})$ of curvature $\frac{1}{3}$ into $S^4(1)$. Two points (x, y, z) and $(-x, -y, -z)$ of $S^2(\frac{1}{3})$ are mapped into the same point. This mapping is the first standard imbedding of the real projective plane $RP^2(\frac{1}{3})$ into $S^4(1)$, known as the *Veronese surface*, which is nothing but the second standard immersion of $S^2(\frac{1}{3})$.

Definition 6.6. Let N^ℓ be a topologically embedded ℓ-dimensional submanifold in a Riemannian m-manifold M^m ($\ell < m - 1$). Denote by $\nu_1(N^\ell)$ the unit normal subbundle of the normal bundle $T^\perp N^\ell$ of N^ℓ in M^m. Then, for a sufficiently small $r > 0$, the mapping

$$\psi : \nu_1(N^\ell) \to M^m : (p, e) \mapsto \exp_p(re)$$

is an immersion which is called the *tube* (or the *tubular hypersurface*) with radius r around N^ℓ. We denote it by $T^r(N^\ell)$.

For 2-type hypersurfaces in $S^4(1)$, we have the following classification result from [Chen (1993b)].

Theorem 6.39. *A closed hypersurface M of $S^4(1) \subset \mathbb{E}^5$ is of 2-type if and only if it is congruent to one of the following two hypersurfaces:*

(a) *a standard embedding $S^1 \times S^2 \subset S^4(1) \subset \mathbb{E}^5$ such that the radii r_1 of S^1 and r_2 of S^2 satisfying $r_1^2 + r_2^2 = 1$ and $(r_1, r_2) \neq (1/\sqrt{3}, \sqrt{2/3}\,)$;*
(b) *a tube $T^r(V^2)$ with radius $r \neq \frac{\pi}{2}$ over the Veronese surface V^2 in $S^4(1)$.*

Proof. If M is a hypersurface of $S^4(1)$ given by (a) or (b), then M is an isoparametric hypersurface with constant scalar curvature and nonzero constant mean curvature. Thus, by Corollary 6.6, M is either of 1-type or of 2-type. If M is given by (a), it is of 2-type. If M is given by (b), then it follows from a direct computation that, at the point $\psi(p, e) = \exp_p(re)$ corresponding to a point $(p, e) \in T^r(V^2)$, the principal curvatures of the tube in $S^4(1)$ are given by

$$\frac{\cot r - \sqrt{3}}{1 + \sqrt{3} \cot r}, \quad \frac{\cot r + \sqrt{3}}{1 - \sqrt{3} \cot r}, \quad \cot r. \tag{6.153}$$

It is direct to verify from (6.153) that the tube is of 1-type if and only if $r = \pi/2$. But this case was excluded.

Conversely, if M is a 2-type closed hypersurface of $S^4(1) \subset \mathbb{E}^5$, then it follows from Theorem 6.35 that M has nonzero constant mean curvature and constant scalar curvature. Therefore, a result of [Chang (1993)] implies that M is an isoparametric hypersurface. Since M is of 2-type, the number g of distinct principal curvatures must either 2 or 3.

If $g = 2$, then Theorem 1 of [Chen (1991c)] implies that M is the direct product of a circle and a 2-sphere as described in case (a).

If $g = 3$, then M lies in the family of parallel hypersurfaces of Cartan's hypersurface. It is known that the Cartan hypersurface is the only minimal one in this family of 3-dimensional isoparametric hypersurfaces. It is also know that such hypersurfaces can be described as tubes of constant radius r about the minimal Veronese embedding of $RP^2(\frac{1}{3})$ into $S^4(1)$. Therefore, if we denote such a tubular hypersurface by $T^r(RP^2)$, then the three principal curvatures are given by (6.153). As we already pointing out above, the tube $T^r(RP^2)$ is of 1-type if and only if $r = \pi/2$. Hence, for $g = 3$, we have case (b) since M is assumed to be of 2-type. $\square$

Let $S_1^3(1) \subset \mathbb{E}_t^4$ be the standard model of a de Sitter spacetime with curvature one. If $\gamma(s)$ is a null curve in $S_1^3(1)$ with an associated Cartan frame $\{A, B, C\}$, i.e., $\{A, B, C\}$ is a pseudo-orthonormal frame of vector fields along $\gamma(s)$. For surfaces M_s with index s in $S_1^3(1)$, we have

$$\langle A, A \rangle = \langle B, B \rangle = \langle A, C \rangle = \langle B, C \rangle = 0, \ \langle A, B \rangle = -1, \ \langle C, C \rangle = 1,$$

such that $\dot{\gamma}(s) = A(s)$, $\dot{C}(s) = -aA(s) - k(s)B(s)$, where a is a nonzero constant and $k(s) \neq 0$ for all s.

The map $\psi : (s, u) \to \gamma(s) + uB(s)$ parametrizes a Lorentzian surface into $S_1^3(1)$ which is called a *B-scroll* (cf. [Dajczer and Nomizu (1981)]).

Theorem 6.40. [Alías et. al. (1994)] *A surface M_s in $S_1^3(1)$ is of 2-type if and only if it is an open piece of $H^1(-r) \times S^1(\sqrt{1+r^2})$, $S_1^1(r) \times S^1(\sqrt{1-r^2})$ or a B-scroll over a null curve.*

6.10 Spherical k-type hypersurfaces with $k \geq 3$

In order to study 3-type spherical hypersurface, we need to compute $\Delta^2 H$. For any vector η normal to a submanifold $M \subset \mathbb{E}^m$, we put

$$\widetilde{A}(\eta) = \sum_{i=1}^{n} h(e_i, A_\eta e_i), \tag{6.154}$$

where h is the second fundamental form and $e_1, \ldots, e_n$ is an orthonormal frame of TM. In particular, $\widetilde{A}(H) = \mathfrak{a}(H) + \|A_\xi\|^2 H$, $\xi = H/|H|$ if $H \neq 0$.

The next lemma extends the formula given in Proposition 4.7.

Lemma 6.10. *Let M be an n-dimensional submanifold of $\mathbb{E}^m$. Then for each normal vector field η of M we have*

$$\Delta \eta = \Delta^D \eta + \widetilde{A}(\eta) + (\Delta \eta)^T, \tag{6.155}$$

where

$$(\Delta \eta)^T = n \sum_{i=1}^{n} \langle D_{e_i} H, \eta \rangle e_i + 2 \sum_{i=1}^{n} A_{D_{e_i} \eta} e_i. \tag{6.156}$$

Proof. This can be proved in the same way as Proposition 4.7. □

When M is a submanifold lying in the unit hypersurface $S^{m-1}(1)$ of $\mathbb{E}^m$ centered at origin, we denote as before the mean curvature vector of M in $S^{m-1}(1)$ and $\mathbb{E}^m$ by H and $\hat{H}$, respectively. In this case, let ξ denote the unit normal vector field in the direction of $\hat{H}$ so that $\hat{H} = \hat{\alpha}\xi$.

We also need the following lemma for later use.

Lemma 6.11. [Chen (1990b); Chen and Li (1991)] *Let M be a submanifold of $\mathbb{E}^m$. If M has parallel mean curvature vector, i.e., $DH = 0$, then*

$$\Delta H = \tilde{A}(H), \tag{6.157}$$

$$\Delta^2 H = \Delta^D(\tilde{A}(H)) + \tilde{A}(\tilde{A}(H)). \tag{6.158}$$

In particular, if M is a hypersurface with constant mean curvature in $S^{n+1}(1) \subset \mathbb{E}^{n+2}$, then we have

$$\Delta H = ||h||^2 \hat{H} - nH^2 \mathbf{x},$$

$$\Delta^2 H = 2A_{\hat{H}}(\nabla\varphi) + (\Delta\varphi + \varphi^2 + n\varphi + n^2 H^2)\hat{H} \tag{6.159}$$

$$- n(\hat{H}^2\varphi + 2n\hat{H}^2 + n)\mathbf{x},$$

where $H^2 = \langle H, H \rangle$, $\hat{H}^2 = \langle \hat{H}, \hat{H} \rangle$ and $\varphi = ||A_\xi||^2$.

Proof. Follows from Proposition 4.7 and Lemma 6.10. $\qquad\square$

For 3-type spherical hypersurfaces we have the following.

Theorem 6.41. [Chen (1990b); Chen and Li (1991)] *Every 3-type hypersurface of a hypersphere $S^{n+1}(r) \subset \mathbb{E}^{n+2}$ has non-constant mean curvature.*

Proof. Assume that M is a 3-type hypersurface of a hypersphere $S^{n+1}(r)$ with constant mean curvature $\hat{\alpha}$. Without loss of generality, we may assume that $S^{n+1}(r)$ is a the unit hypersphere centered at the origin.

Since M is of 3-type, the position vector function of M admits a spectral decomposition:

$$\mathbf{x} = \mathbf{c}_0 + \mathbf{x}_p + \mathbf{x}_r + \mathbf{x}_q, \quad \Delta\mathbf{x}_i = \lambda_i\mathbf{x}_i, \quad i = p, r, q, \tag{6.160}$$

where $\mathbf{c}_0$ is a constant vector and $\lambda_p < \lambda_r < \lambda_q$. By applying Beltrami's formula to (6.160), we find

$$\Delta^2 H = a\Delta H + bH + c(\mathbf{x} - \mathbf{c}_0), \tag{6.161}$$

with

$$a = \lambda_p + \lambda_r + \lambda_q, \quad b = -(\lambda_p\lambda_r + \lambda_p\lambda_q + \lambda_r\lambda_q), \quad c = \frac{\lambda_p\lambda_r\lambda_q}{n}.$$

We also have as before that

$$H = \hat{H} - \mathbf{x}, \quad \hat{H} = \hat{\alpha}\xi, \quad H^2 = \hat{H}^2 + 1. \tag{6.162}$$

Since M has constant mean curvature, it follows from (6.121) that

$$\Delta H = ||h||^2 \hat{H} - nH^2 \mathbf{x}. \tag{6.163}$$

By substituting (6.162) and (6.163) into (6.161), we find

$$\Delta^2 H = (a||h||^2 + b)\hat{H} - (naH^2 + b - c)\mathbf{x} - c\mathbf{c}_0. \tag{6.164}$$

From Lemma 6.11 and (6.164) we obtain

$$2A_{H'}(\nabla\varphi) = -c(\mathbf{c}_0)^T, \tag{6.165}$$

$$n(\varphi + 2n)\hat{H}^2 + n^2 = naH^2 + b - c + c\langle \mathbf{x}, \mathbf{c}_0\rangle, \tag{6.166}$$

$$(\Delta||h||^2 + \varphi^2 + n\varphi + n^2H^2)\hat{a} = (a||h||^2 + b)\hat{a} - c\langle \xi, \mathbf{c}_0\rangle, \tag{6.167}$$

where $(\mathbf{x}_0)^T$ is the tangential component of $\mathbf{x}_0$. Since M has constant mean curvature, by taking the gradient of (6.166) we obtain

$$n\hat{H}^2\nabla\varphi = c\nabla\langle \mathbf{x}, \mathbf{c}_0\rangle = c(\mathbf{c}_0)^T. \tag{6.168}$$

Consequently, it follows from (6.165) and (6.168) that

$$A_{\hat{H}}(\nabla\varphi) = -\frac{1}{2}n\hat{H}^2\nabla\varphi. \tag{6.169}$$

Now we claim that the function $\varphi = ||A_\xi||^2$ is a constant. This can be done as follows: By taking the gradient of (6.167) we get

$$\hat{a}\,\nabla(\Delta\varphi + \varphi^2 + n\varphi) = a\hat{a}\,\nabla\varphi - c\nabla\langle \xi, \mathbf{c}_0\rangle. \tag{6.170}$$

Since $\nabla\langle \xi, \mathbf{c}_0\rangle = -A_\xi(\mathbf{c}_0)^T$, (6.168), (6.169) and (6.170) imply

$$\nabla\left(2\Delta\varphi + 2\varphi^2 + 2(n - a)\varphi + n^2\hat{H}^2\varphi\right) = 0,$$

which is equivalent to

$$2\Delta\varphi + 2\varphi^2 + 2(n - a)\varphi + n^2\hat{H}^2\varphi = constant. \tag{6.171}$$

By applying (6.169), (6.172) and by taking covariant derivatives of (6.159) and (6.164) with respect to $Y \in T(M)$, we obtain

$$(\Delta\varphi + \varphi^2 + n\varphi + n^2H^2 - a\varphi - na - b)A_{\hat{H}}Y$$
$$= (nH^2 + b - c - n\hat{H}^2\varphi - 2n^2\hat{H}^2 - n^2)Y$$
$$\quad - n\hat{H}^2\nabla_Y(\nabla\varphi) - \frac{n}{2}(Y\varphi)\hat{H} \tag{6.172}$$
$$= (Y\varphi)X + h(Y, \nabla\varphi).$$

Since $\sum_{i=1}^n \langle \nabla_{e_i}(\nabla\varphi), e_i\rangle = -\Delta\varphi$, equation (6.172) implies that

$$\hat{H}^2\varphi^2 + (2n - a)\hat{H}^2\varphi + \hat{H}^2(n^2\hat{H}^2 + 3n^2 - 2na - b)$$
$$+ (n^2 - na - b + c) = 0.$$

From the last equation we conclude that φ is constant. Consequently, M is a hypersurface of $S^{n+1}(1)$ with constant mean curvature and constant scalar curvature. Therefore, it follows from Theorem 6.34 that M is not of 3-type, which is a contradiction. $\qquad\square$

This theorem was proved independently in [Nagatomo (1991)] under an additional assumption that M is a closed hypersurface.

The next theorem classifies finite type surfaces of constant curvature in a 3-sphere.

Theorem 6.42. [Chen and Dillen (1990a)] *Standard spheres and products of two planar circles are the only finite type closed surfaces with constant curvature in* $S^3(r) \subset \mathbb{E}^4$.

Proof. Assume that M is a closed surface of constant (Gauss) curvature c in $S^3(1)$. Then we have either $c = 0$ or $c \geq 1$ and in the later case M is totally umbilical (see, e.g., page 139 of [Spivak (1979)]). Thus we only to consider the case $c = 0$. Hence we may assume that M is isometric to a torus $\mathbb{E}^2/\Lambda$, where Λ is a lattice given by

$$\Lambda = \{(2n\pi u, 2m\pi v + 2n\pi w) : n, m \in \mathbb{Z}\},$$

and u, v and w are real numbers with $u, v > 0$.

If x, y are Euclidean coordinates on 2-plane $\mathbb{E}^2$, then the Laplacian is $\Delta = -\frac{\partial^2}{\partial x^2} - \frac{\partial^2}{\partial y^2}$. The eigenfunctions of Δ are given by

$$\left\{ \cos\left(\frac{\ell v - kw}{uv}x + \frac{k}{v}y\right), \sin\left(\frac{\ell v - kw}{uv}x + \frac{k}{v}y\right) : k, \ell \in \mathbb{Z} \right\}$$

and the spectrum of Δ is given by

$$Spec(\mathbb{E}^2/\Lambda) = \left\{ \left(\frac{\ell v - kw}{uv}\right)^2 + \left(\frac{k}{v}\right)^2 : k, \ell \in \mathbb{Z} \right\}.$$

Now, suppose that M is of finite type, say of k-type. Then the position vector field $\mathbf{x}$ of M in $\mathbb{E}^4$ can be written as

$$\mathbf{x} = \sum_{(k,\ell) \in T} \left\{ A_{k\ell} \cos\left(\frac{\ell v - kw}{uv}x + \frac{k}{v}y\right) + B_{k\ell} \sin\left(\frac{\ell v - kw}{uv}x + \frac{k}{v}y\right) \right\},$$

where T is a finite subset of $\mathbb{Z} \times \mathbb{Z}$ and $A_{k\ell}, B_{k\ell}$ are constant vectors in $\mathbb{E}^4$. Let $\gamma_\theta(t) = (t\cos\theta, t\sin\theta)$ be a closed geodesic of M. It is known that there are infinitely many values θ such that γ_θ is closed. Then

$$(\mathbf{x}(\gamma_\theta))(t) = \sum_{(k,\ell) \in T} \left\{ A_{k\ell} \cos\left(\left(\frac{\ell v - kw}{uv}\cos\theta + \frac{k}{v}\sin\theta\right)t\right) \right.$$
$$\left. + B_{k\ell} \sin\left(\left(\frac{\ell v - kw}{uv}\cos\theta + \frac{k}{v}\sin\theta\right)t\right) \right\}. \tag{6.173}$$

Hence $\mathbf{x}(\gamma_\theta)$ is a curve of finite type. Indeed, as the Laplacian of a curve, parameterized by the arc length s is $-\partial^2/\partial s^2$, we see from (6.173) that

$\mathbf{x}(\gamma_\theta)$ is written as a finite sum of eigenfunctions of $-\partial^2/\partial s^2$. Since every closed curve of finite type in $S^3(1)$ is a W-curve according to Theorem 6.15. In particular, it is of 1- or 2-type. Hence the set

$$\left\{ \left(\frac{\ell v - kw}{uv} \right)^2 + \left(\frac{k}{v} \right)^2 : (k, \ell) \in T \right\}$$

contains at most two different elements. Using the fact that this holds for an infinite number of θ's, this implies that T contains at most two elements. Therefore M is of 1-type, and hence congruent to the Clifford torus $M = S^1(1/\sqrt{2}) \times S^1(1/\sqrt{2})$ (cf. [Lawson (1969)]) or of 2-type and hence the product of two circles $S^1(a)$ and $S^1(b)$ with different radii a and b satisfying $a^2 + b^2 = 1$ according to Theorem 6.13.   □

We have the following classification of finite type ruled surfaces S^3.

Theorem 6.43. [Hasanis-Vlachos (1994a)] *A ruled surface in $S^3(r)$ is of finite type if and only if it is an open part of ruled minimal surface of $S^3(r)$ or an open part of a product of two circles of different radii.*

For finite type spherical hypersurfaces the author poses the following (cf. Conjecture 3 in [Chen (1996b)]).

Conjecture 6.3. *Every finite type spherical hypersurface is either of 1-type or of 2-type.*

This conjecture remains open.

6.11   Finite type hypersurfaces in hyperbolic space

Recall that the Minkowski spacetime $\mathbb{E}_1^{n+2}$ is the pseudo-Euclidean space equipped with the Lorentzian metric

$$g = -dx_1^2 + \sum_{j=2}^{n+2} dx_j^2, \tag{6.174}$$

where $\mathbf{x} = (x_1, x_2, \ldots, x_{n+2})$ is the canonical coordinate system on $\mathbb{E}_1^{n+2}$.

Assume that M is a hypersurface of the hyperbolic space $H^{n+1}(-1)$, imbedded standardly in $\mathbb{E}_1^{n+2}$, i.e.,

$$H^{n+1}(-1) = \{\mathbf{x} \in \mathbb{E}_1^{n+2} : \langle \mathbf{x}, \mathbf{x} \rangle = -1, \ x_1 > 0\}. \tag{6.175}$$

Then mean curvature vector H of M in $\mathbb{E}_1^{n+2}$ is given by

$$H = \beta\xi + \mathbf{x}, \tag{6.176}$$

where ξ is a unit normal vector field of M in $H^{n+1}(-1)$ and β is the mean curvature of M in $H^{n+1}(-1)$. By applying Proposition 2.8, we find

$$\Delta H = (\Delta\beta + \beta||A_\xi||^2 - n\beta)\xi + (n\beta^2 - n)\mathbf{x}$$
$$+ \frac{n}{2}\nabla\beta^2 + 2A_\xi(\nabla\beta). \tag{6.177}$$

If M is of 2-type, we have the spectral decomposition:

$$\mathbf{x} = \mathbf{c} + \mathbf{x}_p + \mathbf{x}_q, \quad \Delta\mathbf{x}_p = \lambda_p\mathbf{x}_p, \quad \Delta\mathbf{x}_q = \lambda_q\mathbf{x}_q, \quad \lambda_p < \lambda_q, \tag{6.178}$$

for some vector $\mathbf{c} \in \mathbb{E}_1^{n+2}$. Thus, by applying Beltrami's formula, we find

$$n\Delta H = n(\lambda_p + \lambda_q)H + \lambda_p\lambda_q(\mathbf{x} - \mathbf{c}). \tag{6.179}$$

Because ξ and $\mathbf{x}$ form an orthonormal normal frame of the normal bundle of M in $\mathbb{E}_1^{n+2}$, we have

$$\mathbf{c} = \mathbf{c}_T + \langle\mathbf{c}, \xi\rangle\,\xi - \langle\mathbf{c}, \mathbf{x}\rangle\,\mathbf{x}, \tag{6.180}$$

where $\mathbf{c}_T$ denotes the tangential component of $\mathbf{c}$. Thus we have

$$n\Delta H = \{n(\lambda_p + \lambda_q)\beta - \lambda_p\lambda_q\langle\mathbf{c}, \xi\rangle\}\xi - \lambda_p\lambda_q\mathbf{c}_T$$
$$+ \{n(\lambda_p + \lambda_q) + \lambda_p\lambda_q(1 + \langle\mathbf{c}, \mathbf{x}\rangle)\}\mathbf{x}. \tag{6.181}$$

From (6.177) and (6.181) we find

$$4nA_\xi(\nabla\beta) + n^2\nabla\beta^2 = -2\lambda_p\lambda_q\mathbf{c}_T, \tag{6.182}$$

$$n^2\beta^2 - n^2 = n(\lambda_p + \lambda_q) + \lambda_p\lambda_q(1 + \langle\mathbf{c}, \mathbf{x}\rangle), \tag{6.183}$$

$$n\Delta\beta - n^2\beta + n\beta||A_\xi||^2 = n(\lambda_p + \lambda_q)\beta - \lambda_p\lambda_q\langle\mathbf{c}, \xi\rangle. \tag{6.184}$$

By taking covariant derivative of (6.180), we obtain

$$\nabla\langle\mathbf{c}, \mathbf{x}\rangle = \mathbf{c}_T, \quad \nabla\langle\mathbf{c}, \xi\rangle = -A_\xi(\mathbf{c}_T), \tag{6.185}$$

$$\nabla_X\mathbf{c}_T = \langle\mathbf{c}, \xi\rangle A_\xi X + \langle\mathbf{c}, \mathbf{x}\rangle X, \tag{6.186}$$

for $X \in TN$. From (6.182), (6.183), and (6.185), we derive that

$$2n^2\nabla\beta^2 = \lambda_p\lambda_q\mathbf{c}_T, \tag{6.187}$$

$$4A_\xi(\nabla\beta) = -3n\nabla\beta^2. \tag{6.188}$$

It follows from (6.185), (6.187) and (6.188) that

$$\lambda_p\lambda_q\langle\mathbf{c}, \xi\rangle = (n\beta)^3 + k \tag{6.189}$$

for a constant k. By combining (6.180), (6.183), (6.187) and (6.189), we find

$$\lambda_p^2\lambda_q^2||\mathbf{c}||^2 = 4n^4\beta^2||\nabla\beta||^2 + \{(n\beta)^3 + k\}^2 - \{(n\beta)^2 + b\}^2, \tag{6.190}$$

where $b = -n^2 - n(\lambda_p + \lambda_q) - \lambda_p\lambda_q$ which is a constant.

First we give the following.

Theorem 6.44. [Chen (1992)] *Every 2-type hypersurface of the hyperbolic $(n+1)$-space $H^{n+1}(-1) \subset \mathbb{E}_1^{n+2}$ has nonzero constant mean curvature and constant scalar curvature.*

Proof. Assume that M is a 2-type hypersurface of $H^{n+1}(-1) \subset \mathbb{E}_1^{n+2}$. Put $U = \{p \in M : \nabla \beta^2(p) \neq 0\}$. Suppose that U is non-empty. Let V be a connected component of U. Then V is non-empty open submanifold of M. Now, we will study only on the open submanifold V unless mentioned otherwise.

We choose an orthonormal frame $e_1, \ldots, e_n$ on V such that e_1 is the unit vector field in the direction of $\nabla \beta$. It follows from (6.188) that e_1 is an eigenvector of A_ξ with eigenvalue $\kappa_1 = -\frac{3n}{2}\beta$. We also choose $e_2, \ldots, e_n$ to be eigenvectors of A_ξ with eigenvalues given by $\kappa_2, \ldots, \kappa_n$, respectively. So we have

$$\kappa_1 = -\frac{3n}{2}\beta, \quad \kappa_2 + \cdots + \kappa_n = \frac{5n}{2}\beta. \tag{6.191}$$

By taking the covariant derivative of (6.187) with respect to a tangent vector $X \in TN$ and by applying (6.186), we find

$$2n^2\{(X\beta)\nabla\beta + \beta\nabla_X(\nabla\beta)\} = \lambda_p\lambda_q(\langle \mathbf{c}, \xi \rangle A_\xi X + \langle \mathbf{c}, \mathbf{x} \rangle X). \tag{6.192}$$

In particular, if we choose $X = e_1$, then (6.192) implies $\nabla_{e_1} e_1 = 0$, which shows that the integral curves of e_1 are geodesics in M. Thus the connection forms with respect to $\{e_1, \ldots, e_n\}$ satisfy

$$\omega_1^j(e_1) = 0, \quad j = 1, \ldots, n. \tag{6.193}$$

Since e_1 is in the direction of $\nabla\beta$, we also have

$$d\beta = (e_1\beta)\omega^1. \tag{6.194}$$

If we choose $X = e_i, i = 2, \ldots, n$, then (6.192) implies

$$2n^2\beta \|\nabla\beta\| \omega_1^i(e_i) = \lambda_p\lambda_q(\kappa_i \langle \mathbf{c}, \xi \rangle + \langle \mathbf{c}, \mathbf{x} \rangle). \tag{6.195}$$

On the other hand, it follows from Codazzi's equation that

$$e_1\kappa_i = (\kappa_1 - \kappa_i)\omega_1^i(e_i), \quad i > 1. \tag{6.196}$$

By using (6.191), (6.195) and (6.196) we find

$$\lambda_p\lambda_q\{2\langle \mathbf{c}, \xi \rangle \|A_\xi\|^2 + 3n^2\beta^2 \langle \mathbf{c}, \xi \rangle + (3n+2)n\beta \langle \mathbf{c}, \mathbf{x} \rangle\}$$
$$= -10n^3\beta \|\nabla\beta\|^2. \tag{6.197}$$

Now, by combining (6.183), (6.189) and (6.197), we get

$$2(n^3\beta^3 + k)\|A_\xi\|^2 + 3n^2\beta^2(n^3\beta^3 + k)$$
$$+ (3n+2)n\beta(n^2\beta^2 + b) = -10n^3\beta \|\nabla\beta\|^2. \tag{6.198}$$

Differentiating (6.190) with respect to e_1 and using (6.194), we find

$$4n^2\|\nabla\beta\|^2 + 4n^2\beta e_1 e_1\beta + 3n\beta(n^3\beta^3 + k) = 2n^2\beta^2 + 2b. \tag{6.199}$$

From (6.191), (6.193), (6.194), (6.195) and the definition of Δ, we get

$$\Delta\beta = -e_1 e_1 \beta - \lambda_p \lambda_q \left\{ \frac{5}{4n} \langle \mathbf{c}, \xi \rangle + \frac{n-1}{2n^2\beta} \langle \mathbf{c}, \mathbf{x} \rangle \right\}. \qquad (6.200)$$

On the other hand, by using (6.184) and (6.189), we find

$$\Delta\beta = n\beta - \beta||A_\xi||^2 + (\lambda_p + \lambda_q)\beta - n^2\beta^3 - \frac{k}{n}. \qquad (6.201)$$

By combining (6.200) and (6.201) and using (6.183) and (6.189), we obtain

$$4n e_1 e_1 \beta + 5(n^3\beta^3 + k) + \frac{2(n-1)}{n\beta}(n^2\beta^2 + b) + 4n^2\beta \qquad (6.202)$$
$$= 4n^3\beta^3 + 4k + 4n\beta||A_\xi||^2 - 4n\beta(\lambda_p + \lambda_q).$$

If follows from (6.198) and (6.202) that

$$(n^3\beta^3 + k)\{4n^2\beta e_1 e_1 \beta + 7n^4\beta^4 + (4\lambda_p + 4\lambda_q + 6n - 2)n^2\beta^2$$
$$+ nk\beta + 2(n-1)b\} + 20n^5\beta^3||\nabla\beta||^2 \qquad (6.203)$$
$$+ 2(3n+2)n^3\beta^3(n^2\beta^2 + b) = 0.$$

Now, by combining (6.199) and (6.203), we obtain

$$2n(k - 4n^3\beta^3)||\nabla\beta||^2 - (3n+2)n^2\beta^3(n^2\beta^2 + b)$$
$$= (n^3\beta^3 + k)\{2n^3\beta^4 + (2\lambda_p + 2\lambda_q + 3n)n\beta^2 - k\beta + b\}. \qquad (6.204)$$

Therefore we conclude from (6.190) and (6.204) that the mean curvature β is constant on the open submanifold $V \subset M$, which is a contradiction. Consequently, M has constant mean curvature β in $H^{n+1}(-1)$. Since M is of 2-type, the mean curvature β is nonzero. Because M has nonzero constant mean curvature β, it follows from (6.182) that

$$\lambda_p \lambda_q \mathbf{c}_T = 0. \qquad (6.205)$$

Now, we find from (6.185) and (6.205) that $\lambda_p \lambda_q \langle \mathbf{c}, \xi \rangle$ is a constant. Thus $||A_\xi||$ is a constant by (6.184). Since the scalar curvature τ of M satisfies $2\tau = n^2\beta^2 - ||A_\xi||^2 - n(n-1)$, M has constant scalar curvature too. $\square$

Conversely, we have the following result from [Chen (1992)].

Theorem 6.45. *Let M be a hypersurface of $H^{n+1}(-1) \subset \mathbb{E}_1^{n+2}$. If M has constant mean curvature and constant scalar curvature, then M is either of 1-type or of 2-type.*

Proof. Under the hypothesis, we find $DH = 0$. Thus, by Corollary 4.8,

$$\Delta H = \beta ||h||^2 \xi - n||H||^2 \mathbf{x} \tag{6.206}$$

where β is the mean curvature of M in $H^{n+1}(-1)$. Because $H = \beta\xi + \mathbf{x}$, (6.206) implies

$$\Delta H - ||h||^2 H + (n||H||^2 + ||h||^2)\mathbf{x} = 0,$$

where $||h||$ and $||H||$ are constant. Hence, after applying Theorem 6.3, we conclude that M is either of 1-type or of 2-type. $\qquad\square$

Theorem 6.46. *Every 2-type hypersurface in $H^{n+1}(-1) \subset \mathbb{E}_1^{n+2}$ is not closed.*

Proof. Assume that M is a 2-type closed hypersurface of $H^{n+1}(-1)$, then M has constant scalar curvature and nonzero constant mean curvature in $H^{n+1}(-1)$ by Theorem 6.44. Thus Corollary 4.8 implies

$$\Delta H = c_1 H + c_2 \mathbf{x}, \tag{6.207}$$

where c_1, c_2 are constants satisfying $c_1 = ||A_\xi||^2 - n$, $c_2 = n\beta^2 - ||A_\xi||^2$. Since $\Delta \mathbf{x} = -nH$ by Beltrami's formula and M is closed, Hopf's Lemma and (6.207) imply that

$$c_2 \int_M \mathbf{x}\, dV = 0, \tag{6.208}$$

In particular, this gives $c_2 \int_M x_1\, dV = 0$, where x_1 is the first coordinate function of M in $\mathbb{E}_1^{n+2}$. Since M lies in the hyperbolic space $H^{n+1}(-1) \subset \mathbb{E}_1^{n+2}$, x_1 is always positive. Thus we must have $c_2 = 0$ by (6.208). This together with (6.207) implies that M is of 1-type in $\mathbb{E}_1^{n+2}$ according to Theorem 6.1. But this is a contradiction. $\qquad\square$

Example 6.1. For each positive number r and each integer k, $(2 \leq k \leq n)$, let $N_{k,r}^n$ denote the n-dimensional submanifold of $\mathbb{E}_1^{n+2}$ defined by

$$\left\{ (x_1, x_2, \ldots, x_{n+2}) : x_1^2 - \sum_{i=2}^{k} x_i^2 = 1 + r^2, \ \sum_{j=k+1}^{n+2} x_j^2 = r^2 \right\}.$$

Then $N_{k,r}^n$ is a complete 2-type hypersurface of $H^{n+1}(-1)$. Geometrically, $N_{k,r}^n$ is the direct product of a hyperbolic space and a sphere.

Theorem 6.47. [Chen (1992)] *Let M be a hypersurface of $H^{n+1}(-1) \subset \mathbb{E}_1^{n+2}$. If M has at most two distinct principal curvatures, then M is of 2-type if and only if it is an open piece of $N_{k,r}^n$ for some positive integer k, $2 \leq k \leq n$, and some positive number r.*

Proof. Let M be a 2-type hypersurface of $H^{n+1}(-1) \subset \mathbb{E}_1^{n+2}$. Theorem 6.15 implies that M has constant mean curvature and constant scalar curvature. Thus if M has at most two distinct principal curvature, then M is an open portion of $N_{k,r}^n$ for some positive integer $k, 2 \le k \le n$ by a result of Cartan. $\qquad \square$

If $n = 2$, this implies the following.

Corollary 6.7. *Let M be a surface in the hyperbolic 3-space $H^3(-1) \subset \mathbb{E}_1^4$. Then M is of 2-type if and only if M is a flat surface which is an open portion of $N_{2,r}^2$ for some positive number r.*

Remark 6.13. Similarly, for 2-type hypersurfaces in the Lorentzian space forms, we have the following results.

Theorem 6.48. [Chen (1992)] *We have:*

(a) *Every spacelike 2-type hypersurface of the de Sitter spacetime $S_1^{n+1} \subset \mathbb{E}_2^{n+2}$ has nonzero constant mean curvature and constant scalar curvature.*

(b) *Every spacelike 2-type hypersurface of the anti-de Sitter spacetime $H_1^{n+1}(-1)$ has nonzero constant mean curvature and constant scalar curvature.*

(c) *Every spacelike 2-type hypersurface of the Minkowski spacetime $\mathbb{E}_1^{n+1}$ is of null 2-type if it has constant mean curvature.*

(d) *Hyperplanes, hyperbolic hypersurfaces and null 2-type hypersurfaces are the only spacelike hypersurfaces in the Minkowski spacetime $\mathbb{E}_1^{n+1}$ which have constant mean curvature and constant norm of second fundamental form.*

For surfaces M_s with index s in the anti-de Sitter 3-space $H_1^3(-1)$, we have

Theorem 6.49. [Alías et. al. (1994)] *A surface M_s in $H_1^3(-1)$ is of 2-type if and only if it is an open piece of one of $H_1^1(-r) \times S^1(\sqrt{r^2-1})$, $S_1^1(r) \times H^1(-\sqrt{1+r^2})$ or a non-flat B-scroll over a null curve.*

6.12 2-type spherical surfaces of higher codimension

The products of two circles with different radii are 2-type mass-symmetric spherical surfaces lying in some hypersphere $S^3 \subset \mathbb{E}^4$. Now, we provide examples of mass-symmetric 2-type surfaces lying fully in S^5 and in S^7.

Example 6.2. Let $u, v, w \in \mathbb{R}$ with $u, v > 0$. Consider the lattice

$$\Lambda = \{(2n\pi u, 2m\pi v + 2n\pi w) : n, m \in \mathbb{Z}\}. \tag{6.209}$$

The dual lattice of Λ is given by

$$\Lambda^* = \left\{ \left(\frac{h}{2\pi u} - \frac{kw}{2\pi uv}, \frac{k}{2\pi v} \right) : h, k \in \mathbb{Z} \right\}. \tag{6.210}$$

Let T_{uvw} be the flat torus given by $\mathbb{E}^2/\Lambda$. Then the spectrum of T_{uvw} is

$$\left\{ \left(\frac{h}{u} - \frac{kw}{uv} \right)^2 + \frac{k^2}{v^2} : h, k \in \mathbb{Z} \right\}. \tag{6.211}$$

For any nonzero real number ε and natural numbers h and $\bar{\varepsilon}$ satisfying $\varepsilon \neq 2h\bar{\varepsilon}^2/(\bar{\varepsilon}^2 - 2h^2)$, we put

$$\begin{aligned}
u &= \frac{\sqrt{3}\varepsilon\bar{\varepsilon}}{\sqrt{2\varepsilon^2 + \bar{\varepsilon}^2}}, \quad v = \frac{\bar{\varepsilon}}{\sqrt{2\varepsilon^2 + \bar{\varepsilon}^2}}, \\
w &= \frac{(h - \varepsilon)\bar{\varepsilon}}{\sqrt{2\varepsilon^2 + \bar{\varepsilon}^2}}, \quad e = \frac{\sqrt{2}\varepsilon}{\sqrt{2\varepsilon^2 + \bar{\varepsilon}^2}}.
\end{aligned} \tag{6.212}$$

We define an isometric immersion $\mathbf{y} : \mathbb{E}^2 \to S_0^5(1) \subset \mathbb{E}^6$ by

$$\begin{aligned}
\mathbf{y}(s, t) = \Big(& v \cos \frac{\varepsilon s}{u} \cos \frac{t}{v}, v \cos \frac{\varepsilon s}{u} \sin \frac{t}{v}, \\
& v \sin \frac{\varepsilon s}{u} \cos \frac{t}{v}, v \sin \frac{\varepsilon s}{u} \sin \frac{t}{v}, e \sin \frac{\bar{\varepsilon} s}{u} \Big),
\end{aligned} \tag{6.213}$$

which induces an isometric immersion $\mathbf{x} : T_{uvw} \to S_0^5(1) \subset \mathbb{E}^6$.

Proposition 6.5. [Barros and Chen (1987a)] *For a given real number $\varepsilon \neq 0$ and two natural numbers h and $\bar{\varepsilon}$ satisfying $\varepsilon \neq 2h\bar{\varepsilon}^2/(\bar{\varepsilon}^2 - 2h^2)$, the immersion $\mathbf{x} : T_{uvw} \to S_0^5(1) \subset \mathbb{E}^6$ induced from (6.213) is a stationary mass-symmetric 2-type isometric immersion, where u, v, w are defined by (6.212). Furthermore, we have*

(a) $\mathbf{x}$ *is of 1-type if and only if $\bar{\varepsilon}^2 = 4\varepsilon^2$; in this case $\lambda_p = 2$.*
(b) *Otherwise, $\mathbf{x}$ is of 2-type with the lower order p and the upper order q satisfying $\lambda_p = (\bar{\varepsilon}/u)^2$, $\lambda_q = (\varepsilon/u)^2 + (1/v)^2$.*

Remark 6.14. If $\varepsilon = \bar{\varepsilon} = h = 1$, this example is due to [Ejiri (1982a)].

Example 6.3. For three given real numbers u, v and w with $u, v > 0$, let T_{uvw} be the flat torus given by $\mathbb{E}^2/\Lambda$ defined in Example 6.2. For natural numbers $n, m, \bar{n}, \bar{m}$, we put

$$\varepsilon = n - \frac{mw}{v}, \quad \bar{\varepsilon} = \bar{n} - \frac{\bar{m}w}{v}. \tag{6.214}$$

We define an isometric immersion $\mathbf{y} : \mathbb{E}^2 \to S_0^7(1) \subset \mathbb{E}^6$ by

$$
\mathbf{y}(s,t) = \left(c_1 \cos \frac{\varepsilon s}{u} \cos \frac{mt}{v}, c_1 \cos \frac{\varepsilon s}{u} \sin \frac{mt}{v}, c_1 \sin \frac{\varepsilon s}{u} \cos \frac{mt}{v}, \right.
$$
$$
c_1 \sin \frac{\varepsilon s}{u} \sin \frac{mt}{v}, c_2 \cos \frac{\bar{\varepsilon} s}{u} \cos \frac{\bar{m}t}{v}, c_2 \cos \frac{\bar{\varepsilon} s}{u} \cos \frac{\bar{m}t}{v}, \tag{6.215}
$$
$$
\left. c_2 \sin \frac{\bar{\varepsilon} s}{u} \cos \frac{\bar{m}t}{v}, c_2 \sin \frac{\bar{\varepsilon} s}{u} \sin \frac{\bar{m}t}{v} \right),
$$

where c_1, c_2 are real numbers satisfying

$$c_1^2 + c_2^2 = 1, \quad c_1^2\bar{\varepsilon}^2 + c_2^2\varepsilon^2 = u^2, \quad c_1^2 m^2 + c_2^2 \bar{m}^2 = v^2.$$

The immersion (6.215) induces an isometric immersion from T_{uvw} into $S_0^7(1)$ $\subset \mathbb{E}^8$, denoted by $\mathbf{x} : T_{uvw} \to S_0^7(1) \subset \mathbb{E}^8$.

Proposition 6.6. [Barros and Chen (1987a)] *If $v^2(\bar{\varepsilon}^2 - \epsilon^2) \neq u^2(m^2 - \bar{m}^2)$, then the immersion $\mathbf{x} : T_{uvw} \to S_0^7(1) \subset \mathbb{E}^8$ induced from (6.215) is a mass-symmetric isometric immersion with*

$$\{\lambda_p, \lambda_q\} = \left\{ \left(\frac{\varepsilon}{u}\right)^2 + \left(\frac{m}{v}\right)^2, \left(\frac{\bar{\varepsilon}}{u}\right)^2 + \left(\frac{\bar{m}}{v}\right)^2 \right\}.$$

Moreover, T_{uvw} is a Chen surface of $S_0^7(1)$. Furthermore, the immersion $\mathbf{x}$ is stationary if and only if the following equation holds:

$$2c_1^2 \left[\left(\frac{\varepsilon}{u}\right)^2 - \left(\frac{m}{v}\right)^2 \right]^2 + 2c_2^2 \left[\left(\frac{\bar{\varepsilon}}{u}\right)^2 - \left(\frac{\bar{m}}{v}\right)^2 \right]^2 = \lambda_p \lambda_q.$$

We have the following non-existence result for mass-symmetric 2-type surfaces lying fully in S^4.

Theorem 6.50. [Barros and Chen (1987a)] *There exist no mass-symmetric 2-type surface lying fully in $S_o^4(1) \subset \mathbb{E}^5$.*

Proof. Assume that M is a mass-symmetric 2-type surface lying fully in $S_o^4(1)$. We choose an orthonormal local frame e_1, e_2, ξ_3, ξ_4 such that e_1, e_2 are tangent to M and $\xi_3 = \xi$ is a unit normal vector field in the direction of the mean curvature vector $\hat{H}$ of M in $S_o^4(1)$. Let $\hat{\nabla}$ denote the Levi-Civita connection of $S_o^4(1)$. We put

$$\hat{\nabla} e_i = \sum_j \omega_i^j e_j + \sum_r \omega_i^r \xi_r, \quad \hat{\nabla} \xi_r = \sum_j \omega_r^i e_i + \sum_s \omega_r^s \xi_s, \qquad (6.216)$$

for $i, j = 1, 2$ and $r, s = 3, 4$. By Cartan's lemma we have $\omega_i^r = \sum_j \hat{h}_{ij}^r \omega^j$, with $\hat{h}_{ij}^r = \hat{h}_{ji}^r$, where $\hat{h}_{ij}^r$ are the coefficients of the second fundamental form $\hat{h}$ of M in $S_o^4(1)$ and ω^1, ω^2 are the dual frame of e_1, e_2.

Let $d\hat{H}$ denote the $\mathbb{E}^5$-valued 1-form defined by $(d\hat{H})(X) = \tilde{\nabla}_X \hat{H}$ for $X \in T(M)$, where $\tilde{\nabla}$ denotes the Levi-Civita connection of $\mathbb{E}^5$. In terms of the normal connections D and $\hat{D}$ of M in $\mathbb{E}^5$ and in $S_o^4(1)$, we have

$$\Delta^D H = \Delta^{\hat{D}} \hat{H} = \hat{a}\{\langle \hat{D}\xi, \hat{D}\xi_4 \rangle - \text{Tr}(\nabla \omega_3^4)\}\xi_4 + ||\hat{D}\xi||^2 \hat{H}, \qquad (6.217)$$

where $\text{Tr}(\nabla \omega_3^4) = \sum_{i=1}^2 (\nabla_{e_i} \omega_3^4) e_i$. Since M is mass-symmetric and of 2-type, we have

$$\Delta H = (\lambda_p + \lambda_q) H + \frac{\lambda_p \lambda_q}{n} \mathbf{x}, \qquad (6.218)$$

where p and q are the lower and upper orders of M in $\mathbb{E}^5$. It follows from Proposition 4.7, $H = \hat{H} - \mathbf{x}$, (6.217), (6.218) and Theorem 6.32 that

$$||d\hat{H}||^2 = \frac{1}{n^2}(\lambda_p + \lambda_q - n)\{n(\lambda_p + \lambda_q) - \lambda_p \lambda_q - n^2\}, \qquad (6.219)$$

$$\mathfrak{a}(\hat{H}) = \hat{a}(\text{Tr}(\nabla \omega_3^4) - \langle \hat{D}\xi, \hat{D}\xi_4 \rangle)\xi_4. \qquad (6.220)$$

Since $\langle \hat{D}\xi, \hat{D}\xi_4 \rangle = 0$, (6.220) reduces to

$$\mathfrak{a}(\hat{H}) = \hat{a}\, \text{Tr}(\nabla \omega_3^4)\xi_4. \qquad (6.221)$$

Therefore, after applying the definition of $\mathfrak{a}(H)$, we find

$$\text{Tr}(A_\xi A_{\xi_4}) = \text{Tr}(\nabla \omega_3^4). \qquad (6.222)$$

On the other hand, since M is mass-symmetric and of 2-type, M has constant mean curvature in $S_o^4(1)$ by Theorem 6.31. Thus we obtain from Proposition 4.7 that $\text{Tr} A_{D\xi} = 0$. If we choose e_1, e_2 to be eigenvectors of A_{ξ_4}, then by using $\text{Tr} A_{\xi_4} = 0$, we may put

$$A_{\xi_4} e_1 = \mu e_1, \quad A_{\xi_4}(e_2) = -\mu e_2.$$

Hence, by using $\text{Tr} A_{D\xi} = 0$, we find $\mu \omega_3^4 = 0$. After combining this with (6.222) we get $\text{Tr}(A_\xi A_{\xi_4}) = 0$.

Put $W = \{p \in M : A_{\xi_4} \neq 0 \text{ at } p\}$. Assume that $W \neq \emptyset$ and U is a connected component of W. Then U is an open subset of M and we have

$\hat{D}\xi = \hat{D}\xi_4 = 0$ on U. Let e_1, e_2 be an orthonormal tangent frame on U such that, with respect to e_1, e_2, A_ξ and A_{ξ_4} are given by

$$A_\xi = \begin{pmatrix} \beta & 0 \\ 0 & \gamma \end{pmatrix}, \quad A_{\xi_4} = \begin{pmatrix} c & b \\ b & -c \end{pmatrix}. \tag{6.223}$$

Since $\text{Tr}(A_\xi A_{\xi_4}) = 0$, (6.223) yields $(\beta - \gamma)c = 0$. On the other hand, it follows from Theorem 6.31(a) and (6.219) and (6.220) that $\|A_\xi\|^2 \neq 2\hat{H}^2$. Thus U is not pseudo-umbilical, i.e., $\beta \neq \gamma$. Consequently, we have $c = 0$. Moreover, since $\hat{D}\xi = \hat{D}\xi_4 = 0$ on U, Ricci's equation gives $[A_\xi, A_{\xi_4}] = 0$. Therefore we have $b = 0$ as well. Thus $W = \emptyset$. So, we have $A_{\xi_4} = 0$ on M. Consequently, we find $\omega_1^4 = \omega_2^4 = 0$. Now, by taking exterior differentiation of these equations, we obtain

$$\beta \omega^1 \wedge \omega_3^4 = \gamma \omega^2 \wedge \omega_3^4 = 0. \tag{6.224}$$

Let G denote the Gauss curvature of M. Then we get $G = 1 + \beta\gamma$. Let $V = \{p \in M : G(p) \neq 1\}$. From (6.224) we get $\omega_3^4 = 0$ on V, i.e., $D\xi = 0$ Thus it follows from the constancy of $\hat{\alpha}$ that both β and γ are constant on V. Now, by taking the exterior differentiation of $\omega_1^3 = \beta\omega^1$ and of $\omega_2^3 = \gamma\omega^2$, we find $\omega_1^2 = 0$. Thus $G = 0$. So, by using the continuity of G on M, we obtain $G \equiv 0$ or $G \equiv 1$.

If $G \equiv 0$, then by $A_{\xi_4} = \omega_3^4 = 0$, we conclude that M is a flat surface lying in a great hypersphere of $S_o^4(1)$, which is a contradiction. So we have $G \equiv 1$ on M. Hence $\beta\gamma = 0$. Since $\beta + \gamma$ is a constant, both β and γ are constant. Without loss of generality, we may assume that $\beta = 0$. Since M is of 2-type, we have $\gamma \neq 0$. Thus we find $\omega_1^3 = 0$ and $\omega_2^3 = \gamma\omega^2$ with $\gamma \neq 0$. By taking exterior differentiation of these equations, we obtain $\omega_1^2 = 0$ which implies $G = 0$. This is again a contradiction. $\square$

Similarly, we have the following non-existence result for S^6.

Theorem 6.51. [Hasanis-Vlachos (1996a)] *There exist no mass-symmetric 2-type surfaces which lie fully in S^6.*

Stationary mass-symmetric 2-type spherical surfaces were completely classified in [Barros and Chen (1987a)]. In particular, we have the following.

Theorem 6.52. *Let M be a stationary mass-symmetric 2-type surface in $S_o^{m-1}(1) \subset \mathbb{E}^m$. Then M is a flat Chen surface lying fully in a totally geodesic $S_o^5(1) \subset S_o^{m-1}(1)$ or in a totally geodesic $S_o^7(1) \subset S_o^{m-1}(1)$.*

For every mass-symmetric closed submanifold M of $S^{m-1}(1) \subset \mathbb{E}^m$, we know from Corollary 6.5 that $0 < \lambda_p < n < \lambda_q$, where p and q are the lower and the upper orders of M.

The connection forms of stationary mass-symmetric 2-type surfaces lying fully in $S^5(1)$ were determined by the following.

Theorem 6.53. *If M is a stationary mass-symmetric 2-type surface lying fully in $S_o^5(1)$, then M is flat and $\frac{2}{3} < \lambda_p < 2$. Moreover, with respect to a suitable orthonormal frame $e_1, e_2, \xi_3, \xi_4, \xi_5$, the connection form (ω_A^B) is given by (6.225) if $\frac{2}{3} < \lambda_p \leq \frac{4}{3}$ and by (6.225) or (6.226) if $\frac{4}{3} < \lambda_p < 2$:*

$$
\begin{pmatrix}
0 & 0 & \dfrac{\sqrt{2}(1-c)}{\sqrt{3c-2}}\omega^1 & \dfrac{\sqrt{c}}{\sqrt{3c-2}}\omega^2 & 0 \\[2ex]
0 & 0 & \dfrac{\sqrt{c}}{\sqrt{3c-2}}\omega^2 & \dfrac{\sqrt{c}}{\sqrt{3c-2}}\omega^1 & 0 \\[2ex]
\dfrac{\sqrt{2}(c-1)}{\sqrt{3c-2}}\omega^1 & \dfrac{-\sqrt{2}}{\sqrt{3c-2}}\omega^2 & 0 & 0 & \dfrac{-c}{\sqrt{3c-2}}\omega^1 \\[2ex]
\dfrac{-\sqrt{c}}{\sqrt{3c-2}}\omega^2 & \dfrac{-\sqrt{c}}{\sqrt{3c-2}}\omega^1 & 0 & 0 & \dfrac{-\sqrt{2c}}{\sqrt{3c-2}}\omega^2 \\[2ex]
0 & 0 & \dfrac{c}{\sqrt{3c-2}}\omega^1 & \dfrac{\sqrt{2c}}{\sqrt{3c-2}}\omega^2 & 0
\end{pmatrix}, \quad (6.225)
$$

where $c = \lambda_p$ is a real number satisfying $\frac{2}{3} < c < 2$, or

$$
\begin{pmatrix}
0 & 0 & \dfrac{c-4}{2\sqrt{3c-4}}\omega^1 & \dfrac{\sqrt{c}}{2}\omega^2 & 0 \\[2ex]
0 & 0 & \dfrac{\sqrt{3c-4}}{2}\omega^2 & \dfrac{\sqrt{c}}{2}\omega^1 & 0 \\[2ex]
\dfrac{4-c}{2\sqrt{3c-4}}\omega^1 & \dfrac{-\sqrt{3c-4}}{2}\omega^2 & 0 & 0 & \dfrac{-c}{\sqrt{6c-8}}\omega^1 \\[2ex]
\dfrac{-\sqrt{c}}{2}\omega^2 & \dfrac{-\sqrt{c}}{2}\omega^1 & 0 & 0 & \dfrac{-\sqrt{c}}{\sqrt{2}}\omega^2 \\[2ex]
0 & 0 & \dfrac{c}{\sqrt{6c-8}}\omega^1 & \dfrac{\sqrt{c}}{\sqrt{2}}\omega^2 & 0
\end{pmatrix}, \quad (6.226)
$$

where $c = \lambda_p$ is a real number satisfying $\frac{4}{3} < c < 2$.

The connection forms of stationary mass-symmetric 2-type surfaces lying fully in $S^7(1)$ were also determined in [Barros and Chen (1987a)].

Theorem 6.54. *If M is a stationary mass-symmetric 2-type surface lying fully in $S_o^7(1)$, then M is flat. Moreover, with respect to a suitable* *orthonormal frame* $e_1, e_2, \xi_3, \ldots, \xi_7$, *the connection form* (ω_A^B) *is given by*

$$
\begin{pmatrix}
0 & 0 & \beta\omega^1 & 0 & 0 & b\omega^2 & 0 \\
0 & 0 & \gamma\omega^2 & 0 & 0 & b\omega^1 & 0 \\
-\beta\omega^1 & -\gamma\omega^2 & 0 & \alpha_1\omega^1 & \delta\omega^2 & 0 & 0 \\
0 & 0 & -\alpha_1\omega^1 & 0 & 0 & -\dfrac{\gamma\alpha_1}{b}\omega^2 & \dfrac{\delta}{b}\omega^2 \\
0 & 0 & -\delta\omega^2 & 0 & 0 & -\dfrac{\beta\delta}{b}\omega^1 & \dfrac{\alpha_1}{b}\omega^1 \\
-b\omega^2 & -b\omega^1 & 0 & \dfrac{\gamma\alpha_1}{b}\omega^2 & \dfrac{\beta\delta}{b}\omega^1 & 0 & 0 \\
0 & 0 & 0 & -\dfrac{\delta}{b}\omega^2 & -\dfrac{\alpha_1}{b}\omega^1 & 0 & 0
\end{pmatrix},
\tag{6.227}
$$

where $\alpha, \beta, \gamma, \alpha_1, \delta$ *are constants satisfying*

$$
\beta = \frac{\sqrt{cd}}{2\sqrt{2}} + \frac{1}{2}\sqrt{(2-c)(d-2)}, \quad \gamma = -\frac{\sqrt{cd}}{2\sqrt{2}} + \frac{1}{2}\sqrt{(2-c)(d-2)},
$$

$$
\alpha_1 = \frac{(\gamma - \beta)(2 + 3\beta\gamma - \beta^2)}{2(\beta + \gamma)}, \quad \delta = \frac{(\beta - \gamma)(2 + 3\beta\gamma - \gamma^2)}{2(\beta + \gamma)},
$$

$$
b = \sqrt{1 + \beta\gamma},
$$

for some constants $c = \lambda_p$ *and* $d = \lambda_q$ *such that* $0 < c < 2 < d < \infty$.

We also have the following results from [Barros and Chen (1987a)].

Theorem 6.55. *We have:*

(a) *For each real number* c *with* $\frac{2}{3} < c \leq \frac{4}{3}$, *there is a stationary mass-symmetric 2-type flat torus in* $S_0^5(1)$ *whose connections form is given by* (6.225).

(b) *For each real number* c *with* $\frac{4}{3} < c < 2$, *there are two stationary mass-symmetric 2-type flat tori in* $S_0^5(1)$ *whose connection forms are given by* (6.225) *and* (6.226), *respectively.*

Theorem 6.56. *For any* $d \in (2, \infty)$, *there is a stationary mass-symmetric 2-type flat torus lying fully in* $S_0^7(1)$ *such that* $\lambda_q = d$ *and whose connection form is given by* (6.227).

For mass-symmetric 2-type Chen surfaces we have the next result.

Theorem 6.57. [Garay (1988a)] *If* M *is a mass-symmetric 2-type Chen surface in* $S^{m-1}(1)$, *then* M *is either pseudo-umbilical or flat. Furthermore, if* M *is flat, then it lies fully in a totally geodesic* $S^3(1)$, *or a totally geodesic* $S^5(1)$, *or a totally geodesic* $S^7(1)$ *of* $S^{m-1}(1)$.

M. Kotani investigated mass-symmetric 2-type immersions of 2-spheres into a hypersphere of $\mathbb{E}^m$. She proved the following results.

Theorem 6.58. [Kotani (1990)] *Any mass-symmetric 2-type immersion of a topological 2-sphere into S^{m-1} is the diagonal immersion of two 1-type immersions.*

Theorem 6.59. [Kotani (1990)] *If a 2-sphere admits a mass-symmetric 2-type isometric immersion into S^9, then the 2-sphere is of constant curvature.*

Theorem 6.60. [Kotani (1990)] *There do not exist mass-symmetric and 2-type immersions from a topological 2-sphere into hyperbolic spaces.*

Remark 6.15. The last result follows immediately from Theorem 6.45.

Y. Miyata investigated mass-symmetric 2-type immersions of surfaces of constant curvature into spheres. He obtained the next six results.

Theorem 6.61. [Miyata (1988)] *Let $\phi : M(c) \to S^{m-1}(1)$ be a mass-symmetric 2-type immersion of a surface of constant positive curvature c into $S^{m-1}(1)$. Then ϕ is a diagonal sum of two different standard minimal immersions of $M(c)$ into spheres.*

Theorem 6.62. [Miyata (1988)] *There do not exist mass-symmetric 2-type immersions of a surface of constant negative curvature into a sphere.*

Theorem 6.63. [Miyata (1988)] *Let D be a small disk about the origin in the Euclidean plane $\mathbb{E}^2$ and $\phi : D \to S^N(1)$ be a mass-symmetric 2-type full immersion. Then we have:*

(1) *N is odd,*
(2) *ϕ extends uniquely to a mass-symmetric 2-type immersion of $\mathbb{E}^2$ into $S^N(1)$,*
(3) *ϕ can be written in terms of a suitable complex coordinate z on $\mathbb{C} \simeq \mathbb{E}^2$ in the form*

$$f(z) = \sqrt{A_1} \sum_{k=1}^{m} \{P_k \exp \frac{\sqrt{c}}{2}(\mu_k z - \overline{\mu}_k \overline{z}) + \overline{P}_k \exp \frac{\sqrt{c}}{2}(-\mu_k z + \overline{\mu}_k \overline{z})\}$$

$$+ \sqrt{A_2} \sum_{\dot{g}=1}^{m'} \{Q_j \exp \frac{\sqrt{d}}{2}(\eta_j z - \overline{\eta}_j \overline{z}) + \overline{Q}_j \exp \frac{\sqrt{d}}{2}(-\eta_j z + \overline{\eta}_j \overline{z})\},$$

where $c, d \in \mathbb{R}$, $A_1 = \frac{d-2}{d-c}$, $A_2 = \frac{2-c}{d-c}$, $\{\pm\mu_k\}_{k=1}^m$ *are* $2m$ *(resp.,* $\{\pm\eta_j\}_{j=1}^{m'}$ *are* $2m'$*) distinct complex numbers of norm* 1, $N = 2(m+m') - 1$, *and* P_k, $Q_j \in (\mathbb{E}^{N+1})^{\mathbb{C}}$ *are nonzero vectors satisfying*

$$\langle P_k, P_\ell \rangle = 0 \ \forall k, \ell, \ \ \langle P_k, \overline{P}_\ell \rangle = 0 \ \forall k \neq \ell,$$

$$\langle Q_j, Q_i \rangle = 0 \ \forall j, i, \ \ \langle Q_j, \overline{Q}_i \rangle = 0 \ \forall j \neq i,$$

$$\langle P_k, Q_j \rangle = \langle P_k, \overline{Q}_j \rangle = 0 \ \forall k, j, \ \ \sum_{k=1}^m \langle P_k, \overline{P}_k \rangle = \frac{1}{2},$$

$$\sum_{j=1}^{m'} \langle Q_j, \overline{Q}_j \rangle = \frac{1}{2}, \ \ cA_1 \sum_{k=1}^m \mu_k^2 \langle P_k, \overline{P}_k \rangle + dA_2 \sum_{j=1}^{m'} \eta_j^2 \langle Q_j, \overline{Q}_j \rangle = 0.$$

Remark 6.16. Theorem 6.62 says that $\phi(M)$ is an orbit of an abelian subgroup of $SO(2(m + m'))$. Let G be the abelian group of parallel displacements of $\mathbb{E}^2$. Then ϕ is G-equivariant.

Corollary 6.8. [Miyata (1988)] *If* $\phi : M(c) \to S^n(1)$ *is a mass-symmetric 2-type full immersion from a surface of constant curvature* c *into* $S^n(1)$, *then* n *is odd.*

Theorem 6.64. [Miyata (1988)] *Let* M *be a flat surface and* ϕ *be a full mass-symmetric 2-type Chen immersion of* M *into* $S^N(1)$. *If* $N \geq 9$, *then* ϕ *is a diagonal sum of two different minimal immersions into spheres.*

Let $(S^6(1), J, \tilde{g})$ be a nearly Kähler manifold with a canonical almost complex structure on S^6. The automorphism group is the compact simple Lie group G_2. For two imbedded submanifolds M_1 and M_2 of S^6, we say that $M_1 \sim M_2$ if there exists an element φ of G_2 satisfying $\varphi(M_1) = M_2$. Then the relation $\sim$ is an equivalence relation.

Definition 6.7. A submanifold M of a Hermitian manifold $(\tilde{M}, J, \tilde{g})$ is called *totally real* is $J(T_x(M)) \subset T_x^\perp(M)$ for each $x \in M$. And M is called *Lagrangian* if $J(T_x(M)) = T_x^\perp(M)$, $x \in M$ (cf. [Chen and Ogiue (1974a)]).

Let T be a maximal torus of G_2. Since G_2 is of rank 2, any T-orbit in S^6 is a flat surface or a circle or a point. We put

$\mathfrak{F} = \{T$-orbit which is totally real mass-symmetric 2-type$\}$,

$\mathfrak{F}_3 = \{T$-orbit which is totally real mass-symmetric 2-type and imbedded fully into a totally geodesic $S^8(1)\}$,

$\mathfrak{F}_6 = \{T$-orbit which is totally real mass-symmetric 2-type and imbedded fully into a totally geodesic $S^6(1)\}$.

Y. Miyata also proved the following results.

Theorem 6.65. [Miyata (1988)] *We have:*

(1) *If $M \in \mathfrak{F}$, then M is a Chen surface of $S^6(1)$.*
(2) *$\mathfrak{F}_8/\sim$ and $\mathfrak{F}_5/\sim$ form 1-parameter families.*
(3) *$\mathfrak{F}/\sim$ is a disjoint union of $\mathfrak{F}_3/\sim$ and $\mathfrak{F}_5/\sim$.*
(4) *If $M \in \mathfrak{F}_3$, then M lies fully in a totally real and totally geodesic 3-sphere $S^3(1)$ of $S^6(1)$.*
(5) *Suppose that $M \in \mathfrak{F}$ and denote by H the mean curvature vector field of M in $S^6(1)$. Then JH is a normal vector field of M if and only if $M \in \mathfrak{F}_3$ and JH is a tangent vector field of M if and only if $M \in \mathfrak{F}_5$.*
(6) *If $M \in \mathfrak{F}_5$, then M is not stationary.*

Theorem 6.66. [Miyata (1988)] *Suppose that M is a complete totally real mass-symmetric 2-type Chen surface in $S^6(1)$ which is imbedded by isometric imbedding ϕ. Denote by H the mean curvature vector of ϕ.*

(1) *If JH is a normal vector field of M, then $M \in \mathfrak{F}_3$ and*
(2) *if JH is a tangent vector field of M, then $M \in \mathfrak{F}_5$.*

The local version of this result also holds.

For pseudo-umbilical 2-type spherical surfaces, we have the following.

Theorem 6.67. [Garay (1990b)] *Every pseudo-umbilical 2-type surface in a hypersphere of a Euclidean space has constant mean curvature. Moreover, the dimension of the hypersphere is at least 5.*

Theorem 6.68. [Hasanis-Vlachos (1996a)] *Let M be a mass-symmetric 2-type surface lying fully in S^{m-1}. If M has flat normal connection, then m is even unless M is pseudo-umbilical.*

Theorem 6.69. [Kim and Kim (1995b)] *Every closed pseudo-umbilical 2-type surface in $S^{m-1} \subset \mathbb{E}^m$ is mass-symmetric.*

For further results on pseudo-umbilical 2-type submanifolds, see [Hasanis-Vlachos (1996a); Kim and Kim (1995b)].

In views of the above results on mass-symmetric 2-type spherical surfaces, the author proposed the following (cf. [Chen (2014)]).

Conjecture 6.4. *If $\phi : M \to S^{m-1}(r) \subset \mathbb{E}^m$ is a mass-symmetric 2-type isometric full immersion, then m is even.*

Chapter 7

Biharmonic Submanifolds and Biharmonic Conjectures

A pseudo-Riemannian submanifold M of a pseudo-Euclidean m-space $\mathbb{E}_s^m$ is called *biharmonic* if its position vector field $\mathbf{x}$ satisfies

$$\Delta^2 \mathbf{x} = 0, \tag{7.1}$$

i.e., each pseudo-Euclidean coordinate function of M is a biharmonic function on (M, g), where g is the induced metric on the submanifold. Clearly, it follows from Hopf's lemma that biharmonic submanifolds in a Euclidean space are non-closed.

The study of biharmonic submanifolds was initiated by the author in the middle of 1980s in his program of understanding submanifolds of finite type in Euclidean spaces as well as in pseudo-Euclidean spaces. Independently, biharmonic submanifolds as biharmonic maps were also investigated by G.-Y. Jiang in [Jiang (1986a,b)] for his study of Euler-Lagrange's equation of bienergy functional. It was shown independently by the author and Jiang in [Jiang (1987)] that there are no biharmonic surfaces in $\mathbb{E}^3$ except the minimal ones. This non-existence result was generalized by I. Dimitric in his doctoral thesis [Dimitric (1989)] at Michigan State University. Later, biharmonic submanifolds have been investigated by many mathematicians.

Obvious, minimal submanifolds of $\mathbb{E}_s^m$ are trivially biharmonic. On the other hand, the following *biharmonic conjecture* is well-known.

Conjecture 7.1. [Chen (1991b)] *Minimal submanifolds are the only biharmonic submanifolds in Euclidean spaces.*

The study of biharmonic submanifolds is nowadays a very active research subject. In particular, since 2000 biharmonic submanifolds have been receiving a growing attention and have become a popular research subject of study with many important progresses. In this chapter we provide a comprehensive account of the latest updated and new results on biharmonic submanifolds and biharmonic conjectures.

7.1   Necessary and sufficient conditions

Definition 7.1. Let $\phi : M \to \mathbb{E}_s^m$ be an isometric immersion of a pseudo-Riemannian n-manifold M into a pseudo-Euclidean space. Then M is called a *biharmonic submanifold* if and only if

$$\Delta^2 \phi = 0, \quad \text{equivalently } \Delta H = 0, \tag{7.2}$$

holds identically, where H is the mean curvature vector of the submanifold and Δ is the Laplacian on the submanifold.

Theorem 7.1. [Chen (1979b, 1986b)] *Let $\phi : M \to \mathbb{E}_s^m$ be an immersion of a pseudo-Riemannian n-manifold M into a pseudo-Euclidean space $\mathbb{E}_s^m$. Then M is biharmonic if and only if it satisfies the following PDE system:*

$$\Delta^D H + \sum_{i=1}^{n} \epsilon_i h(A_H e_i, e_i) = 0, \tag{7.3}$$

$$n \nabla \langle H, H \rangle + 4 \operatorname{Tr} A_{DH} = 0. \tag{7.4}$$

Proof. This is an immediate consequence of Proposition 4.7 after taking the tangential and normal components from the formula of ΔH.   □

The next two corollaries are easy consequences of Theorem 7.1.

Corollary 7.1. *Every pseudo-Riemannian biharmonic submanifold M of $\mathbb{E}_s^m$ with parallel mean curvature vector satisfies $\operatorname{Tr} A_H^2 = 0$.*

In particular, every pseudo-Riemannian biharmonic hypersurface of $\mathbb{E}_s^{n+1}$ with constant mean curvature and diagonalizable shape operator is minimal.

Proof. Under the hypothesis, we have $DH = 0$. Thus (7.3) reduces to

$$\sum_{i=1}^{n} \epsilon h(A_H e_i, e_i) = 0, \tag{7.5}$$

with $n = \dim M$. Hence, by taking the inner product of (7.5) with H, we obtain $\operatorname{Tr} A_H^2 = 0$.

If M is a pseudo-Riemannian biharmonic hypersurface with diagonalizable shape operator, then we may choose $e_1, \ldots, e_n$ such that $A_H(e_i) = \kappa_i e_i$ for $i = 1, \ldots, n$. Hence it follows from $\operatorname{Tr} A_H^2 = 0$ that $\sum_i \kappa_i^2 = 0$, which implies $H = 0$. Thus M is minimal.   □

Corollary 7.2. *Let $\phi : M \to \mathbb{E}_s^m$ be a biharmonic immersion of a pseudo-Riemannian manifold M into $\mathbb{E}_s^m$. If $\phi(M)$ lies in a pseudo hypersphere $S_s^{m-1}(r) \subset \mathbb{E}_s^m$, then M is either minimal or quasi-minimal in $\mathbb{E}_s^m$.*

Proof. Under the hypothesis, we may assume without loss of generality that $S_s^{m-1}(r)$ is centered at the origin and $r = 1$. Let $\hat{H}$ denotes the mean curvature vector of M in $S_s^{m-1}(r)$ as before. Then $H = \hat{H} - \phi$. Hence we get $\Delta H = \Delta^{\hat{D}} \hat{H}$ which is tangent to $S_s^{m-1}(r)$. Thus, after taking the inner product of (7.3) with ϕ, we find

$$0 = \sum_i \epsilon_i \langle A_H e_i, A_\phi e_i \rangle = -\sum_i \epsilon \langle A_H e_i, e_i \rangle = -n \langle H, H \rangle,$$

which implies the corollary. □

The next result is due to [Dimitric (1989, 1992a)]).

Proposition 7.1. *Every finite type biharmonic pseudo-Riemannian submanifold in $\mathbb{E}_s^m$ is minimal.*

Proof. If M is a finite type biharmonic pseudo-Riemannian submanifold of $\mathbb{E}_s^m$, then the immersion of M admits finite spectral decomposition:

$$\phi = \phi_0 + \phi_{t_1} + \cdots + \phi_{t_k},$$
$$\Delta\phi_0 = 0, \quad \Delta\phi_{t_i} = \lambda_{t_i}\phi_{t_i} \tag{7.6}$$

where $\lambda_{t_1}, \ldots, \lambda_{t_k}$ are nonzero distinct eigenvalues. By applying Δ^r to (7.6) we find

$$0 = \Delta^r \phi = \lambda_{t_1}^r \phi_{t_1} + \cdots + \lambda_{t_k}^r \phi_{t_k},$$
$$r = 2, 3, \ldots. \tag{7.7}$$

Since $\lambda_{t_1}, \ldots, \lambda_{t_k}$ are distinct, system (7.7) is impossible unless $k = 0$. So, $\phi = \phi_0$, which implies that M is minimal. □

Consequently, we have the following.

Corollary 7.3. *Every biharmonic pseudo-Riemannian submanifold of $\mathbb{E}_s^m$ is either minimal or of infinite type.*

For hypersurface of $\mathbb{E}^{n+1}$, Theorem 7.1 reduces to the following.

Corollary 7.4. *Let $\phi : M \to \mathbb{E}_s^{n+1}$ be a pseudo-Riemannian hypersurface of $\mathbb{E}_s^{n+1}$. Then M is biharmonic if and only if it satisfies*

$$\Delta\alpha + \epsilon\alpha \operatorname{Tr} A^2 = 0, \tag{7.8}$$

$$A(\nabla\alpha) = -\frac{n}{2}\epsilon\alpha\nabla\alpha, \tag{7.9}$$

where $A = A_\xi$, $H = \alpha\xi$ and $\epsilon = \langle \xi, \xi \rangle = \pm 1$.

7.2   Biharmonic curves and surfaces in pseudo-Euclidean space

First, we have the following.

Lemma 7.1. *Every unit speed biharmonic curve in a pseudo-Euclidean m-space $\mathbb{E}_r^m$ lies in a linear 3-subspace of $\mathbb{E}_r^m$.*

Proof. Let $\gamma : I \to E_r^m$ be an isometric immersion of an open interval I into $\mathbb{E}_r^m$. Let s be an arclength function of γ. The Laplacian Δ of γ is then given by $-d^2/ds^2$ and the mean curvature vector is $H = \gamma''(s)$. The condition that γ being biharmonic is equivalent to $\gamma^{(iv)}(s) = 0$. Thus if γ is biharmonic, then there exist constant vectors $c_1, c_2, c_3, c_4 \in \mathbb{E}_r^m$ such that

$$\gamma(s) = c_1 s^3 + c_2 s^2 + c_3 s + c_4. \tag{7.10}$$

Consequently, γ lies in a 3-dimensional linear subspace of $\mathbb{E}_r^m$. $\square$

The next result classifies biharmonic curves in $\mathbb{E}_r^3$ for any index r.

Theorem 7.2. [Chen and Ishikawa (1991)] *Let $\gamma : I \to \mathbb{E}_r^3$ be a unit speed curve in $\mathbb{E}_r^3$. Then γ is biharmonic if and only if one of the following three statements holds:*

(a) *$r \in \{0, 1, 2\}$ and γ lies in a line.*
(b) *$r = 1$ and, up to rigid motions of $\mathbb{E}_1^3$, γ is given either by*

$$\gamma(s) = (\delta s^3 + \bar{\delta} s^2, \delta s^3 + \bar{\delta} s^2, s) \tag{7.11}$$

for some constants δ and $\bar{\delta}$ satisfying $\delta^2 + \bar{\delta}^2 \neq 0$ or by

$$\gamma(s) = \left(\frac{a^2 s^3}{6}, \frac{a s^2}{2}, s - \frac{a^2 s^3}{6} \right) \tag{7.12}$$

for some nonzero constant a.
(c) *$r = 2$ and, up to rigid motions of $\mathbb{E}_2^3$, γ is given by*

$$\gamma(s) = \left(\frac{a^2 s^3}{6}, \frac{a s^2}{2}, s + \frac{a^2 s^3}{6} \right) \tag{7.13}$$

for some nonzero constant a.

Proof. If γ is given by (a), (b) or (c), then it is easy to verify that γ is biharmonic.

Conversely, suppose that $\gamma : I \to \mathbb{E}^3_r$ is biharmonic. Then (7.10) holds. So, by using the condition that s is the arclength of γ we may obtain

$$\langle c_1, c_1 \rangle = \langle c_1, c_2 \rangle = \langle c_2, c_3 \rangle = 0, \tag{7.14}$$

$$3 \langle c_1, c_3 \rangle + 2 \langle c_2, c_2 \rangle = 0, \quad \langle c_3, c_3 \rangle = 1. \tag{7.15}$$

Without loss of generality, we may assume

$$c_3 = (0, 0, 1). \tag{7.16}$$

It is clear that the mean curvature vector $H = 0$ if and only if γ is linear, so it is a part of a line. Next, we assume $H \neq 0$.

Case (1): $\langle H, H \rangle = 0$. We have $\langle c_2, c_2 \rangle = \langle c_1, c_3 \rangle = 0$. Thus it follows from (7.14), (7.15) and (7.16) that $r = 1$. Moreover, up to rigid motions of $\mathbb{E}^3_1$, c_1 and c_2 are given by $c_1 = (\delta, \delta, 0)$, $c_2 = (\bar{\delta}, \bar{\delta}, 0)$ for some constants δ and $\bar{\delta}$ satisfying $\delta^2 + \bar{\delta}^2 \neq 0$. Thus γ is a curve given by (7.11).

Case (2): $\langle H, H \rangle \neq 0$. In this case, (7.10) implies $H = 6c_1 s + 2c_2$. Thus (7.14) gives $\langle H, H \rangle = 4 \langle c_2, c_2 \rangle$ which is a nonzero constant. Let N denote the unit vector in the direction of H so that $H = \kappa N$ with $\kappa > 0$. We treat the cases $\langle N, N \rangle = 1$ and $\langle N, N \rangle = -1$, separately.

Case (2.1): $\langle N, N \rangle = 1$. We have $\kappa^2 = 4 \langle c_2, c_2 \rangle = const. \neq 0$ and $T = \gamma'$ and N are orthonormal spacelike vectors in $\mathbb{E}^3_r$. Put $Y = N' + \kappa T$. Then Y is perpendicular to T and N. Since

$$Y = \left(\frac{6}{\kappa} + 3\kappa s^2 \right) c_1 + 2\kappa s c_2 + \kappa c_3,$$

equations (7.14) and (7.15) imply that $\langle Y, Y \rangle = -8 \langle c_2, c_2 \rangle + \kappa^2 = -\kappa^2 < 0$. Thus $r = 1$ and Y is timelike; also $T, N, \kappa^{-1} Y$ are orthonormal. Hence we obtain $\langle c_2, c_2 \rangle = \kappa^2/4$ and $\langle c_1, c_3 \rangle = -\kappa^2/6$. From these and (7.14)-(7.16), we may choose a coordinate system on $\mathbb{E}^3_1$ such that

$$c_1 = \left(\frac{a^2}{6}, 0, -\frac{a^2}{6} \right), \quad c_2 = \left(0, \frac{a}{2}, 0 \right).$$

Therefore, up to rigid motions of $\mathbb{E}^3_1$, γ is given by (7.12).

Case (2.2): $\langle N, N \rangle = -1$. In this case, we have $\kappa^2 = -4 \langle c_2, c_2 \rangle$. Thus we find form (7.14) and (7.15) that

$$\langle c_2, c_2 \rangle = -\frac{\kappa^2}{4}, \quad \langle c_1, c_3 \rangle = \frac{\kappa^2}{6}.$$

Since $\kappa \neq 0$, the vector c_2 is timelike. Because T and N are orthonormal, we find form a direct computation that the vector field $Y = N' - \kappa T$ satisfies

$$\langle Y, Y \rangle = 8 \langle c_2, c_2 \rangle + \kappa^2 = -\kappa^2.$$

Thus Y is timelike. Since $T, N, \kappa^{-1} Y$ are orthonormal, we have $r = 2$. Moreover, we obtain (7.13) from (7.14), (7.15) and $r = 2$. $\square$

The following corollary is an immediate consequence of Lemma 7.1 and Theorem 7.2.

Corollary 7.5. [Dimitric (1989)] *Every biharmonic curve of a Euclidean space is an open portion of a straight line.*

For further results on biharmonic curves in a Minkowski 3-space, see [Inoguchi (2003, 2006)].

For biharmonic surfaces we have the following nonexistence result.

Theorem 7.3. *Every biharmonic immersion of a spacelike surface M into $\mathbb{E}_r^3$ ($r = 0, 1$) is minimal.*

Proof. Let $\phi : M \to \mathbb{E}_r^3$ be a biharmonic isometric immersion from a Riemannian 2-manifold M into $\mathbb{E}_r^3$. Put $H = \alpha e_3$. Then we have

$$\langle H, H \rangle = \alpha^2 \epsilon_3, \quad \epsilon_3 = \langle e_3, e_3 \rangle = \pm 1.$$

From Corollary 7.4 we find

$$\Delta \alpha + \epsilon_3 \|A_3\|^2 = 0, \tag{7.17}$$

$$A_3(\nabla \alpha) = -\epsilon_3 \alpha (\nabla \alpha). \tag{7.18}$$

Let e_1, e_2 be an orthonormal frame which diagonalize A_3. Thus we have

$$h(e_1, e_1) = \epsilon_3 \beta e_3, \quad h(e_1, e_2) = 0, \quad h(e_2, e_2) = \epsilon_3 \gamma e_3 \tag{7.19}$$

for some functions β, γ. It follows from Codazzi's equation and (7.19) that

$$\begin{aligned} e_2 \beta &= (\beta - \gamma)\omega_1^2(e_1), \\ e_1 \gamma &= (\beta - \gamma)\omega_1^2(e_2). \end{aligned} \tag{7.20}$$

Let $U = \{x \in M : \nabla \alpha^2 \neq 0 \text{ at } x\}$. Then U is an open subset. Assume $U \neq \emptyset$. Then (7.18) implies that $\nabla \alpha$ is an eigenvector of A_3 with eigenvalue $-\epsilon_3 \alpha$ on U. Let us choose e_1 in the direction of $\nabla \alpha$ on U. Then we have $e_2 \alpha = 0$ on U. Moreover, the shape operator A_3 satisfies

$$\beta = -\epsilon_3 \alpha, \quad \gamma = 3\epsilon_3 \alpha, \quad \|A_3\|^2 = 10\alpha^2. \tag{7.21}$$

Since $e_2 \alpha = 0$, (7.20) and (7.21) imply

$$\omega_1^2(e_1) = 0, \quad d\omega^1 = 0. \tag{7.22}$$

Thus locally $\omega^1 = du$ for some function u. Since $d\alpha \wedge \omega^1 = d\alpha \wedge du = 0$, α is a function of u. Let α' and α'' denote the first and the second derivatives of α with respect to u. From (7.20), (7.21) and (7.22), we find

$$4\alpha \omega_1^2 = -3\alpha' \omega^2. \tag{7.23}$$

Also from $e_2\alpha = 0$ and (7.23) we have

$$4\alpha\Delta\alpha = 3(\alpha')^2 - 4\alpha\alpha''. \tag{7.24}$$

From (7.17), (7.21) and (7.24) we get

$$4\alpha\alpha'' - 3(\alpha')^2 - 40\epsilon_3\alpha^4 = 0.$$

Therefore we obtain

$$(\alpha')^2 = 8\epsilon_3\alpha^4 + C\alpha^{3/2}, \quad C \in \mathbb{R}. \tag{7.25}$$

On the other hand, the equation of Gauss, (7.19) and (7.23) yield

$$\alpha\alpha'' - (\alpha')^2 + 4\epsilon_3\alpha^4 = 0. \tag{7.26}$$

From (7.25) and (7.26) we obtain $(\alpha')^2 = 56\epsilon_3\alpha^4$. By combining this with (7.26) we conclude that α is constant on U which is a contradiction. Thus we have $U = \emptyset$. So, M has constant mean curvature. Therefore, by (7.17), we get $\alpha = 0$. Consequently, ϕ is a minimal immersion. $\qquad\square$

Remark 7.1. Theorem 7.3 is an extension of an unpublished result done in 1985. This pseudo-Riemannian version as stated in Theorem 7.3 was written while S. Ishikawa was working on his doctoral thesis in 1988 at Michigan State University (see [Chen and Ishikawa (1991)]). In a quite different approach, Theorem 7.2 with $r = 0$ was also obtained independently by Jiang in his study of Euler-Lagrange's equation of bienergy functional (cf. [Jiang (1987)]).

Next, we provide some examples of spacelike biharmonic surfaces in $\mathbb{E}_r^4$ with nonzero constant mean curvature.

Example 7.1. Let M be a spacelike surface in $\mathbb{E}_2^4$ defined by

$$\mathbf{x}(u, v) = \left(\frac{1}{6}a^2u^3, \frac{1}{2}au^2, u + \frac{1}{6}a^2u^3, v\right), \quad 0 \neq a \in \mathbb{R}. \tag{7.27}$$

Then $H = \frac{1}{2}(a^2u, a, a^2u, 0)$ and $\Delta H = 0$. Hence M is a biharmonic surface satisfying $\langle H, H \rangle = -\frac{1}{4}a^2$.

Example 7.2. Let M be a spacelike surface in $\mathbb{E}_1^4$ defined by

$$\mathbf{x}(u, v) = \left(\frac{1}{6}a^2u^3, \frac{1}{2}au^2, u - \frac{1}{6}a^2u^3, v\right), \quad 0 \neq a \in \mathbb{R}. \tag{7.28}$$

Then $H = \frac{1}{2}(a^2u, a, -a^2u, 0)$ and $\Delta H = 0$. Thus M is a biharmonic surface satisfying $\langle H, H \rangle = \frac{1}{4}a^2$.

Theorem 7.4. [Chen and Ishikawa (1991)] *Let M be a spacelike surface in $\mathbb{E}_r^4$. Then M is a biharmonic surface with nonzero constant mean curvature if and only if one of the following two cases occurs:*

(1) $r = 2$ *and M is congruent to a surface given by* (7.27) *in Example 7.1.*
(2) $r = 1$ *and M is congruent to a surface given by* (7.28) *in Example 7.2.*

Proof. Assume that M is a spacelike biharmonic surface in $\mathbb{E}_r^4$ with constant $\langle H, H \rangle \neq 0$. Let e_3 be the unit normal vector field in the direction of H so that $H = \alpha e_3$ with constant $\alpha > 0$. It follows from Theorem 7.1 that

$$\operatorname{Tr} A_4 = 0, \tag{7.29}$$

$$\epsilon_4 \|\omega_4^3\|^2 + \|A_3\|^2 = 0, \tag{7.30}$$

$$\epsilon_4 \operatorname{Tr}(A_H A_4) = \alpha \operatorname{Tr}(\nabla \omega_3^4), \tag{7.31}$$

$$\omega_3^4(e_1) A_4 e_1 + \omega_3^4(e_2) A_4 e_2 = 0, \tag{7.32}$$

where $\operatorname{Tr}(\nabla \omega_3^4) = \sum (\nabla_{e_i} \omega_3^4)(e_i)$. Let us put

$$\begin{aligned}
\omega_2^1 &= -\omega_1^2 = f_1 \omega^1 + f_2 \omega^2, \\
\omega_3^4 &= -\epsilon_3 \epsilon_4 \omega_4^3 = g_1 \omega^1 + g_2 \omega^2.
\end{aligned} \tag{7.33}$$

It follows from (7.29), (7.32) and (7.33) that

$$\|A_4\| \, \|\omega_3^4\| = 0. \tag{7.34}$$

If $\omega_3^4 = 0$, (7.30) gives $A_3 = 0$. Thus $\alpha = 0$ which is a contradiction. So,

$$\omega_3^4 \neq 0 \quad \text{and} \quad A_4 \equiv 0. \tag{7.35}$$

From (7.31) and (7.35) we find $\operatorname{Tr}(\nabla \omega_3^4) = 0$. Combining this with (7.33) gives

$$e_1 g_1 + e_2 g_2 + f_1 g_2 - f_2 g_1 = 0. \tag{7.36}$$

From the equation of Ricci and (7.35) we find

$$e_1 g_2 - e_2 g_1 = f_1 g_1 + f_2 g_2. \tag{7.37}$$

Since $A_4 = 0$, we may choose e_1, e_2 such that

$$h(e_1, e_1) = \epsilon_3 \beta e_3, \quad h(e_1, e_2) = 0, \quad h(e_2, e_2) = \epsilon_3 \gamma e_3 \tag{7.38}$$

for some functions β, γ. From the equation of Codazzi, (7.33) and (7.36), we obtain

$$\beta g_2 = 0, \quad e_2 \beta = f_1(\gamma - \beta), \quad \gamma g_1 = 0, \quad e_1 \gamma = f_2(\gamma - \beta). \tag{7.39}$$

Since $A_3 \neq 0$ and $\omega_3^4 \neq 0$, (7.39) implies that exactly one of β, γ is zero and also exactly one of g_1, g_2 is zero. Without loss of generality, we may put

$$g_1 \neq 0, \quad g_2 = 0, \quad \beta \neq 0, \quad \gamma = 0. \tag{7.40}$$

In particular, this implies $\beta - \gamma \neq 0$ and β is constant. Thus, from (7.39) and (7.40), we get

$$f_1 = f_2 = 0, \quad i.e., \quad \omega_1^2 = \omega_2^1 = 0, \tag{7.41}$$

from which we conclude that M is flat. Moreover, from (7.36), (7.37), (7.40) and (7.41), we get

$$\omega_1^2 = \omega_2^3 = \omega_1^4 = \omega_2^4 = 0, \quad \omega_1^3 = \epsilon_3 \beta \omega^1, \quad \omega_3^4 = \delta \omega^1, \tag{7.42}$$

where β and δ are nonzero constants.

On the other hand, it follows from (7.30), (7.42) and the constancy of $\alpha \neq 0$ that $\epsilon_4 \delta^2 + \beta^2 = 0$. Thus we get $\epsilon_4 = -1$, $\delta^2 = \beta^2$. Hence we have $s > 0$. Without loss of generality, we may choose e_1, e_2, e_3, e_4 such that $\delta = \beta$. Thus we obtain

$$\omega_1^2 = \omega_2^3 = \omega_1^4 = \omega_2^4 = 0, \, \omega_1^3 = \epsilon_3 \beta \omega^1$$

and $\omega_3^4 = \beta \omega^1$ From these we get $\tilde{\nabla} e_2 = 0$ and $d\omega^1 = d\omega^2 = 0$. Hence e_2 is a constant spacelike unit vector in $\mathbb{E}_r^4$.

Let γ_1 and γ_2 be integral curves of e_1 and e_2 through a given point in M, respectively. Then γ_2 is an open portion of a line L and γ_1 is a curve lying in a $\mathbb{E}_r^3$ perpendicular to L.

Since M is locally the Riemannian product of γ_1 and γ_2, the biharmonicity of M implies the biharmonicity of γ_1 in $\mathbb{E}_r^3$. Furthermore, the mean curvature vector H_1 of γ_1 is $\epsilon_3 \beta e_3$ which satisfies $\langle H_1, H_1 \rangle = \epsilon_3 \beta^2 \neq 0$. Thus, by applying Theorem 7.1, we conclude that either $r = 1$ and, up to rigid motions, γ_1 is given by (7.12) or $r = 2$ and, up to rigid motions, γ_1 is given by (7.13). Consequently, the biharmonic surface is congruent either to (7.27) or to (7.28).

The converse is easy to verify. $\quad\square$

Recall that spacelike surfaces with lightlike mean curvature vector in a spacetime are called *marginally trapped surfaces*. We give the following examples of marginally trapped biharmonic surfaces.

Example 7.3. Let D be an open domain of $\mathbb{E}^2$ with Euclidean coordinates (x, y). Let $\varphi(x, y)$ be a function satisfying $\Delta \varphi \neq 0$ and $\Delta^2 \varphi = 0$ and let

$z = u(x, y) + iv(x, y)$ be a holomorphic function in $z = x + iy$ satisfying $\|\nabla u\| = 1$. Consider the map $\phi : D \to \mathbb{E}_1^4$ defined by

$$\phi(x, y) = (\varphi(x, y), u(x, y), v(x, y), \varphi(x, y)). \tag{7.43}$$

By applying the Cauchy-Riemann condition, it is easy to verify that ϕ defines a marginally trapped biharmonic surface with lightlike mean curvature vector given by $H = -\frac{1}{2}(\Delta\varphi, 0, 0, \Delta\varphi)$.

Theorem 7.5. [Chen and Ishikawa (1991)] *Let M be a marginally trapped surface in $\mathbb{E}_1^4$. Then M is biharmonic if and only if it is congruent to a surface defined by (7.43) in Example 7.3.*

Proof. Let $\phi : M \to \mathbb{E}_1^4$ be a marginally trapped surface. We choose e_3 and e_4 such that e_3 is spacelike and e_4 is timelike, i.e., $\epsilon_3 = 1, \epsilon_4 = -1$. Because H is lightlike, we may assume that

$$H = \delta(e_3 + e_4) \tag{7.44}$$

for some function $\delta \neq 0$. By a direct computation we find

$$\sum h(A_H e_i, e_i) = \delta(\|A_3\|^2 + \text{Tr}(A_3 A_4))e_3 \tag{7.45}$$
$$- \delta(\|A_4\|^2 + \text{Tr}(A_3 A_4))e_4,$$
$$\Delta^D H = \{\Delta\delta - \delta\|\omega_3^4\|^2 - \delta\,\text{Tr}(\nabla\omega_3^4) - 2\omega_3^4(\nabla\delta)\}(e_3 + e_4), \tag{7.46}$$
$$\text{Tr}\, A_{DH} = \sum_i \{e_i\delta + \delta\omega_3^4(e_i)\}(A_3 e_i + A_4 e_i). \tag{7.47}$$

From Theorem 7.1, (7.45), (7.46) and (7.47) we get

$$\delta\{\|A_3\|^2 + \text{Tr}(A_3 A_4) - \text{Tr}(\nabla\omega_3^4) - \|\omega_3^4\|^2\} + \Delta\delta = 2\omega_3^4(\nabla\delta), \tag{7.48}$$
$$\delta\{\|A_4\|^2 + \text{Tr}(A_3 A_4) + \text{Tr}(\nabla\omega_3^4) + \|\omega_3^4\|^2\} - \Delta\delta = -2\omega_3^4(\nabla\delta). \tag{7.49}$$

Summing up (7.48) and (7.49) gives $\|A_3 + A_4\|^2 = 0$. Therefore we obtain $A_3 = -A_4$ and $A_H = 0$. Hence the second fundamental form satisfies

$$h(e_1, e_1) = \beta(e_3+e_4), \; h(e_1, e_2) = \mu(e_3+e_4), \; h(e_2, e_2) = \gamma(e_3+e_4) \tag{7.50}$$

for some functions β, γ, μ. It follows from the equation of Gauss, (7.50) and $\langle H, H \rangle = 0$ that M is flat. Therefore we may choose e_1, e_2 such that $\omega_1^2 = 0$. Moreover, from the equation of Ricci and (7.50) that $R^D = 0$. So M has flat normal connection as well. Thus we may choose e_3, e_4 such that $\omega_3^4 = 0$. Thus, by applying $A_{e_3+e_4} = 0$, we obtain $\tilde{\nabla}(e_3 + e_4) = 0$. Consequently, $e_3 + e_4$ is a constant lightlike vector. Hence we may choose a pseudo-Euclidean coordinate system on $\mathbb{E}_1^4$ so that $e_3 + e_4$ is given by

$$e_3 + e_4 = (\epsilon, 0, 0, \epsilon) \tag{7.51}$$

for a constant $\epsilon \neq 0$. Since M is flat, M is locally an open portion of $\mathbb{E}^2$. Suppose that the position vector field ϕ of M is given by

$$\phi(x, y) = (\varphi(x, y), u(x, y), v(x, y), \eta(x, y)), \tag{7.52}$$

where (x, y) is a Euclidean coordinate system of $\mathbb{E}^2$.

It follows from

$$\langle e_1, e_3 + e_4 \rangle = \langle e_2, e_3 + e_4 \rangle = 0$$

that $e_i \langle \phi, e_3 + e_4 \rangle = 0$ for $i = 1, 2$. Thus if we put $e_1 = \partial/\partial u$ and $e_2 = \partial/\partial v$, then we obtain from (7.51) and (7.52) that

$$\frac{\partial(\varphi - \eta)}{\partial u} = \frac{\partial(\varphi - \eta)}{\partial v} = 0. \tag{7.53}$$

Without loss of generality we may assume $\eta = \varphi$. Thus (7.52) becomes

$$\phi(x, y) = (\varphi(x, y), u(x, y), v(x, y), \varphi(x, y)). \tag{7.54}$$

Hence we get $e_1 = (\varphi_x, u_x, v_x, \varphi_x)$ and $e_2 = (\varphi_y, u_y, v_y, \varphi_y)$. Because e_1 and e_2 are orthonormal, we find

$$u_x^2 + v_x^2 = 1, \quad u_y^2 + v_y^2 = 1, \quad u_x u_y + v_x v_y = 0. \tag{7.55}$$

Case (a): $u_x = 0$. In this case, (7.55) implies $v_x^2 = u_y^2 = 1$ and $v_y = 0$. Thus, after applying suitable rigid motion of $\mathbb{E}_1^4$, we obtain (7.43) with $u = x$ and $v = u$.

Case (b): $u_y = 0$, or $v_x = 0$, or $v_y = 0$. This can be done case (a).

Case (c): $u_x u_y v_x v_y \neq 0$. It follows from $u_x \neq 0$ and the last equation of (7.47) that $u_y = -v_x v_y / u_x$. By substituting this into the second equation in (7.55) and applying the first equation we find $u_x = \pm v_y$. Similarly, we also have $u_y = \pm v_x$.

Because we have $u_x u_y v_x v_y \neq 0$, the last equation in (7.55) implies that we have either $(u_x, u_y) = (v_y, -v_x)$ or $(u_x, u_y) = (-v_y, v_x)$. Without loss of generality, we may assume that u, v satisfy Cauchy-Riemann's equations:

$$u_x = v_y, \quad u_y = -v_x, \tag{7.56}$$

From (7.38) we have $\Delta u = \Delta v = 0$, i.e., u and v are harmonic functions. Moreover, from it follows from (7.55) and (7.56) that $\|\nabla u\| = \|\nabla v\| = 1$. Consequently, the marginally trapped surfaces is given by Example 7.3.

The converse of this is easy to verify. $\qquad\square$

Recall that Lorentzian surfaces in a pseudo-Riemannian manifold with lightlike mean curvature vector are called quasi-minimal.

Biharmonic quasi-minimal surfaces in $\mathbb{E}_2^4$ were classified in the following theorem.

Theorem 7.6. [Chen (2008)] *A quasi-minimal surface is biharmonic in* $\mathbb{E}_2^4$ *if and only if it is congruent to one of the following:*

(1) *a surface given by*

$$\frac{1}{\sqrt{2}}\left(\varphi(x,y), x+y, x-y, \varphi(x,y)\right),$$

where $\varphi(x,y)$ *is a function on an open domain* $U \subset \mathbb{E}_1^2$ *which satisfies* $\varphi_{xy} \neq 0$ *and* $\varphi_{xxyy} = 0$;

(2) *a surface given by*

$$\phi(x,y) = z(x)y + w(x),$$

where z *is a null curve in the light cone* $\mathcal{LC}$ *and* w *is a null curve such that* $\langle z', w' \rangle = 0$ *and* $\langle z, w' \rangle = -1$.

For biharmonic surfaces with constant Gauss curvature in $\mathbb{E}_r^4$ we have:

Theorem 7.7. [Chen and Ishikawa (1991)] *Let* M *be a non-minimal space-like surface with constant Gauss curvature in* $\mathbb{E}_r^4$. *Then* M *is a biharmonic surface with flat normal connection if and only if it is congruent to a surface given by Example 7.1, or a surface given by Example 7.2, or a surface given by Example 7.3.*

For biharmonic submanifolds with parallel mean curvature, we have the next three results.

Theorem 7.8. [Fu (2013b)] *Let* M *be a non-minimal spacelike biharmonic submanifold of* $\mathbb{E}_r^m$ *with index* $r \geq 1$. *If* M *has parallel mean curvature vector, then* M *is a marginally trapped surface congruent to*

$$(f, \psi, f),$$

where f *is a function satisfying* $\Delta f = a$ *for some real number* $a \neq 0$ *and* $\psi : M \to \mathbb{E}_{r-1}^{m-2}$ *is a minimal immersion.*

Theorem 7.9. [Fu (2013b)] *Let* M *be a non-minimal biharmonic Lorentzian surface in a pseudo-Euclidean m-space* $\mathbb{E}_r^m$ *with index* $r \geq 1$. *If* M *has parallel mean curvature vector, then* M *it is marginally trapped.*

Theorem 7.10. [Fu (2013b)] *Let M be a non-minimal biharmonic Lorentzian surfaces in $\mathbb{E}_r^m$ with index $r \geq 1$. If M has parallel mean curvature vector, then M is congruent to the following one of the following surfaces:*

(1) *a surface given by*

$$(f, \psi, f),$$

where f is a function satisfying $\Delta f = a$ for some real number $a \neq 0$ and $\psi : M \to \mathbb{E}_{r-1}^{m-2}$ is a minimal immersion;

(2) *a flat marginally trapped ruled surface defined by*

$$\phi(s, t) = z(t)s + w(t),$$

where $z(t)$ is a null curve in the light cone $\mathcal{LC}$ and $w(t)$ is a null curve in $\mathbb{E}_r^m$ satisfying $\langle z', w' \rangle = 0$ and $\langle z, w' \rangle = -1$.

For further results concerning spacelike and Lorentzian biharmonic surfaces in $\mathbb{E}_r^4$, see [Chen and Ishikawa (1991, 1998); Ishikawa (1992b); Sasahara (2007a)].

Remark 7.2. The author would like to point out that author's biharmonic conjecture, i.e., Conjecture 7.1, remains open for surfaces in $\mathbb{E}^4$. (The lowest dimension case for the conjecture remaining open.)

7.3 Biharmonic hypersurfaces in pseudo-Euclidean space

As an extension of Theorem 7.3 with $s = 0$, Dimitric proved the following result.

Theorem 7.11. [Dimitric (1989, 1992a)] *Let M be a biharmonic hypersurface of a Euclidean space with at most two principal curvatures, then M is a minimal submanifold.*

Since every conformally flat hypersurface of $\mathbb{E}^{n+1}$ with $n \geq 4$ is quasi-umbilical, Theorem 7.11 implies immediately the following.

Corollary 7.6. [Dimitric (1989); Ferrández et al. (1991b)] *Every conformally flat biharmonic hypersurface M of $\mathbb{E}^{n+1}$ with $n \neq 3$ is minimal.*

Remark 7.3. O. J. Garay proved that every conformally flat biharmonic hypersurface of a Euclidean 4-space is minimal [Garay (1994a)].

For biharmonic hypersurfaces of $\mathbb{E}_s^4$, we have

Theorem 7.12. *Every pseudo-Riemannian biharmonic hypersurface with diagonalizable shape operator in a pseudo-Euclidean 4-space is a minimal hypersurface.*

Proof. Assume that M is a non-minimal biharmonic hypersurface of $\mathbb{E}_s^4$. Then M has non-constant mean curvature according to Corollary 7.1. Without loss of generality, we can choose e_1 in the direction of $\nabla\alpha$. Then by applying the corollary, the diagonalized shape operator of M takes the form

$$A = \begin{pmatrix} -\frac{3}{2}\epsilon\alpha & & \\ & \mu_2 & \\ & & \mu_3 \end{pmatrix}. \tag{7.57}$$

We choose e_1, e_2, e_3 consisting of eigenvectors of A so that e_1 is the direction of $\nabla\alpha$. So we have

$$e_1\alpha \neq 0, \quad e_2\alpha = e_3\alpha = 0. \tag{7.58}$$

Case (1): M has three different principal curvatures. Since all three principal curvatures are mutually different, we have

$$\mu_2, \mu_3 \neq -\frac{3}{2}\epsilon\alpha, \quad \mu_2 \neq \mu_3. \tag{7.59}$$

Writing $\nabla_{e_i} e_j = \sum_k \omega_{ij}^k e_k$, the equation of Codazzi gives

$$\omega_{12}^1 = \omega_{13}^1 = 0, \quad \omega_{21}^2 = \frac{-2e_1\mu_2}{3\epsilon\alpha + 2\mu_2}, \quad \omega_{31}^3 = \frac{-2e_1\mu_3}{3\epsilon\alpha + 2\mu_3}, \tag{7.60}$$

$$\omega_{23}^2 = \frac{e_3\mu_2}{\mu_3\mu_2}, \quad \omega_{31}^3 = \frac{e_2\mu_3}{\mu_2 - \mu_3}. \tag{7.61}$$

Also, in view of (7.58), we have that $[e_2, e_3]\alpha = 0$. Together with Codazzi's equation we find

$$\omega_{13}^2 - \omega_{21}^3 = \omega_{32}^1 = 0. \tag{7.62}$$

By applying the equation of Gauss, we find

$$e_3\left(\frac{2e_1\mu_2}{3\epsilon\alpha + 2\mu_2}\right) = \frac{e_3\mu_2}{\mu_2 - \mu_3}\left(\frac{-2e_1\mu_3}{3\epsilon\alpha + 2\mu_3} + \frac{2e_1\mu_2}{3\epsilon\alpha + 2\mu_2}\right), \tag{7.63}$$

$$e_2\left(\frac{2e_1\mu_3}{3\epsilon\alpha + 2\mu_3}\right) = \frac{e_2\mu_3}{\mu_3 - \mu_2}\left(\frac{-2e_1\mu_2}{3\epsilon\alpha + 2\mu_2} + \frac{2e_1\mu_3}{3\epsilon\alpha + 2\mu_3}\right). \tag{7.64}$$

On the other hand, in view of (7.58), equation (7.8) takes the form:

$$e_1^2\alpha - \left(\frac{2e_1\mu_2}{3\epsilon\alpha + 2\mu_2} + \frac{2e_1\mu_3}{3\epsilon\alpha + 2\mu_3}\right)e_1\alpha - \epsilon\epsilon_1\left(\frac{45}{2}\alpha^2 - 2\mu_2\mu_3\right) = 0. \tag{7.65}$$

Acting with e_2 and e_3 respectively on (7.65) and combining with (7.63)-(7.65) gives

$$e_2\left(\frac{2e_1\mu_2}{3\epsilon\alpha + 2\mu_2}\right) = \frac{e_2\mu_3}{\mu_2 - \mu_3}\left(\frac{-2e_1\mu_2}{3\epsilon\alpha + 2\mu_2} + \frac{2e_1\mu_3}{3\epsilon\alpha + 2\mu_3}\right) \qquad (7.66)$$
$$+ \frac{2\epsilon\alpha}{\epsilon_1 e_1\alpha}(\mu_2 - \mu_3)e_2\mu_3,$$

$$e_3\left(\frac{2e_1\mu_3}{3\epsilon\alpha + 2\mu_3}\right) = \frac{e_3\mu_2}{\mu_3 - \mu_2}\left(\frac{-2e_1\mu_3}{3\epsilon\alpha + 2\mu_3} + \frac{2e_1\mu_2}{3\epsilon\alpha + 2\mu_2}\right) \qquad (7.67)$$
$$+ \frac{2\epsilon\alpha}{\epsilon_1 e_1\alpha}(\mu_3 - \mu_2)e_3\mu_2.$$

From Gauss' equation for $\langle R(e_1, e_2)e_1, e_2\rangle$ and $\langle R(e_3, e_1)e_1, e_3\rangle$ we get

$$e_1\left(\frac{-2e_1\mu_2}{3\epsilon\alpha + 2\mu_2}\right) + \left(\frac{2e_1\mu_2}{3\epsilon\alpha + 2\mu_2}\right)^2 = \frac{3}{2}\epsilon\epsilon_1\alpha\mu_2, \qquad (7.68)$$

$$e_1\left(\frac{-2e_1\mu_3}{3\epsilon\alpha + 2\mu_3}\right) + \left(\frac{2e_1\mu_3}{3\epsilon\alpha + 2\mu_3}\right)^2 = \frac{3}{2}\epsilon\epsilon_1\alpha\mu_3. \qquad (7.69)$$

Using (7.60)-(7.62) we obtain

$$[e_1, e_2] = \frac{2e_1\mu_2}{3\epsilon\alpha + 2\mu_2}e_2. \qquad (7.70)$$

In addition, we take into account the relation

$$e_1\left(\frac{e_2\mu_3}{\mu_2 - \mu_3}\right) = \frac{2(e_1\mu_3)(e_2\mu_3)}{(3\epsilon\alpha + \mu_3)(\mu_2 - \mu_3)} \qquad (7.71)$$

which follows from Gauss' equation for $\langle R(e_3, e_1)e_2, e_3\rangle$. Applying both sides of (7.70) on $2e_1\mu_2/(3\epsilon\alpha + 2\mu_2)$ and using (7.66), (7.68), (7.69) and (7.71), we deduce that

$$0 = \frac{e_2\mu_3}{\mu_2 - \mu_3}\left[\left(\frac{2e_1\mu_2}{3\epsilon\alpha + 2\mu_2} - \frac{2e_1\mu_3}{3\epsilon\alpha + 2\mu_3}\right)^2 + \frac{\epsilon}{\epsilon_1}e_1\left(\frac{\alpha}{e_1\alpha}\right)(\mu_2 - \mu_3)^2\right.$$
$$\left.- \frac{\epsilon\alpha(\mu_2 - \mu_3)}{\epsilon_1 e_1\alpha}\left(\left(\frac{6e_1\mu_2}{3\epsilon\alpha + 2\mu_2} - \frac{e_1\mu_3}{3\epsilon\alpha + 2\mu_3}\right)(\mu_2 - \mu_3) - 2e_1(\mu_2 - \mu_3)\right)\right].$$

This equation shows that either $e_2\mu_3$ or the expression between square brackets has to vanish. Now prove that $e_2\mu_3$ necessarily has to be zero, since the assumption that $e_2\mu_3 \neq 0$ runs into contradiction. Indeed, suppose that $e_2\mu_3 \neq 0$ then the factor between square brackets has to vanish:

$$e_1\left(\frac{\alpha}{e_1\alpha}\right) = \frac{\alpha}{e_1\alpha}\left(\left(\frac{6e_1\mu_2}{3\epsilon\alpha + 2\mu_2} - \frac{2e_1\mu_3}{3\epsilon\alpha + 2\mu_3}\right) - \frac{2e_1(\mu_2 - \mu_3)}{\mu_2 - \mu_3}\right)$$
$$- \frac{\epsilon_1}{\epsilon(\mu_2 - \mu_3)^2}\left(\frac{2e_1\mu_2}{3\epsilon\alpha + 2\mu_2} - \frac{2e_1\mu_3}{3\epsilon\alpha + 2\mu_3}\right)^2.$$

Acting with e_2 on both sides of the last equation, in view of (7.64), (7.66) and (7.70), gives

$$\frac{\epsilon}{\epsilon_1 e_1 \alpha}(\mu_2 - \mu_3)^2 = 2\left(\frac{2e_1\mu_2}{3\epsilon\alpha + 2\mu_2} - \frac{2e_1\mu_3}{3\epsilon\alpha + 2\mu_3}\right).$$

Applying e_2 again on the previous equation, gives in addition

$$\frac{2\epsilon}{\epsilon_1 e_1 \alpha}(\mu_2 - \mu_3)^2 = \left(\frac{2e_1\mu_2}{3\epsilon\alpha + 2\mu_2} - \frac{2e_1\mu_3}{3\epsilon\alpha + 2\mu_3}\right)$$

from which, it follows that $\mu_2 = \mu_3$, which is a contradiction. Therefore we conclude that $e_2\mu_3 = 0$. Analogously, from (7.60) we have

$$[e_1, e_3] = \frac{2e_1\mu_3}{3\epsilon\alpha + 2\mu_3}e_3 \tag{7.72}$$

both sides of which we apply to $2e_1\mu_3/(3\epsilon\alpha - 2\mu_3)$. In view of the equation of Gauss for $\langle R(e_1, e_2)e_3, e_2\rangle$ we find

$$e_1\left(\frac{e_3\mu_2}{\mu_3 - \mu_2}\right) = \frac{2(e_1\mu_2)(e_3\mu_2)}{(3\epsilon\alpha + 2\mu_3)(\mu_3 - \mu_2)}e_3. \tag{7.73}$$

Now, by using (7.67)-(7.69) and (7.73) and the result of former action, we deduce that

$$0 = \frac{e_3\mu_2}{\mu_3 - \mu_2}\left[\left(\frac{2e_1\mu_2}{3\epsilon\alpha + 2\mu_2} - \frac{2e_1\mu_3}{3\epsilon\alpha + 2\mu_3}\right)^2 + \frac{\epsilon}{\epsilon_1}e_1\left(\frac{\alpha}{e_1\alpha}\right)(\mu_3 - \mu_2)^2 \right.$$
$$\left. + \frac{\epsilon\alpha(\mu_3 - \mu_2)}{\epsilon_1 e_1 \alpha}\left(\left(\frac{2e_1\mu_2}{3\epsilon\alpha + 2\mu_2} - \frac{e_1\mu_3}{6\epsilon\alpha + 2\mu_3}\right)(\mu_3 - \mu_2) + 2e_1(\mu_3 - \mu_2)\right)\right].$$

In a similar way as above, one can show that $e_3\mu_2$ necessarily has to vanish. Indeed, following the same line of proof, the assumption that $e_3\mu_2 \neq 0$ runs into contradiction.

Summarizing, we can state that we have proved independently that $e_2\mu_3$ and $e_3\mu_2$ have to vanish separately. Hence we conclude that

$$e_2\mu_3 = e_3\mu_2 = 0. \tag{7.74}$$

In view of (7.74), the Gauss equation for $\langle R(e_2, e_3)e_2, e_3\rangle$ gives the following relation:

$$\frac{4\epsilon_1(e_1\mu_2)(e_1\mu_3)}{(3\epsilon\alpha + 2\mu_2)(3\epsilon\alpha + 2\mu_3)} + \mu_2\mu_3 = 0. \tag{7.75}$$

By calculating $e_1^2\alpha$ from (7.68) and (7.69), we find

$$e_1^2\alpha = -9\epsilon_1\alpha^3 + \frac{1}{3}\left(\frac{2e_1\mu_2}{3\epsilon\alpha + 2\mu_2} - \frac{2e_1\mu_3}{3\epsilon\alpha + 2\mu_3}\right)e_1\alpha$$
$$+ \frac{4}{9}\epsilon\left(\left(\frac{2e_1\mu_2}{3\epsilon\alpha + 2\mu_2}\right)e_1\mu_2 + \left(\frac{2e_1\mu_3}{3\epsilon\alpha + 2\mu_3}\right)e_1\mu_3\right) + \frac{2}{3}\epsilon_1\alpha\mu_2\mu_3.$$

Substituting this expression in (7.65) and using (7.75), we get

$$\frac{4}{3}\epsilon\epsilon_1\left(\left(\frac{2e_1\mu_2}{3\epsilon\alpha + 2\mu_2}\right)e_1\mu_3 + \left(\frac{2e_1\mu_3}{3\epsilon\alpha + 2\mu_3}\right)e_1\mu_2\right) - \frac{135\epsilon + 54}{2}\alpha^3 \qquad (7.76)$$

$$+ 4\epsilon_1\left(\frac{2e_1\mu_2}{3\epsilon\alpha + 2\mu_2} + \frac{2e_1\mu_3}{3\epsilon\alpha + 2\mu_3}\right)e_1\alpha + (2 + 6\epsilon)\alpha\mu_2\mu_3 = 0.$$

Using (7.75), equation (7.76) reduces to

$$4\epsilon_1\left(\frac{2e_1\mu_2}{3\epsilon\alpha + 2\mu_2} + \frac{2e_1\mu_3}{3\epsilon\alpha + 2\mu_3}\right)e_1\alpha + 6(2 + \epsilon)\alpha\mu_2\mu_3 - \frac{54 + 135\epsilon}{2}\alpha^3 = 0$$

and finally we get

$$\left(\frac{2e_1\mu_2}{3\epsilon\alpha + 2\mu_2} + \frac{2e_1\mu_3}{3\epsilon\alpha + 2\mu_3}\right)e_1\alpha = \frac{54 + 135\epsilon}{8\epsilon_1}\alpha^3 - \frac{6 + 3\epsilon}{2\epsilon_1}\alpha\mu_2\mu_3. \qquad (7.77)$$

Thus (7.65) gives

$$e_1^2\alpha = \frac{54 + 135\epsilon}{8\epsilon_1}\alpha^3 - \frac{6 + 3\epsilon}{2\epsilon_1}\alpha\mu_2\mu_3. \qquad (7.78)$$

Acting with e_1 on both sides of (7.77) and using (7.68), (7.69) and (7.78), we get

$$\left((378 + 486\epsilon)\alpha^2 - 4(10 + 3\epsilon)\mu_2\mu_3\right)e_1\alpha$$

$$= \left(\frac{2e_1\mu_2}{3\epsilon\alpha + 2\mu_2} + \frac{2e_1\mu_3}{3\epsilon\alpha + 2\mu_3}\right)\left((324 + 558\epsilon)\alpha^3 - 4(6 + 7\epsilon)\alpha\mu_2\mu_3\right). \qquad (7.79)$$

Applying again e_1 on (7.79), we take

$$\frac{1026 + 1053\epsilon}{4}\alpha(e_1\alpha)^2 + \frac{117 + 104\epsilon}{2\epsilon_1}\alpha\mu_2^2\mu_3^2$$

$$+ \left((12 + 14\epsilon)\alpha\mu_2\mu_3 - \frac{54 + 103\epsilon}{2}\alpha^3\right)\left(\frac{2e_1\mu_2}{3\epsilon\alpha + 2\mu_2} + \frac{2e_1\mu_3}{3\epsilon\alpha + 2\mu_3}\right) \qquad (7.80)$$

$$+ \frac{6426 + 4500\epsilon}{8\epsilon_1}\alpha^3\mu_2\mu_3 - \frac{155763 + 132678\epsilon}{32\epsilon_1}\alpha^5 = 0$$

Multiplying (7.77) and (7.79) we obtain

$$\{(378 + 486\epsilon)\alpha^2 - 4(10 + 3\epsilon)\mu_2\mu_3\}(e_1\alpha)^2$$

$$= \left(\frac{54 + 135\epsilon}{8\epsilon_1}\alpha^3 - \frac{6 + 3\epsilon}{2\epsilon_1}\alpha\mu_2\mu_3\right)\{(324 + 558\epsilon)\alpha^3 - (6 + 7\epsilon)\alpha\mu_2\mu_3\}.$$

If we put $Y = \mu_2\mu_2/\alpha^2$, then for the Lorentzian case $\epsilon = 1$ by using (7.77) and the last equation, relation (7.80) is reduced to the following algebraic equation of 4th degree in Y:

$$F(Y) = 140608Y^4 - 6157008Y^3 + 59355504Y^2$$

$$+ 485815806Y - 6863560515 = 0,$$

which shows that even in case of the existence of a real solution, $\mu_2\mu_3$ and α^2 have to be proportional with a numerical factor. Hence

$$\mu_2\mu_3 = \lambda\alpha^2, \tag{7.81}$$

where p is a root of F. In view of (7.81), equation (7.77) combined with equation (7.79) gives

$$8\epsilon_1(432 - 26\lambda)(e_1\alpha)^2 = (441 - 26\lambda)(36\lambda - 189)\alpha^4. \tag{7.82}$$

By taking the derivative of this equation with respect to e_1 and using (7.78) and (7.80), we find

$$(1612\lambda^2 - 36819\lambda + 163053)\alpha^3 = 0.$$

In order to obtain the contradiction and finish the proof, we have merely to check that neither of the roots of $1612\lambda^2 - 36819\lambda + 163053$ is also a root of $F(Y)$. Since this is true we conclude that $\alpha = 0$, which is a contradiction. In a similar way as above, we can work for the Riemannian case $\epsilon = -1$ and take the same result.

Case (2): *M has two different principal curvatures.* Under this assumption, we will show that the assumption runs into contradiction. As we mentioned earlier $\nabla\alpha$ is an eigenvector of A with corresponding eigenvalue $-\frac{3}{2}\epsilon\alpha$. We now choose a local orthonormal frame e_1, e_2, e_3 consisting of eigenvectors of A and such that e_1 is a unit vector in the direction of $\nabla\alpha$. With respect to this local frame, A takes the form:

$$A = \begin{pmatrix} -\frac{3}{2}\epsilon\alpha & & \\ & \mu & \\ & & \mu \end{pmatrix}. \tag{7.83}$$

Moreover, the relations (7.58) are valid.

Combining the relation $\operatorname{Tr} A = 3\epsilon\alpha$ and (7.83), we deduce that $\mu = 9\epsilon\alpha/4$. Hence $\operatorname{Tr} A^2 = 99\alpha^2/8$. Writing $\nabla_{e_i}e_j = \sum_{k=1}^{3} \omega_{ij}^k e_k$ as before, we find from Codazzi's equation for $\langle(\nabla_{e_1}A)e_2, e_2\rangle$ and $\langle(\nabla_{e_1}A)e_3, e_3\rangle$ that

$$\omega_{21}^2 = \omega_{31}^3 = -\frac{3}{5} \cdot \frac{e_1\alpha}{\alpha}. \tag{7.84}$$

Next, the Gauss equation for $\langle R(e_1, e_2)e_1, e_2\rangle$ shows that

$$e_1\omega_{21}^2 = \frac{27}{8}\epsilon_1\alpha^2 - (\omega_{21}^2)^2. \tag{7.85}$$

On the other hand, in view of (7.58), (7.84) and (7.85), we have

$$\epsilon_1 e_1^2\alpha + 2\epsilon_1\omega_{21}^2 e_1\alpha - \frac{99}{8}\epsilon\alpha^3 = 0. \tag{7.86}$$

Acting on (7.84) with e_1 and using (7.85) yields

$$e_1^2 \alpha = \frac{40}{9} \alpha (\omega_{21}^2)^2 - \frac{45}{8} \epsilon_1 \alpha^3.$$

Hence equation (7.86) reduces to

$$\left[\frac{10}{9} \epsilon_1 (\omega_{21}^2)^2 - \frac{45 + 99\epsilon}{8} \alpha^2 \right] \alpha = 0.$$

The expression between the bracket has to vanish, since $\alpha \neq 0$. Thus

$$\frac{10}{9} \epsilon_1 (\omega_{21}^2)^2 - \frac{45 + 99\epsilon}{8} \alpha^2 = 0.$$

Now, by applying e_1 on this relation, we get

$$\frac{10}{9} \epsilon_1 (\omega_{21}^2)^2 - \frac{945 + 165\epsilon}{8} \alpha^2 = 0.$$

By combining the last two relations we obtain $\alpha = 0$ which is a contradiction. $\qquad\Box$

Remark 7.4. Theorem 7.12 was first proved in [Hasanis-Vlachos (1995a)] for hypersurfaces in $\mathbb{E}^4$. It was extended to pseudo-Riemannian biharmonic hypersurfaces of $\mathbb{E}_s^4$ in [Defever et al. (2006)].

Made use of the fact that there are four canonical forms of the shape operator of any hypersurface M_2^3 with index 2 in $\mathbb{E}_2^4$ with respect to suitable tangent frames, the following result was proved along the same lines as Theorem 7.9, treating the remaining three cases of canonical forms for the shape operator.

Theorem 7.13. [Papantoniou and Petoumenos (2012)] *Every pseudo-Riemannian biharmonic hypersurfaces M_2^3 in $\mathbb{E}_2^4$ is minimal.*

7.4 Recent developments on biharmonic conjecture

All of the results mentioned so far support author's Biharmonic Conjecture, Conjecture 7.1. In this section we present more recent results which provide further supports of this conjecture.

Definition 7.2. An immersed submanifold M of a Riemannian manifold $\tilde{M}$ is said to be *properly immersed* if the immersion is a proper map, i.e., the preimage of each compact set in $\tilde{M}$ is compact in M.

Notice that the properness of the immersion implies the completeness of the immersed submanifold.

Theorem 7.14. [Akutagawa and Maeta (2013)] *Every properly immersed biharmonic submanifold M in $\mathbb{E}^m$ is minimal.*

Proof. We need the following.

Lemma 7.2. *Let M be a biharmonic immersed submanifold in $\mathbb{E}^m$. Then the squared mean curvature H^2 of M satisfies*

$$\Delta H^2 \leq -2nH^4. \tag{7.87}$$

Proof. Under the hypothesis, equation (7.3) of Theorem 7.1 implies that, at each point $x \in M$, we have When $H(x) \neq 0$, set $e_m = H(x)/\|H(x)\|$. Then we find from (7.3) that

$$\Delta H^2 = 2\langle \Delta^D H, H \rangle - 2\sum_i \langle D_{e_i} H, D_{e_i} H \rangle \tag{7.88}$$

$$\leq -\langle h(A_H e_i, e_i), H \rangle = -2\|A_H\|^2.$$

When $H(x) \neq 0$, we put $e_N = H(x)/\|H(x)\|$. Then we have

$$\Delta H^2 \leq -2H^2 \sum_i \langle A_{e_N} e_i, A_{e_N} e_i \rangle$$

$$= -2H^2 \|A_{e_N}\|^2 \leq -2nH^4.$$

Even when $H(x) = 0$, the above inequality (7.87) still holds at x. $\square$

Assume that M is non-minimal and $H(x_o) \neq 0$. Set $u(x) = H^2(x)$ for $x \in M$. For each $\rho > 0$, consider the function

$$F(x) = F_\rho(x) := \left(\rho^2 - \|\phi(x)\|^2 \right)^2 u(x), \quad x \in M \cap \phi^{-1}(\bar{B}_\rho),$$

where $\bar{B}_\rho(x_o)$ is the closed ball of radius ρ centered at x_o in M and ϕ is the immersion of M in $\mathbb{E}^m$. Then, there exists $\rho_o > 0$ such that $x_o \in \phi^{-1}(B_{\rho_o})$. For each $\rho \geq \rho_o$, $F = F_\rho$ is a nonnegative function which is not identically zero on $M \cap \phi^{-1}(\bar{B}_\rho)$. Take any $\rho \geq \rho_o$ and fix it. Since M is properly immersed in $\mathbb{E}^m$, $M \cap \phi^{-1}(\bar{B}_\rho)$ is compact. By this fact combined with $F = 0$ on $M \cap \phi^{-1}(\bar{B}_\rho)$, there exists a maximum point $p \in M \cap \phi^{-1}(\bar{B}_\rho)$ of $F = F_\rho$ such that $F(p) > 0$. So we have $\nabla F = 0$ at p, and hence

$$\frac{\nabla u}{u} = \frac{2\nabla \|\phi(x)\|^2}{\rho^2 - \|\phi(x)\|^2} \quad \text{at } p. \tag{7.89}$$

We also have that $\Delta F \geq 0$ at p. Combining this with (7.89) we obtain

$$\frac{\nabla u}{u} \leq \frac{g \nabla ||\phi(x)||_M^2}{\rho^2 - ||\phi(x)||^2} - \frac{2\Delta ||\phi(x)||^2}{\rho^2 - ||\phi(x)||^2} \quad \text{at } p. \tag{7.90}$$

From $\Delta^2 \phi = 0$ we note that

$$\Delta ||\phi(x)||^2 = -2 \sum_i |\nabla_{e_i} \phi(x)|^2 + 2 \langle \Delta\phi(x), \phi(x) \rangle$$

$$\geq 2n + 2n ||H|| \, |\phi(x)|, \tag{7.91}$$

$$||\nabla ||\phi(x)||^2 ||_M^2 \leq 4n ||\phi(x)||^2.$$

It then follows from (7.87), (7.90) and (7.91) that

$$u(p) \leq \frac{12 ||\phi(p)||^2}{(\rho^2 - ||\phi(p)||^2)^2} + \frac{2(1 + \sqrt{u(p)} ||\phi(p)||)}{\rho^2 - ||\phi(p)||^2},$$

and hence

$$F(p) \leq 12 ||\phi(p)||^2 + 2(\rho^2 - ||\phi(p)||^2) + 2\sqrt{F(p)} ||\phi(p)||.$$

Therefore there exists a positive constant $c > 0$ which is independent of ρ such that $F(p) \leq c\rho^2$. Since $F(p)$ is the maximum of $F = F_\rho$, we have

$$F(x) \leq F(p) \leq c\rho^2$$

for $x \in M \cap \phi^{-1}(\bar{B}_\rho)$. Hence

$$||H(x)||^2 = u(x) \leq \frac{c\rho^2}{(\rho^2 - ||\phi(x)||^2)^2} \tag{7.92}$$

for $x \in M \cap \phi^{-1}(\bar{B}_\rho)$ and $\rho \geq \rho_0$. Letting $\rho \nearrow \infty$ in (7.92) for $x = x_o$, we find $||H(x_o)||^2 = 0$, which is a contradiction. $\qquad\square$

A hypersurface of a Riemannian manifold is called *weakly convex* if it has non-negative principal curvatures.

Theorem 7.15. [Luo (2013a)] *Every weakly convex biharmonic hypersurface of $\mathbb{E}^{n+1}$ is minimal.*

Proof. Assume that M is a convex biharmonic hypersurface in $\mathbb{E}^{n+1}$. Put $H = \alpha\xi$, where ξ is a unit normal vector field. Without loss of generality, we may assume that $\alpha \geq 0$.

Suppose that M is non-minimal, then there is a non-empty connected open subset V of M on which $\alpha > 0$. According to Corollary 7.1, α is non-constant on V.

Let $e_1, \ldots, e_n$ be an orthonormal frame consisting of eigenvectors of A_ξ with corresponding eigenvalues $\kappa_1, \ldots, \kappa_n$. For each $k \in \{1, \ldots, n\}$ we have

$$\sum_i \langle A_{D_{e_i}} H e_i, e_k \rangle = \sum_i \langle h(e_i, e_k), D_{e_i} H \rangle = \langle h(e_k, e_k), D_{e_k} H \rangle$$

on V. Thus, by (7.12) of Theorem 7.1, we obtain

$$
\begin{aligned}
0 &= n \nabla_{e_k} H^2 + 4 \sum_i \langle A_{D_{e_i}} H e_i, e_k \rangle \\
&= n \nabla_{e_k} H^2 + 4 \langle h(e_k, e_k), D_{e_k} H \rangle \\
&= 2 n \alpha \nabla_{e_k} \alpha + 4 \kappa_k \langle \xi, D_{e_k} H \rangle \\
&= (2 n \alpha + 4 \kappa_k) e_k \alpha.
\end{aligned}
$$

Since $2 n \alpha + 4 \kappa_k > 0$ by weakly convexity, one gets $e_1 \alpha = \cdots = e_n \alpha = 0$. Thus α is constant, which is a contradiction. $\qquad\square$

We have the following result for $\delta(2)$ and $\delta(3)$-ideal biharmonic hypersurfaces.

Theorem 7.16. [Chen and Munteanu (2013)] *Every $\delta(2)$-ideal and $\delta(3)$-ideal biharmonic hypersurface in $\mathbb{E}^{n+1}$ is minimal.*

It follows from Theorem 4.6 that a $\delta(k)$-ideal hypersurface has at most $k + 1$ distinct principal curvatures. Biharmonic hypersurfaces with three distinct principal curvatures in a Euclidean space were classified very recently by Y. Fu as follows.

Theorem 7.17. [Fu (2014a,b)] *Every biharmonic hypersurfaces with three distinct principal curvatures in a Euclidean space is minimal.*

Recall that a submanifold of $\mathbb{E}^m$ is called *pseudo-umbilical* if its shape operator A_H with respect to its mean curvature vector is proportional to the identity map.

Theorem 7.18. [Dimitric (1989, 1992a)] *Let M be a pseudo-umbilical biharmonic submanifold of a Euclidean space. If the dimension of M is different from 4, then it is a minimal submanifold.*

Remark 7.5. Y.-L. Ou constructed in [Ou (2009)] a two-parameter family of non-minimal conformal biharmonic immersions of a cylinder into $\mathbb{E}^3$. Ou's examples show that author's Biharmonic Conjecture cannot be generalized to biharmonic conformal submanifolds in Euclidean spaces.

A special case of author's biharmonic conjecture is the following.

Biharmonic Conjecture for Hypersurfaces: *Every biharmonic hypersurface of Euclidean spaces is minimal.*

The global version of author's Biharmonic Conjecture can be found in [Akutagawa and Maeta (2013)].

Global Version of Author's Biharmonic Conjecture: *Every complete biharmonic submanifold of a Euclidean space is minimal.*

Theorems 7.14 and 7.15 support this global version of the biharmonic conjecture.

7.5 Harmonic, biharmonic and k-biharmonic maps

In the following, we only consider smooth maps. Given a map

$$\phi : (M, g) \to (\tilde{M}, \tilde{g}) \tag{7.93}$$

between two Riemannian manifolds. The *energy density* $e(\phi)$ of ϕ is a non-negative function defined by

$$e(\phi)_u = \frac{1}{2} ||d\phi_u||^2 = \frac{1}{2} \operatorname{Tr}(\phi^* h)_u, \quad u \in M, \tag{7.94}$$

where $d\phi$ is the differential of the map ϕ. The *energy* $E(\phi)$ of ϕ is defined to be the functional:

$$E(\phi) = \int_\Omega e(\phi) dV, \tag{7.95}$$

for any compact domain $\Omega \subseteq M$.

Harmonic maps $\phi : (M, g) \to (\tilde{M}, \tilde{g})$ between Riemannian manifolds are the critical point of the energy functional $E(\phi)$. Thus they are the solutions of the corresponding Euler-Lagrange equation for the energy.

The Euler-Lagrangian associated with the energy functional is given by the *tension field* $\tau_1(\phi)$ which is defined by (cf. [Eells and Sampson (1964)])

$$\tau_1(\phi) = \operatorname{div}(d\phi). \tag{7.96}$$

Thus the map ϕ is harmonic if and only if its tension field $\tau_1(\phi)$ vanishes identically. Harmonic maps have broad applications to many areas in science and engineering including the robotic mechanics (cf. [Brockett and Park (1994); Urakawa (2011)]).

J. Eells and L. Lemaire proposed in [Eells and Lemaire (1983)] the problem to consider the *k-harmonic maps*, which are the critical maps of the *k-energy functional*

$$E_k(\phi) = \frac{1}{2} \int_M e_k(\phi) \, dV_g, \tag{7.97}$$

for $k = 2, 3, \ldots$, where $e_k(\phi)$ is defined by

$$e_k(\phi) = \frac{1}{2} \|(d + \delta)^k \phi\|^2. \tag{7.98}$$

Definition 7.3. A k-harmonic map is called *proper* if it is not a harmonic map. In particular, a biharmonic submanifold is called *proper* if it is non-minimal.

The Euler-Lagrange equation of the bienergy functional was computed by G.-Y. Jiang in [Jiang (1986b)]. The first variational formula for the bienergy functional is given by

$$\frac{dE_2(\phi_t)}{dt}\bigg|_{t=0} = \int_\Omega \langle \tau_2(\phi), \eta \rangle \, dV_g, \tag{7.99}$$

where η is the variational vector field along ϕ and $\tau_2(\phi)$ is defined by

$$\tau_2(\phi) = -\Delta^\phi \tau_1(\phi) - \mathrm{Tr}_g \tilde{R}(d\phi(\cdot), \tau_1(\phi)) d\phi(\cdot), \tag{7.100}$$

with

$$\Delta^\phi \tau_1(\phi) = \mathrm{Tr}_g(\nabla^\phi_{\nabla^M} - \nabla^\phi \nabla^\phi) \tau_1(\phi)$$

and $\tilde{R}$ being the curvature tensor of $(\tilde{M}, \tilde{g})$. Consequently, ϕ is a biharmonic map if and only if its bitension field $\tau_2(\phi)$ vanishes identically.

The first variational formula of the k-energy functional $E_k(\phi)$ with $k \geq 3$ is given by

$$\frac{dE_k(\phi_t)}{dt}\bigg|_{t=0} = \int_\Omega \langle \tau_k(\phi), \eta \rangle \, dV_g, \tag{7.101}$$

where η is the variational vector field along ϕ and the k-tension field $\tau_k(\phi)$ have been computed in [Wang (1989)] (cf. [Maeta (2014b)]). In particular, if $\phi : (M, g) \to \mathbb{E}^m$ is a (smooth) map, then the Euler-Lagrange equation of E_k is

$$\tau_k(\phi) = (-1)^{k-1} \Delta^{k-1} \tau_1(\phi) = 0. \tag{7.102}$$

When $\iota : (M, g) \to \mathbb{E}^m$ is an isometric immersion, the tension field of ϕ is given by $\tau_1(\phi) = nH$. Hence, by (7.102), M is a biharmonic submanifold of $\mathbb{E}^m$ if and only if $\Delta H = 0$ holds. Consequently, for submanifolds of a Euclidean space, the notion of biharmonic maps coincides with the notion of biharmonic submanifolds introduced by the author.

The following results are easy consequences of (7.100).

Proposition 7.2. [Jiang (1986b)] *Let $\phi : (M,g) \to (\tilde{M}, \tilde{g})$ be a map from a Riemannian manifold (M,g) into a Riemannian manifold with non-positive curvature. If M is closed and orientable, then ϕ is biharmonic if and only if it is harmonic.*

Proof. Assume $\phi : (M,g) \to (\tilde{M}, \tilde{g})$ is a biharmonic map, then (7.100) yields $\Delta^\phi \tau_1(\phi) = -\mathrm{Tr}\, \tilde{R}(d\phi, \tau_1(\phi))d\phi$. Thus

$$\int_M ||\nabla \tau_1||^2 dV_g = \int_M \langle \Delta^\phi \tau_1, \tau_1 \rangle \, dV_g$$

$$= -\int_M \mathrm{Tr}\, \langle \tilde{R}(d\phi, \tau_1)d\phi, \tau_1 \rangle \, dV_g \leq 0, \qquad (7.103)$$

which implies that $\nabla \tau_1 = 0$. From

$$\mathrm{div}\, \langle \tau_1, d\phi \rangle = \mathrm{Tr}\, \langle \nabla \tau_1, d\phi \rangle + ||\tau_1||^2,$$

we obtain

$$0 = \int_M \mathrm{div}\, \langle \tau_1, d\phi \rangle \, dV_g = \int_M ||\tau_1||^2 dV_g.$$

Hence $\tau_1 = 0$, i.e., ϕ is harmonic.

The converse is trivial. $\qquad \square$

Corollary 7.7. [Oniciuc (2002)] *Every biharmonic submanifold with constant mean curvature in a Riemannian manifold of non-positive curvature is minimal.*

Proof. First, let us assume that $\phi : M \to \tilde{M}$ is a biharmonic map of a Riemannian manifold M into a Riemannian manifold $\tilde{M}$ of non-positive curvature. Then (7.100) implies $\Delta^\phi \tau_1(\phi) = -\mathrm{Tr}\, \tilde{R}(d\phi, \tau_1(\phi))d\phi$. Replacing in the Weitzenböak formula

$$\frac{1}{2}\Delta^\phi ||\tau_1||^2 = \langle \Delta^\phi \tau_1, \tau_1 \rangle - ||\nabla \tau_1||^2$$

and using $||\tau_1|| = const.$, we obtain $||\nabla \tau_1||^2 \leq 0$. Thus $\nabla \tau_1 = 0$.

Next, for an isometric immersion, we have $||\tau_1||^2 = -\mathrm{Tr}\, \langle d\phi, \nabla \tau_1 \rangle$. Therefore $\tau_1 = 0$, i.e., ϕ is a minimal immersion. $\qquad \square$

If the ambient space is a real space form, we obtain the following.

Corollary 7.8. *An n-dimensional submanifold of a Riemannian manifold $R^m(c)$ of constant curvature c is biharmonic if and only if*

$$\Delta^D H + \sum_{i=1}^{n} h(A_H e_i, e_i) = ncH, \tag{7.104}$$

$$n \nabla H^2 + 4 \operatorname{Tr} A_{DH} = 0,$$

where H is the mean curvature vector of M in $R^m(c)$ and $H^2 = \langle H, H \rangle$ is the squared mean curvature.

Proof. For an n-dimensional submanifold of a Riemannian manifold $R^m(c)$ of constant curvature c, the bitension field $\tau_2(\iota)$ of the inclusion map of M reduces to

$$\tau_2(\iota) = \operatorname{Tr}(\nabla d\tau_1(\iota)) + n\tau_1(\iota) = n\{\operatorname{Tr}(\nabla dH) + nH\}, \tag{7.105}$$

where $\tau_1(\iota)$ is the tension field of ι.

To prove this result one can choose, without loss of generality, a system of normal coordinate $u_1, \ldots, u_n$ around an arbitrary point $x \in M$. If we put $e_i = \partial/\partial u_i$ at x, then we have

$$\begin{aligned}
\operatorname{Tr}(\nabla dH) &= \sum_{i=1}^{n} \tilde{\nabla}_{e_i} \tilde{\nabla}_{e_i} H = \sum_{i=1}^{n} \tilde{\nabla}_{e_i} \{\tilde{D}_{e_i} H - A_H(e_i)\} \\
&= -\Delta^{\tilde{D}} H - \sum_{i=1}^{n} h(A_H e_i, e_i) \\
&\quad - \sum_{i=1}^{n} \{A_{\tilde{D}_{e_i}} e_i + \nabla_{e_i} A_H e_i\} \\
&= -\Delta^{\tilde{D}} H - \sum_{i=1}^{n} h(A_H e_i, e_i) - 2\operatorname{Tr} A_{DH} + \frac{n}{2} \nabla H^2,
\end{aligned} \tag{7.106}$$

where $\tilde{\nabla}$ is the Levi-Civita connection of $R^m(c)$ and $\tilde{D}$ is the normal connection of M in $R^m(c)$. Now, by taking the tangential and normal components of $\tau_2(\iota)$ and by applying (7.105) and (7.106), we obtain the corollary. $\square$

Remark 7.6. Corollary 7.8 is due to [Oniciuc (2002)] when $c \neq 0$. When $c = 0$, it follows trivially from formula (4.13) of [Chen (1983b)].

In particular, we have the following.

Corollary 7.9. *A hypersurface M of a Riemannian manifold $R^{n+1}(c)$ of constant sectional curvature c is biharmonic if and only if its mean curvature function α is a solution of the following PDEs:*

$$\Delta \alpha - \alpha \|A\|^2 + nc\alpha = 0,$$
$$2A(\nabla \alpha) + n\alpha \nabla \alpha = 0. \tag{7.107}$$

The following relationship between the bitension fields of a submanifold in $S^m(1)$ and in $\mathbb{E}^{m+1}$ can be found in [Caddeo et al. (2002)].

Proposition 7.3. *If $\phi : (M, g) \to S^m(1)$ is an isometric immersion and $\varphi = \iota \circ \phi$, where $\iota : S^m(1) \to \mathbb{E}^{m+1}$ is the canonical inclusion, then*

$$\tau_2(\phi) = \tau_2(\varphi) + 2n\tau_1(\varphi) + \{2n^2 - |\tau_1(\varphi)|^2\}\varphi.$$

7.6 Equations of biharmonic hypersurfaces

The following theorem from [Ou (2010)] is an extension of Corollary 7.9.

Theorem 7.19. *Let M be a hypersurface of a Riemannian $(n+1)$-manifold N^{n+1}. Then ϕ is biharmonic if and only if the following two equations hold:*

$$\Delta \alpha + \alpha \|A\|^2 - \alpha \, Ric^N(\xi, \xi) = 0,$$
$$2A(\nabla \alpha) + n\alpha \nabla \alpha - 2\alpha \left(Ric^N(\xi)\right)^T = 0, \tag{7.108}$$

where α is the mean curvature, $Ric^N : T_x N \to T_x N$ is the Ricci operator of N^{n+1} defined by $\langle Ric^N(Z), W \rangle = Ric^N(Z, W)$ and A is the shape operator with respect to a unit normal vector field ξ.

Proof. Let $\phi : M \to N^{n+1}$ is the isometric immersion and $H = \alpha\xi$ is the mean curvature vector. Choose a local orthonormal frame $e_1, \ldots, e_n$ on M so that $\{d\phi(e_1), \ldots, d\phi(e_n), \xi\}$ is an adapted orthonormal frame defined on M. Identifying $d\phi(X) = X$, $\nabla_X^\phi W = \nabla_X^N W$ and noting that the tension field of ϕ is $\tau_1(\phi) = n\alpha\xi$ we can compute the bitension field of ϕ as:

$$\tau_2(\phi) = \sum_i \left\{ n\nabla_{e_i}^\phi \nabla_{e_i}^\phi H - n\nabla_{\nabla_{e_i} e_i}^\phi H - nR^N(d\phi(e_i), H)d\phi(e_i) \right\}$$
$$= n\sum_i \left\{ (e_i e_i \alpha)\xi + 2(e_i\alpha)\nabla_{e_i}^N \xi + \alpha\nabla_{e_i}^N \nabla_{e_i}^N \xi - (\nabla_{e_i} e_i)\alpha\xi \right. \tag{7.109}$$

$$- \alpha \nabla^N_{\nabla_{e_i} e_i} \xi \Big\} - n\alpha \sum_i R^N(d\phi(e_i), \xi) d\phi(e_i)$$

$$= -n(\Delta\alpha)\xi - 2nA(\nabla\alpha) + n\alpha\Delta^\phi\xi - n\alpha \sum_i R^N(d\phi(e_i), \xi) d\phi(e_i),$$

where R^N denote the curvature tensor of N^{n+1}.

To find the tangential and normal parts of the bitension field we first compute the tangential and normal components of the curvature term to have

$$\sum_{i,k} \langle R^N(d\phi(e_i), \xi) d\phi(e_i), e_k \rangle e_k = -(Ric(\xi))^T, \qquad (7.110)$$

$$\sum_i \langle R^N(d\phi(e_i), \tau_1(\phi)) d\phi(e_i), \xi \rangle = -n\alpha\, Ric^N(\xi, \xi). \qquad (7.111)$$

To find the normal part of $\Delta^\phi\xi$ we compute:

$$\langle \Delta^\phi\xi, \xi \rangle = \sum_i \langle \nabla^N_{e_i} \nabla^N_{e_i}\xi - \nabla^N_{\nabla_{e_i} e_i}\xi, \xi \rangle = -\sum_i \langle \nabla^N_{e_i}\xi, \nabla^N_{e_i}\xi \rangle. \qquad (7.112)$$

On the other hand, we have

$$||A||^2 = \sum_{i,j} \langle \nabla^N_{e_i}\xi, e_j \rangle^2 = \sum_{i,j} \langle \nabla^N_{e_i}\xi, \langle \nabla^N_{e_i}\xi, e_j \rangle e_j \rangle = \sum_i \langle \nabla^N_{e_i}\xi, \nabla^N_{e_i}\xi \rangle,$$

which, together with (7.112), implies that

$$(\Delta^\phi\xi)^\perp = \langle \Delta^\phi\xi, \xi \rangle\xi = -\sum_i \langle \nabla^N_{e_i}\xi, \nabla^N_{e_i}\xi \rangle\xi = -||A||^2\xi. \qquad (7.113)$$

A straightforward computation gives the tangential part of $\Delta^\phi\xi$ as

$$(\Delta^\phi\xi)^T = \sum_{i,k} \langle \nabla^N_{e_i}\nabla^N_{e_i}\xi - \nabla^N_{\nabla_{e_i} e_i}\xi, e_k \rangle e_k$$

$$= -\sum_{i,k} \langle \nabla^N_{e_i} Ae_i - A(\nabla_{e_i} e_i), e_k \rangle e_k = -\sum_{i,k} ((\nabla_{e_i} b)(e_k, e_i))\, e_k, \qquad (7.114)$$

where $h(X, Y) = b(X, Y)\xi$. Substituting Codazzi's equation for a hypersurface into (7.114) and using the normal coordinates at a point we find

$$(\Delta^\phi\xi)^T = -\sum_{i,k} [(\nabla_{e_i} b)(e_k\, e_i)] e_k = -n\nabla\alpha + \sum_k (Ric(\xi, e_k)) e_k.$$

Therefore, by collecting all the tangent and normal parts of the bitension field separately, we obtain

$$(\tau_2(\phi))^\perp = -n \left(\Delta\alpha + \alpha||A||^2 - \alpha\, Ric^N(\xi, \xi) \right) \xi, \qquad (7.115)$$

$$(\tau_2(\phi))^T = -n \left(2A(\nabla\alpha) + n\alpha\nabla\alpha - 2\alpha(Ric(\xi))^T \right), \qquad (7.116)$$

from which the theorem follows. $\qquad\square$

An immediate consequence of Theorem 7.19 is the following.

Corollary 7.10. [Ou (2010)] *A constant mean curvature hypersurface in a Riemannian manifold is biharmonic if and only if either it is minimal or,* $Ric^N(\xi, \xi) = ||A||^2$ *and* $(Ric^N(\xi))^T = 0$.

The following is also a consequence of Theorem 7.19.

Corollary 7.11. *A hypersurface in an Einstein space* N^{n+1} *is biharmonic if and only if its mean curvature function* α *satisfies*

$$\Delta\alpha + \alpha ||A||^2 - \frac{2\alpha}{n+1}\tau = 0,$$
$$2A(\nabla\alpha) + n\alpha\nabla\alpha = 0,$$

(7.117)

where τ *is the scalar curvature of* N^{n+1}.

Another easy consequence of Theorem 7.19 is the following.

Theorem 7.20. [Ou (2010)] *A totally umbilical hypersurface in an Einstein space with non-positive scalar curvature is biharmonic if and only if it is minimal.*

Proof. Take an orthonormal frame $\{e_1, \ldots, e_m, \xi\}$ of N^{n+1} adapted to the hypersurface M such that $Ae_i = \kappa_i e_i$ for $i = 1, \ldots, n$. Since M is supposed to be totally umbilical, i.e., all κ_i at any point $p \in M$ are equal to the same number $\kappa(p)$. It follows that $\alpha = \kappa$, $||A||^2 = n\kappa^2$, $A(\nabla\alpha) = \kappa\nabla\kappa$. The biharmonic hypersurface equation (7.117) becomes

$$\Delta\kappa + n\kappa^3 - \frac{2\kappa}{n+1}\tau = 0, \quad \kappa\nabla\kappa = 0.$$

After solving these equations we have either $\kappa = 0$ and hence $\alpha = 0$, or $\kappa = \pm\sqrt{\tau/(n(n+1))}$ is a constant and this happens only if the scalar curvature is nonnegative, from which we obtain the theorem. $\square$

Consequently, we have the following.

Corollary 7.12. [Ou (2010)] *Any totally umbilical biharmonic hypersurface in a Ricci flat manifold is minimal.*

Proof. Follows from Theorem 7.20 and the fact that a Ricci flat manifold is an Einstein space with zero scalar curvature. $\square$

7.7 Biharmonic submanifolds in sphere

The following lemma is an easy consequence of Proposition 7.3.

Lemma 7.3. [Caddeo et al. (2002)] *Let $\gamma : I \to S^m(1) \subset \mathbb{E}^{m+1}$ be a unit speed curve in $S^m(1)$. Then γ is biharmonic in $S^m(1)$ if and only if*

$$\gamma^{iv} + 2\gamma'' + (1 - k_g^2)\gamma = 0, \tag{7.118}$$

where k_g is the first Frenet curvature of γ in $S^m(1)$.

Example 7.4. Unit speed proper biharmonic curves in spheres were classified in [Caddeo et al. (2002)]. By using Lemma 7.3 it was proved that such curves in $S^m(1) \subset \mathbb{E}^{m+1}$ have constant first Frenet curvature κ_g in $S^m(1)$.

When $\kappa_g = 1$, they are the 1-type curves parametrized by

$$\gamma(s) = c_0 + c_1 \cos \sqrt{2}s + c_2 \sin \sqrt{2}s,$$

where c_0, c_1, c_2 are mutually orthogonal constant vectors in $\mathbb{E}^{m+1}$ with $||c_0||^2 = ||c_1||^2 = ||c_2||^2 = \frac{1}{2}$.

When $0 < \kappa_g < 1$, they are 2-type W-curves parametrized by

$$\gamma(s) = c_1 \cos as + c_2 \sin as + c_3 \cos bs + c_4 \sin bs, \quad a^2 \neq b^2,$$

where c_1, c_2, c_3, c_4 are mutually orthogonal constant vectors in $\mathbb{E}^{m+1}$ with $||c_1||^2 = ||c_2||^2 = ||c_3||^2 = ||c_4||^2 = \frac{1}{2}$ and $a^2 + b^2 = 2$.

Contrast to biharmonic submanifolds in Euclidean spaces, there are proper biharmonic submanifolds in spheres. Here we give such examples.

Example 7.5. [Jiang (1986a,b)] Let n_1, n_2 be two positive integers and let a, b be two positive real numbers such that $a^2 + b^2 = 1$. Then there are two cases:

(1) If $n_1 \neq n_2$, then $S^{n_1}(a) \times S^{n_2}(b)$ is a proper biharmonic hypersurface of $S^{n_1+n_2+1}(1)$ if and only if $a = b = 1/\sqrt{2}$;

(2) If $n_1 = n_2 = q$, then the following statements are equivalent:

(2.1) $S^q(a) \times S^q(b)$ is a biharmonic hypersurface of $S^{2q+1}(1)$.
(2.2) $S^q(a) \times S^q(b)$ is a harmonic hypersurface of $S^{2q+1}(1)$.
(2.3) $a = b = \frac{1}{\sqrt{2}}$.

In the case of S^3 the above example gives the minimal Clifford torus.

Example 7.6. [Caddeo et al. (2001a)] Let $S^n(a) \times \{b\}$ be the submanifold of $\mathbb{E}^{n+2}$ defined by

$$\left\{ (x_1, \ldots, x_{n+1}, b) \in \mathbb{E}^{n+2} : \sum_{i=1}^{n+1} x_i^2 = a^2, a^2 + b^2 = 1, 0 < a < 1 \right\}.$$

Then M is a proper biharmonic hypersurface of $S^{n+1}(1)$ if and only if $a = 1/\sqrt{2}$ and $b = \pm 1/\sqrt{2}$.

Example 7.7. [Caddeo et al. (2002)] The map

$$\left(\frac{x_2 x_3}{\sqrt{6}}, \frac{x_1 x_3}{\sqrt{6}}, \frac{x_1^2 - x_2^2}{2\sqrt{6}}, \frac{x_1^2 + x_2^2 - 2x_3^2}{6\sqrt{2}}, \frac{1}{\sqrt{2}} \right), \quad \sum_{i=1}^{3} x_i^2 = 1,$$

defines a proper biharmonic imbedding of RP^2 into $S^5(1)$.

Definition 7.4. A *Hopf cylinder* of S^3 is the inverse image of a spherical curve γ in S^2 via the Hopf vibration $\pi : S^3 \to S^2$.

In particular, if γ is a closed curve in S^2, then the Hopf cylinder $\pi^{-1}(\gamma)$ is called a *Hopf torus*.

For biharmonic spherical hypersurfaces, the following results are known.

Theorem 7.21. [Inoguchi (2004)] *There is no proper biharmonic Hopf cylinder in S^3.*

Theorem 7.22. [Chen (1993)] *Let M be a biharmonic oriented closed hypersurface of $S^{n+1}(1)$. If the norm $||h||$ of the second fundamental form satisfies $0 \leq ||h||^2 \leq n$, then either $||h||^2 = 0$ or $||h||^2 = n$ and the mean curvature of M is constant.*

Definition 7.5. A non-minimal hypersurface with constant mean curvature vector is called a *CMC hypersurface*.

For biharmonic hypersurfaces with 3 distinct principal curvatures, we have the following.

Theorem 7.23. [Balmuş et al. (2010b)] *There do not exist closed proper biharmonic CMC hypersurfaces with three distinct principal curvatures in $S^{n+1}(1)$*

Proof. By Corollary 7.13, a proper biharmonic hypersurface M with constant mean curvature in S^{n+1} must have constant scalar curvature. Suppose that M is a closed hypersurface with 3 distinct principal curvatures and with constant mean and scalar curvatures. Then a result of [Chang (1993)] implies that M is an isoparametric hypersurface. Thus, by a result of É. Cartan, the three distinct principal curvatures have the same multiplicity $m = 2^r$, $r = 0, 1, 2, 3$, and they are given respectively by

$$k_1 = \cot\theta, \; k_2 = \cot\left(\theta + \frac{\pi}{3}\right) = \frac{k_1 + \sqrt{3}}{1 + \sqrt{3}k_1},$$

$$k_3 = \cot\left(\theta + \frac{2\pi}{3}\right) = \frac{k_1 + \sqrt{3}}{1 - \sqrt{3}k_1}, \quad \theta \in (0, \pi/3).$$

Thus the squared norm of the shape operator is

$$|A|^2 = 2^q(k_1^2 + k_2^2 + k_3^2) = 2^q \frac{9k_1^6 + 45k_1^2 + 6}{(1 - 3k_1^2)^2}. \tag{7.119}$$

On the other hand, since M is biharmonic and of constant mean curvature, Corollary 7.11 implies that $|A|^2 = 3 \cdot 2^q$. Hence (7.119) implies that k_1 is a solution of $3k_1^6 - 9k_1^4 + 21k_1^2 + 1 = 0$, which has no real roots. $\qquad\square$

Remark 7.7. Recently, Y. Fu proved that biharmonic hypersurfaces of $S^{n+1}(1)$ with at most three distinct principal curvature must have constant mean curvature [Fu (2014c)]. Consequently, the condition of CMC in Theorem 7.23 can be removed.

For biharmonic spherical submanifolds, the following results are known.

Theorem 7.24. [Balmuş et al. (2008a)] *Let M be a closed submanifold of $S^m(1)$ with constant squared mean curvature given by $\hat{H}^2 = k$. Then M is proper biharmonic if and only if*

(1) $\hat{H}^2 = 1$ *and M is a 1-type submanifold in $\mathbb{E}^{m+1}$ with $\lambda_p = 2n$,*
(2) $\hat{H}^2 = k \in (0, 1)$ *and M is a 2-type submanifold with $\lambda_p = n(1 - \sqrt{k})$ and $\lambda_q = n(1 + \sqrt{k})$ in $\mathbb{E}^{m+1}$,*

where p is the lower order and q is the upper order of M in $\mathbb{E}^{m+1}$.

Proof. We apply directly Corollary 7.8. Denote by $\phi : M \to S^m(1)$ the inclusion of M in $S^m(1)$ and by $\iota : S^m(1) \to \mathbb{E}^{m+1}$ the canonical inclusion. Let $\varphi : M \to \mathbb{E}^{n+1}$, $\varphi = \iota \circ \phi$, be the inclusion of M in $\mathbb{E}^{m+1}$. Denote by $\hat{H}$ the mean curvature vector field of M in $S^m(1)$ and by H the mean curvature vector field of M in $\mathbb{E}^{m+1}$. Then we have $H = \hat{H} - \varphi$.

It follows from Proposition 7.3 that $\tau_2(\phi) = 0$ if and only if

$$\Delta H - 2nH + n(\hat{H}^2 - 1)\varphi = 0. \tag{7.120}$$

There are two situations to be analyzed.

Case (1): $\hat{H}^2 = 1$. In this case we have $\Delta H = 2nH$. Thus Theorem 6.1 implies that M is a 1-type submanifold of $\mathbb{E}^{n+1}$ with eigenvalue $\lambda_p = 2n$.

Case (2): $\hat{H}^2 = k \in (0,1)$. Equation (7.120) and Beltrami's formula imply

$$0 = \Delta^2 H - 2n\Delta H - n^2(k-1)H.$$

Thus the monic polynomial with distinct roots described in Theorem 6.1, which provides the type of M is $P(\Delta) = \Delta^2 - 2n\Delta - n^2(k-1)\Delta$, so M is of 2-type with $\lambda_p = n(1 - \sqrt{k})$ $\lambda_q = n(1 + \sqrt{k})$.

For the converse, let us assume that M has constant mean curvature 1 in $S^m(1)$. If it is of 1-type in $\mathbb{E}^{m+1}$ with $\lambda_p = 2n$, then we have $\Delta\varphi = 2n\varphi$ and $\Delta H = 2nH$. Thus M satisfies (7.120). So it is biharmonic in $S^m(1)$.

If $\hat{H}^2 = k \in (0,1)$ and if M is 2-type in $\mathbb{E}^{m+1}$ with $\lambda_{p,q} = n(1 \pm \sqrt{k})$, then we have $\varphi = x_p + x_q$ with $\Delta x_p = \lambda_p x_p$ and $\Delta x_q = \lambda_q x_q$. Thus after applying the Laplacian we obtain

$$H = -\{x_p + x_q + \sqrt{k}(x_q - x_q)\} = -\varphi - \sqrt{k}(x_p - x_q),$$
$$\Delta H = -n\{(k+1)\varphi + 2(-\varphi - H)\} = -n\{(k-1)\varphi - 2H\}.$$

Hence, by using (7.120), we conclude that M is biharmonic in $S^m(1)$. $\square$

Corollary 7.13. *Let M be a proper biharmonic hypersurface with constant mean curvature in $S^{n+1}(1)$. Then M has constant scalar curvature.*

Proof. Follows from Theorems 6.33 and 7.24. $\square$

For pseudo-umbilical spherical submanifolds, we have the following.

Theorem 7.25. [Balmuş et al. (2008a)] *Let M be a pseudo-umbilical submanifold of $S^m(1)$ with $\dim M \neq 4$. If M is biharmonic, then it has constant mean curvature and constant scalar curvature.*

Theorem 7.26. [Balmuş et al. (2008a)] *Let M be a pseudo-umbilical biharmonic submanifold of $S^{n+2}(1)$ with $n = \dim M \neq 4$. Then the scalar curvature τ of M satisfies $\tau \leq n(n-1)$, with the equality holding if and only if it is open in $S^n(1/\sqrt{2})$.*

Theorem 7.27. [Balmuş et al. (2008a)] *Let M be a pseudo-umbilical submanifold of $S^{n+2}(1)$ with $n = \dim M \neq 4$. Then M is proper biharmonic if and only if it is minimal in $S^{n+1}(1/\sqrt{2})$.*

Theorem 7.28. [Balmuş et al. (2008a)] *If M is a proper biharmonic hypersurface with at most two distinct principal curvatures in $S^{n+1}(1)$, then M is congruent to an open portion of $S^n(1/\sqrt{2})$ or of $S^k(1/\sqrt{2}) \times S^{n-k}(1/\sqrt{2})$ with $0 < k < n$.*

Theorem 7.29. [Balmuş et al. (2008a)] *If M is a proper biharmonic hypersurface in $S^{n+1}(1)$ with $n \geq 3$, then the following three statements are equivalent:*

(a) *M is quasi-umbilical.*
(b) *M is conformally flat.*
(c) *M is an open part of $S^n(1/\sqrt{2})$ or of $S^1(1/\sqrt{2}) \times S^{n-1}(1/\sqrt{2})$.*

Theorem 7.30. [Balmuş et al. (2008a)] *Let M be a proper biharmonic hypersurface of a hypersphere $S^{n+1}(a) \subset S^{n+2}(1)$ with $a \in (0,1)$. Then $a \geq 1/\sqrt{2}$. Moreover,*

(a) *if $a = 1/\sqrt{2}$, then M is minimal in $S^{n+1}(1/\sqrt{2})$,*
(b) *if $a > 1/\sqrt{2}$, then M is an open part of $S^n(1/\sqrt{2})$.*

The following methods for constructing biharmonic submanifolds of codimension greater than 1 in $S^m(1)$ were given in [Caddeo et al. (2002)].

Theorem 7.31 (Composition property). *Let M be a minimal submanifold of $S^{m-1}(a) \subset S^m(1)$. Then M is proper biharmonic in $S^m(1)$ if and only if $a = 1/\sqrt{2}$.*

Theorem 7.32 (Product composition property). *Let $M_1^{n_1}$ and $M_2^{n_2}$ be two minimal submanifolds of $S^{m_1}(r_1)$ and $S^{m_2}(r_2)$, respectively, where $m_1 + m_2 = m - 1$, $r_1^2 + r_2^2 = 1$. Then $M_1^{n_1} \times M_2^{n_2}$ is proper biharmonic in $S^m(1)$ if and only if $r_1 = r_2 = 1/\sqrt{2}$ and $n_1 \neq n_2$.*

Remark 7.8. By using the composition property given in Theorem 7.31 and applying Lawson's examples of closed orientable imbedded minimal surfaces of arbitrary genus in S^3, it was shown in [Caddeo et al. (2002)] that there exist closed orientable imbedded proper biharmonic surfaces of arbitrary genus in S^4.

For proper biharmonic spherical surfaces with parallel mean curvature vector, we have the next result.

Theorem 7.33. [Balmuş et al. (2008a)] *Let M be a proper biharmonic surface with parallel mean curvature vector in $S^m(1)$. Then it is minimal in $S^{n-1}(1/\sqrt{2})$.*

For biharmonic surface in $S^4(1)$, the condition of parallel mean curvature vector can be weaken to constant mean curvature as follows.

Theorem 7.34. [Balmuş and Oniciuc (2009)] *If M is a proper biharmonic surface in $S^4(1)$ with constant mean curvature, then M is a minimal surface in $S^{n-1}(1/\sqrt{2})$.*

Definition 7.6. A submanifold M of a Riemannian manifold is said to have *parallel normalized mean curvature vector* (or simply PNMC) if the unit vector field in the direction of the mean curvature vector is parallel in the normal bundle (cf. [Chen (1980)]).

The next three results were proved in [Chen (1980)].

Theorem 7.35. *Let M be a surface of class C^ω in a complete simply-connected real space form $R^m(c)$ of constant curvature c. If M has PNMC, then either M is a minimal surface of $R^m(c)$ or M lies in a totally geodesic submanifold $R^4(c)$ of $R^m(c)$.*

Theorem 7.36. *Let M be a flat surface of class C^ω in $\mathbb{E}^m$. If M has PNMC, then it is one of the following surfaces:*

(a) *a flat minimal surface of a hypersphere of $\mathbb{E}^m$;*
(b) *an open piece of the product surface of two plane circles; or*
(c) *a developable surface lying in a linear 3-subspace $\mathbb{E}^3 \subset \mathbb{E}^m$.*

Theorem 7.37. *Let M be a Riemann sphere immersed in $\mathbb{E}^m$. If M has PNMC, then either it is a minimal surface of a hypersphere of $\mathbb{E}^m$ or it lies in a linear 3-subspace $\mathbb{E}^3 \subset \mathbb{E}^m$.*

Biharmonic spherical submanifolds with PNMC were studied in [Balmuş et al. (2013)]. The following results were obtained.

Theorem 7.38. *Let $\phi : M \to S^m(1)$ be a PNMC biharmonic submanifold with at most two distinct principal curvatures in the direction of the mean curvature vector. Then either ϕ induces a minimal immersion of M in $S^{m-1}(1/\sqrt{2})$ or locally $\phi(M) = M_1^{n_1} \times M_2^{n_2} \subset S^{m_1}(1/\sqrt{2}) \times S^{m_2}(1/\sqrt{2}) \subset S^m(1)$, where M_i is a minimal imbedded submanifold of $S^{m_i}(1/\sqrt{2})$ for $i = 1, 2$, and $n_1 + n_2 = n$, $n_1 \neq n_2$ and $m_1 + m_2 = m - 1$.*

Theorem 7.39. *There exist no PNMC biharmonic 3-type submanifolds in any sphere. In particular, there do not exist biharmonic 3-type hypersurfaces in $S^{n+1} \subset \mathbb{E}^{n+2}$.*

Theorem 7.40. *Let $\phi : M \to S^m(1)$ be a proper biharmonic submanifold. If M is of finite type, mass-symmetric and linearly independent, then it is of 2-type in $\mathbb{E}^{m+1}$.*

The following two conjectures were made in [Caddeo et al. (2002)].

Conjecture 7.2. *Any biharmonic submanifold of S^m has constant mean curvature.*

Conjecture 7.3. *The only proper biharmonic hypersurfaces in $S^{n+1}(1)$ are the open parts of hyperspheres $S^n(1/\sqrt{2})$ and of the generalized Clifford tori $S^k(1/\sqrt{2}) \times S^{n-k}(1/\sqrt{2})$, $0 < k < n$.*

All results given in this section support these two conjectures.

7.8 Biharmonic submanifolds in hyperbolic space and generalized biharmonic conjecture

There are several non-existence results for biharmonic submanifolds lying in the hyperbolic m-space $H^m(-1)$ of constant sectional curvature -1.

First we mention the following results.

Theorem 7.41. [Caddeo et al. (2002)] *Let M be a pseudo-umbilical submanifold of $H^m(-1)$ with $\dim M \neq 4$. Then M is biharmonic if and only if it is minimal.*

Proof. Assume that M is a pseudo-umbilical submanifold of the hyperbolic m-space $H^m(-1)$ with $\dim M \neq 4$. Let $u_1, \ldots, u_n$ be a system of normal coordinates around an arbitrary point $x \in M$ and let $e_i = \partial/\partial u_i$ be the corresponding coordinate vector fields. Denote by $H, D,$ and A the mean curvature vector, the normal connection, and the shape operator of M in $H^m(-1)$. At x we have

$$\operatorname{Tr} A_{DH} = \sum_i \nabla_{e_i} A_H e_i - \frac{n}{2} \nabla H^2.$$

Since M is pseudo-umbilical, the first term in the right-hand side is

$$\sum_i \nabla_{e_i} A_H(e_i) = \sum_i \nabla_{e_i}(H^2 e_i) = \nabla H^2$$

at x and therefore

$$\operatorname{Tr} A_{DH} = \left(1 - \frac{n}{2}\right)\nabla H^2.$$

Substituting this into the second equation of (7.104) gives $(4-n)\nabla H^2 = 0$. Thus, for $n \neq 4$, the mean curvature of M is constant. Since any biharmonic submanifold with constant mean curvature in a Riemannian manifold with nonpositive sectional curvature is minimal according to Corollary 7.7, we obtain the theorem. $\qquad\square$

In particular, we have

Corollary 7.14. [Caddeo et al. (2002)] *Let $\gamma : I \to H^m(-1)$ be a unit speed curve in the hyperbolic m-space $H^m(-1)$. Then γ is biharmonic if and only if it is a geodesic.*

Using the same method as the proof of Theorem 7.3, we have the following result.

Theorem 7.42. [Caddeo et al. (2002)] *Let M be a surface of $H^3(-1)$. Then M is biharmonic if and only if it is minimal.*

Proof. Assume that M is a biharmonic surface of $H^3(-1)$. Suppose that M is proper. We shall prove that the mean curvature is constant, which implies that M is minimal according to Corollary 7.7.

Let $\{e_1, e_2\}$ be a local orthonormal frame field on M and let ξ be a unit normal vector field. Assume that $H = \alpha\xi$. In this case conditions (7.104) become

$$\Delta\alpha = (-2 - ||A||^2)\alpha, \quad A(\nabla\alpha) + \alpha\nabla\alpha = 0. \tag{7.121}$$

Put $U = \{x \in M | (\nabla\alpha^2)(x) \neq 0\}$. We claim that $U = \emptyset$. Assume that $U \neq \emptyset$ and let $e_1 = \nabla\alpha/|\nabla\alpha|$. Then we have

$$e_2\alpha = 0, \quad \nabla\alpha = (e_1\alpha)e_1, \tag{7.122}$$

and the second fundamental form h of M is given by

$$h(e_1, e_1) = -\alpha\xi, \quad h(e_1, e_2) = 0, \quad h(e_2, e_2) = 3\alpha\xi. \tag{7.123}$$

Thus

$$||A||^2 = 10\alpha^2. \tag{7.124}$$

Since $H^3(-1)$ has constant sectional curvature and M is a hypersurface, Codazzi's equation gives

$$e_2 f\alpha - 4\alpha\omega_2^1(e_1), \quad 3e_1\alpha = -4\alpha\omega_1^2(e_2), \tag{7.125}$$

where ω^1, ω^2 are the 1-forms dual of e_1, e_2 and ω_i^j are the connection forms given by $\nabla e_i = \omega_i^j e_j$.

Now, equations (7.122) and (7.125) imply that $\omega_2^1(e_1) = 0$ and $d\omega^1 = 0$. Thus locally ω^1 is a exact 1-form. Hence $\omega^1 = du$ for some local function u. Since we have $d\alpha = (e_1\alpha)\omega^1 + (e_2\alpha)\omega^2$ and $e_2\alpha = 0$, we find $d\alpha \wedge \omega^1 = 0$, which means that α is a function of u. Denoting by α' and α'' the first and second derivatives of α with respect to u, then the second formula of (7.125) implies that

$$4\alpha\omega_1^2 = -3\alpha'\omega^2. \tag{7.126}$$

Again, (7.122) and (7.125) give

$$4\alpha\Delta f = 3(\alpha')^2 - 4\alpha\alpha'', \tag{7.127}$$

and, from (7.121) and (7.124), we find

$$4\alpha\alpha'' - 3(\alpha')^2 - 8\alpha^2 - 40\alpha^4 = 0. \tag{7.128}$$

If we put $y = (\alpha')^2$, then the last equation gives

$$2\alpha\frac{dy}{\alpha} - 3y = 40\alpha^4 + 8\alpha^2, \tag{7.129}$$

which implies

$$(\alpha')^2 = 8\alpha^4 + 8\alpha^2 + C\alpha^{\frac{3}{2}}, \tag{7.130}$$

for some constant C. On the other hand, the Gauss equation gives

$$G = -1 - 3\alpha^2, \quad d\omega_1^2 = -G\omega^1 \wedge \omega^2 \tag{7.131}$$

where G is the Gauss curvature. From (7.123), (7.126) and (7.131), we get

$$4\alpha\alpha'' - 7(\alpha')^2 + 16\alpha^4 + \frac{16}{3}\alpha^2 = 0. \tag{7.132}$$

But (7.128) and (7.132) imply

$$(\alpha')^2 = 14\alpha^4 + \frac{10}{3}\alpha^2. \tag{7.133}$$

Summing up (7.130) and (7.133) shows that α satisfy a polynomial equation with constant coefficients. Hence M has constant mean curvature. $\square$

Analogous to Theorem 7.11, we have the next result for biharmonic hypersurfaces in $H^{n+1}(-1)$ with at most two distinct principal curvatures.

Theorem 7.43. [Balmuş et al. (2008a)] *Any biharmonic hypersurface of $H^{n+1}(-1)$ with at most two distinct principal curvatures is minimal.*

By investigating biharmonic hypersurfaces in H^4 case by case according to the number of distinct principal curvatures, Theorem 7.42 was extended in [Balmuş et al. (2010b)] to the following.

Theorem 7.44. *If M is a biharmonic hypersurface of a hyperbolic 4-space H^4, then M is minimal in H^4.*

Based on the results mentioned above, Caddeo, Montaldo and Oniciuc proposed in [Caddeo et al. (2001a)] the following.

Generalized Chen's (Biharmonic) Conjecture: *Every biharmonic submanifold in a Riemannian manifold of non-positive sectional curvature is minimal.*

7.9 Recent development on generalized biharmonic conjecture

There exist numerous recent results concerning Generalized Biharmonic Conjecture. In this section, we provide a list of results in this respect.

Theorem 7.45. [Ou (2010)] *Every totally umbilical biharmonic hypersurface in an Einstein space with non-positive scalar curvature is minimal.*

Proof. Follows immediately from Theorem 7.20. □

Theorem 7.46. [Nakauchi and Urakawa (2011)] *If M is a biharmonic hypersurface with finite total mean curvature, i.e.,*

$$\int_M H^2 dV < \infty,$$

in a Riemannian manifold of non-positive Ricci curvature, then M is a minimal hypersurface.

Theorem 7.47. [Nakauchi and Urakawa (2013a)] *Let M be a biharmonic submanifold in a Riemannian manifold $\tilde{M}$ of non-positive sectional curvature. If M has finite total mean curvature, then M is a minimal submanifold.*

Proof. Let φ denote the biharmonic immersion of M. Then, according to (7.100), the biharmonic map equation $\tau_2(\varphi) = 0$ is equivalent to

$$\Delta H - \sum_{i=1}^n \tilde{R}(H, d\varphi(e_i))d\varphi(e_i) = 0, \qquad (7.134)$$

where H is the mean curvature vector. Take any point x_0 in M, and for every $r > 0$, let us consider the following cut-off function λ on M:

$$0 \le \lambda(x) \le 1 \ (x \in M), \quad \lambda(x) = 1 \ (x \in B_r(x_0)),$$

$$\lambda(x) = 0 \ (x \notin B_{2r}(x_0)), \quad |\nabla \lambda| \le \frac{2}{r} \ (\text{on } M),$$

where $B_r(x_0) := \{x \in M : d(x, x_0) < r\}$ and d is the distance of M. In both sides of (7.134), taking inner product with $\lambda^2 H$, and integrate them over M, we have

$$\int_M \langle \Delta H, \lambda^2 H \rangle \, dV = \int_M \sum_{i=1}^n \langle \tilde{R}(H, d\varphi(e_i)) d\varphi(e_i), H \rangle \lambda^2 \, dV. \quad (7.135)$$

Since the sectional curvature of $\tilde{M}$ is non-positive, $\tilde{g}(\tilde{R}(u, v)v, u) \le 0$ for $u, v \in T_y \tilde{M}$, $y \in \tilde{M}$, the right hand side of (7.135) is non-positive, i.e.,

$$\int_M \langle \Delta H, \lambda^2 H \rangle \, dV \le 0. \quad (7.136)$$

On the other hand, the right hand side coincides with

$$\int_M \langle \tilde{\nabla} H, \tilde{\nabla}(\lambda^2 H) \rangle dV = \int_M \sum_{i=1}^n \langle \tilde{\nabla}_{e_i} H, \tilde{\nabla}_{e_i}(\lambda^2 H) \rangle dV$$

$$= \int_M \lambda^2 \sum_{i=1}^n |\tilde{\nabla}_{e_i} H|^2 dV + 2 \int_M \sum_{i=1}^n \lambda \, (e_i \lambda) \, \langle \tilde{\nabla}_{e_i} H, H \rangle dV,$$

since $\tilde{\nabla}_{e_i}(\lambda^2 H) = \lambda^2 \tilde{\nabla}_{e_i} H + 2\lambda(e_i \lambda) H$. Therefore we have

$$\int_M \lambda^2 \sum_{i=1}^n |\tilde{\nabla}_{e_i} H|^2 dV \le -2 \int_M \sum_{i=1}^n \langle \lambda \, \tilde{\nabla}_{e_i} H, (e_i \lambda) H \rangle dV. \quad (7.137)$$

Now apply with $V := \lambda \tilde{\nabla}_{e_i} H$, and $W := (e_i \lambda) H$, to Young's inequality: for all $V, W \in \Gamma(\varphi^{-1} TN)$ and $\epsilon > 0$,

$$\pm 2 \langle V, W \rangle \le \epsilon |V|^2 + \frac{1}{\epsilon} |W|^2,$$

the right hand side of (7.137) is smaller than or equal to

$$\epsilon \int_M \lambda^2 \sum_{i=1}^n |\tilde{\nabla}_{e_i} H|^2 dV + \frac{1}{\epsilon} \int_M H^2 \sum_{i=1}^n |e_i \lambda|^2 dV. \quad (7.138)$$

By taking $\epsilon = \frac{1}{2}$, we obtain

$$\int_M \lambda^2 \sum_{i=1}^n |\tilde{\nabla}_{e_i} H|^2 dV \le \frac{1}{2} \int_M \lambda^2 \sum_{i=1}^n |\tilde{\nabla}_{e_i} H|^2 dV$$

$$+ 2 \int_M H^2 \sum_{i=1}^n |e_i \lambda|^2 dV.$$

Thus we have

$$\int_M \lambda^2 \sum_{i=1}^n |\tilde{\nabla}_{e_i} H|^2 dV \le 4 \int_M H^2 \sum_{i=1}^n |e_i \lambda|^2 dV$$

$$\le \frac{16}{r^2} \int_M H^2 dV < \infty.$$

Since (M, g) is complete, we can tend r to infinity, and then the left hand side goes to $\int_M \sum_{i=1}^n |\tilde{\nabla}_{e_i} H|^2 dV$, we obtain

$$\int_M \sum_{i=1}^n |\tilde{\nabla}_{e_i} H|^2 dV \le 0.$$

Hence we have $\tilde{\nabla}_X H = 0$ for any X tangent to M. Therefore it follows from Weingarten's formula that $A_H = 0$. Consequently, we conclude that $H = 0$. $\qquad\square$

Theorem 7.46 was extended by S. Maeta to the following.

Theorem 7.48. [Maeta (2014a)] *If M is a complete biharmonic submanifold M in a Riemannian manifold of non-positive sectional curvature whose mean curvature vector satisfies $\int_M H^\alpha dv < \infty$ for some finite $\alpha \ge 2$, then M is minimal.*

Theorem 7.49. [Nakauchi et al. (2014)] *If M is a complete biharmonic submanifold with finite bi-energy in a non-positively curved Riemannian manifold, then M is minimal.*

Let M is a complete, non-closed Riemannian manifold. A function f on M is called an *L^q-function* if

$$||f||_{L^q} := \left(\int_M |f|^q dV \right)^{\frac{1}{q}}$$

converges. Denote by $L^q(M)$ the space of L^q functions on M.

Theorem 7.50. [Alías et. al. (2013)] *Let M be a complete orientable biharmonic hypersurface in a Riemannian manifold with non-positive Ricci tensor whose mean curvature $||H||$ satisfying $||H|| \in L^2(M)$. Then M is a minimal hypersurface.*

Theorem 7.51. [Maeta (2014a)] (see also [Jiang (1986b)]) *If M is a biharmonic closed submanifold in a Riemannian manifold with non-positive sectional curvature, then M is minimal.*

For a complete Riemannian manifold M and a real number $\alpha \geq 0$, if the sectional curvature K of M satisfies

$$K(x) \geq -L\big(1 + \mathrm{dist}_N(x, x_0)^2\big)^{\alpha/2} \qquad (7.139)$$

for some $L > 0$ and $x_0 \in M$, then we call that K has a *polynomial growth bound of order* α *from below*.

Theorem 7.52. [Maeta (2014c)] *If M is a biharmonic properly immersed submanifold in a complete Riemannian manifold with non-positive sectional curvature whose sectional curvature has polynomial growth bound of order less than 2 from below, then M is minimal.*

Let M be a complete Riemannian manifold and $x_0 \in M$. Then M is said to have *at most polynomial volume growth*, if there exists a nonnegative integer s such that $\mathrm{vol}(B_\rho(x_0)) \leq C\rho^s$, where $B_\rho(x_0)$ is the geodesic ball centered at x_0 with radius ρ and C is a positive constant independent of ρ.

Theorem 7.53. [Luo (2013b)] *If M is a complete biharmonic submanifold (resp., hypersurface) in a Riemannian manifold whose sectional curvature (resp., Ricci curvature) is non-positive with at most polynomial volume growth, then M is minimal.*

Theorem 7.54. [Luo (2013b)] *If M is a complete biharmonic submanifold (resp., hypersurface) in a Riemannian manifold whose sectional curvature (resp., Ricci curvature) is smaller that $-\epsilon$ for some constant $\epsilon > 0$, then M is minimal.*

Theorem 7.55. [Luo (2013b)] *If M is a complete biharmonic submanifold (resp., hypersurface) M in a Riemannian manifold of non-positive sectional (resp., Ricci) curvature which satisfies $\int_M \|H\|^p dV < \infty$ for some $p > 0$, then M is minimal.*

Let M be a submanifold of a Riemannian manifold with inner product $\langle\ ,\ \rangle$. Then M is called ϵ-*superbiharmonic* if it satisfies

$$\langle \Delta H, H \rangle \geq (1 - \epsilon)\|\nabla H\|^2, \qquad (7.140)$$

where $\epsilon \in [0, 1]$ is a constant.

Theorem 7.56. [Wheeler (2013)] *If M is a proper ϵ-superbiharmonic submanifold in a complete Riemannian manifolds satisfying the decay condition at infinity*

$$\lim_{\rho \to \infty} \frac{1}{\rho^2} \int_{\phi^{-1}(B_\rho)} H^2 dV = 0,$$

where ϕ is the immersion, B_ρ is a geodesic ball of N with radius ρ, then M is minimal.

Theorem 7.56 was extended in [Luo (2013b)] to the next two results.

Theorem 7.57. *Let M be a proper ϵ-superbiharmonic submanifold with $\epsilon > 0$ in a complete Riemannian manifold $\tilde{M}$ whose mean curvature satisfying the curvature growth condition*

$$\lim_{\rho \to \infty} \frac{1}{\rho^2} \int_{\phi^{-1}(B_\rho)} H^{2+a} dv = 0,$$

where ϕ is the immersion, B_ρ is a geodesic ball of $\tilde{M}$ with radius ρ, and $a \geq 0$ is a constant. Then M is minimal.

Theorem 7.58. *If M is a complete ϵ-superbiharmonic submanifold with $\epsilon > 0$ in a complete Riemannian manifold $\tilde{M}$ satisfying $\int_M H^{2+a} dv < \infty$, where $a \geq 0$ is a constant, then M is minimal.*

A Riemannian manifold is called a *Cartan-Hadamard manifold* if it is complete, simply-connected and of non-positive sectional curvature.

Theorem 7.59. [Asserda and Kassi (2013)] *Let $\tilde{M}$ be a Cartan-Hadamard manifold whose Ricci curvature satisfies $Ric^{\tilde{M}}(x) \geq -G(d_{\tilde{M}}(x, x_0))$ for some function $G : [0, \infty) \to [0, \infty)$ satisfying $G(0) \geq 1$, $G' \geq 0$, $G^{-1/2} \notin L^1([0, +\infty))$, then every biharmonic submanifold of $\tilde{M}$ is minimal.*

The following result was obtained in [Baird et al. (2010)].

Theorem 7.60. *Every biharmonic immersion from a complete non-closed manifold of nonnegative Ricci curvature into a manifold of non-positive sectional curvature with finite bi-energy is minimal.*

All of the results of this section given above provide partial solutions to the generalized Chen's biharmonic conjecture.

On the other hand, Y.-L. Ou and L. Tang discovered the following counter-examples to this conjecture.

Theorem 7.61. [Ou and Tang (2012)] *Let $A, B > 0$, c be constants, $\mathbb{R}^5_+$ be the upper-half space defined by $\mathbb{R}^5_+ = \{(x_1, \ldots, x_4, z) \in \mathbb{R}^5 : z > 0\}$, and $f(z) = (Az + B)^t$. Then, for any $t \in (0, 1/2)$ and any $(a_1, a_2, a_3, a_4) \in S^3(\sqrt{2t/(1-2t)})$, the isometric immersion*

$$\varphi : \mathbb{R}^4 \to \left(\mathbb{R}^5_+, h = f^{-2}(z) \left\{ \sum_{i=1}^{4} dx_i{}^2 + dz^2 \right\} \right) \tag{7.141}$$

with $\varphi(x_1, \ldots, x_4) = (x_1, \ldots, x_4, \sum_{i=1}^{4} a_i x_i + c)$ *gives a proper biharmonic hypersurface into a conformally flat space with strictly negative sectional curvature.*

Theorem 7.61 gives infinitely many counter-examples to the generalized Chen's biharmonic conjecture. These examples are incomplete Riemannian manifolds. In views of the results given above, it is very interesting to know whether the generalized Chen's biharmonic conjecture still holds true for every *complete* biharmonic submanifold in a Riemannian manifold of nonpositive sectional curvature. The following is global version of generalized Chen's biharmonic conjecture (see [Maeta (2014a)]).

Conjecture 7.4. *Any complete biharmonic submanifold in a Riemannian manifold with nonpositive sectional curvature is minimal.*

Remark 7.9. There are many studies of biharmonic submanifolds in various model spaces other than real space forms. Here, we provide a list of such studies for readers' convenience.

(1) For biharmonic submanifolds in complex space forms, see [Fetcu et al. (2010); Fu (2013a); Sasahara (2007a, 2013)].
(2) For biharmonic submanifolds in Sasakian space forms, see [Cho et al. (2007); Fetcu (2008); Fetcu and Oniciuc (2009a,b,d, 2012); Kocayigit and Hacsalihoglu (2012); Sasahara (2008, 2013)].
(3) For biharmonic submanifolds in other contact metric manifolds, see [Arslan et. al. (2005); Inoguchi (2004); Markellos and Papantoniou (2011); Perkas et al. (2011)].
(4) For biharmonic submanifolds in product and warped product manifolds, see [Balmuş et al. (2007); Roth (2103)].
(5) For biharmonic submanifolds in 3-dimensional geometries, see [Fetcu and Oniciuc (2007, 2009c); Loubeau and Montaldo (2005); Ou and Wang (2011); Sasahara (2012a)].
(6) For conformal biharmonic submanifolds, see [Ou (2009)].

7.10   Biminimal immersions

E. Loubeau and S. Montaldo defined the notion of biminimal immersions.

Definition 7.7. [Loubeau and Montaldo (2008)] An immersion $\phi :$ $(M, g) \to (\tilde{M}, \tilde{g})$ between Riemannian manifolds, or its image, is called

λ-*biminimal* if it is a critical point of the λ-*bienergy*

$$E_{2,\lambda}(\phi) = E_2(\phi) + \lambda E(\phi), \quad \lambda \in \mathbb{R}, \qquad (7.142)$$

for any smooth variation of the map $\phi_t : (-\epsilon, +\epsilon) \times M \to \tilde{M}$, $\phi_0 = \phi$, such that $V = \frac{d\phi_t}{dt}\big|_{t=0}$ is normal to $\phi(M)$.

Using the Euler-Lagrange equations for harmonic and biharmonic maps, we see that an immersion is λ-biminimal if and only if

$$[\tau_{2,\lambda}]^\perp = [\tau_2]^\perp + \lambda[\tau_1]^\perp = 0, \qquad (7.143)$$

for $\lambda \in \mathbb{R}$, where $[\,,\,]^\perp$ is the normal component of $[\,,\,]$. In the instance of an isometric immersion $\phi : M \to \tilde{M}$, the condition is given by

$$\{\Delta^\phi H - \operatorname{Tr} \tilde{R}(H, d\phi)d\phi\}^\perp + \lambda H = 0. \qquad (7.144)$$

Definition 7.8. An immersion is called *free biminimal* or simply *biminimal* if it is λ-biminimal with $\lambda = 0$.

Clearly, every biharmonic immersion is biminimal, but the converse is not always true.

The following result provides many examples of biminimal curves.

Theorem 7.62. [Loubeau and Montaldo (2008)] *Let (M, g) be a Riemannian manifold. Fix a point $p \in M$ and choose a function f depending only on the geodesic distance from p. Then any geodesic on (M, g) going through p is a biminimal curve on $(M, \overline{g} = e^{2f}g)$.*

Proof. Let γ be a geodesic of (M, g) and let T be the unit tangent vector field along γ. Since the function f depends only on the geodesic distance from p and γ is a geodesic on (M, g), we have $\nabla^\gamma_{\partial/\partial t} T = 0$, and the tension field of γ with respect to the metric $\overline{g} = e^{2f}g$ is:

$$\overline{\tau_1}(\gamma) = \overline{\nabla}^\gamma_{\frac{\partial}{\partial t}} d\gamma\left(\frac{\partial}{\partial t}\right) = \overline{\nabla}^\gamma_{\frac{\partial}{\partial t}} T = \nabla^\gamma_{\frac{\partial}{\partial t}} T + 2(Tf)T - \nabla f,$$

where in this last equality we have used (5.30). Thus $\overline{\tau_1}(\gamma) = (Tf)T$.

With respect to $\overline{g}$, the bitension field of γ is

$$\overline{\tau_2}(\gamma) = \overline{\nabla}^\gamma_{\frac{\partial}{\partial t}} \overline{\nabla}^\gamma_{\frac{\partial}{\partial t}} (Tf)T) - \overline{\nabla}^\gamma_{\overline{\nabla}^\gamma_{\frac{\partial}{\partial t}} \frac{\partial}{\partial t}}((T(f)T) + \overline{R}(T, (Tf)T)T$$

$$= \overline{\nabla}^\gamma_{\frac{\partial}{\partial t}} \overline{\nabla}^\gamma_{\frac{\partial}{\partial t}}((Tf)T)$$

$$= \overline{\nabla}^\gamma_{\frac{\partial}{\partial t}}((T^2f)T + (T(f)^2)T)$$

$$= (T^3f)T + (T^2f)(Tf)T + 2(T^2f)(Tf)T + (Tf)^3T$$

$$= [T^3f + 3(T^2f)Tf + (Tf)^3]T.$$

Therefore $\overline{\tau_2}(\gamma)$ has no normal component and the curve γ is biminimal on $(M, \overline{g} = e^{2f}g)$. $\qquad \square$

Corollary 7.15. [Loubeau and Montaldo (2008)] *Let r denote the geodesic distance from a point $p \in (M, g)$, and $f(r) = \ln(ar^2 + br + c)$, $a, b, c \in \mathbb{R}$. Then a geodesic on (M, g) through p is a biharmonic map on $(M, \bar{g} = e^{2f}g)$.*

Proof. It follows from the proof of Theorem 7.62 that a geodesic on (M, g) through p becomes a biharmonic map on $(M, \bar{g})$ if f is a solution of the ordinary differential equation: $f'''(r) + 3f''(r)f'(r) + f'(r)^3 = 0$. To solve this equation, put $y = f'$ to obtain

$$y'' + 3y'y + y^3 = 0. \tag{7.145}$$

Then, using the transformation $y = x'/x$, equation (7.145) becomes $x'''/x = 0$ which has the solution $x(r) = \bar{a}r^2 + \bar{b}r + \bar{c}$, $\bar{a}, \bar{b}, \bar{c} \in \mathbb{R}$. Finally, from $f(r) = \ln(d\, x(r))$, $d \in \mathbb{R}$, we find the desired f. □

Remark 7.10. Biminimal curves in $\mathbb{E}^2$, $S^2(1)$ and $H^2(-1)$ were completely classified in [Inoguchi and Lee (2012b)] in terms of the curvature function of the biminimal curves.

Proposition 7.4. [Loubeau and Montaldo (2008)] *Let $\phi : M \to \tilde{M}$ be an isometric immersion of codimension one with mean curvature vector $H = \alpha\xi$ and mean curvature α. Then ϕ is λ-biminimal if and only if*

$$\Delta\alpha + \alpha||h||^2 = (Ric(\xi) - \lambda)\alpha. \tag{7.146}$$

Proof. In a local orthonormal frame $e_1, \ldots, e_n$ on M, the tension field of ϕ is $\tau_1(\phi) = nH$ and its bitension field is

$$\tau_2(\phi) = n\sum_{i=1}^{n} \left[\nabla^\phi_{e_i}\nabla^\phi_{e_i}H + \nabla^\phi_{\nabla_{e_i}e_i}H - \tilde{R}(d\phi(e_i), H)d\phi(e_i) \right]$$

$$= n\sum_{i=1}^{n} \left[\nabla^\phi_{e_i}\left(e_i(\alpha)\xi + \alpha\nabla^\phi_{e_i}\xi\right) - (\nabla_{e_i}e_i)(\alpha)\xi \right.$$

$$\left. - \alpha\nabla^\phi_{\nabla_{e_i}e_i}\xi + \alpha\tilde{R}(d\phi(e_i), \xi)d\phi(e_i) \right]$$

$$= n\sum_{i=1}^{n} \left[(e_i^2\alpha)\xi + 2e_i\alpha\nabla^\phi_{e_i}\xi + \alpha\nabla^\phi_{e_i}\nabla^\phi_{e_i}\xi - (\nabla_{e_i}e_i)\alpha\xi \right.$$

$$\left. - \alpha\nabla^\phi_{\nabla_{e_i}e_i}\xi \right] + n\alpha\sum_{i=1}^{n}\tilde{R}(d\phi(e_i), \xi)d\phi(e_i)$$

$$= n(\Delta\alpha)\xi + 2n\sum_{i=1}^{n}e_i\alpha\nabla^\phi_{e_i}\xi - n\alpha\Delta^\phi\xi + n\alpha\,\text{Tr}\,\tilde{R}(d\phi, \xi)d\phi.$$

But (1) $\langle \nabla^\phi_{e_i} \xi, \xi \rangle = 0$ and (2) $\langle \sum_{i=1}^n \tilde{R}(d\phi(e_i), \xi) d\phi(e_i), \xi \rangle = Ric(\xi)$.
For $\langle \Delta^\phi \xi, \xi \rangle$, first we have

$$\langle \Delta^\phi \xi, \xi \rangle = \sum_{i=1}^n \langle -\nabla^\phi_{e_i} \nabla^\phi_{e_i} \xi + \nabla^\phi_{\nabla_{e_i} e_i} \xi, \xi \rangle = ||\nabla^\phi \xi||^2.$$

Thus if h is the second fundamental form of ϕ, then we have

$$||\nabla^\phi_{e_i} \xi||^2 = \langle \nabla^\phi_{e_i} \xi, \nabla^\phi_{e_i} \xi \rangle = \sum_{j=1}^n \langle \nabla^\phi_{e_i} \xi, e_j \rangle^2, \quad \forall i = 1, \ldots, n,$$

which implies that $\sum_{i=1}^n \langle \nabla^\phi_{e_i} \xi, \nabla^\phi_{e_i} \xi \rangle = ||h||^2$. Consequently, we obtain

$$-\langle \tau_{2,\lambda}(\phi), \xi \rangle = n \left(\Delta \alpha + \alpha ||h||^2 - \alpha Ric(\xi) + \lambda \alpha \right),$$

which implies the proposition. □

Corollary 7.16. [Loubeau and Montaldo (2008)] *A hypersurface M in a real space form of constant curvature c is λ-biminimal if and only if*

$$\Delta \alpha + \alpha \left(n^2 \alpha^2 - 2\tau + n(n-2)c + \lambda \right) = 0, \tag{7.147}$$

where α is the mean curvature and τ is the scalar curvature of M.

Moreover, an isometric immersion $\phi : M \to R^3(c)$ from a surface to a 3-dimensional real space form is λ-biminimal if and only if

$$\Delta \alpha + \alpha \left(4\alpha^2 - 2G + \lambda \right) = 0. \tag{7.148}$$

Proof. Let $e_1, \ldots, e_n$ be an orthonormal frame of M corresponding to the principal curvatures $\kappa_1, \ldots, \kappa_n$ and h its second fundamental form, then

$$||h||^2 = \kappa_1^2 + \cdots + \kappa_n^2 = n^2 H^2 - 2\tau + n(n-1)c,$$

where $K(e_i, e_j)$ is the sectional curvature on M^n of the plane spanned by e_i and e_j, and $\tau = \sum_{i<j} K(e_i, e_j)$ is the scalar curvature of M. Since $Ric(\xi) = nc$, the immersion ϕ is λ-biminimal if and only if (7.148) holds. □

By using Corollary 7.16, the following examples were given in [Loubeau and Montaldo (2008)].

Example 7.8. Consider the orthogonal projection $\pi : \mathbb{E}^3 \to \mathbb{E}^2$ given by $\pi(x, y, z) = (x, y)$. Then π is a Riemannian submersion with totally geodesic fibres and integrable horizontal distribution. Then a vertical cylinder over a free biminimal curve $\mathbb{E}^2$ is a free biminimal surface; e.g., one can consider the cylinder on the logarithmic spiral.

Example 7.9. Let $\pi : S^3(1) \to S^2(1/2)$ denote the Hopf fibration defined by $\pi(z, w) = (2z\bar{w}, |z|^2 - |w|^2)$, where we have identified

$$S^3(1) = \{(z, w) \in \mathbb{C}^2 : |z|^2 + |w|^2 = 1\},$$
$$S^2(\tfrac{1}{2}) = \{(z, t) \in \mathbb{C} \times \mathbb{R} : |z|^2 + t^2 = \tfrac{1}{4}\}.$$

It is direct to verify that a Hopf cylinder $\pi^{-1}(\gamma)$ is λ-biminimal in $S^3(1)$ if and only if γ is $(\lambda + 4)$-biminimal in $S^2(1/2)$.

Example 7.10. Let $\pi : H^3(-1) \to H^2(-1)$ be defined by

$$\pi(x, y, z) = \left(x, 0, \sqrt{y^2 + z^2}\right).$$

For a unit speed curve γ in $H^2(-1)$, $\pi^{-1}(\gamma)$ is of Gauss curvature -1 and it is λ-biminimal in $H^3(-1)$ with $\lambda = -2$.

By applying the same argument as the proof of Theorem 7.14, S. Maeta obtains the following non-existence result.

Theorem 7.63. [Maeta (2012b)] *Every complete λ-biminimal submanifold with $\lambda \geq 0$ in a Euclidean space is minimal.*

We need the following.

Lemma 7.4. [Maeta (2014c)] *Let M be a λ-biminimal submanifold with $\lambda \geq 0$ in a Riemannian manifold $\tilde{M}$ of non-positive sectional curvature. Then we have*

$$\Delta H^2 \leq -2n\,H^4, \quad n = \dim M. \tag{7.149}$$

Proof. Let $e_1, \ldots, e_n$ be a local orthonormal frame on M. Denote by φ the immersion of M in $\tilde{M}$. Then (7.144) implies, at each $x \in M$, that

$$-\Delta H^2 = 2 \sum_{i=1}^{n} \langle D_{e_i} H, D_{e_i} H \rangle - 2 \langle \Delta^D H, H \rangle$$

$$= 2 \sum_{i=1}^{n} \langle D_{e_i} H, D_{e_i} H \rangle + 2 \sum_{i=1}^{n} \langle h(A_H e_i, e_i), H \rangle$$

$$\qquad - 2 \sum_{i=1}^{n} \left\langle \tilde{R}(H, d\varphi(e_i)) d\varphi(e_i), H \right\rangle + 2\lambda \langle H, H \rangle \tag{7.150}$$

$$\geq 2 \sum_{i=1}^{n} \langle A_H e_i, A_H e_i \rangle$$

$$\geq 2n H^4,$$

where the first inequality follows from $\tilde{K} \leq 0$ and $\lambda \geq 0$. The last inequality follows from $\|A_{e_{n+1}}\|^2 \geq n H^4$. $\qquad\square$

Proposition 7.5. [Maeta (2014c)] *Let M be a compact λ-biminimal submanifold with $\lambda \geq 0$ in a Riemannian manifold with non-positive sectional curvature. Then M is minimal.*

Proof. Applying the standard maximum principle to the elliptic inequality (7.149) gives $H = 0$. □

Proposition 7.6. [Maeta (2014c)] *Let M be a λ-biminimal submanifold with $\lambda \geq 0$ in a Riemannian manifold with non-positive sectional curvature. If the mean curvature is constant, then M is minimal.*

Proof. Since M has constant mean curvature, $\Delta H^2 = 0$. Thus (7.149) implies $H = 0$. □

Proposition 7.7. [Maeta (2014c)] *Let M be a complete λ-biminimal submanifold with $\lambda \geq 0$ in a Riemannian manifold with non-positive sectional curvature. If the Ricci curvature of M is bounded from below, then M is minimal.*

Proof. Follow immediately from Theorem 2.4 and Lemma 7.4. □

For λ-biminimal hypersurfaces, S. Maeta proved the following.

Lemma 7.5. [Maeta (2014c)] *Let M be a λ-biminimal hypersurface with $\lambda \geq 0$ in a Riemannian manifold with non-positive Ricci curvature. Then we have*

$$\Delta H^2 \leq -2nH^4, \quad n = \dim M. \tag{7.151}$$

Proof. Let ξ denote the unit normal vector field in the direction of the mean curvature vector. Denote by $\tilde{R}$ and $\widetilde{Ric}$ the curvature tensor and the Ricci tensor of $\tilde{M}$, respectively. Since

$$\sum_{i=1}^{n} \langle \tilde{R}(H, d\varphi(e_i))d\varphi(e_i), H \rangle = H^2 \widetilde{Ric}(\xi, \xi) \leq 0,$$

the same argument as in Lemma 7.4 shows this lemma. □

Proposition 7.8. [Maeta (2014c)] *Let M be a compact λ-biminimal hypersurface with $\lambda \geq 0$ in a Riemannian manifold with non-positive Ricci curvature. Then M is a minimal hypersurface.*

Proof. Applying the standard maximum principle to the elliptic inequality (7.151) gives $H = 0$. □

Proposition 7.9. [Maeta (2014c)] *Let M be a complete λ-biminimal hypersurface with $\lambda \geq 0$ in a Riemannian manifold with non-positive Ricci curvature. If the Ricci curvature of M is bounded from below, then M is a minimal hypersurface.*

Proof. Follows from Theorem 2.4 and Lemma 7.5.   □

Proposition 7.10. [Maeta (2014c)] *Let M be a λ-biminimal hypersurface with $\lambda \geq 0$ in a Riemannian manifold with non-positive Ricci curvature. If the mean curvature is constant, then M is minimal.*

Proof. Since M has constant mean curvature, we have $\Delta H^2 = 0$. Hence (7.151) implies $H = 0$.   □

S. Maeta also proved the following result in [Maeta (2014c)].

Theorem 7.64. *Let $\tilde{M}$ be a complete Riemannian manifold with non-positive sectional curvature. If the sectional curvature of $\tilde{M}$ has a polynomial growth bound of order less than 2 from below, then any λ-biminimal properly immersed hypersurface with $\lambda \geq 0$ in $\tilde{M}$ is minimal.*

The next two results were due to [Luo (2014)].

Theorem 7.65. *Let $\phi : M \to \tilde{M}$ be a complete λ-biminimal submanifold (resp. hypersurface) with $\lambda \geq 0$ in a Riemannian manifold $\tilde{M}$ with non-positive sectional curvature (resp. Ricci curvature). Then it is minimal.*

Theorem 7.66. *Let M be an n-dimensional complete λ-biminimal submanifold with $\lambda \geq 0$ in a real space form $R^m(c)$ with constant sectional curvature $c \leq 0$. If $\lambda > nc$, then it is minimal.*

Remark 7.11. Example 7.8 and Example 7.10 show that Theorem 7.65 and Theorem 7.66 are sharp, respectively.

For λ-biminimal hypersurfaces in $\mathbb{E}^{n+1}$, (7.147) reduces to

$$\Delta\alpha + \alpha(||h||^2 + \lambda) = 0. \tag{7.152}$$

The next two corollaries follows easily from (7.152).

Corollary 7.17. [Maeta (2012b)] *Let $\phi : M \to \mathbb{E}^{n+1}$ be a codimension one isometric immersion with harmonic mean curvature. If M is biminimal, then it is minimal.*

Corollary 7.18. [Maeta (2012b)] *Let* $\phi : M \to \mathbb{E}^{n+1}$ *be a codimension one isometric immersion with harmonic mean curvature. Then M is non-trivial biminimal if and only if* $||h||^2 = -\lambda$ *for some real number* $\lambda < 0$.

By applying Corollary 7.18, S. Maeta provides the following non-trivial biminimal example with $\lambda < 0$.

Example 7.11. For each $\lambda < 0$, the inclusion of $S^n\left(\sqrt{n}/\sqrt{-\lambda}\right)$ in $\mathbb{E}^{n+1}$ is λ-biminimal.

Theorem 7.63 was later extended to the following.

Theorem 7.67. [Maeta (2014c)] *Let* $(\tilde{M}, \tilde{g})$ *be a complete Riemannian manifold with non-positive sectional curvature. If the sectional curvature $\tilde{K}$ has a polynomial growth bound of order less than 2 from below, then every λ-biminimal properly immersed submanifolds with $\lambda \geq 0$ into $(\tilde{M}, \tilde{g})$ is minimal.*

Several rigidity theorems for biminimal hypersurfaces in $S^{n+1}(1)$ were established as follows.

Theorem 7.68. [Cao (2011)] *The only totally umbilical biminimal hypersurfaces in $S^{n+1}(1)$ are either $S^n(1)$ or $S^n(1/\sqrt{2})$.*

Theorem 7.69. [Cao (2011)] *Let M be a closed biminimal hypersurface in $S^{n+1}(1)$. If the squared mean curvature of M satisfies $H^2 \geq 1$, then $M = S^n(1/\sqrt{2})$.*

Theorem 7.70. [Cao (2011)] *Let M be a closed biminimal hypersurface in $S^{n+1}(1)$. If the squared mean curvature H^2 and the squared norm $||h||^2$ of the second fundamental form of M satisfies*

$$n \leq ||h||^2 \leq n + \frac{n^3 H^2}{2(n-1)} - \frac{n(n-2)}{2(n-1)}\sqrt{n^2 H^4 + 4(n-1)H^2},$$

then one of the following three cases occurs:

(a) *M is a Clifford minimal hypersurface $S^k\left(\sqrt{k}/\sqrt{n}\right) \times S^{b-k}\left(\sqrt{n-k}/\sqrt{n}\right)$ for some $k = 1, \ldots, n-1$.*
(b) *$H^2 = 1$ and $M = S^n(1/\sqrt{2})$.*
(c) *$H^2 = \left(\frac{n-2}{n}\right)^2$ and $S^{n-1}(1/\sqrt{2}) \times S^1(1/\sqrt{2})$.*

Remark 7.12. T. Sasahara classified several special families of biminimal immersions, such as biminimal Legendrian surfaces in 5-dimensional

Sasakian space forms, biminimal Lagrangian surfaces in complex space forms under the condition that the integral curves of the Maslov vector field are geodesics, as well as biminimal Lagrangian surfaces of non-zero constant mean curvature in complex 2-dimensional complex space forms (cf. [Sasahara (2005b, 2009a,b, 2010a)]).

For instance, T. Sasahara proved the following.

Theorem 7.71. [Sasahara (2009b, 2010a)] *Let* $\phi : M \to \tilde{M}^2(4c)$ *be a biminimal Lagrangian immersion into a 2-dimensional complex space form of constant holomorphic curvature* $4c$, *where* $c \in \{-1, 0, 1\}$. *Then the mean curvature of* M^2 *is nonzero constant if and only if* $c = 1$ *and* $\phi(M)$ *is congruent to*

$$\pi \left(\sqrt{\frac{\mu^2}{\mu^2+1}} e^{-\frac{i}{\mu}x}, \sqrt{\frac{1}{\mu^2+1}} e^{i\mu x} \cos y, \sqrt{\frac{1}{\mu^2+1}} e^{i\mu x} \sin y \right) \quad (7.153)$$

in $CP^2(4)$, *where* $\mu = \sqrt{7 \pm \sqrt{41}}/2$ *and* $\pi : S^5(1) \to CP^2(4)$ *is the Hopf fibration.*

It is well-known that there do not exist totally umbilical Lagrangian submanifolds in a complex space form $\tilde{M}^n(4c)$ with $n \geq 2$ other than the totally geodesic ones ([Chen and Ogiue (1974b)]). For this reason the author introduced in [Chen (1997a,b)] the notion of *Lagrangian H-umbilical submanifolds* as the simplest Lagrangian submanifolds next to totally geodesic ones in complex space forms.

Definition 7.9. A Lagrangian H-umbilical submanifold of a Kähler manifold is a Lagrangian submanifold whose second fundamental form takes the following form:

$$\begin{aligned}
h(e_1, e_1) &= \zeta J e_1, \\
h(e_2, e_2) &= \cdots = h(e_n, e_n) = \mu J e_1, \\
h(e_1, e_j) &= \mu J e_j, \\
h(e_j, e_k) &= 0, \quad j \neq k, \quad j, k = 2, \ldots, n,
\end{aligned} \quad (7.154)$$

for suitable functions ζ and μ with respect to some suitable orthonormal frame field $e_1, \ldots, e_n$.

All the examples of non-minimal biminimal Lagrangian surfaces obtained so far are Lagrangian H-umbilical surfaces. Since it is important to study the case of higher dimension, thus it is natural and interesting to investigate non-minimal biminimal Lagrangian H-umbilical submanifolds in complex space forms $\tilde{M}^n(4c)$ with $n \geq 3$.

In this respect, T. Sasahara proved the following.

Theorem 7.72. [Sasahara (2012a)] *Let $\phi : M \to \tilde{M}^n(4c)$ be a biminimal Lagrangian H-umbilical immersion into a complex space form of constant sectional curvature $4c$, where $n \geq 3$ and $c \in \{-1, 0, 1\}$. Then the mean curvature of M is nonzero constant if and only if $c = 1$ and $\phi(M)$ is congruent to*

$$\pi\left(\sqrt{\tfrac{\mu^2}{\mu^2+1}}e^{-\frac{i}{\mu}x}, \sqrt{\tfrac{1}{\mu^2+1}}e^{i\mu x}y_1, \ldots, \sqrt{\tfrac{1}{\mu^2+1}}e^{i\mu x}y_n\right) \subset CP^n(4), \quad (7.155)$$

where $\mu = \sqrt{(n+5 \pm \sqrt{n^2+6n+25})/2n}$ and $y_1^2 + \cdots + y_n^2 = 1$.

Remark 7.13. J.-I. Inoguchi studied biminimal Legendre curves in a 3-dimensional Sasakian space form in [Inoguchi (2007)]. For further results on biminimal curves, see [Arslan et. al. (2009); Maeta (2012b); Turhan and Körpinar (2010, 2011)].

Remark 7.14. There exist many studies on biminimal submanifolds in various model spaces other than real space forms.

(1) For biminimal surfaces in 3-dimensional Lorentzian space forms, see [Sasahara (2012a)].
(2) For biminimal submanifolds in complex space forms, see [Sasahara (2009b, 2010a, 2014b)].
(3) For biminimal submanifolds in Sasakian space forms, see [Sasahara (2009a, 2014a, 2007b)].
(4) For biminimal surfaces in contact 3-manifolds, see [Inoguchi (2007)].
(5) For biharmonic submanifolds in Thurston's 3-dimensional geometries, see [Loubeau and Montaldo (2005)].

7.11 Biconservative immersions

As described by David Hilbert in [Hilbert (1924)], the *stress-energy tensor* associated to a variational problem is a symmetric 2-covariant tensor $\hat{S}$ conservative at critical points, i.e., with div $\hat{S} = 0$.

In the context of harmonic maps $\phi : (M, g) \to (\tilde{M}, \tilde{g})$ between two Riemannian manifolds, that by definition are critical points of the energy

$$E(\phi) = \frac{1}{2}\int_M |d\phi|^2 \, dV_g,$$

the stress-energy tensor was studied in detail by Baird and Eells in [Baird and Eells (1981)] and Sanini in [Sanini (1983)].

Indeed, the Euler-Lagrange equation associated to the energy is equivalent to the vanishing of the tension field $\tau_1(\phi) = \operatorname{Tr}\nabla d\phi$ (see [Eells and Sampson (1964)]), and the tensor

$$\hat{S} = \frac{1}{2}|d\phi|^2 g - \phi^*\tilde{g}$$

satisfies $\operatorname{div}\hat{S} = -\langle\tau_1(\phi), d\phi\rangle$. Hence $\operatorname{div}\hat{S} = 0$ when the map is harmonic. In the case of isometric immersions, the condition $\operatorname{div}\hat{S} = 0$ is always satisfied, since $\tau_1(\phi)$ is normal.

The study of the stress-energy tensor for the bienergy was initiated in [Jiang (1987)] and afterwards developed in [Loubeau et al. (2008)]. Its expression is

$$\hat{S}_2(X, Y) = \frac{1}{2}|\tau_1(\phi)|^2\langle X, Y\rangle + \langle d\phi, \nabla\tau_1(\phi)\rangle\langle X, Y\rangle$$
$$-\langle d\phi(X), \nabla_Y\tau_1(\phi)\rangle - \langle d\phi(Y), \nabla_X\tau_1(\phi)\rangle,$$

and it satisfies the condition

$$\operatorname{div}\hat{S}_2 = -\langle\tau_2(\phi), d\phi\rangle, \tag{7.156}$$

thus conforming to the principle of a stress-energy tensor for the bienergy.

If $\phi : (M, g) \to (\tilde{M}, \tilde{g})$ is an isometric immersion then (7.156) becomes

$$\operatorname{div}\hat{S}_2 = -\tau_2(\phi)^T. \tag{7.157}$$

Based on these, R. Caddeo, S. Montaldo, C. Oniciuc and P. Piu defined the notion of *biconservative submanifolds* in [Caddeo et al. (2014)] as follows.

Definition 7.10. A submanifold of a Riemannian manifold is called *biconservative* if it satisfies $\operatorname{div}\hat{S}_2 = 0$.

Remark 7.15. Biconservative hypersurfaces in a Euclidean space are also known as *H-hypersurfaces* in [Hasanis-Vlachos (1995a)].

It follows from (7.157) that isometric immersions with $\operatorname{div}\hat{S}_2 = 0$ correspond to immersions satisfying

$$(\Delta H)^T = 0, \tag{7.158}$$

where H is the mean curvature vector. Consequently, we have the following lemma from Proposition 4.7 and Corollary 7.8 (see also [Caddeo et al. (2014)]).

Lemma 7.6. *An n-dimensional submanifold of a real space form is biconservative if and only if its mean curvature vector H satisfies*

$$\operatorname{Tr} A_{DH} = -\frac{n}{4}\nabla\langle H, H\rangle. \tag{7.159}$$

Corollary 4.8 and Lemma 7.6 imply the following (see also [Caddeo et al. (2014)]).

Corollary 7.19. *Every submanifold of a real space form with parallel mean curvature vector is biconservative. In particular, every CMC hypersurface of real space form is biconservative.*

Due to the last corollary, the main interest on biconservative hypersurfaces in real space forms are the classification of non-CMC biconservative hypersurfaces.

Obviously, biharmonic submanifolds are always biconservative. Hence biconservative submanifolds form a much bigger family of submanifolds including biharmonic submanifolds.

The next theorem classifies biconservative hypersurfaces in $\mathbb{E}^4$.

Theorem 7.73. [Hasanis-Vlachos (1995a)] *Let M be a biconservative hypersurface of the Euclidean 4-space $\mathbb{E}^4$. Then M is congruent to one of the following hypersurfaces:*

(a) *a CMC hypersurface;*
(b) *a CMC rotational hypersurface generated by a unit speed plane curve $(f(s), g(s))$ such that f satisfies $3ff'' = 2(1 - (f')^2)$;*
(c) *a generalized cylinder on a surface of revolution in $\mathbb{E}^3$ with non-constant mean curvature given by*

$$x(u, v, s) = \big(f(s) \cos u, f(s) \sin u, v, g(s)\big),$$

where $(f(s), g(s))$ is a unit speed curve with non-constant curvature satisfying $3ff'' = 1 - (f')^2$;
(d) *a $SO(2) \times SO(2)$-invariant hypersurface in $\mathbb{E}^4$ with non-constant mean curvature given by*

$$\big(f(s) \cos u, f(s) \sin u, g(s) \cos v, g(s) \sin v\big),$$

where $(f(s), g(s))$ is a unit speed curve with non-constant curvature satisfying

$$f'g'' - f''g' = \frac{1}{3}\left(\frac{f'}{g} - \frac{g'}{f}\right).$$

Remark 7.16. It follows from [Hasanis-Vlachos (1995a); Caddeo et al. (2014)] that every biconservative surface in $\mathbb{E}^3$ is either a CMC surface or a surface of revolution.

For biconservative surfaces in $\mathbb{E}^3$, we also have the following.

Theorem 7.74. [Caddeo et al. (2014)] *Let M^2 be a biconservative surface of revolution in $\mathbb{E}^3$ with non constant mean curvature. Then, locally, the surface can be parametrized by $\phi_c(\rho, v) = \big(\rho \cos v, \rho \sin v, u(\rho)\big)$, where*

$$u(\rho) = \frac{3}{2c} \left(\rho^{1/3} \sqrt{c\rho^{2/3} - 1} + \frac{1}{\sqrt{c}} \ln \left[2(c\rho^{1/3} + \sqrt{c^2 \rho^{2/3} - c}) \right] \right),$$

with c a positive constant and $\rho \in (c^{-3/2}, \infty)$. The parametrization ϕ_c consists of a family of biconservative surfaces of revolution any two of which are not locally isometric.

Biconservative $\delta(2)$-ideal hypersurfaces of a Euclidean space were classified in the following.

Theorem 7.75. [Chen and Munteanu (2013)] *A $\delta(2)$-ideal biconservative hypersurface in a Euclidean $(n + 1)$-space $\mathbb{E}^{n+1}$, $n \geq 3$, is either minimal or a spherical hypercylinder.*

We also have the next two results for biconservative hypersurfaces of a Euclidean space.

Theorem 7.76. [Montaldo et al. (2013)] *There exists an infinite family of proper $SO(p+1) \times SO(q+1)$-invariant biconservative hypersurfaces (cones) in $\mathbb{E}^m$ ($m = p + q + 2$).*

Their corresponding profile curves $\gamma(s)$ tend asymptotically to the profile of a minimal cone. If $p+q \leq 17$, at infinity the profile curves γ intersect the profile of the minimal cone at infinitely many points, while, if $p + q \geq 18$, at infinity the profile curves γ do not intersect the profile of the minimal cone.

Theorem 7.77. [Montaldo et al. (2013)] *There exists an infinite family of complete, proper $SO(p + 1)$-invariant biconservative hypersurfaces in a Euclidean m-space $\mathbb{E}^m$ with $m = p + 2$. Their corresponding profile curves are of "catenary" type.*

In these two classes of invariant families given in Theorems 7.76 and 7.77, there exists no proper biharmonic immersion.

For further classification results on biconservative surfaces, see [Caddeo et al. (2014); Fu (2013c, 2014d); Fu and Li (2013); Sasahara (2012c)].

7.12 Iterated Laplacian and polyharmonic submanifolds

A map $\phi : M \to \mathbb{E}^m$ is *k-harmonic* if $\Delta^k \phi = 0$ holds identically for some integer $k \geq 1$. In this section we present results related to k-harmonic maps and k-harmonic submanifolds.

Proposition 7.11. [Chen (2007)] *If map $\phi : M \to \mathbb{E}^m$ of a Riemannian manifold M into $\mathbb{E}^m$ is k-harmonic for some $k \in \{1, 2, \ldots\}$, then either ϕ is harmonic, i.e., $\Delta \phi = 0$, or ϕ is of infinite type.*

Proof. Let $\phi : M \to \mathbb{E}^m$ be a map satisfying $\Delta^k \phi = 0$ for some integer $k \geq 1$. If ϕ is of r-type ($r < \infty$), then ϕ admits a spectral decomposition:
$$\phi = c + \phi_1 + \cdots + \phi_r, \tag{7.160}$$
where c is a constant vector and $\phi_1, \ldots, \phi_r$ are non-constant maps satisfying $\Delta \phi_i = \lambda_{t_i} \phi_i$ for some distinct eigenvalues $\lambda_{t_1} < \cdots < \lambda_{t_r}$. After applying Δ^{k+j} to (7.160) we obtain
$$0 = \lambda_{t_1}^{k+j} \phi_1 + \cdots + \lambda_{t_r}^{k+j} \phi_r \tag{7.161}$$
for $j = 0, 1, 2, \ldots$, which is impossible unless $r = 1$, $\lambda_{t_1} = 0$. Thus either ϕ is harmonic or ϕ is of infinite type. $\qquad\square$

An immediate consequence of Proposition 7.11 is the following.

Corollary 7.20. *Every non-harmonic map $\phi : M \to \mathbb{E}^m$ of a Riemannian manifold into $\mathbb{E}^m$ satisfying $\Delta^k \phi = 0$ for some $k \geq 2$ is of infinite type.*

Another immediate consequence of Proposition 7.11 is the following.

Corollary 7.21. *Every map $\phi : M \to \mathbb{E}^m$ from a Riemannian manifold into $\mathbb{E}^m$ with nonzero constant tension field is of infinite type.*

The following are examples satisfying $\Delta^k \phi = 0$ but $\Delta^{k-1} \phi \neq 0$.

Example 7.12. [Chen (2007)] For an integer $k \geq 2$, the map $\phi_k : \mathbb{E}^3 \to \mathbb{E}^3$:
$$\phi_k(x, y, z) = (x^2 + y^2 + z^2)^{k-1} (x, y, z) \tag{7.162}$$
satisfies $\Delta^k \phi = 0$ and $\Delta^{k-1} \phi \neq 0$. Hence ϕ_k is of infinite type according to Corollary 7.20.

Example 7.13. [Chen (2007)] Consider the *inversion* $\psi : \mathbb{E}^4 \to \mathbb{E}^4$ about $S^3(1)$ centered at the origin given by
$$\psi(x_1, \ldots, x_4) = \left(\frac{x_1}{x_1^2 + \cdots + x_4^2}, \ldots, \frac{x_4}{x_1^2 + \cdots + x_4^2} \right). \tag{7.163}$$
A straight-forward computation shows that ψ is a non-harmonic map which satisfies $\Delta^2 \psi = 0$. Thus by Corollary 7.20 ψ is a map of infinite type.

Proposition 7.11 extends a result of [Dimitric (1989)]. For further results in this respect, see [Chen (2007); Maeta (2012a); Wei (2008)].

The next corollary is also an immediate consequence of Proposition 7.11.

Corollary 7.22. *The only k-harmonic curves in* $\mathbb{E}^m$ *are open parts of lines.*

Proof. If $\gamma : I \to \mathbb{E}^m$ is a k-harmonic curve, then γ is a line segment or of infinite type according to Proposition 7.11. But the second case is impossible by Theorem 6.2. $\qquad\qquad\qquad\qquad\qquad\qquad\qquad\qquad\qquad\qquad$ $\square$

Remark 7.17. Corollary 7.22 was rediscovered in [Maeta (2012a)]

Obviously, there do not exist closed k-harmonic submanifolds in any Euclidean space.

Based on Corollary 7.22 S. Maeta made another generalized biharmonic conjecture as follows.

Conjecture 7.5. *The only k-harmonic submanifolds of a Euclidean space are the minimal ones.*

Theorem 7.78. [Nakauchi and Urakawa (2013c)] *If* $\phi : M \to \mathbb{E}^m$ *is a k-harmonic isometric immersion from a complete domain into* $\mathbb{E}^m$ *with finite j-energy for all* $j = 1, \ldots, 2k - 2$, *then it is minimal.*

An isometric immersion is called *triharmonic* if it is 3-harmonic. For triharmonic submanifolds we have the following (see also [Maeta (2014b)]).

Theorem 7.79. [Maeta (2012a)] *Let* $\phi : M \to \tilde{M}$ *be an isometric immersion of a closed Riemannian manifold* M *into a Riemannian manifold* $\tilde{M}$ *with non-positive constant sectional curvature. Then triharmonic is harmonic.*

Both Theorem 7.78 and Theorem 7.79 support Conjecture 7.5.

Remark 7.18. Triharmonic submanifolds in a hyperbolic space have been investigated in [Maeta et al. (2013)]. In particular, Maeta, Nakauchi and Urakawa proved that when the domain is complete and both the 4-energy of the immersion ϕ and the L^4-norm of the tension field $\tau_1(\phi)$ are finite, then such a triharmonic immersion is always minimal.

Remark 7.19. Every harmonic map is a k-harmonic map for $k = 1, 2, \ldots$. On the other hand, a k-harmonic map need not be an s-harmonic for $s > k$ (cf. [Maeta (2012a)]).

Chapter 8

λ-biharmonic and Null 2-type Submanifolds

8.1 (k, ℓ, λ)-harmonic maps and submanifolds

We extend k-harmonic maps to (k, ℓ, λ)-harmonic maps as follows.

Definition 8.1. For two given integers k, ℓ with $k > \ell \geq 1$ and a real number λ, a map $\phi : (M, g) \to (\tilde{M}, \tilde{g})$ from a Riemannian manifold into another is called (k, ℓ, λ)-*harmonic* if it is a critical point of the (k, ℓ, λ)-*energy functional* $E_{k\ell\lambda}(\phi)$ defined by

$$E_{k\ell\lambda}(\phi) = E_k(\phi) + \lambda E_\ell(\phi) \tag{8.1}$$

for variations $\phi_t : (-\epsilon, \epsilon) \times M \to \tilde{M}$ satisfying $\phi_0 = \phi$, where $E_1(\phi) = E(\phi)$. If we restrict the variations ϕ_t of ϕ only to the normal variations, then the critical points of the (k, ℓ, λ)-energy functional are called (k, ℓ, λ)-*minimal*.

The $(2, 1, \lambda)$-harmonic maps are called λ-*biharmonic maps* and $(2, 1, \lambda)$-minimal maps are called λ-*biminimal maps*.

It follows from (7.102) and Definition 8.1 that a map $\phi : (M, g) \to \mathbb{E}^m$ is (k, ℓ, λ)-harmonic if and only if it satisfies

$$(-1)^{k-1}\Delta^k \phi + (-1)^{\ell-1}\lambda \Delta^\ell \phi = 0. \tag{8.2}$$

Consequently, we have the following.

Lemma 8.1. *Let $\phi : (M, g) \to \mathbb{E}^m$ be a map. Then ϕ is (k, ℓ, λ)-harmonic if and only if it satisfies*

$$\Delta^k \phi = (-1)^{k-\ell-1}\lambda \Delta^\ell \phi. \tag{8.3}$$

In view of Lemma 8.1, we give the next two definitions.

Definition 8.2. A map $\phi : (M, g) \to \mathbb{E}_s^m$ from a Riemannian manifold (M, g) into $\mathbb{E}_s^m$ is called (k, ℓ, λ)-*harmonic* if ϕ satisfies (8.3) for integers $k > \ell \geq 1$ and real number λ.

Definition 8.3. An isometric immersion $\phi : (M, g) \to \mathbb{E}_s^m$ of a pseudo-Riemannian manifold (M, g) into $\mathbb{E}_s^m$ is called (k, ℓ, λ)-*harmonic* if the immersion ϕ satisfies

$$\Delta^k \phi = (-1)^{k-\ell-1} \lambda \Delta^\ell \phi \tag{8.4}$$

for integers $k > \ell \geq 1$ and real number λ. In particular, $(2, 1, \lambda)$-harmonic submanifolds of $\mathbb{E}_s^m$ are simply called λ-*biharmonic submanifolds*.

Obviously, 0-biharmonic submanifolds are nothing but biharmonic submanifolds, (k, ℓ, λ)-harmonic immersions are automatically (k, ℓ, λ)-minimal, and $(2, 1, \lambda)$-minimal immersions are exactly λ-biminimal immersions.

Example 8.1. If $\phi : (M, g) \to \mathbb{E}^m$ is a 1-type map of order $\{p\}$, then after applying a suitable translation we have $\Delta \phi = \lambda_p \phi = -\tau_1(\phi)$. Thus

$$\Delta^t \phi = -\lambda_p^{t-1} \tau_1(\phi)$$

for any integer $t \geq 1$. Therefore, by Lemma 8.1, ϕ is (k, ℓ, λ)-harmonic for any integers k, ℓ satisfying $k > \ell \geq 1$ and constant λ. Hence every 1-type $\phi : (M, g) \to \mathbb{E}^m$ is (k, ℓ, λ)-harmonic for $k > \ell \geq 1$.

Similarly, by a similar argument, we also know that every null 2-type map into $\mathbb{E}^m$ is (k, ℓ, λ)-harmonic for any $k > \ell \geq 1$.

Example 8.2. For a positive integer p, consider the isometric immersion $\phi : (S^1(1) \times \mathbb{R}, g_0) \to \mathbb{E}^3$ from $(S^1(1) \times \mathbb{R}, g_0 = ds^2 + dt^2)$ into $\mathbb{E}^3$ given by

$$\phi(s, t) = \left(\frac{\cos\left(\sqrt[2p]{2}\, s\right)}{\sqrt[2p]{2}}, \frac{\sin\left(\sqrt[2p]{2}\, s\right)}{\sqrt[2p]{2}}, t \right). \tag{8.5}$$

It is direct to verify that the mean curvature vector H of ϕ is

$$H = -\frac{\sqrt[2p]{2}}{2} \left(\cos\left(\sqrt[2p]{2}\, s\right), \sin\left(\sqrt[2p]{2}\, s\right), 0 \right)$$

and it satisfies $\Delta^t H = 2^{t/p} H$, $t = 1, 2, \ldots$. Hence ϕ is (k, ℓ, λ)-harmonic with $\lambda = 2^{(k-\ell)/p}$ for $k > \ell \geq 1$. Because $\sqrt[2p]{2}$ is an irrational number, ϕ is of infinite type.

Example 8.3. Consider the parabola $\phi : (\mathbb{R}, g = ds^2) \to \mathbb{E}^2$ defined by $\phi(s) = (s, s^2)$. The tension field of ϕ is $\tau_1(\phi) = (0, 2)$ which is a nonzero constant vector. Consequently, ϕ is (k, ℓ, λ)-harmonic for any $k > \ell > 1$ and any $\lambda \in \mathbb{R}$.

Similarly, the map $\psi : (\mathbb{E}^2, g = ds^2 + dt^2) \to \mathbb{E}^3$ defined by

$$\psi(s, t) = (s, s^2, t)$$

has constant tension field given by $(0, 2, 0)$. Thus ψ is a (k, ℓ, λ)-harmonic map for any $k > \ell > 1$ and any constant λ.

We have the following general result for (k, ℓ, λ)-harmonic maps.

Proposition 8.1. *Let $\phi : (M, g) \to \mathbb{E}_s^m$ be a map of a Riemannian manifold into $\mathbb{E}_s^m$. If ϕ is (k, ℓ, λ)-harmonic, then it is either of 1-type, of null 2-type, or of infinite type.*

Proof. Let $\phi : M \to \mathbb{E}_s^m$ be a (k, ℓ, λ)-harmonic. Then it satisfies (8.3) for some integers $k > \ell \geq 1$ and some real number λ. If ϕ is of finite type, then ϕ admits a spectral decomposition:

$$\phi = c + \phi_1 + \cdots + \phi_r, \tag{8.6}$$

where c is a constant vector and $\phi_1, \ldots, \phi_r$ are non-constant maps satisfying

$$\Delta \phi_i = \lambda_{t_i} \phi_i, \quad i = 1, \ldots, r, \tag{8.7}$$

for some distinct eigenvalues $\lambda_{t_1} < \cdots < \lambda_{t_r}$.

It follows from (8.3), (8.6), (8.7), and Beltrami's formula that

$$0 = \lambda_{t_1}^\ell (\lambda_{t_1}^{k-\ell} - (-1)^{k-\ell} \lambda) \phi_1 + \cdots + \lambda_{t_r}^\ell (\lambda_{t_r}^{k-\ell} - (-1)^{k-\ell} \lambda) \phi_r. \tag{8.8}$$

After applying Δ^j to (8.8), we obtain

$$0 = \lambda_{t_1}^{\ell+j} (\lambda_{t_1}^{k-\ell} - (-1)^{k-\ell} \lambda) \phi_1 + \cdots + \lambda_{t_r}^{\ell+j} (\lambda_{t_r}^{k-\ell} - (-1)^{k-\ell} \lambda) \phi_r$$

for $j = 0, 1, 2, \ldots$, which is impossible unless

$$\lambda_{t_1}^\ell (\lambda_{t_1}^{k-\ell} - (-1)^{k-\ell} \lambda) = \cdots = \lambda_{t_r}^{\ell+j} (\lambda_{t_r}^{k-\ell} - (-1)^{k-\ell} \lambda) = 0$$

for all $j = 0, 1, 2, \ldots$. But this can happen only when either (a) $r = 1$ or (b) $r = 2$ and one of $\lambda_{t_1}, \lambda_{t_2}$ is zero. Consequently, M is either of 1-type or of null 2-type. $\qquad \square$

For pseudo-Riemannian submanifolds of $\mathbb{E}_s^m$, we have the following.

Proposition 8.2. *Let $\phi : M \to \mathbb{E}_s^m$ be a finite type isometric immersion of a pseudo-Riemannian manifold into $\mathbb{E}_s^m$. If M is (k, ℓ, λ)-harmonic, then it is either of 1-type or of null 2-type.*

Proof. This can be done in the same way as Proposition 8.1. $\qquad \square$

An important consequence of Proposition 8.2 is the following.

Theorem 8.1. *Every (k, ℓ, λ)-harmonic spacelike closed submanifold M of $\mathbb{E}_s^m$ is of 1-type. Moreover, M is a minimal submanifold of a pseudo hypersphere S_s^{m-1} in $\mathbb{E}_s^m$.*

Proof. Let M be a (k, ℓ, λ)-harmonic spacelike closed submanifold of $\mathbb{E}_s^m$. Proposition 8.2 implies that M is of 1-type, null 2-type or infinite type.

If it is of 1-type, it follows from Theorem 6.16 that M is a minimal submanifold of $\mathbb{E}_s^m$, or a minimal submanifold of a pseudo hypersphere, or a minimal submanifold of a pseudo hyperbolic space. Because M is a closed manifold, Corollaries 4.6 and 4.7 imply that M must be a minimal submanifold of a pseudo hypersphere.

Since M is a closed manifold, M cannot be of null 2-type according to Proposition 6.1. Finally, because the mean curvature vector satisfies (8.3) and M is closed, M cannot be of infinite type by Theorem 6.1. □

Corollary 8.1. *Every (k, ℓ, λ)-harmonic closed submanifold of $\mathbb{E}^m$ is a minimal submanifold of a hypersphere.*

Remark 8.1. Example 8.2 shows that Theorem 8.1 and Corollary 8.1 are false for non-closed submanifolds.

Proposition 8.2 also provides the following classification of (k, ℓ, λ)-harmonic curves.

Corollary 8.2. *If γ is a (k, ℓ, λ)-harmonic curve in $\mathbb{E}^m$, then it is an open part of a line, of a circle, or of a right circular helix which is congruent to*

$$\left(as, b\cos s, b\sin s, 0, \ldots, 0 \right), \quad b = \sqrt{1 - a^2}, \ a \in (0, 1).$$

Proof. Follows from Corollary 6.1 and Propositions 6.3 and 8.1. □

Since maps with constant tension are (k, ℓ, λ)-harmonic for $k > \ell > 1$; in particular, it is k-harmonic, it is interesting to study such maps.

Proposition 8.3. [Chen (2007)] *If a non-harmonic map $\phi : M \to \mathbb{E}^m$ satisfies $\Delta^{k-1}\tau_1 = 0$ for some integer $k \geq 2$, then ϕ is of infinite type.*

Proof. If $\phi : M \to \mathbb{E}^m$ satisfies $\Delta^{k-1}\tau_1 = 0$ for an integer $k \geq 2$, then it is $(k, 1, 0)$-biharmonic. Thus Proposition 8.1 implies that ϕ is either of 1-type, or null 2-type or infinite type. If it is of 1-type, we have

$$\phi = c + \phi_1, \tag{8.9}$$

where $c \in \mathbb{E}^m$ and ϕ is a non-constant map satisfying $\Delta\phi_1 = \lambda_p\phi_1$ with eigenvalue λ_p. Applying Δ^k to (8.9) gives $\lambda_p^k\phi_1 = 0$. Consequently, we have $\Delta\phi = 0$, which is a contradiction.

If ϕ is of null 2-type, it admits a spectral decomposition $\phi = \phi_1 + \phi_2$ so that $\Delta\phi_1 = 0$ and $\Delta\phi_2 = \lambda_p\phi_p$ with $\lambda_p \neq 0$. By applying Δ^k to this spectral decomposition we find $\lambda_p^k \phi_2 = 0$, which gives $\lambda_p = 0$. Consequently, this case is also impossible. $\square$

Proposition 8.3 implies immediately the following.

Corollary 8.3. *If a non-harmonic function on a Riemannian manifold is k-harmonic for some $k \geq 2$, then it is of infinite type.*

Another immediate consequence of Proposition 8.3 is the following.

Corollary 8.4. [Chen (2007)] *Every map $\phi : M \to \mathbb{E}^m$ of a Riemannian manifold into $\mathbb{E}^m$ with nonzero constant tension field is of infinite type.*

Example 8.3 shows that there are maps ϕ with nonzero constant tension field. The next result shows that this cannot occur when ϕ is isometric.

Corollary 8.5. *If an isometric immersion $\phi : M \to \mathbb{E}^m$ has constant mean curvature vector (or equivalently, ϕ satisfies $\Delta \phi = B$, $B \in \mathbb{E}^m$), then the immersion is minimal.*

Proof. Since M has constant mean curvature vector H, Weingarten's equation implies that $A_H = 0$. This implies that $H = 0$. $\square$

8.2 Null 2-type hypersurfaces

It follows from Beltrami's formula that minimal submanifolds of $\mathbb{E}_s^m$ are constructed from harmonic functions. According to J. Douglas and T. Rado's solutions to the famous Plateau problem, there are ample examples of minimal surfaces in Euclidean spaces. The study of minimal surfaces has attracted many mathematicians during the last two centuries.

According to Corollary 6.1, null 2-type curves are right circular helices lying fully in $\mathbb{E}^3$. It is easy to see that circular cylinders in $\mathbb{E}^3$ is of null 2-type. The position vector fields of such curves and surfaces admit the following simple spectral decomposition:

$$\phi = \phi_0 + \phi_p, \quad \Delta \phi_0 = 0, \quad \Delta \phi_p = \lambda \phi_p, \tag{8.10}$$

for some non-constant maps ϕ_0 and ϕ_p and nonzero eigenvalue λ.

The following are some examples of null 2-type submanifolds.

Example 8.4. Let $\iota_1 : S^p(1) \to \mathbb{E}^{p+1}$, $0 < p < n$, be the inclusion map of a hypersphere $S^p(1)$ in $\mathbb{E}^{p+1}$ and let $\iota_2 : \mathbb{E}^{n-p} \to \mathbb{R}^{n-p}$ be the identity map. Then $(\iota_1, \iota_2) : S^p(1) \times \mathbb{E}^{n-p} \to \mathbb{E}^{n+1}$ defines a null 2-type hypersurface.

Example 8.5. Let $\iota_1 : H^p(-1) \to \mathbb{E}_1^{p+1}$, $0 < p < n$, be the inclusion map of a hyperbolic hypersurface in $\mathbb{E}_1^{p+1}$ and $\iota_2 : \mathbb{E}^{n-p} \to \mathbb{E}^{n-p}$ be the identity map. Then $(\iota_1, \iota_2) : H^p(-1) \times \mathbb{E}^{n-p} \to \mathbb{E}_1^{n+1}$ defines a spacelike null 2-type hypersurface in the Minkowski space $\mathbb{E}_1^{n+1}$.

Example 8.6. Let $\varphi = \varphi(x_1, \ldots, x_n)$ be a harmonic function defined on an open domain $D \subset \mathbb{E}^n$ and $f = f(x_1, \ldots, x_n)$ an eigenfunction of the Laplacian $\Delta = -\sum_{i=1}^n \partial^2 / \partial x_i^2$ with a nonzero eigenvalue defined on D. Then

$$\phi(x_1, \ldots, x_n) = \left(\varphi + f, x_1, \ldots, x_n, \varphi + f \right)$$

is a marginally trapped null 2-type immersion of D into $\mathbb{E}_1^{n+2}$.

It follows from Beltrami's formula and (8.10) that the mean curvature vector H of every null 2-type submanifold of $\mathbb{E}_s^m$ satisfies

$$\Delta H = \lambda H \tag{8.11}$$

for some nonzero real number λ. Hence null 2-type submanifolds are critical points of the $(2, 1, \lambda)$-energy functional. Thus they are $(2, 1, \lambda)$-harmonic, i.e., λ-biharmonic. Obviously, every null 2-type submanifold of a Euclidean space is also biconservative.

Submanifolds of $\mathbb{E}_s^m$ are said to have *proper mean curvature vector* if their mean curvature vector satisfies condition (8.11).

The author posed the following geometric question in the 1980s (see e.g., [Chen (1988a,b, 1996b)]).

Question 8.1. *Determine submanifolds which are constructed from eigenfunctions of Δ with two eigenvalues 0 and $\lambda \neq 0$. In other words, to classify null 2-type submanifolds.*

The next result provides the first answer to this question.

Theorem 8.2. [Chen (1988a)] *A surface M in $\mathbb{E}^3$ is of null 2-type if and only if it is an open portion of a circular cylinder.*

Proof. Let M be a surface in $\mathbb{E}^3$, ξ a unit local normal vector field, and $\{e_1, e_2\}$ an orthonormal local frame of M. If we put $H = \alpha \xi$ and $A = A_\xi$, then $\operatorname{Tr} A_{DH} = A(\nabla \alpha)$. Also, we have

$$\Delta^D H = (\Delta \alpha)\xi, \quad \sum_{i=1}^2 h(A_H e_i, e_i) = \alpha \|A\|^2 \xi. \tag{8.12}$$

Thus formulas (4.46) and (4.47) give

$$\Delta H = \{\Delta \alpha + \alpha \|A\|^2\}\xi + \nabla \alpha^2 + 2 A(\nabla \alpha). \tag{8.13}$$

If M is of null 2-type, then Theorem 8.2 implies that $\Delta H = \lambda H$ for some constant $\lambda \neq 0$. Comparing this with (8.13) gives

$$\Delta \alpha = \alpha(\lambda - ||A||^2), \quad A(\nabla \alpha^2) = -\alpha \nabla \alpha^2. \tag{8.14}$$

If we put $U = \{x \in M : \nabla \alpha^2 \neq 0 \text{ at } x\}$, then $\nabla \alpha^2$ is an eigenvector of A with eigenvalue $-\alpha$ on U. The other eigenvalue is 3α. Let us choose e_1 in the direction of $\nabla \alpha^2$. Then $e_2 \alpha = 0$. Also, we have

$$h(e_1, e_1) = -\alpha \xi, \quad h(e_1, e_2) = 0, \quad h(e_2, e_2) = 3\alpha \xi. \tag{8.15}$$

It follows from (8.15) and Codazzi's equation that

$$3e_1 \alpha = -4\alpha \omega_1^2(e_2), \quad \omega_1^2(e_1) = 0 \text{ on } U. \tag{8.16}$$

Let ω^1, ω^2 be the dual 1-forms of e_1, e_2. We find from $\omega_1^2(e_1) = 0$ and Cartan's structure equation that $d\omega^1 = 0$. Thus, by Poincaré lemma, we have $\omega^1 = du$ for some local function u. Combining this with $d\alpha = (e_1 \alpha)\omega^1$ yields $d\alpha \wedge du = 0$. Thus α is a function of u. Hence, by (8.16), we get

$$d\alpha = \alpha'(u)du, \quad 4\alpha \omega_1^2 = -3\alpha'(u)\omega^1. \tag{8.17}$$

Exterior differentiating the second equation in (8.17) gives

$$4\alpha \alpha'' - 7(\alpha')^2 + 16\alpha^4 = 0. \tag{8.18}$$

Let $y = (\alpha')^2$. Then we obtain from (8.17) that

$$y = (\alpha')^2 = c\alpha^{\frac{7}{2}} - 16\alpha^4, \quad c \in \mathbb{R}. \tag{8.19}$$

On the other hand, it follows from (8.14) and (8.15) that

$$\alpha \Delta \alpha = (\lambda - 10\alpha^2)\alpha^2. \tag{8.20}$$

By applying (8.16) we get from (8.20) that

$$4\alpha \alpha'' - 3(\alpha')^2 + 4(\lambda - 10\alpha^2)\alpha^2 = 0. \tag{8.21}$$

From (8.18) and (8.21) we find $(\alpha')^2 = 14\alpha^4 - \lambda \alpha^2$. By combining this with (8.18) we conclude that α^2 is constant on U. Hence U is empty. Thus M has constant mean curvature. Now, we derive from (8.14) that $||A||^2$ is the constant λ. Consequently, M has constant mean curvature and Gauss curvature. Since M is of null 2-type, M is an open portion of a circular cylinder. The converse is trivial. $\qquad \square$

The following is a simple characterization of null 2-type hypersurfaces (cf. Theorem 5 of [Chen and Lue (1988)]).

Theorem 8.3. *Let M be a CMC hypersurface of $\mathbb{E}^{n+1}$. Then M has constant scalar curvature if and only if it is of null 2-type unless it is an open portion of a hypersphere.*

Proof. Let M be a CMC hypersurface of $\mathbb{E}^{n+1}$. Then Corollary 4.8 gives

$$\Delta H = ||A_H||^2 \xi, \tag{8.22}$$

where ξ is the unit normal vector in the direction of the mean curvature vector H. If M is of null 2-type, we have $\Delta H = \lambda H$ for some nonzero real number λ. Comparing this with (8.22) gives $||A_H||^2 = \lambda ||H||$. Thus $||A_H||^2$ is a nonzero constant. Consequently, M has constant scalar curvature by Gauss' equation.

Conversely, if M is a CMC hypersurface with constant scalar curvature, then (8.22) yields

$$\Delta H = \lambda H \tag{8.23}$$

where $\lambda = ||A_H||^2/||H||$ is a nonzero constant. If we put $\phi_p = \lambda^{-1}\Delta\phi$ and $\phi_0 = \phi - \phi_p$, then it follows from Beltrami's formula and (8.23) that

$$\Delta\phi_p = -\frac{n}{\lambda}\Delta H = -nH = \Delta\phi = \lambda\phi_p, \quad \Delta\phi_0 = 0.$$

Hence we obtain (8.10). If ϕ_0 is a constant vector, then M is of 1-type. Since M is non-minimal, it must be spherical by Corollary 6.3. Therefore M is of null 2-type, unless it is an open portion of a hypersphere.   □

Null 2-type $\delta(2)$-ideal hypersurfaces were determined as follows.

Theorem 8.4. [Chen and Garay (2012)] *A null 2-type hypersurface of $\mathbb{E}^{n+1}$ is an open portion of a spherical hypercylinder $S^{n-1} \times \mathbb{R}$ if and only if it is $\delta(2)$-ideal.*

It is important to determine when a null 2-type hypersurface has constant mean curvature. In this respect, the following results are known.

For null 2-type hypersurfaces we have the following.

Theorem 8.5. [Ferrández and Lucas (1991)] *Every null 2-type Dupin hypersurface in $\mathbb{E}^{n+1}$ has constant mean curvature.*

Theorem 8.6. [Defever (1997)] *Every null 2-type hypersurface in $\mathbb{E}^4$ has constant mean curvature.*

Very recently, It was proved in [Fu (2014e); Liu and Yang (2014)] that every null 2-type hypersurface with at most three principal curvatures in a Euclidean space has constant mean curvature.

For null 2-type hypersurfaces in Minkowski spaces we have the following.

Theorem 8.7. [Chen (1992)] *Every spacelike 2-type hypersurface of $\mathbb{E}_1^{n+1}$ is of null 2-type if it has constant mean curvature.*

Theorem 8.8. [Chen (1992); Ferrández and Lucas (1992a)] *Let M be a null 2-type spacelike hypersurface of $\mathbb{E}_1^{n+1}$. If M has at most two distinct principal curvatures, then it has constant mean curvature and constant scalar curvature. Locally, such a hypersurface is the product of a linear subspace $\mathbb{E}^k$ and a hyperbolic space H^{n-k} with $0 < k < n$.*

For null 2-type pseudo-Riemannian hypersurfaces M_r^n of index r in a pseudo-Euclidean $(n + 1)$-space $\mathbb{E}_s^{n+1}$ of index s, we have the following.

Theorem 8.9. [Ferrández and Lucas (1992b)] *A Lorentzian surface in $\mathbb{E}_1^3$ is of null 2-type if and only if it is an open piece of a Lorentzian cylinder or a B-scroll.*

Theorem 8.10. [Arvanitoyeorgos et al. (2009b)] *Let M_1^3 be a nondegenerate Lorentzian hypersurface of the Minkowski 4-space $\mathbb{E}_1^4$. If M_1^3 is of null 2-type, then it has constant mean curvature.*

Theorem 8.11. [Arvanitoyeorgos and Kaimakamis (2013)] *Let M_2^3 be a nondegenerate Lorentzian null 2-type hypersurface of $\mathbb{E}_2^4$. Then M_2^3 has constant mean curvature.*

Theorem 8.12. [Arvanitoyeorgos et al. (2007a)] *Let M_r^3 $(r = 0, 1, 2, 3)$ be a nondegenerate hypersurface in $\mathbb{E}_s^4$ with diagonalizable shape operator. If M_r^3 is of null 2-type, then it has constant mean curvature.*

Remark 8.2. For the classification of null 2-type surfaces in S_1^3 and in H_1^3, see [Houh (1988a); Alías et. al. (1994)].

The following conjecture was posed in [Chen (1991b, 1996b)].

Conjecture 8.1. *The only complete non-closed hypersurfaces of finite type in a Euclidean space are either minimal or of null 2-type.*

All of the above results support Conjecture 8.1. However, this conjecture remains open.

8.3 Null 2-type submanifolds with parallel mean curvature

For null 2-type submanifolds with parallel mean curvature, we have

Theorem 8.13. [Chen and Lue (1988)] *Let M be a non-spherical submanifold of $\mathbb{E}^m$ with parallel mean curvature vector. Then M is of null 2-type if and only if it is a Chen submanifold with constant $\|A_H\| \neq 0$.*

Proof. Let $\phi : M \to \mathbb{E}^m$ be a submanifold with parallel mean curvature vector in $\mathbb{E}^m$. Then Corollary 4.8 gives

$$\Delta H = ||A_H||^2 \xi + \mathfrak{a}(H), \tag{8.24}$$

where ξ is the unit in the direction of H. If M is of null 2-type, we have $\Delta H = \lambda H$ for some nonzero real number λ. Comparing this with (8.24) gives $\mathfrak{a}(H) = 0$ and $||A_H||^2 = \lambda ||H||$. Therefore M is a Chen submanifold and $||A_H||^2$ is a nonzero constant.

Conversely, if M is a non-spherical Chen submanifold such that $||A_H||$ is a nonzero constant, then M is non-minimal in $\mathbb{E}^m$. Thus (8.24) yields

$$\Delta H = \lambda H \tag{8.25}$$

where $\lambda = ||A_H||^2/||H||$ is a nonzero constant. If we put $\phi_p = \lambda^{-1}\Delta\phi$ and $\phi_0 = \phi - \phi_p$, then it follows from Beltrami's formula and (8.25) that

$$\Delta\phi_p = -\frac{n}{\lambda}\Delta H = -nH = \Delta\phi = \lambda\phi_p, \quad \Delta\phi_0 = 0.$$

Hence we obtain (8.10). If ϕ_0 is a constant vector, then M is of 1-type. Since M is non-minimal, it is spherical according to Corollary 6.3. This is a contradiction. Therefore M must be of null 2-type. $\square$

The following result was obtained in [Chen (1992)].

Theorem 8.14. *Let M be a 2-type pseudo-Riemannian submanifold of $\mathbb{E}^m_s$. If M has parallel mean curvature vector, then one of the following four cases occurs:*

(a) *M is of null 2-type.*
(b) *M is non-null and, up to translations and dilations, it lies in a pseudo-hypersphere of $\mathbb{E}^m_s$.*
(c) *M is non-null and, up to translations and dilations, it lies in a pseudo-hyperbolic $(m-1)$-space of $\mathbb{E}^m_s$.*
(d) *M is non-null and, up to translations, it lies in the light cone $\mathcal{LC}$ of $\mathbb{E}^m_s$.*

Proof. Let M be a pseudo-Riemannian submanifold of $\mathbb{E}^m_s$ with parallel mean curvature vector. Then Corollary 4.8 gives

$$\Delta H = \sum_{r=n+1}^{m} \epsilon_r \operatorname{Tr}(A_H A_{e_r}) e_r, \tag{8.26}$$

where $e_{n+1}, \ldots, e_m$ is a pseudo-orthonormal frame of the normal bundle of M. If M is of 2-type, then up to translations the position vector ϕ of M admits the following spectral decomposition

$$\phi = \phi_p + \phi_q,$$
$$\Delta\phi_p = \lambda_p \phi_p, \quad \Delta\phi_q = \lambda_q \phi_q, \quad \lambda_p < \lambda_q.$$

Thus we find

$$n\Delta H = n(\lambda_p + \lambda_q)H + \lambda_p\lambda_q\phi. \tag{8.27}$$

It follows from (8.26) and (8.27) that either (i) M is of null 2-type or (ii) M is non-null and ϕ is normal to M at each point on M. The later case implies $\langle\phi,\phi\rangle = c$, which is constant. Consequently, up to translations and dilations, M lies in $S_s^{m-1}(1)$, or in $H_{s-1}^{m-1}(-1)$, or in the light cone $\mathcal{L}C$ according to $c > 0$, $c < 0$ or $c = 0$, respectively. $\qquad\square$

The next three corollaries follow from Theorem 8.14 and Corollary 8.12.

Corollary 8.6. *Every 2-type submanifold of $\mathbb{E}^m$ with parallel mean curvature vector is of null 2-type.*

Corollary 8.7. *Every 2-type nondegenerate hypersurface of $\mathbb{E}_s^{n+1}$ with constant mean curvature is of null 2-type.*

Proof. Due to the fact that pseudo-hyperspheres and pseudo-hyperbolic hypersurfaces are of 1-type and the light-cone $\mathcal{L}C$ are degenerate. $\qquad\square$

Corollary 8.8. [Chen and Lue (1988)] *Every 2-type closed hypersurface of $\mathbb{E}^{n+1}$ has non-constant mean curvature.*

For 2-type surfaces with parallel mean curvature vector, we have the following complete classification.

Theorem 8.15. [Chen and Lue (1988)] *Let M be a surface with parallel mean curvature vector in $\mathbb{E}^m$. Then M is of 2-type if and only if it is an open portion of one of the following two surfaces:*

(a) *a circular cylinder $\mathbb{R} \times S^1(r) \subset \mathbb{E}^3 \subset \mathbb{E}^m$, which is a null 2-type surface;*
(b) *the product of two circles $S^1(a) \times S^1(b) \subset \mathbb{E}^4 \subset \mathbb{E}^m$ with different radii.*

Proof. Let M be a surface with $DH = 0$ in $\mathbb{E}^m$. Then M lies either in a totally geodesic $\mathbb{E}^3$ or in a hypersphere S^3 of a totally geodesic $\mathbb{E}^4$ (cf. e.g., [Chen (1973b), page 106]). Assume that M is of 2-type. If it lies in a totally geodesic $\mathbb{E}^3$, then it is of null 2-type by Corollary 8.6. Thus by Theorem 8.2 we obtain case (a). If it lies in a S^3, we have case (b) according to Theorem 6.38. $\qquad\square$

Recall that a submanifold is called *PNMC* if it has parallel normalized mean curvature vector.

Assume that M is a null 2-type surface in $\mathbb{E}^4$ satisfying $\Delta H = \lambda H$. We choose an orthonormal normal frame e_3, e_4 such that e_3 is in the direction of H so that $H = \alpha e_3$, where α is the mean curvature. It follows from Proposition 4.7 that

$$\Delta\alpha = \alpha\left(\lambda - \|\omega_3^4\|^2 - \|A_3\|^2\right), \tag{8.28}$$

$$\alpha\,\mathrm{Tr}(A_3 A_4) = 2\omega_3^4(\nabla\alpha) + \alpha\,\mathrm{Tr}(\nabla\omega_3^4), \tag{8.29}$$

$$A_3(\nabla\alpha) = -\alpha\nabla\alpha - \alpha\omega_3^4(e_1)A_4(e_1) - \alpha\omega_3^4(e_2)A_4(e_2). \tag{8.30}$$

For simplicity, we put

$$\omega_2^1 = f_1\omega^1 + f_2\omega^2, \quad \omega_4^3 = g_1\omega^1 + g_2\omega^2. \tag{8.31}$$

We choose e_1, e_2 which diagonalize A_3. So we may put

$$A_3 = \begin{pmatrix} \beta & 0 \\ 0 & \gamma \end{pmatrix}, \quad A_4 = \begin{pmatrix} \mu & \delta \\ \delta & -\mu \end{pmatrix}. \tag{8.32}$$

From the definition of $\mathrm{Tr}(\nabla\omega_4^3)$ and (8.31) we have

$$\mathrm{Tr}(\nabla\omega_4^3) = e_1 g_1 + e_2 g_2 + f_1 g_2 - f_2 g_1. \tag{8.33}$$

We find from (8.29), (8.32) and (8.33) that

$$\alpha\mu(\gamma - \beta) = 2g_1 e_1\alpha + 2g_2 e_2\alpha + \alpha(e_1 g_1 + e_2 g_2 + f_1 g_2 - f_2 g_1). \tag{8.34}$$

It follows from (8.30), (8.31) and (8.32) that

$$\begin{aligned}
(3\beta + \gamma)e_1\alpha &= 2\alpha(g_1\mu + g_2\delta), \\
(\beta + 3\gamma)e_2\alpha &= 2\alpha(g_1\delta - g_2\mu).
\end{aligned} \tag{8.35}$$

By exterior differentiating the first equation in (8.31) we have

$$e_1 f_2 - e_2 f_1 = f_1^2 + f_2^2 + \beta\gamma - \mu^2 - \delta^2. \tag{8.36}$$

Similarly, by exterior differentiating the second equation in (8.31) we find

$$e_1 g_2 - e_2 g_1 = (\beta - \gamma)\delta + f_1 g_1 + f_2 g_2. \tag{8.37}$$

From (8.32) we obtain

$$\omega_1^3 = \beta\omega^2, \quad \omega_2^3 = \gamma\omega^2, \quad \omega_1^4 = \mu\omega^1 + \delta\omega^2, \quad \omega_2^4 = \delta\omega^1 - \mu\omega^2. \tag{8.38}$$

Taking exterior differentiation of (8.38) we find

$$e_2\beta = (\gamma - \beta)f_1 + \delta g_1 - \mu g_2, \tag{8.39}$$

$$e_1\gamma = (\gamma - \beta)f_2 + \mu g_1 + \delta g_2, \tag{8.40}$$

$$e_1\delta - e_2\mu = 2\mu f_1 + 2\delta f_2 - \beta g_2, \tag{8.41}$$

$$e_1\mu + e_2\delta = -2\delta f_1 + 2\mu f_2 - \gamma g_1. \tag{8.42}$$

The first result on null 2-type PNMC surfaces is the following.

Theorem 8.16. [Chen (1988b)] *A surface M in $\mathbb{E}^4$ is an open portion of a circular cylinder lying in a hyperplane of $\mathbb{E}^4$ if and only if it is a null 2-type PNMC surface.*

Proof. If M is a null 2-type surface in $\mathbb{E}^4$, then it is not minimal. Assume that M has parallel normalized mean curvature vector e_3. Then $\omega_3^4 = 0$. Thus it follows from (8.29) that $\text{Tr}(A_3 A_4) = 0$, from which we get

$$(\beta - \gamma)\mu = 0. \tag{8.43}$$

Let us put $V = \{x \in M : \beta \neq \gamma \text{ at } x\}$. If $W \neq \emptyset$, then we have $\mu \neq 0$ on V. Further, from (8.30) we find

$$A_3(\nabla \alpha) = -\alpha \nabla \alpha. \tag{8.44}$$

Let $\tilde{V} = \{x \in V : \nabla \alpha^2 \neq 0 \text{ at } x\}$. Then $\nabla \alpha$ is an eigenvector of A_3 with eigenvalue $-\alpha$ on $\tilde{V}$. Thus we may put $\beta = -\alpha$ and $\gamma = 3\alpha$ on $\tilde{V}$. Hence we obtain from (8.37) that

$$0 = (\beta - \gamma)\delta = -4\alpha\delta. \tag{8.45}$$

From this we find $\delta = 0$. By combining this with $\mu = 0$, we have $A_4 = 0$. Thus, by applying $De_3 = 0$, we conclude that each connected component of $\tilde{V}$ lies in a hyperplane of $\mathbb{E}^4$. Consequently, by Theorem 8.2, we have $\nabla \alpha^2 = 0$, which is impossible unless $\tilde{V} = \emptyset$.

If V is not dense in M, then $interior(M \setminus W)$ is nonempty. Let Z be a connected component of $interior(M \setminus W)$. Hence (8.35) implies that $e_1 \alpha^2 = e_2 \alpha^2 = 0$ on Z, from which we obtain $\nabla \alpha^2 = $ on Z. By combining this with the constancy of α on V, we conclude that M has constant mean curvature. Thus M has parallel mean curvature vector. Consequently, by Theorem 8.15, M is an open portion of a circular cylinder.

The converse is trivial. □

Combining Theorems 7.35 and 8.16 gives the following.

Corollary 8.9. *A surface M of class C^ω in $\mathbb{E}^m$ is an open portion of a circular cylinder lying in a totally geodesic $\mathbb{E}^3 \subset \mathbb{E}^m$ if and only if it is a null 2-type PNMC surface.*

Remark 8.3. It was proved in [Li (1994)] that Corollary 8.9 is also true for surfaces of class C^∞ as well.

Theorem 8.17. [Dursun (2005)] *Let M be a 3-dimensional null 2-type PNMC submanifold of $\mathbb{E}^5$. If M has two distinct principal curvatures in the mean curvature vector direction and if it has constant norm of second fundamental form, then locally it is congruent to one of $\mathbb{R} \times S^2 \subset \mathbb{E}^4 \subset \mathbb{E}^5$, $\mathbb{E}^2 \times S^1 \subset \mathbb{E}^4 \subset \mathbb{E}^5$, and $\mathbb{R} \times S^1(a) \times S^1(a)$.*

8.4   Null 2-type submanifolds with constant mean curvature

In view of Theorem 8.3, in order to study null 2-type submanifolds with constant mean curvature it suffices to investigate such submanifolds with codimension ≥ 2. The first result on such submanifolds is the following.

Theorem 8.18. [Chen (1988b)] *A surface with constant mean curvature in $\mathbb{E}^4$ is of null 2-type if and only if it is congruent to an open part of a helical cylinder defined by*

$$(u, a\cos v, a\sin v, bv), \tag{8.46}$$

where a, b are real numbers with $a \neq 0$.

Proof. Let M be a null 2-type surface in $\mathbb{E}^4$. Then we have $\Delta H = \lambda H$ for some nonzero constant λ. Assume that M has constant mean curvature α. Then $\alpha \neq 0$, otherwise it is of 1-type. By (8.35) we find

$$\mu g_1 + \delta g_2 = 0, \quad \delta g_1 - \mu g_2 = 0, \tag{8.47}$$

which implies

$$(\mu^2 + \delta^2)g_1 = (\mu^2 + \delta^2)g_2 = 0. \tag{8.48}$$

Put $W = \{x \in M : \mu^2 + \delta^2 \neq 0 \text{ at } x\}$. If $W \neq \emptyset$, we have $De_3 = 0$, so $g_1 = g_2 = 0$ on W. Therefore, by Theorem 8.16, each connected component of W lies in a hyperplane of $\mathbb{E}^4$. But this is impossible since $A_4 = 0$ in this case. Therefore we obtain $\mu = \delta = 0$ identically on M. Now, it follows from (8.41) and (8.42) that

$$\beta g_2 = \gamma g_1 = 0. \tag{8.49}$$

Since $\beta + \gamma = 2\alpha \neq 0$, at least one of β and γ is nonzero. If both are nonzero, M has parallel normalized mean curvature vector. In this case, Theorem 8.16 implies that M is an open part of a circular cylinder which in turn implies $\beta\gamma = 0$. Thus exactly one of β, γ is nonzero. Without loss of generality, we may assume $\beta = 0$ and γ is a nonzero constant. Hence, by (8.49), we have

$$\beta = g_1 = 0, \quad \gamma = 2\alpha \neq 0. \tag{8.50}$$

By combining (8.39), (8.40) and (8.50), we find $f_1 = f_2 = 0$. Hence, by (8.34) and (8.37), g_2 is constant. Consequently, we have

$$\omega_1^2 = \omega_1^3 = \omega_1^4 = \omega_2^4 = 0, \quad \omega_2^3 = 2\alpha\omega^2, \quad \omega_3^4 = g_2\omega^2,$$

for some constants α and g_2. Therefore M is a flat surface whose connection forms coincide with that of a helical cylinder given by (8.46) for some nonzero constants a, b. Consequently, by applying fundamental theorem of submanifolds, we conclude that M is an open part of a helical cylinder.

The converse is easy to verify. $\square$

The pseudo-Riemannian version of Theorem 8.18 are the following.

Theorem 8.19. [Chen and Song (1989a)] *A spacelike surface with constant mean curvature in $\mathbb{E}_1^4$ is of null 2-type if and only if it is congruent to an open part of one of the following surfaces:*

(1) *a helical cylinder given by*
$$\left(\frac{bv}{c}, a\cos\frac{v}{c}, a\sin\frac{v}{c}, u\right),$$
  where a, b, c are real numbers with $a^2 > b^2$, $c = \sqrt{a^2 - b^2}$;

(2) *a helical cylinder given by*
$$\left(a\cosh\frac{v}{c}, a\sinh\frac{v}{c}, \frac{bv}{c}, u\right),$$
  where a, b, c are real numbers with $a > 0$, $c = \sqrt{a^2 + b^2}$.

Theorem 8.20. [Chen and Song (1989a)] *A spacelike surface M with constant mean curvature in $\mathbb{E}_2^4$ is of null 2-type if and only if it is congruent to an open part of one of the following surfaces:*

(1) *A helical cylinder given by*
$$\left(b\sin\frac{v}{c}, b\cos\frac{v}{c}, \frac{av}{c}, u\right),$$
  where a, b are real numbers with $a^2 > b^2 > 0$ and $c = \sqrt{a^2 - b^2}$;

(2) *a helical cylinder given by*
$$\left(\frac{bv}{c}, a\cosh\frac{v}{c}, a\sinh\frac{v}{c}, u\right),$$
  where a, b are real numbers with $a^2 > b^2 > 0$ and $c = \sqrt{a^2 - b^2}$.

For null 2-type Chen surfaces we have the following.

Theorem 8.21. [Li (1995)] *Let M be a null 2-type Chen surface in $\mathbb{E}^m$ with constant mean curvature. If M is not pseudo-umbilical, then it is a flat surface lying fully in a totally geodesic $\mathbb{E}^r \subset \mathbb{E}^m$ with $r = 3, 4, 5$ or 6.*

Remark 8.4. Circular cylinders in $\mathbb{E}^3$ and helical cylinders in $\mathbb{E}^4$ are null 2-type Chen surfaces. The following are examples of null 2-type Chen surfaces with constant mean curvature lying fully in $\mathbb{E}^5$ and $\mathbb{E}^6$ [Li (1995)].

Example 8.7. Let M be a surface defined by
$$\left(au, b\cos u\cos v, b\cos u\sin v, b\sin u\cos v, b\sin u\sin v\right).$$
Then M is a null 2-type Chen surface with constant mean curvature lying fully in $\mathbb{E}^5$.

Example 8.8. Let M be a surface defined by
$$\left(au, cv, b\cos u\cos v, b\cos u\sin v, b\sin u\cos v, b\sin u\sin v\right).$$
Then M is a null 2-type Chen surface with constant mean curvature lying fully in $\mathbb{E}^6$.

Assume that $\gamma(s)$ is a null curve in $\mathbb{E}^4_1$ and let $\{A, B, C, D\}$ be a pseudo-orthonormal frame of vector fields along $\gamma(s)$ such that
$$\langle A, A\rangle = \langle B, B\rangle = 0, \quad \langle A, B\rangle = -1,$$
$$\langle A, C\rangle = \langle A, D\rangle = \langle B, C\rangle = \langle B, D\rangle = 0, \tag{8.51}$$
$$\langle C, C\rangle = \langle D, D\rangle = 1, \quad \langle C, D\rangle = 0, \quad \gamma'(s) = A(s).$$

Let $X(s)$ denote the matrix $(A(s)\,B(s)\,C(s)\,D(s))$ consisting of column vectors with respect to the standard coordinates of $\mathbb{E}^4_1$. Then $X(s)$ satisfies
$$X^T(s)EX(s) = T, \tag{8.52}$$
where $E = \mathrm{Diag}(-1, 1, 1, 1)$, X^T is the transpose of X, and
$$T = \begin{pmatrix} 0 & -1 & 0 & 0 \\ -1 & 0 & 0 & 0 \\ 0 & 0 & 1 & 0 \\ 0 & 0 & 0 & 1 \end{pmatrix}.$$

Consider the following system of ordinary differential equations:
$$X'(s) = X(s)M(s), \quad M = \begin{pmatrix} \kappa_1 & 0 & -a & 0 \\ 0 & -\kappa_1 & -\kappa_2 & -\kappa_3 \\ -\kappa_2 & -a & 0 & 0 \\ -\kappa_3 & 0 & 0 & 0 \end{pmatrix}. \tag{8.53}$$

For a given initial condition $X(0) = (A(0)\,B(0)\,C(0)\,D(0))$ satisfying
$$X^T(0)EX(0) = T,$$
the system (8.53) with the initial condition has a unique solution. Because T is symmetric and MT is skew-symmetric, we get $(X(s)TX^T(s))' = 0$. Therefore $X(s)TX^T(s) = E$. Therefore $\{A(s), B(s), C(s), D(s)\}$ form a null frame field along γ in $\mathbb{E}^4_1$. The map
$$\phi : (s, t) \mapsto \gamma(s) + tB(s)$$
defines a Lorentzian surface of null 2-type with constant mean curvature a^2 in $\mathbb{E}^4_1$ which is called an *extended B-scroll* over the null curve γ.

For Lorentzian null 2-type CMC surfaces in $\mathbb{E}^4_1$, we have the following.

Theorem 8.22. [Kim and Kim (1996a)] *Let M be a Lorentzian surface in $\mathbb{E}^4_1$. Then M is a CMC surface of null 2-type if and only if, up to rigid motions of $\mathbb{E}^4_1$, M is an open portion of one of the following surfaces:*

(1) $\mathbb{E}_1^1 \times S^1(r) \subset \mathbb{E}_1^3 \subset \mathbb{E}_1^4$;

(2) $\mathbb{E}^1 \times S_1^1(r) \subset \mathbb{E}_1^3 \subset \mathbb{E}_1^4$;

(3) a Lorentzian surface defined by

$$\Big(acu, b\cos cu, b\sin cu, v \Big),$$

where a, b are constants with $a^2 > b^2 > 0$ and $c = 1/\sqrt{a^2 - b^2}$;

(4) a Lorentzian surface defined by

$$\Big(b\sinh cu, b\cosh cu, acu, v \Big),$$

where a, b are constants with $b \neq 0$, $b^2 > a^2$, and $c = 1/\sqrt{b^2 - a^2}$;

(5) an extended B-scroll on a null curve in $\mathbb{E}_1^4$.

Analogous to the last theorem, null 2-type Lorentzian surfaces with constant mean curvature in $\mathbb{E}_2^4$ are classified in [Kim (1998)].

8.5 Marginally trapped null 2-type submanifolds

Recall that a spacelike submanifold in a Lorentzian manifold is called marginally trapped if its mean curvature vector is lightlike at each point.

Null 2-type marginally trapped surfaces in the Minkowski 4-space were classified in the following theorem.

Theorem 8.23. [Chen and Song (1989b)] *A marginally trapped surface in* $\mathbb{E}_1^4$ *is of null 2-type if and only if it is a flat surface congruent to*

$$\Big(f(u, v) + \varphi(u, v), u, v, f(u, v) + \varphi(u, v) \Big)$$

where φ *is a harmonic function and* f *is an eigenfunction of the Laplacian* Δ *with nonzero eigenvalue.*

Proof. Assume that M is a marginally trapped surface in $\mathbb{E}_1^4$. It follows from Proposition 4.7 that

$$\Delta H = \Delta^D H + \sum_{i=1}^{2} h(A_H e_i, e_i) + 2\operatorname{Tr} A_{DH} \qquad (8.54)$$

for any orthonormal frame e_1, e_2 of M. Since H is lightlike, we may choose a pseudo-orthonormal frame e_3, e_4 such that e_3 is spacelike, e_4 is timelike, and

$$H = \delta(e_3 + e_4), \quad \delta \neq 0, \qquad (8.55)$$

for some function δ. By a direct computation we have

$$\sum_{i=1}^{2} h(A_H e_i, e_i) = \delta(||A_3||^2 + \text{Tr}(A_3 A_4))e_3$$
$$- \delta(||A_4||^2 + \text{Tr}(A_3 A_4))e_4, \tag{8.56}$$

and

$$\Delta^D H = \{\Delta\delta - \delta||\omega_3^4||^2 - \delta\,\text{Tr}(\nabla\omega_3^4) - 2\omega_3^4(\nabla\delta)\}(e_3 + e_4), \tag{8.57}$$

where $\text{Tr}(\nabla\omega_3^4) = \sum_i (\nabla\omega_3^4)(e_i)$. Moreover, it follows from (8.55) that

$$\text{Tr}\,A_{DH} = \sum_i \{e_i\delta + \delta\omega_3^4(e_i)\}\{A_3(e_i) + A_4(e_i)\}. \tag{8.58}$$

Since M is of null 2-type, we have

$$\Delta H = \lambda H \tag{8.59}$$

for some constant $\lambda \neq 0$. Thus, by using (8.54)-(8.57) and (8.59), we find

$$\text{Tr}\,A_{DH} = 0,$$
$$\delta\lambda + \delta||\omega_3^4||^2 + \delta\,\text{Tr}(\nabla\omega_3^4) + 2\omega_3^4(\nabla\delta) - \Delta\delta$$
$$= \delta(||A_3||^2 + \text{Tr}(A_3 A_4)), \tag{8.60}$$
$$\delta\lambda + \delta||\omega_3^4||^2 + \delta\,\text{Tr}(\nabla\omega_3^4) + 2\omega_3^4(\nabla\delta) - \Delta\delta$$
$$= -\delta(||A_4||^2 + \text{Tr}(A_3 A_4)).$$

It follows from the last two equations in (8.60) that $\text{Tr}(A_3 + A_4)^2 = 0$. Hence we find from (8.55) that

$$A_H = 0, \quad A_3 = -A_4. \tag{8.61}$$

So, the second fundamental form of M satisfies

$$h(e_1, e_1) = \beta(e_3 + e_4),$$
$$h(e_1, e_2) = \mu(e_3 + e_4), \tag{8.62}$$
$$h(e_2, e_2) = \gamma(e_3 + e_4),$$

for some functions β, γ, μ. By applying the equation of Gauss and (8.62), we see that M is a flat surface. Hence we may choose e_1, e_2 such that

$$\nabla e_1 = \nabla e_2 = \omega_1^2 = 0. \tag{8.63}$$

Also, it follows from (8.61) that M has flat normal connection. Hence we may choose e_3, e_4 in such way that $De_1 = De_2 = \omega_3^4 = 0$. Therefore we obtain from (8.63) that $\tilde{\nabla}(e_3 + e_4) = 0$. Consequently, $e_3 + e_4$ is a constant lightlike vector in $\mathbb{E}^4_1$. Thus, without loss of generality we may put

$$e_3 + e_4 = (1, 0, 0, 1). \tag{8.64}$$

Since M is flat, we can assume that M is covered by an open domain $\Omega \subset \mathbb{E}^2$ and that M is represented by

$$\phi(s, t) = (x_1(s, t), x_2(s, t), x_3(s, t), x_4(s, t)). \tag{8.65}$$

Let us choose e_1, e_2 with $e_1 = \partial\phi/\partial s$ and $e_2 = \partial\phi/\partial t$. It follows from (8.64) and (8.65) and $\langle e_1, e_3 + e_4 \rangle = \langle e_2, e_3 + e_4 \rangle = 0$ that

$$\frac{\partial(x_1 - x_4)}{\partial s} = \frac{\partial(x_1 - x_4)}{\partial t} = 0.$$

Thus, after applying a suitable translation on $\mathbb{E}^4_1$, we obtain

$$\phi(s, t) = (\varphi(s, t), v(s, t), w(s, t), \varphi(s, t)) \tag{8.66}$$

for some functions φ, v, w. Since $e_1 = \partial\phi/\partial s$, $e_2 = \partial\phi/\partial t$ are orthonormal, we have

$$v_s^2 + w_s^2 = v_t^2 + w_t^2 = 1, \quad v_s v_t + w_s w_t = 0. \tag{8.67}$$

Hence

$$v_s v_{ss} + w_s w_{ss} = 0, \quad v_t v_{tt} + w_t w_{tt} = 0, \tag{8.68}$$

$$v_s v_{st} + w_s w_{st} = 0, \quad v_t v_{st} + w_t w_{st} = 0, \tag{8.69}$$

$$v_s v_{st} + v_t v_{ss} + w_s w_{st} + w_t w_{ss} = 0, \tag{8.70}$$

$$v_s v_{tt} + v_t v_{st} + w_s w_{tt} + w_t w_{st} = 0. \tag{8.71}$$

Moreover, since $\Delta\phi = -2H$, (8.55), (8.64) and (8.66) give $\Delta v = \Delta w = 0$. Thus we find

$$v_{ss} = -v_{tt}, \quad w_{ss} = -w_{tt}. \tag{8.72}$$

From (8.68) and (8.72) we obtain

$$(v_s w_t - v_t w_s) v_{tt} = 0. \tag{8.73}$$

If $v_s = 0$, then (8.67) gives $w_s^2 = 1$, $w_t = 0$ and $v_t^2 = 1$. Hence $v_t = \pm 1$. Thus we find

$$v = \pm t + c_1, \quad w = \pm s + c_2$$

for some real numbers c_1, c_2. Therefore, up to rigid motions of $\mathbb{E}^4_1$, the immersion ϕ takes the following form:

$$\phi(s, t) = (\varphi, s, t, \varphi). \tag{8.74}$$

If $v_s \neq 0$, we find from (8.67) that

$$v_t = -\frac{w_s w_t}{v_s}. \tag{8.75}$$

Thus (8.73) and (8.75) imply $w_t v_{tt} = 0$. Since $v_s \neq 0$, we find $w_t \neq 0$ from (8.67). Hence we obtain from $w_t v_{tt} = 0$ that $v_{tt} = 0$. So, we find from (8.68), (8.72) and $v_{tt} = 0$ that

$$v_{ss} = w_{ss} = w_{tt} = 0. \tag{8.76}$$

Now, by (8.70), (8.71) and (8.76) we have

$$v_s v_{st} + w_s w_{st} = 0, \quad v_t v_{st} + w_t w_{st} = 0. \tag{8.77}$$

Hence we get

$$(v_s w_t - v_t w_s)v_{st} = 0. \tag{8.78}$$

If $v_{st} \neq 0$, we get $w_t = v_t w_s / v_s$. By combining this with (8.75), we find $w_t = 0$ and $v_t^2 = 1$. Hence we have $v_{st} = 0$ by virtue of (8.69), which is a contradiction. Therefore we must have $v_{st} = 0$.

Similarly, we also have $w_{st} = 0$ from (8.75) and (8.77). Consequently, both $v = v(s,t)$ and $w = w(s,t)$ are linear in s and t. Consequently, after applying a suitable translation on $\mathbb{E}_1^4$, we have

$$\phi(s,t) = \big(\varphi, (s,t), s, t, \varphi(s,t)\big). \tag{8.79}$$

It follows from $\Delta\phi = -2H$ and $\Delta H = \lambda H$ and (8.79) that $\Delta^2 \varphi = \lambda \Delta \varphi$. Therefore, if we put

$$u = \varphi - \frac{1}{\lambda}\Delta\varphi, \quad f = \frac{1}{\lambda}\Delta\varphi,$$

we obtain $\varphi = f + u, \Delta f = \lambda f$ and $\Delta u = 0$. Consequently, M is a flat surface such that, up to rigid motions, M is given by $(f + \varphi, u, v, f + \varphi)$.

The converse is easy to verify.   □

Theorem 8.23 implies immediately the following.

Corollary 8.10. *There exist no marginally trapped null 2-type surfaces in* $\mathbb{E}_1^4$ *which lies either in the de Sitter spacetime* $S_1^3(1)$ *or in the hyperbolic* $H^3(-1)$.

Recall that a pseudo-Riemannian submanifold M with positive index in a pseudo-Riemannian manifold is called quasi-minimal if M has lightlike mean curvature vector at each point.

Analogous to Theorem 8.23, null 2-type quasi-minimal surfaces in $\mathbb{E}_2^4$ are classified as follows.

Theorem 8.24. [Kim and Kim (1996c)] *Let M be a quasi-minimal surface in $\mathbb{E}_2^4$. Then M is of null 2-type if and only if it is a flat surface congruent to an open portion of*

$$\left(u, f(u,v) + \varphi(u,v), f(u,v) + \varphi(u,v), v \right)$$

with f is an eigenfunction of the Laplacian Δ with nonzero eigenvalue and φ is a harmonic function.

This theorem was proved in a similar way as Theorem 8.23.

8.6   λ-biharmonic submanifolds of $\mathbb{E}_s^m$

The following simple and natural geometric question was proposed by the author in [Chen (1988a,b)].

Question 8.2. *Determine all submanifolds in a Euclidean space (or, more generally, in a pseudo-Euclidean space) which have proper mean curvature vector. In other word, classify λ-biharmonic submanifolds in Euclidean and pseudo-Euclidean spaces.*

The next result provides a primary answer to this question.

Theorem 8.25. [Chen (1988a, 2011b)] *Let M be a pseudo-Riemannian submanifold of $\mathbb{E}_s^m$. Then M is λ-biharmonic if and only if it is one of the following submanifolds:*

(a) *a biharmonic submanifold;*
(b) *a 1-type submanifold;*
(c) *a null 2-type submanifold.*

Proof. Assume that $\phi : M \to \mathbb{E}_s^m$ is an isometric immersion of a pseudo-Riemannian n-manifold into a pseudo-Euclidean m-space $\mathbb{E}_s^m$. If $\Delta H = \lambda H$ with $\lambda = 0$, then M is a biharmonic submanifold.

Next, assume that the condition $\Delta H = \lambda H$ holds for some nonzero real number λ. Let us put

$$\phi_p = \frac{1}{\lambda}\Delta\phi; \text{ and } \phi_0 = \phi - \phi_p.$$

Then we have

$$\Delta\phi_p = -\frac{n}{\lambda}\Delta H = \Delta\phi = \lambda\phi_p,$$

$$\Delta\phi_0 = \Delta\phi - \Delta\phi_p = 0.$$

Hence M is either of 1-type or of null 2-type, depending on ϕ_0 is a constant vector or a non-constant map.

The converse is easy to verify.   □

By combining Theorem 8.10, Theorem 8.11 and the minimality of bi-harmonic surfaces in $\mathbb{E}^3$ from Theorem 7.5, we have the following complete classification of all λ-biharmonic surfaces in $\mathbb{E}^3$.

Theorem 8.26. *Let M be a surface in $\mathbb{E}^3$. Then M is λ-biharmonic if and only if it is either a minimal surface or an open part of a circular cylinder.*

Theorems 7.12, 8.25 and 8.3 imply the following.

Theorem 8.27. [Defever (1997)] *Every λ-biharmonic hypersurface of $\mathbb{E}^4$ has constant mean curvature.*

The next two theorems are pseudo-Riemannian versions of Theorem 8.27.

Theorem 8.28. [Arvanitoyeorgos et al. (2009b)] *Every λ-biharmonic non-degenerate Lorentzian hypersurface of $\mathbb{E}_1^4$ has constant mean curvature.*

Theorem 8.29. [Arvanitoyeorgos et al. (2007a)] *Every λ-biharmonic non-degenerate hypersurfaces M_r^3 ($r = 0, 1, 2, 3$) with diagonal shape operator in $\mathbb{E}_s^4$ has constant mean curvature.*

Theorem 8.30. [Liu and Yang (2014)] *Let M_r^n ($n \geq 4$) be a nondegenerate hypersurface of $\mathbb{E}_s^{n+1}$ satisfying $\Delta H = \lambda H$. If M_r^n has diagonalizable shape operator with at most three distinct principal curvatures, then it has constant mean curvature.*

8.7 λ-biharmonic submanifolds in H^m

The following result classifies all submanifolds of $H^m(-1) \subset \mathbb{E}_1^{m+1}$ which are λ-biharmonic in $\mathbb{E}_1^{m+1}$.

Theorem 8.31. [Chen (1994a)] *Let $\phi : M \to \mathbb{E}_1^{m+1}$ be a λ-biharmonic space-like submanifold of $\mathbb{E}_1^{m+1}$. Then $\phi(M)$ lies in the hyperbolic m-space $H^m(-1) \subset \mathbb{E}_1^{m+1}$ if and only if either*

(a) M is a minimal submanifold of $H^m(-1)$, or

(b) M lies in a totally umbilical hypersurface of $H^m(-1)$ as minimal submanifold.

Proof. Let $\phi : M \to H^m(-1) \subset \mathbb{E}_1^{m+1}$ be an isometric immersion. We choose an orthonormal frame $e_1, \ldots, e_n, e_{n+1}, \ldots, e_{m+1}$ such that $e_1, \ldots, e_n$ are tangent to M and $e_{m+1} = \phi$. Then we have

$$H = H' + \phi, \quad A_\phi = -I, \tag{8.80}$$

where H' is the mean curvature vector of M in $H^m(-1)$. Thus formula (4.46) reduces to

$$\Delta H = \Delta^D H + 2\operatorname{Tr} A_{DH} - nH' + n(\langle H', H' \rangle - 1)\phi$$
$$+ \frac{n}{2} \nabla\langle H, H \rangle + \sum_{r=n+1}^{m} \operatorname{Tr}(A_{H'} A_r)e_r. \tag{8.81}$$

If $\Delta H = \lambda H$ for some constant λ, then (8.80) and (8.81) give

$$\langle H', H' \rangle = 1 + \frac{\lambda}{n}, \quad \langle H, H \rangle = \frac{\lambda}{n}. \tag{8.82}$$

Thus M has constant mean curvature in $H^m(-1)$ and in $\mathbb{E}_1^{m+1}$.

Case (1): $H' = 0$. In this case we have $H = \phi$. Hence Beltrami's formula yields $\Delta H = -nH$. Thus H is an eigenvector of Δ with eigenvalue $-n$.

Case (2): If $H' \neq 0$. Let us choose e_{n+1} such that

$$H' = ae_{n+1}, \quad a = \sqrt{1 + \lambda/n}, \quad \lambda > -n. \tag{8.83}$$

Put $De_r = \sum_{t=n+1}^{m+1} \omega_r^t e_t$. Then $\omega_{n+1}^{m+1} = \cdots = \omega_m^{m+1} = 0$. Hence we have

$$\Delta^D H = a \sum_{r=n+2}^{m} \left\{ \sum_{i=1}^{n} \sum_{t=n+1}^{m} \omega_{n+1}^t(e_i)\omega_r^t(e_i) - \operatorname{Tr}(\nabla \omega_{n+2}^r) \right\} e_r$$
$$+ \|De_{n+1}\|^2 H'. \tag{8.84}$$

Also, $\operatorname{Tr} A_{n+1}^2 = n + \lambda - \|De_{n+1}\|^2$ holds. From this and (8.82) we find

$$n \operatorname{Tr} A_{n+1}^2 - (\operatorname{Tr} A_{n+1})^2 = -n\|De_{n+1}\|^2. \tag{8.85}$$

On the other hand, we also have

$$n \operatorname{Tr} A_{n+1}^2 - (\operatorname{Tr} A_{n+1})^2 = \sum_{i<j}(\kappa_i - \kappa_j)^2, \tag{8.86}$$

where $\kappa_1, \ldots, \kappa_n$ are the eigenvalues of A_{n+1}. Therefore (8.83), (8.84) and (8.86) imply that

$$A_{n+1} = aI, \quad De_{n+1} = DH' = DH = 0, \tag{8.87}$$

which show that M is pseudo-umbilical in $H^m(-1)$ with nonzero parallel mean curvature vector. Consequently, by Proposition 4.18, M lies in a totally umbilical hypersurface of $H^m(-1)$ as a minimal submanifold.

Conversely, if M is a minimal submanifold of a totally umbilical hypersurface of $H^m(-1)$, then M is pseudo-umbilical with nonzero parallel mean curvature vector. Hence (8.87) holds for some nonzero real number a. Thus it follows from (8.84) that $\Delta^D H = 0$. Also, it follows from (8.87) that

$$\sum_{r=n+1}^{m} \text{Tr}(A_{H'} A_r) e_r = n \langle H', H' \rangle e_{n+1}. \tag{8.88}$$

Thus we get from (8.84), (8.87) and (8.88) that $\Delta H = n(a^2 - 1)H$. $\qquad \square$

Definition 8.4. A complete flat totally umbilical hypersurface of a hyperbolic space is called a *horosphere*.

The next result from [Chen (1994a)] relates closely to Theorem 8.31.

Theorem 8.32. *If M is a non-minimal submanifold of $H^m(-1) \subset \mathbb{E}_1^{m+1}$, then the following four statements are equivalent:*

(1) *M is a biharmonic submanifold of $\mathbb{E}_1^{m+1}$.*
(2) *M is a marginally trapped submanifold of $\mathbb{E}_1^{m+1}$.*
(3) *M is a pseudo-umbilical submanifold with unit parallel mean curvature vector in $H^m(-1)$.*
(4) *M is a minimal submanifold of a horosphere of $H^m(-1)$.*

Proof. Assume that M is a non-minimal submanifold of the hyperbolic space $H^{m-1}(-1) \subset \mathbb{E}_1^{m+1}$.

(1) $\Rightarrow$ (2) & (3): If M is a biharmonic submanifold of $\mathbb{E}_1^{n+1}$, then we have $\Delta H = 0$. Hence (8.82) implies that $\langle H, H \rangle = 0$ and $\langle H', H' \rangle = 1$. Since $H = H' + \phi$ is nonzero, H is a lightlike vector field. Moreover, by Theorem 8.31, M is a pseudo-umbilical submanifold with parallel mean curvature vector in $H^m(-1)$. Thus we have statement (3). Moreover, it follows from (8.80), (8.82), (8.87) and $\Delta H = 0$ that $A_H = 0$. Hence we may conclude from Weingarten's formula and (8.87) that H is a constant vector in $\mathbb{E}_1^{m+1}$. This gives statement (2).

(2) $\Rightarrow$ (1) is obvious.

(3) $\Rightarrow$ (1): If M is a pseudo-umbilical submanifold with unit parallel mean curvature vector in $H^m(-1)$, then Theorem 8.31 implies $\Delta H = \lambda H$

for some $\lambda \in \mathbf{R}$. On the other hand, (8.82) and $\langle H', H' \rangle = 1$ yield $\lambda = 0$. Thus M is a biharmonic submanifold of $\mathbb{E}_1^{m+1}$.

(3) $\Rightarrow$ (4): If M is pseudo-umbilical with unit parallel mean curvature vector in $H^m(-1)$, then $H = H' + \phi$ is a lightlike constant vector by statement (2). Thus we have $\langle c, \phi \rangle = -1$. Let N denote the hyperplane section of $H^m(-1)$ defined by

$$N = \{x \in \mathbb{E}_1^{m+1} : \langle x, x \rangle = -1 \ \langle c, x \rangle = -1\}.$$

If we put $\xi = c - \phi$, then $\langle \xi, \phi \rangle = 0$ and $\langle \xi, \xi \rangle = 1$. So ξ is a unit normal vector field of N in $H^m(-1)$. From the definition of ξ, it is clear that N is a totally umbilical hypersurface of $H^m(-1)$ with constant mean curvature one. Thus it follows from Gauss' equation that N is flat. Because the mean curvature vector H of M in $\mathbb{E}_1^{m+1}$ is c which is equal to $\xi - \phi$, H is also normal to N. Therefore M lies in a horosphere of $H^m(-1)$ as a minimal submanifold. This proves statement (4).

(4) $\Rightarrow$ (3): If M lies in a horosphere N of $H^{m-1}(-1)$ as a minimal submanifold, then M is a pseudo-umbilical submanifold with parallel mean curvature vector in $H^m(-1)$. Because N is flat and $H^m(-1)$ has constant curvature -1, Gauss' equation implies that M has constant mean curvature one in $H^m(-1)$. This proves statement (3). $\square$

When $m = n + 1$, Theorem 8.32 implies immediately the following.

Corollary 8.11. [Chen (1994a)] *Let M be a hypersurface of $H^{n+1}(-1) \subset \mathbb{E}_1^{n+2}$. Then M is an open portion of a horosphere if and only if it is biharmonic in $\mathbb{E}_1^{n+2}$.*

Remark 8.5. Another simple characterization of horosphere in terms of ΔH can be found in [Dimitric (2009)].

The next result can be found in [Chen (1994a)].

Theorem 8.33. *Let M be an n-dimensional non-minimal submanifold of the hyperbolic space $H^m(-1) \subset \mathbb{E}_1^{n+1}$. If M is a biharmonic submanifold of $\mathbb{E}_1^{m+1}$, then we have:*

(a) *The Ricci tensor of M is negative semi-definite.*
(b) *M is Ricci flat if and only if M lies in an $(n+1)$-dimensional totally geodesic $H^{n+1}(-1) \subset H^m(-1)$.*
(c) *M is Ricci flat if and only if M is flat.*

(d) *The mean curvature vector of M in $\mathbb{E}_1^{m+1}$ is a constant vector, say c; and M lies in the de Sitter spacetime $S_1^{m-1}(c, 1)$ with c as its center. Moreover, the positive vector of M in $\mathbb{E}_1^{m+1}$ is the mean curvature vector of M in the de Sitter spacetime.*

(e) *M lies in the hyperplane L of $\mathbb{E}_1^m$ defined by $\langle \phi, c \rangle = -1$, where c is the center of $S_1^{m-1}(c, 1)$ given by (d).*

8.8 λ-biharmonic submanifolds in S^m and S_1^m

For spherical submanifolds with proper mean curvature vector, we have the following result.

Theorem 8.34. [Chen (1995)] *Let $\phi : M \to S^m(1) \subset \mathbb{E}^{m+1}$ be an isometric immersion of a Riemannian manifold M into $S^m(1) \subset \mathbb{E}^{m+1}$. If M is λ-biharmonic in $\mathbb{E}^{m+1}$, then it is either a minimal submanifold of $S^m(1)$ or a minimal submanifold of a totally umbilical hypersurface of $S^m(1)$.*

Proof. Under the hypothesis, after comparing $\Delta H = \lambda H$ using the formula of ΔH, we find

$$\lambda = n + ||A_\xi||^2 + ||\hat{D}\xi||^2, \quad \lambda = n||H||^2, \tag{8.89}$$

where ξ is a unit normal vector in the direction of the mean curvature vector $\hat{H}$ of M in $S^{m-1}(1)$ and $\hat{D}$ is the normal connection of M in $S^{m-1}(1)$. These two expressions of λ given in (8.89) imply that M is a pseudo-umbilical submanifold of $S^{m-1}(1)$ with parallel mean curvature vector. Therefore, by applying Proposition 4.17, we obtain the desired result. $\quad\square$

Theorem 8.2 and Theorem 8.34 imply the following.

Corollary 8.12. *We have:*

(1) *Every null 2-type submanifold M of $\mathbb{E}^m$ is non-spherical, i.e., M does not lie in any hypersphere of $\mathbb{E}^m$.*

(2) *Every biharmonic submanifold in a Euclidean space is non-spherical.*

Next, we present some classification results for submanifolds in S_1^m with proper mean curvature vector.

Theorem 8.35. [Chen (1995)] *Let M be a spacelike submanifold of the de Sitter spacetime $S_1^{m-1}(1) \subset \mathbb{E}_1^m$. Then M is a λ-biharmonic submanifold in $\mathbb{E}_1^m$ with $\lambda < n = \dim M$ if and only if M lies in a spacelike non-totally geodesic totally umbilical hypersurface of $S_1^{m1}(1)$ as a minimal submanifold.*

Theorem 8.35 implies the following.

Corollary 8.13. *If M is an n-dimensional spacelike submanifold of the de Sitter spacetime $S_1^{m-1}(1) \subset \mathbb{E}_1^{n+1}$ whose mean curvature vector M in $\mathbb{E}_1^m$ satisfies $\Delta H = \lambda H$ with $\lambda < n$, then M is of 1-type.*

Theorem 8.36. [Chen (1995)] *If $\phi : M \to S_1^m(1) \subset \mathbb{E}_1^{m+1}$ is an isometric immersion of a Riemannian n-manifold M into $S_1^m(1)$, then M is λ-biharmonic in $\mathbb{E}_1^m$ with $\lambda = n$ if and only if ether*

(a) *M is a minimal submanifold of $S_1^{m-1}(1)$, or*
(b) *ϕ is congruent to $(f + \varphi, x_3, \ldots, x_m, f + \varphi)$, where φ is a harmonic function, f is an eigenfunction of Δ with eigenvalue n, and $\Phi = (x_3, \ldots, x_m) : M \to S^{m-3}(1) \subset \mathbb{E}^{m-2}$ is a minimal immersion of M into $S^{m-3}(1)$.*

Theorem 8.37. [Chen (1995)] *Let M be a Riemannian n-manifold, λ be a real number $> n$, $c \in \mathbb{E}_1^{m+1}$ be a spacelike vector satisfying $\langle c, c \rangle = 1 - n/\lambda > 0$, and $N_1(c)$ be the hypersurface of $S_1^m(1)$ defined by*

$$N_1(c) = \{ x \in S_1^m(1) : \langle c, x \rangle = \langle c, c \rangle \}.$$

Then we have:

(a) *$N_1(c)$ is a non-totally geodesic totally umbilical hypersurface of $S_1^m(1)$.*
(b) *If M is a spacelike minimal submanifold of $N_1(c)$, then*

(b.1) *M is a pseudo-umbilical submanifold of $S_1^m(1)$;*
(b.2) *the mean curvature vector of M in $S_1^m(1)$ is a nonzero parallel normal vector field; and*
(b.3) *M is a λ-biharmonic submanifold of $\mathbb{E}_1^{m+1}$ with $\lambda > n$.*

Conversely, if $\phi : M \to S_1^m(1) \subset \mathbb{E}_1^{m+1}$ is an isometric immersion from a Riemannian n-manifold M into $S_1^m(1)$ such that the mean curvature vector of M in $\mathbb{E}_1^{m+1}$ satisfies conditions (b.2) and (b.3), then M lies in $N_1(c)$ as a minimal submanifold, where c is a spacelike vector in $\mathbb{E}_1^m$ with $\langle c, c \rangle = 1 - n/\lambda$.

Theorem 8.37 provides many examples of spacelike submanifolds of $S_1^m(1)$ satisfying the condition $\Delta H = \lambda H$ for $\lambda > n$. The following gives more examples.

Theorem 8.38. [Chen (1995)] *Let $m \geq 5$, $r \in (0, 1)$, and let*

$$y = (y_4, \ldots, y_{m+1}) : M \to S^{m-3}(r) \subset \mathbb{E}^{m-2}$$

be a minimal immersion of a Riemannian n-manifold into $S^{m-4}(r) \subset \mathbb{E}^{m-3}$ centered at the origin and with radius r. Then, for any non-constant harmonic function φ on M and any eigenfunction f of Δ with eigenvalue n/r^2, the mapping $\phi : M \to \mathbb{E}_1^{m+1}$ given by

$$\phi = \left(f + \varphi, \sqrt{1 - r^2}, y_4, \ldots, y_m, f + \varphi \right) \tag{8.90}$$

is an isometric immersion of M into $S_1^m(1)$ satisfying

(a) $\Delta H = \lambda H$ with $\lambda = n/r^2 > n$;
(b) $DH = \omega \xi$ for some 1-form $\omega \neq 0$ and constant light–like vector $\xi \neq 0$.

Conversely, if $\phi : M \to S_1^m(1) \subset \mathbb{E}_1^{m+1}$ is an isometric immersion of a Riemannian n-manifold satisfying (a) and (b), then ϕ is congruent to (8.90), where

$$(y_4, \ldots, y_{m+1}) : M \to S^{m-3}(r) \subset \mathbb{E}^{m-2}$$

is a minimal immersion, $r \in (0, 1)$, φ is harmonic function, and f satisfies $\Delta f = (n/r^2)f$.

Theorem 8.38 implies that, for each $m \geq 5$, there exist many spacelike submanifolds in $S_1^m(1)$ which satisfy the λ-biharmonic condition:

$$\Delta H = \lambda H, \quad \lambda > n = \dim M, \quad DH \neq 0.$$

The next result shows that there do not exist such surfaces in $S_1^m(1)$ for $m \leq 4$.

Theorem 8.39. [Chen (1995)] *Let $\phi : M \to S_1^4(1) \subset \mathbb{E}_1^5$ be an isometric immersion of a surface M into $S_1^4(1)$ such that the mean curvature vector H of M in $\mathbb{E}_1^5$ satisfies $\Delta H = \lambda H$ for some $\lambda > 2$. Then H is parallel in the normal bundle and M lies in a non-totally geodesic totally umbilical hypersurface of $S_1^4(1)$ as a minimal submanifold.*

Remark 8.6. L. J. Alías, A. Ferrández and P. Lucas classified in [Alías et. al. (1995a)], λ-biharmonic pseudo-Riemannian hypersurfaces M_t^n in S_1^{n+1} and H_1^{n+1} for the following three cases:

(1) $n = 2$;
(2) M_t^n is non-minimal and with diagonalizable shape operator;
(3) the shape operator of M_t^n have no complex eigenvalues.

Remark 8.7. For the classifications of λ-biharmonic and quasi-minimal Lagrangian surfaces in Lorentzian complex space forms, see [Sasahara (2005a)].

Chapter 9

Applications of Finite Type Theory

In this chapter we present numerous applications of finite type theory to several important subjects in differential geometry for Riemannian submanifolds and maps via their spectral decompositions.

9.1 Total mean curvature and order of submanifolds

In this section we provide general optimal relations between total mean curvature and the order of Euclidean submanifolds via finite type theory.

We need the following.

Lemma 9.1. *Let* $\phi : M \to \mathbb{E}^m$ *be an isometric immersion of an oriented closed Riemannian n-manifold M into $\mathbb{E}^m$. Then* $\langle d\phi, d\phi \rangle = n$.

Proof. Let $e_1, \ldots, e_n$ be an orthonormal basis of $T_p N$, $x \in M$. Then

$$\langle d\phi, d\phi \rangle = \sum_{i=1}^{n} \langle d\phi(e_i), d\phi(e_i) \rangle = \sum_{i=1}^{n} \langle e_i, e_i \rangle = n. \qquad \square$$

The following is known as Minkowski-Hsiung's formula.

Proposition 9.1. *Let* $\phi : M \to \mathbb{E}^m$ *be an isometric immersion of a oriented closed Riemannian n-manifold M into $\mathbb{E}^m$. Then*

$$\int_M \{1 + \langle \phi, H \rangle\} dV = 0. \qquad (9.1)$$

Proof. From the formula of Beltrami, we have $\Delta \phi = -nH$. Hence, by Proposition 3.1 and Lemma 9.1, we find

$$n \int_M \langle \phi, H \rangle \, dV = -(\phi, \Delta \phi) = -(d\phi, d\phi) = -n \int_M dV. \qquad \square$$

The following problem was proposed in [Chen (1975)].

Problem 9.1. *Let (M, g) be a closed Riemannian manifold. What are the relationship between the total mean curvature of an isometric immersion $\phi : M \to \mathbb{E}^m$ and the Riemannian structure of (M, g)?*

Recall that we simply denote $||H||^k$ by H^k for $k \neq 0$. By applying finite type theory we provide several solutions to Problem 9.1.

Theorem 9.1. *[Chen (1979b)] Let $\phi : M \to \mathbb{E}^m$ be an isometric immersion of a closed Riemannian n-manifold M into $\mathbb{E}^m$. Then, for $k \geq 2$, we have*

$$\int_M H^k dV \geq \left(\frac{\lambda_p}{n}\right)^{k/2} \text{vol}(M), \tag{9.2}$$

where p is the lower order of M in $\mathbb{E}^m$. The equality sign of (9.2) holds for some real number $k \geq 2$ if and only if M is of 1-type.

Proof. Let p and q denote the lower and upper order of M in $\mathbb{E}^m$. Then we have the spectral decomposition:

$$\phi = \phi_0 + \sum_{t=p}^{q} \phi_t, \quad \Delta\phi_t = \lambda_t \phi_t. \tag{9.3}$$

Thus

$$n^2 \int_M H^2 dV = n^2 (H, H) = (\Delta\phi, \Delta\phi) = \sum_{t=p}^{q} \lambda_t^2 ||\phi_t||^2. \tag{9.4}$$

On the other hand, it follows from (9.2) and (9.3) that

$$n \int_M dV = -n(\phi, H) = (\phi, \Delta\phi) = \sum_{t=p}^{q} \lambda_t ||\phi_t||^2. \tag{9.5}$$

Hence, by combining (9.4) and (9.5), we find

$$n^2 \int_M H^2 dV - n\lambda_p \int_M dV = \sum_{t=p+1}^{q} \lambda_t (\lambda_t - \lambda_p) ||\phi_t||^2 \geq 0.$$

Therefore we obtain

$$\int_M H^2 dV \geq \left(\frac{\lambda_p}{n}\right) \text{vol}(M), \tag{9.6}$$

with the equality holding if and only if M is of 1-type.

Now, by using Hölder's inequality, we find

$$\left(\frac{\lambda_p}{n}\right) \text{vol}(M) \leq \int_M H^2 dV \leq \left(\int_M H^{2r} dV\right)^{1/r} \left(\int_M dV\right)^{1/s},$$

with $r^{-1} + s^{-1} = 1$, $r, s > 1$. By choosing $r = k/2$, we obtain inequality (9.2) for $k > 2$.

The remaining part is easy to verify. $\qquad\qquad\square$

Remark 9.1. Theorem 9.1 generalizes a result of [Reilly (1977)]:

$$\int_M H^2 dV \geq \frac{\lambda_1}{n} \text{vol}(M),$$ (9.7)

with the equality holding if and only if M is a minimal submanifold of a hypersphere.

The next result gives a sharp relationship between total mean curvature and the upper order of the submanifold.

Theorem 9.2. [Chen (1983a)] *Let $\phi : M \to \mathbb{E}^m$ be an isometric immersion of a closed Riemannian n-manifold M into $\mathbb{E}^m$. Then, for any real number k, $0 < k \leq 4$, we have*

$$\int_M H^k dV \leq \left(\frac{\lambda_q}{n}\right)^{k/2} \text{vol}(M),$$ (9.8)

where q is the upper order of M. The equality sign of (9.8) holds for some $k \in (0, 4]$ if and only if M is of 1-type of order $\{q\}$.

Proof. From Proposition 2.8 we have

$$\Delta H = \Delta^D H + \sum_{i=1}^{n} h(A_H e_i, e_i) + \frac{n}{2} \nabla \langle H, H \rangle + 2 \text{Tr} A_{DH}.$$

Thus

$$\begin{aligned} \langle \Delta H, H \rangle &= \langle \Delta^D H, H \rangle + \sum_{i=1}^{n} \langle h(A_H e_i, e_i), H \rangle \\ &= \langle \Delta^D H, H \rangle + \|A_H\|^2. \end{aligned}$$ (9.9)

Moreover, it follows from (9.3)-(9.5) that

$$n \int_M dV = \sum_{t=p}^{q} \lambda_t \|\phi_t\|^2,$$ (9.10)

$$n^2 \int_M H^2 dV = \sum_{t=p}^{q} \lambda_t^2 \|\phi_t\|^2,$$ (9.11)

$$n^2 \int_M \langle H, \Delta H \rangle dV = \sum_{t=p}^{q} \lambda_t^3 \|\phi_t\|^2.$$ (9.12)

If $q = \infty$, inequality (9.8) is trivial. So we may assume $q < \infty$. It we put

$$\Lambda = n^2 \int_M \langle H, \Delta H \rangle dV - n^2 (\lambda_p + \lambda_q) \int_M H^2 dV + n \lambda_p \lambda_q \int_M dV,$$ (9.13)

then we find from (9.10)-(9.12) that

$$\Lambda = \sum_{t=p+1}^{q-1} \lambda_t(\lambda_t - \lambda_q)(\lambda_t - \lambda_q)||\phi_t||^2 \leq 0, \tag{9.14}$$

with the equality holding if and only if M is of 1-type or 2-type. Combining (9.9), (9.13) and (9.14) gives

$$n^2 \int_M \langle H, \Delta^D H \rangle dV + n^2 \int_M ||A_H||^2 dV$$
$$- n^2(\lambda_p + \lambda_q) \int_M H^2 dV + n\lambda_p\lambda_q \int_M dV = 0. \tag{9.15}$$

Since M is a closed manifold, Hopf's lemma implies

$$\int_M \langle H, \Delta H \rangle dV = \int_M ||DH||^2 dV. \tag{9.16}$$

Let $\kappa_1, \ldots, \kappa_n$ be the eigenvalues of A_H. It is easy to verify that

$$||A_H||^2 = nH^4 + \frac{1}{n}\sum_{i<j}(\kappa_i - \kappa_j)^2. \tag{9.17}$$

By combining (9.15), (9.16) and (9.17), and applying Schwartz's inequality, we find

$$0 \geq n^2 \int_M ||DH||^2 dV + n^3 \int_M H^4 dV + n \sum_{i<j} \int_M (\kappa_i - \kappa_j)^2 dV$$
$$- n^2(\lambda_p + \lambda_q) \int_M H^2 dV + n\lambda_p\lambda_q \int_M dV$$
$$\geq n^2 \int_M ||DH||^2 dV + \frac{n^3 \left(\int_M H^2 dV\right)^2}{\mathrm{vol}(M)} + n\lambda_p\lambda_q \int_M dV$$
$$+ n \sum_{i<j} \int_M (\kappa_i - \kappa_j)^2 dV - n^2(\lambda_p + \lambda_q) \int_M H^2 dV. \tag{9.18}$$

Thus

$$0 \geq n\,\mathrm{vol}(M) \int_M ||DH||^2 dV + \mathrm{vol}(M) \sum_{i<j} \int_M (\kappa_i - \kappa_j)^2 dV$$
$$+ \left(n \int_M H^2 dV - \lambda_p\mathrm{vol}(M)\right)\left(n \int_M H^2 dV - \lambda_q\mathrm{vol}(M)\right). \tag{9.19}$$

Now, by combining Theorem 9.1 and (9.19), we have

$$\int_M H^2 dV \leq \left(\frac{\lambda_q}{n}\right) \mathrm{vol}(M). \tag{9.20}$$

Substituting (9.20) into the first inequality of (9.18), we find (9.8) for $k = 4$. Thus, by using Hölder's inequality, we derive that

$$\int_M H^k dV \leq \left(\int_M H^4 dV \right)^{k/4} (\text{vol}(M))^{1-k/4}$$

$$= \left(\int_M (H^k)^p dV \right)^{1/p} (\text{vol}(M))^{1-1/p}$$

with $p = 4/k$ for any real number $k \in (0, 4)$. Hence, by combining this with the inequality (9.8) with $k = 4$ obtained above, we also have inequality (9.8) for any $k \in (0, 4)$. Consequently, we obtain (9.8) for any $k \in (0, 4]$.

If the equality sign of (9.8) holds for a real number $k \in (0, 4]$, then all of the inequalities in (9.14)-(9.20) become equalities. Therefore M is pseudo-umbilical with parallel mean curvature vector. Hence, by applying Proposition 4.16 and Theorem 7.2, we conclude that M is of 1-type.

The converse is easy to verify. $\qquad\square$

By combining Theorems 9.1 and 9.2 we have the following.

Theorem 9.3. [Chen (1979b, 1983a)] *Let* $\phi : M \to \mathbb{E}^m$ *be an isometric immersion of a closed Riemannian n-manifold M into $\mathbb{E}^m$. Then*

$$\left(\frac{\lambda_p}{n} \right) \text{vol}(M) \leq \int_M H^2 dV \leq \left(\frac{\lambda_q}{n} \right) \text{vol}(M), \qquad (9.21)$$

where p and q are the lower and the upper orders of M. Either equality sign in (9.21) holds if and only if M is of 1-type.

An immediate consequence of Theorem 9.3 is the following.

Corollary 9.1. *Let M be an n-dimensional closed submanifold of $\mathbb{E}^m$. If M has constant mean curvature, then $\lambda_p \leq n H^2 \leq \lambda_q$. Either equality holds if and only if M is of 1-type.*

9.2   Conformal property of $\lambda_1 \text{vol}(M)$

A 1-type immersion $\phi : M \to \mathbb{E}^m$ with order $\{p\}$ is called an *immersion of order p*. The following *conformal property of $\lambda_1 \text{vol}(M)$* as an application of Reilly's inequality (9.7) was first proved in [Chen (1979a)].

Theorem 9.4. *Let M be a closed Riemannian surface which admits an order 1 isometric embedding into $\mathbb{E}^m$. Then, for any closed surface $\overline{M}$ in $\mathbb{E}^m$ which is conformally equivalent to M, we have*

$$\lambda_1 \text{vol}(M) \geq \bar{\lambda}_1 \text{vol}(\overline{M}), \qquad (9.22)$$

with equality sign of (9.22) *holding if and only if* $\overline{M}$ *is also of order* 1.

Proof. Let $\phi : N \to \mathbb{E}^m$ be an order 1 isometric embedding of a closed Riemannian surface. Then, by Reilly's result, we have

$$\int_M H^2 dS_M = \frac{\lambda_1}{2} \text{vol}(M). \tag{9.23}$$

Since the total mean curvature is a conformal invariant, we also have

$$\int_{\overline{M}} \overline{H}^2 dS_{\overline{M}} = \int_M H^2 dS_M, \tag{9.24}$$

where the left-hand side is the corresponding quantity for $\overline{M}$.

On the other hand, we have

$$\int_{\overline{M}} \overline{H}^2 dS_{\overline{M}} \geq \overline{\lambda}_1 \text{vol}(\overline{M}). \tag{9.25}$$

After combining (9.23), (9.24) and (9.25), we obtain (9.22). If the equality sign of (9.22) holds, then the equality sign of (9.25) holds. Hence $\overline{M}$ is of order 1. The converse is clear. $\square$

Let $T_a^2 = S^1(a) \times S^1(a)$ be a square torus. Then $\lambda_1 \text{vol}\,(T_a^2) = 4\pi^2$. The standard embedding of T_a^2 into $\mathbb{E}^4 \subset \mathbb{E}^m$ is of order 1.

Definition 9.1. A closed surface M in $\mathbb{E}^m$ is called a *conformal square torus* if it can be obtained from the standard embedding of a square torus via some conformal mappings of Euclidean space.

Theorem 9.4 implies immediately the following [Chen (1979a)].

Corollary 9.2. *If* M *is a conformal square torus in* $\mathbb{E}^m$, *then*

$$\lambda_1 \text{vol}(M) \leq 4\pi^2,$$

with the equality holding if and only if M *admits an order 1 imbedding.*

9.3 Total mean curvature and λ_1, λ_2

For an isometric immersion $\phi : M \to \mathbb{E}^m$ of a closed Riemannian manifold M into $\mathbb{E}^m$ with center of mass at ϕ_0, the *moment* of M is defined by

$$\mathcal{M} = \int_M ||\phi - \phi_0||^2 dV. \tag{9.26}$$

The next result gives a sharp relationship between moment, total mean curvature and the first and second nonzero eigenvalues λ_1, λ_2 of Δ.

Proposition 9.2. [Chen (1987b)] *Let $\phi : M \to \mathbb{E}^m$ be an isometric immersion of a closed Riemannian n-manifold M into $\mathbb{E}^m$. Then we have*

$$\int_M H^2 dV \geq \frac{1}{n}(\lambda_1 + \lambda_2)\text{vol}(M) - \frac{\lambda_1 \lambda_2}{n^2}\mathcal{M}, \tag{9.27}$$

where $\mathcal{M}$ is the moment of M. With the equality holding if and only if M is either of 1-type with order 1 or order 2, or it is of 2-type with order $\{1, 2\}$.

Proof. Let $\phi : M \to \mathbb{E}^m$ be an isometric immersion of a closed Riemannian n-manifold. Then we have the spectral decomposition (9.3) of ϕ. Thus

$$\mathcal{M} = \langle \phi - \phi_0, \phi - \phi_0 \rangle = \sum_{t=p}^{q}(\phi_t, \phi_t). \tag{9.28}$$

From Corollary 5.1 and (9.3), we have

$$n\,\text{vol}(M) = (d\phi, d\phi) = (\phi, \delta d\phi) = (\phi, \Delta\phi) = \sum_{t=p}^{q}\lambda_t(\phi_t, \phi_t), \tag{9.29}$$

$$n^2 \int_M H^2 dV = (\Delta\phi, \Delta\phi) = \sum_{t=p}^{q}\lambda_t^2(\phi_t, \phi_t). \tag{9.30}$$

By combining (9.28), (9.29) and (9.30), we find

$$
\begin{aligned}
n^2 \int_M H^2 dV &- n(\lambda_1 + \lambda_2)\text{vol}(M) + \lambda_1\lambda_2\mathcal{M} \\
&= \sum_{t=p}^{q}(\lambda_t - \lambda_1)(\lambda_t - \lambda_2)(\phi_t, \phi_t) \geq 0,
\end{aligned}
\tag{9.31}
$$

which proves inequality (9.27). If the equality sign of (9.27) holds, then inequality (9.31) becomes equality. Thus M must be of order 1, 2 or $\{1, 2\}$.
The converse is clear. $\square$

Proposition 9.3. *Let $\phi : M \to S^{m-1}(1) \subset \mathbb{E}^m$ be an isometric immersion of a closed Riemannian n-manifold into a unit hypersphere of $\mathbb{E}^m$. If M is mass-symmetric in $S^{m-1}(1)$, then*

$$\int_M H^2 dV \geq \frac{1}{n^2}\{n(\lambda_1 + \lambda_2) - \lambda_1\lambda_2\}\text{vol}(M), \tag{9.32}$$

with the equality holding if and only if M is of order 1, 2 or $\{1, 2\}$.

Proof. Under the hypothesis, the moment $\mathcal{M}$ is nothing but the volume of M. Hence, the proposition follows from Proposition 9.2. □

By applying Proposition 9.3 we have the next sharp relationship between λ_1, λ_2 of closed minimal submanifolds in projective spaces.

Theorem 9.5. [Chen (1987b)] *Let M be a closed minimal Riemannian n-manifold. If M admits a minimal isometric immersion into $RP^m(1)$, $CP^m(4)$ or $QP^m(4)$, then the first two nonzero eigenvalues of Δ satisfy*

$$\frac{m}{2(m+1)}\lambda_1\lambda_2 \geq n(\lambda_1 + \lambda_2 - 2n - 2d). \tag{9.33}$$

where $d = 1, 2$ or 4 according to $\mathbb{F} = \mathbb{R}, \mathbb{C}$ or $\mathbb{Q}$. Moreover, we have

(1) *For $QP^m(4)$, the equality sign of (9.33) holds if and only if $n = 4k$ and M is a totally geodesic $QP^k(4)$;*

(2) *For $CP^m(4)$, the equality holding if and only if M is one of the compact Hermitian symmetric spaces:*

$$CP^k(4),\ CP^k(2),\ SO(2+k)/SO(2) \times SO(k),\ CP^k(4) \times CP^k(4),$$

$$U(2+k)/U(2) \times U(k),\ (k > 2),\ SO(10)/U(5)\ \text{and}\ E_6/\text{Spin}(10) \times T,$$

with appropriate metric and k.

For Kähler submanifolds, Theorem 9.5 is due to [Ros (1984a)] (see also [Udagawa (1986b)]).

9.4 Total mean curvature and circumscribed radii

Let $B_o(r)$ denote the open ball in $\mathbb{E}^m$ of radius r centered at the origin o. Put $c = ||\phi_0||$, where ϕ_0 is the center of mass of M in $\mathbb{E}^m$.

The total mean curvature can also be estimated in terms of the order of immersions and circumscribed radii as follows.

Theorem 9.6. [Chen and Jiang (1995)] *Let M be an n-dimensional closed submanifold of $\mathbb{E}^m$. The we have*

(a) *If M lies in a closed ball $\overline{B_o(R)}$ with radius R, then*

$$\int_M H^k dV \geq \frac{\text{vol}(M)}{(R^2 - c^2)^{k/2}} \tag{9.34}$$

for a real number $k \geq 2$, with any one of the equalities holding if and only if M is of 1-type and M lies in the boundary $\partial(B_o(R))$ of $B_o(R)$.

(b) *If M lies in $\mathbb{E}^m - B_o(r)$, then*

$$\left(\frac{\lambda_p}{n}\right)^2 (r^2 - c^2)\mathrm{vol}(M) \leq \int_M H^2 dV \leq \left(\frac{\lambda_q}{n}\right)^2 (R^2 - c^2)\mathrm{vol}(M), \quad (9.35)$$

where p and q are the lower and the upper orders of M in $\mathbb{E}^m$. Either equality of (9.35) holding if and only if M is of type 1 and it lies in the boundary $\partial(B_o(r))$.

Proof. For a given integer $\nu \geq p$, we put

$$u^\nu = (||\phi_p||, \ldots, ||\phi_\nu||), \quad v^\nu = (\lambda_p||\phi_p||, \ldots, \lambda_\nu||\phi_\nu||).$$

Thus we obtain $\langle u^\nu, u^\nu \rangle = \sum_{t=p}^\nu ||\phi_t||^2$, $\langle u^\nu, v^\nu \rangle = \sum_{t=p}^\nu \lambda_t ||\phi_t||^2$ and $\langle v^\nu, v^\nu \rangle = \sum_{t=p}^\nu \lambda_t^2 ||\phi_t||^2$. By the Schwartz inequality, we find

$$\left(\sum_{t=p}^\nu ||\phi_t||^2\right)\left(\sum_{t=p}^\nu \lambda_t^2 ||\phi_t||^2\right) \geq \left(\sum_{t=p}^\nu \lambda_t ||\phi_t||^2\right)^2. \quad (9.36)$$

On the other hand, we have

$$\langle u^\nu, u^\nu \rangle \to (\phi, \phi) - c^2 \mathrm{vol}(M),$$

$$\langle v^\nu, v^\nu \rangle \to n^2 \int_M H^2 dV,$$

$$\langle u^\nu, v^\nu \rangle \to -n \int_M \langle \phi, H \rangle \, dV = n \, \mathrm{vol}(M)$$

as $\nu \to \infty$. Thus (9.36) yields

$$((\phi, \phi) - c^2 \mathrm{vol}(M))\int_M H^2 dV \geq \mathrm{vol}(M)^2. \quad (9.37)$$

Since $(\phi, \phi) \leq R^2 \mathrm{vol}(M)$, (9.37) implies (9.34) for $k = 2$.

By applying Hölder's inequality, we find

$$\left(\int_M H^{2r} dV\right)^{1/r}\left(\int_M dV\right)^{1/s} \geq \int_M H^2 dV \geq \frac{\mathrm{vol}(M)}{R^2 - c^2}$$

with $r^{-1} + s^{-1} = 1, r, s > 1$. If we choose $r = k/2$, we obtain (9.34).

It is easy to verify that the equality sign of (9.34) holds for some k if and only if M is of 1-type and it lies in the boundary $\partial(B_o(R))$ of $B_o(R)$. This gives statement (a). For statement (b) we consider

$$n^2((\phi, \phi) - c^2 \mathrm{vol}(M))\int_M H^2 dV \geq n^2 \mathrm{vol}(M)^2 = \left(\sum_{t \geq p} \lambda_t ||\phi_t||^2\right)^2$$

$$\geq \lambda_p^2\left(\sum_{t \geq p} ||\phi_t||^2\right)^2 = \lambda_p^2((\phi, \phi) - c^2 \mathrm{vol}(M))^2,$$

which implies the first inequality of (9.35). The second inequality of (9.35) can be proved in a similar way.

The remaining part can be verified easily. $\square$

Remark 9.2. Inequalities (9.34) and (9.34) improve the main result of [Rotondaro (1993)] and inequality (14) in page 210 of [Burago and Zalgaller (1988)], respectively.

Theorem 9.6 implies immediately the following.

Corollary 9.3. *Let M be an n-dimensional closed submanifold of $\mathbb{E}^m$.*

(1) *If M lies in the closed ball $\overline{B_o(R)}$, then $\max H^2 \geq (R^2 - c^2)^{-1}$, with equality holding if and only if M is a minimal submanifold of a hypersphere of $\mathbb{E}^m$;*

(2) *If M has constant mean curvature and lies in $\mathbb{E}^m - B_o(r)$, then*

$$\left(\frac{\lambda_p}{n}\right)^2 (r^2 - c^2) \leq H^2 \leq \left(\frac{\lambda_q}{n}\right)^2 (R^2 - c^2),$$

with either equality holding if and only if M is of 1-type.

The next result provides a sharp estimate of λ_1 for arbitrary closed Euclidean submanifolds in term of its circumscribed radii.

Proposition 9.4. [Chen and Jiang (1995)] *Let M be an n-dimensional closed submanifold of $\mathbb{E}^m$.*

(1) *If M is contained in $\mathbb{E}^m - B_o(r)$, then we have $\lambda_1 \leq n/(r^2 - c^2)$, with equality sign holding if and only if M is of 1-type and it lies in the boundary $\partial(B_o(r))$ of $B_o(r)$;*

(2) *If M is contained in the closed ball $\overline{B_o(R)}$, we have*
$$\lambda_p(r^2 - c^2) \leq n \leq \lambda_q(R^2 - c^2),$$
where p and q are the lower and upper orders of M. Either equality sign holding if and only if M is of 1-type and it lies in the boundary $\partial(B_o(R))$ of $B_o(R)$.

Proof. Follows from the following two equations:
$$n \operatorname{vol}(M) = \sum_{t \geq p} \lambda_t \|\phi_t\|^2 \geq \lambda_p((\phi, \phi) - c^2 \operatorname{vol}(M)) > \lambda_1(r^2 - c^2)\operatorname{vol}(M),$$
$$n \operatorname{vol}(M) = \sum_{t \geq p} \lambda_t \|\phi_t\|^2 \leq \lambda_q((\phi, \phi) - c^2 \operatorname{vol}(M)) \leq \lambda_q(R^2 - c^2)\operatorname{vol}(M). \quad \square$$

Proposition 9.4 gives the following sharp estimates on λ_1 for ellipsoids.

Corollary 9.4. *Let M be the ellipsoid in $\mathbb{E}^3$ defined by*
$$\frac{x^2}{a^2} + \frac{y^2}{b^2} + \frac{z^2}{c^2} = 1, \quad a \leq b \leq c.$$
Then $\lambda_1 \leq 2/a^2$, with equality sign holding if and only if M is a sphere.

Proposition 9.4 can also be applied to obtain the following estimate of λ_1 for closed tubes [Chen and Jiang (1995)].

Proposition 9.5. *Let σ be a closed curve in $\mathbb{E}^3$. Denote by σ_0 the center of mass of σ and by $T^\epsilon(\sigma)$ the tube around σ with a sufficiently small radius ϵ. If σ is contained in $\mathbb{E}^3 - B_{\sigma_0}(r)$ and $\epsilon < r$, then the first nonzero eigenvalue λ_1 of the Laplacian of $T^\epsilon(\sigma)$ satisfies $\lambda_1 < 2/(r - \epsilon)^2$.*

Proof. Without loss of generality, we may assume that $\sigma : [0, \ell] \to \mathbb{E}^3$ is a unit speed closed curve whose center of mass is the origin. Denote by $\{\mathbf{T}, \mathbf{N}, \mathbf{B}\}$ the Frenet frame of σ. Then the tube $T^\epsilon(\sigma)$ is given by

$$\phi(s, \theta) = \sigma(s) + \epsilon \cos \theta \, \mathbf{N} + \epsilon \sin \theta \, \mathbf{B}. \tag{9.38}$$

From (9.38) we find $dV = \epsilon(1 - \epsilon\kappa \cos \theta)d\theta \wedge ds$, where κ is the curvature function of σ. Thus, by using (9.38) we have

$$
\begin{aligned}
\phi_0 &= \frac{\epsilon}{\text{vol}(M)} \int_0^\ell \int_0^{2\pi} \phi(s, \theta)(1 - \epsilon\kappa \cos \theta)d\theta \wedge ds \\
&= \frac{\epsilon}{\text{vol}(M)} \left(2\pi \int_0^\ell \sigma(s)ds - \epsilon\pi \int_0^\ell H_\sigma ds \right),
\end{aligned}
\tag{9.39}
$$

where H_σ is the mean curvature vector of σ. Since $\int_0^\ell H_\sigma ds = 0$, (9.39) implies that the center of mass of is the center of mass of σ. Therefore, by Proposition 9.4, we find $(r - \epsilon)^2\lambda_1 \leq 2$. If $(r - \epsilon)^2\lambda_1 = 2$, the tube is of 1-type which is impossible. Therefore we obtain the lemma. $\square$

The following result is a generalization of Proposition 9.3.

Theorem 9.7. [Chen and Jiang (1995)] *Let M be an n-dimensional closed submanifold of $\mathbb{E}^m$ and $c = \|\phi_0\|$ with ϕ_0 being the center of mass.*

(1) *If M is contained in the closed ball $\overline{B_o(R)}$, then*

$$\int_M H^2 dV \geq \frac{1}{n^2}\{n(\lambda_1 + \lambda_2) + (c^2 - R^2)\lambda_1\lambda_2\}\text{vol}(M),$$

with equality sign holding if and only if M is contained in the boundary $\partial(B_o(R))$; moreover, either M is of 1-type with order 1 or 2, or of 2-type and of order $\{1, 2\}$.

(2) *If M is contained in $\mathbb{E}^m - B_o(r)$ and M is of finite type with lower order p and upper order q, then*

$$\int_M H^2 dV \leq \frac{1}{n^2}\{n(\lambda_p + \lambda_q) + (c^2 - r^2)\lambda_p\lambda_q\}\text{vol}(M),$$

with the equality holding if and only if M lies in the boundary $\partial(B_o(r))$ and M is either of 1-type or of 2-type.

This can be proved in the same way as Proposition 9.3.

Analogous to Theorem 9.6(b), we also have the next two results.

Theorem 9.8. [Chen and Jiang (1995)] *Let M be an n-dimensional closed submanifold of $\mathbb{E}^m$. Then we have*

$$\left(\frac{\lambda_p^2}{n}\right)\mathrm{vol}(M) \le \int_M \|\nabla H\|^2 dV \le \left(\frac{\lambda_q^2}{n}\right)\mathrm{vol}(M), \qquad (9.40)$$

where p and q are the lower and upper orders of M, respectively. Either equality of (9.40) holds if and only if M is of 1-type.

Furthermore, if M is contained in the closed ball $\overline{B_o(R)}$ and lies in outside of the open ball $B_o(r)$, then we also have

$$\left(\frac{\lambda_p^3}{n^2}\right)(r^2 - c^2)\mathrm{vol}(M) \le \int_M \|\nabla H\|^2 dV \le \left(\frac{\lambda_q^3}{n^2}\right)(R^2 - c^2)\mathrm{vol}(M).$$

$$(9.41)$$

Either one of the equality signs of (9.41) holds if and only if M is lies on a hypersphere centered at the origin and it is of 1-type.

Theorem 9.9. [Chen and Jiang (1995)] *Let M be an n-dimensional closed submanifold of $\mathbb{E}^m$. Then we have*

$$\left(\frac{\lambda_p^3}{n}\right)\mathrm{vol}(M) \le \int_M \|\Delta H\|^2 dV \le \left(\frac{\lambda_q^3}{n}\right)\mathrm{vol}(M). \qquad (9.42)$$

Either equality of (9.42) holds if and only if M is of 1-type.

Furthermore, if M is contained in the closed ball $\overline{B_o(R)}$ and lies outside of the open ball $B_o(r)$, then we also have

$$\left(\frac{\lambda_p^2}{n}\right)^2 (r^2 - c^2)\mathrm{vol}(M) \le \int_M \|\Delta H\|^2 dV \le \left(\frac{\lambda_q^2}{n}\right)^2 (R^2 - c^2)\mathrm{vol}(M).$$

Either one of the equality sign holding if and only if M is lies in a hypersphere centered at the origin and it is of 1-type.

9.5 Spectra of spherical submanifolds

Now, we apply the finite type theory to derive sharp estimates of eigenvalues of Δ for spherical closed submanifolds.

Theorem 9.10. [Chen (1983a)] *If M is an n-dimensional closed submanifold of a unit hypersphere $S^m(1)$ of $\mathbb{E}^{m+1}$, then we have*

(a) *If M is mass-symmetric in $S^m(1)$, then $\lambda_1 \leq \lambda_p \leq n$; and $\lambda_p = n$ holds if and only if M is minimal in $S^m(1)$.*

(b) *If M is of finite type, then $\lambda_q \geq n$, with the equality sign holding if and only if M is of 1-type,*

where p and q denotes the lower and the upper orders of M in $\mathbb{E}^{m+1}$.

Proof. Without loss of generality, we may assume that $S^{m-1}(1)$ is centered at the origin. Denote by H and $\hat{H}$ the mean curvature vectors of M in $\mathbb{E}^{m+1}$ and in $S^m(1)$, respectively. Then we have $H^2 = \hat{H}^2 + 1$.

For statement (a), let us assume that M is mass-symmetric in $S^m(1)$ and $\phi = \sum_{t=p}^q \phi_t$ is the spectral decomposition of M in $\mathbb{E}^{m+1}$. Then it follows from Beltrami's formula and Proposition 9.1 that

$$n\,\mathrm{vol}(M) = -n(\phi, H) = (\phi, \Delta\phi) = \sum_{t=p}^q \lambda_t ||\phi_t||^2 \geq \lambda_p ||\phi||^2. \qquad (9.43)$$

Since $\langle \phi, \phi \rangle = 1$, we have $||\phi||^2 = (\phi, \phi) = \mathrm{vol}(M)$. Thus we obtain from (9.43) that $\lambda_p \leq n$, with the equality holding if and only if M is of 1-type.

For statement (b), we assume that M is of finite type, i.e., q is finite. Thus, by Theorem 9.2, we have

$$\mathrm{vol}(M) \leq \int_M H^2 dV \leq \left(\frac{\lambda_q}{n}\right) \mathrm{vol}(M), \qquad (9.44)$$

which gives $\lambda_q \geq n$. If $\lambda_q = n$, it follows from (9.44) and Theorem 9.2 that M is minimal in $S^m(1)$. The converse is clear. $\qquad\square$

The following is a consequence of Theorem 9.10 (cf. Corollary 6.5).

Corollary 9.5. *Let M be a mass-symmetric closed submanifold of $S^m(1) \subset \mathbb{E}^{m+1}$. Then $\lambda_1 \leq n = \dim N$, with the equality sign holds if and only if M is of order 1.*

9.6 The first standard imbedding of projective spaces

In this section we present the first standard imbedding of projective spaces. Such an imbedding have been considered in several articles, see for instance [Tai (1968); Little (1976); Sakamoto (1977); Ros (1983); Chen (1983c)].

Let $\mathbb{F}$ denote the field of $\mathbb{R}$ of real numbers, $\mathbb{C}$ of complex numbers or $\mathbb{Q}$ of quaternions. In a natural way, we have $\mathbb{R} \subset \mathbb{C} \subset \mathbb{Q}$.

For each $z = z_0 + z_1 i + z_2 j + z_3 k \in \mathbb{Q}$, $z_0, z_1, z_2, z_3 \in \mathbb{R}$, the conjugate of z is $\bar{z} = z_0 - z_1 i - z_2 j - z_3 k$. If $z \in \mathbb{C}$. So $\bar{z}$ coincides with the ordinary complex conjugate of z. If $z \in \mathbb{R}$, we have $\bar{z} = z$. It is convenient to define $d = d(\mathbb{F})$ to $1, 2$ or 4, according to $\mathbb{F}$ is $\mathbb{R}$, $\mathbb{C}$ or $\mathbb{Q}$, respectively.

For a matrix A over $\mathbb{F}$, denote by A^T and $\bar{A}$ the transpose and the conjugate of A, respectively. Let $z = (z_i) \in \mathbb{F}^{m+1}$ be a column vector. A matrix $A = (a_{ij}), 0 \leq i, j, \leq m$, operates on z by the rule

$$Az = \begin{pmatrix} a_{00} & \cdots & a_{0m} \\ & \cdots & \\ a_{m0} & \cdots & a_{mm} \end{pmatrix} \begin{pmatrix} z_0 \\ \vdots \\ z_m \end{pmatrix}. \tag{9.45}$$

We will use the following notations:

$$M(m+1; \mathbb{F}) = \text{the space of } (m+1) \times (m+1) \text{ matrices over } \mathbb{F},$$
$$H(m+1; \mathbb{F}) = \{A \in M(m+1; \mathbb{F}) : A^* = A\}$$
$$= \text{the space of } (m+1) \times (m+1) \text{ Hermitian matrices over } \mathbb{F},$$
$$U(m+1; \mathbb{F}) = \{A \in M(m+1; \mathbb{F}) : A^*A = I\},$$

where $A^* = \bar{A}^T$ and I is the identity matrix. If $A \in H(m+1; \mathbb{F})$, then A is a symmetric matrix. Therefore, we have $U(m+1; \mathbb{R}) = O(m+1)$, $U(m+1; \mathbb{C}) = U(m+1)$ and $U(m+1; \mathbb{Q}) = Sp(m+1)$.

$\mathbb{F}^{m+1}$ is an $(m+1)d$-dimensional vector spar over $\mathbb{R}$ endowed with the usual Euclidean inner product $\langle z, w \rangle = Re(z^*w)$. And $M(m+1; \mathbb{F})$ is an $(m+1)^2 d$-dimensional Euclidean space with the inner product:

$$\langle A, B \rangle = \frac{1}{2} Re \, \text{Tr}(AB^*). \tag{9.46}$$

For $A, B \in H(m+1; \mathbb{F})$, we have $\langle A, B \rangle = \frac{1}{2} \text{Tr}(AB)$.

Let $\mathbb{F}P^m$ denote the projective m-space over $\mathbb{F}$ regarded as the quotient space of $S^{(m+1)d-1}(1) = \{z \in \mathbb{F}^{m+1} : z^*z = 1\}$ obtained by identifying z with $z\lambda$, where z is a column vector and $\lambda \in \mathbb{F}$ such that $|\lambda| = 1$. The metric g_0 on $\mathbb{F}P^m$ is the invariant metric so that $\pi : S^{(m+1)d-1}(1) \to \mathbb{F}P^m$ is a Riemannian submersion. The holomorphic sectional curvature of CP^m and the quaternion sectional curvature of QP^m are both 4.

Via (9.45) we have an action of $U(m+1; \mathbb{F})$ on $S^{(m+1)d-1}(1)$. Such an action induces an action of $U(m+1; \mathbb{F})$ on $\mathbb{F}P^m$. Denote by o the point in $\mathbb{F}P^m$ whose homogenous coordinates (z_i) are $z_0 = 1, z_1 = \cdots = z_m = 0$. Then the isotropy subgroup at o is $U(1; \mathbb{F}) \times U(m; \mathbb{F})$, which gives the following well-known isometry:

$$\mu : \mathbb{F}P^m \cong U(m+1; \mathbb{F})/(U(1; \mathbb{F}) \times U(m; \mathbb{F})). \tag{9.47}$$

Define a mapping $\tilde{\varphi} : S^{(m+1)d-1}(1) \to H(m+1; \mathbb{F})$ by

$$\tilde{\varphi}(z) = zz^* = \begin{pmatrix} |z_0|^2 & z_0\bar{z}_1 & \cdots & z_0\bar{z}_m \\ & \cdots & \cdots & \\ z_m\bar{z}_0 & z_m\bar{z}_1 & \cdots & |z_m|^2 \end{pmatrix}, \quad z = (z_i) \in S^{(m+1)d-1}(1). \quad (9.48)$$

Then $\tilde{\varphi}$ induces a mapping of $\mathbb{F}P^m$ into $H(m+1; \mathbb{F})$:

$$\varphi(\pi(z)) = \tilde{\varphi}(z) = zz^*. \quad (9.49)$$

We denote $\varphi(\pi(z))$ by $\varphi(z)$ if there is no confusion. Define a hyperplane $H_1(m+1, \mathbb{F})$ by

$$H_1(m+1, \mathbb{F}) = \{A \in H(m+1; \mathbb{F}) : \operatorname{Tr} A = 1\}.$$

Then $\dim H_1(m+1, \mathbb{F}) = m + m(m+1)d/2$. It follows from (9.48) that

$$\varphi(\mathbb{F}P^m) = \{A \in H(m+1; \mathbb{F}) : A^2 = A \text{ and } \operatorname{Tr} A = 1\}. \quad (9.50)$$

Let $U(m+1; \mathbb{F})$ act on $M(m+1; \mathbb{F})$ by

$$P(A) = PAP^{-1}, \quad P \in U(m+1; \mathbb{F}), \quad A \in M(m+1; \mathbb{F}). \quad (9.51)$$

We have

$$\langle P(A), P(B) \rangle = \langle A, B \rangle. \quad (9.52)$$

Hence the action of $U(m+1; \mathbb{F})$ preserves the inner product of $M(m+1; \mathbb{F})$. Moreover, we also have $\varphi(Pz) = P(\varphi(z)) \in \varphi(\mathbb{F}P^m)$ for $z \in \mathbb{F}P^m$ and $P \in U(m+1; \mathbb{F})$. Consequently, we have the following.

Lemma 9.2. [Tai (1968)] *The imbedding* $\varphi : \mathbb{F}P^m \to H(m+1; \mathbb{F})$ *given by* (9.49) *is equivariant and invariant under the action of* $U(m+1; \mathbb{F})$.

Now, we claim that $\varphi : \mathbb{F}P^m \to H(m+1; \mathbb{F})$ is the first standard imbedding. Let $A \in \varphi(\mathbb{F}P^m)$. Consider a curve $A(t)$ with $A(0) = A$, $A'(0) = X \in T_A(\mathbb{F}P^m)$. It follows from $A^2(t) = A(t)$ that $XA + AX = X$. Because $\dim\{X \in H(m+1; \mathbb{F}) : XA + AX = X\} = md$, we obtain

$$T_A(\mathbb{F}P^m) = \{X \in H(m+1; \mathbb{F}) : XA + AX = X\}. \quad (9.53)$$

There is another expression of $T_A(\mathbb{F}P^m)$ as follows: For $u, v \in \mathbb{F}^{m+1}$, we define $\alpha(u, v) = u^*v$. Let $z \in S^{(m+1)d-1}(1)$ and v be a horizontal vector in $T_z(S^{(m+1)d-1}(1))$. We identify v with its image $\pi_*(v) \in T_{\pi(z)}(\mathbb{F}P^m)$. Let $a(t)$ be a curve in $S^{(m+1)d-1}(1)$ satisfying $a(0) = z$ and $a'(0) = v$. Then $A(t) = a(t)a(t)^*$ is a curve in $\varphi(\mathbb{F}P^m)$ through $A = zz^*$. From these we find $\varphi_*(v) = vz^* + zv^*$. Hence

$$T_A(\mathbb{F}P^m) = \{vz^* + zv^* : v \in \mathbb{F}^{m+1} \text{ and } \alpha(z, v) = 0\} \quad (9.54)$$

where $A = zz^*$, $z \in S^{(m+1)d-1}(1)$. A vector $\xi \in H(m+1;\mathbb{F})$ is normal to FP^m at A if and only if $\langle X, \xi \rangle = 0$ for all $X \in T_A(FP^m)$. Hence $\xi \in T_A^\perp(FP^m)$ if and only if $\mathrm{Tr}(X\xi) = 0$, $X \in T_A(FP^m)$. Thus (9.53) gives

$$T_A^\perp(\mathbb{F}P^m) = \{\xi \in H(m+1;\mathbb{F}) : A\xi = \xi A\}. \qquad (9.55)$$

For each $A \in \varphi(\mathbb{F}P^m)$, we have

$$\left\langle A - \frac{1}{m+1}I, A - \frac{1}{m+1}I \right\rangle = \frac{1}{2}\mathrm{Tr}\left(A - \frac{1}{m+1}I\right)^2 = \frac{m}{2(m+1)}.$$

Therefore $\mathbb{F}P^m$ is imbedded in a hypersphere $S(r)$ of $H(m+1;\mathbb{F})$ centered at $(1/(m+1))I$ with radius $r = \sqrt{m/2(m+1)}$.

Let $X \in T_A(\mathbb{F}P^m)$ and Y a vector field tangent to $\mathbb{F}P^m$. Consider a curve $A(t)$ in $\varphi(\mathbb{F}P^m)$ so that $A(0) = A$ and $A'(0) = X$. Denote by $Y(t)$ the restriction of Y to $A(t)$. Since $Y(t) \in T_{A(t)}(\mathbb{F}P^m)$, (9.53) implies

$$A(t)Y(t) + Y(t)A(t) = Y(t), \qquad (9.56)$$

from which we find

$$\tilde{\nabla}_X Y = Y'(0) = A(\tilde{\nabla}_X Y) + (\tilde{\nabla}_X Y)A + XY + YX, \qquad (9.57)$$

where $\tilde{\nabla}$ is the Levi-Civita connection of $H(m+1;\mathbb{F})$. Using (9.53) we get

$$AXY = XYA. \qquad (9.58)$$

Thus

$$A(XY + YX)(I - 2A) = -XYA - YXA = (XY + YX)(I - 2A)A.$$

Hence we find from (9.55) that

$$(XY + YX)(I - 2A) \in T_A^\perp(\mathbb{F}P^m). \qquad (9.59)$$

On the other hand, by multiplying A to (9.56) from the right, we get

$$(XY + YX)A + A(\tilde{\nabla}_X Y)A = 0. \qquad (9.60)$$

Thus we derive from (9.50), (9.58) and (9.59) that

$$2(XY + YX)A + A(\tilde{\nabla}_X Y) + (\tilde{\nabla}_X Y)A \in T_A(\mathbb{F}P^m). \qquad (9.61)$$

Now, by combining (9.57), (9.59) and (9.60), we obtain

$$\tilde{h}(X, Y) = (XY + YX)(I - 2A), \qquad (9.62)$$

$$\nabla_X Y = 2(XY + YX)A + A(\tilde{\nabla}_X Y) + (\tilde{\nabla}_X Y)A, \qquad (9.63)$$

where $\tilde{h}$ is the second fundamental form of $\mathbb{F}P^m$ in $H(m+1;\mathbb{F})$ at A and ∇ is the induced connection on $\mathbb{F}P^m$. It is direct to verify that the second fundamental form $\tilde{h}$ of $\mathbb{F}P^m$ is parallel, i.e., $\tilde{\nabla}\tilde{h} = 0$.

From (9.62) we conclude that the mean curvature vector $\tilde{H}$ of $\mathbb{F}P^m$ in $H(m+1;\mathbb{F})$ at A is given by

$$\tilde{H} = \frac{2}{m}\{I - (m+1)A\}, \tag{9.64}$$

which is parallel to the radius vector $A - I/(m+1)$. Hence $\mathbb{F}P^m$ is imbedded in the hypersphere $S(r)$ as a minimal submanifold. Thus the imbedding is constructed with eigenfunctions with eigenvalue $\lambda = 2(m+1)d$. Since $2(m+1)d$ is exactly the first nonzero eigenvalue λ_1 of Δ on $\mathbb{F}P^m$, we conclude that φ is first standard imbedding.

We summarize these results as the following well-known theorem.

Theorem 9.11. *The isometric imbedding $\varphi : \mathbb{F}P^m \to H(m+1;\mathbb{F})$ defined by (9.49) is the first standard imbedding of $\mathbb{F}P^m$ into $H(m+1;\mathbb{F})$. Moreover, the second fundamental form $\tilde{h}$ and the mean curvature vector $\tilde{H}$ of $\mathbb{F}P^m$ in $H(m+1;\mathbb{F})$ are (9.62) and (9.64). Also, $\mathbb{F}P^m$ lies in a hypersphere of $H(m+1;\mathbb{F})$ centered at $I/(m+1)$ and with radius $r = \sqrt{m/2(m+1)}$.*

Let $A = zz^*$ be a point in $\varphi(\mathbb{F}P^m)$. For each $X \in T_A(\mathbb{F}P^m)$, there is a vector in $\mathbb{F}^{m+1}$ such that $\alpha(z,v) = 0$ and $X = vz^* + zv^*$. If $\mathbb{F} = \mathbb{C}$, we put

$$JX = viz^* - ziv^*. \tag{9.65}$$

The J defines the complex structure of $\varphi(CP^m)$. Similarly, we may define the quaternionic structure $\{J_1, J_2, J_3\}$ on $\varphi(QP^m)$. Let $X = uz^* + zu^*, Y = vz^* + zv^*$ be two vectors in $T_A(CP^m)$, where $A = zz^*, \alpha(v,z) = \alpha(u,z) = 0$, and $z \in S^{2(m+1)-1}(1)$. Then $JX = uiz^* - ziu^*$, $JY = viz^* - ziv^*$. Thus

$$(JX)(JY) = uv^* + zu^*vz^* = XY. \tag{9.66}$$

Consequently, by using (9.62) we obtain

$$\tilde{h}(JX, JY) = \tilde{h}(X,Y) \ \ X,Y \in T_A(CP^m). \tag{9.67}$$

A similar formula holds for QP^m in $H(m+1;\mathbb{Q})$.

Let $z_0 = \{1, 0, \ldots, 0\}^T$ and $A_0 = z_0 z_0^*$. Then (9.54) implies

$$T_{A_0}(\mathbb{F}P^m) = \left\{ X = \begin{pmatrix} 0 & b^* \\ b & 0 \end{pmatrix} : b \in \mathbb{F}^m \right\}. \tag{9.68}$$

Using (9.46) we see that a vector $X \in T_{A_0}(\mathbb{F}P^m)$ is a unit vector if and only if X takes the form: $X = \begin{pmatrix} 0 & b^* \\ b & 0 \end{pmatrix}$, with $b^*b = 1$. Thus we derive via (9.46) and (9.62) that

$$\tilde{h}(X,X) = 2 \begin{pmatrix} -1 & 0 \\ 0 & bb^* \end{pmatrix}. \tag{9.69}$$

Since $\mathbb{F}P^m$ is equivariantly imbedded in $H(m+1;\mathbb{F})$, (9.69) yields

$$\|\tilde{h}(X,X)\| = 2 \text{ for any unit vector } X \in T_A(\mathbb{F}P^m), \ A \in \varphi(\mathbb{F}P^m). \quad (9.70)$$

We need the following.

Lemma 9.3. *Let $\mathbb{F}P^m$ be imbedded by φ into $H(m+1;\mathbb{F})$. Then for any orthonormal vectors X, Y tangent to $\mathbb{F}P^m$ the sectional curvature $\tilde{K}$ of $\mathbb{F}P^m$ satisfies*

$$\langle \tilde{h}(X,X), \tilde{h}(Y,Y) \rangle = \frac{1}{3}\{4 + 2\tilde{K}(X,Y)\}.$$

Proof. Since X, Y are orthonormal, (9.70) implies

$$32 = \langle \tilde{h}(X+Y, X+Y), \tilde{h}(X+Y, X+Y) \rangle$$
$$+ \langle \tilde{h}(X-Y, X-Y), \tilde{h}(X-Y, X-Y) \rangle$$
$$= 16 + 4\langle \tilde{h}(X,X), \tilde{h}(Y,Y) \rangle + 8\langle \tilde{h}(X,Y), \tilde{h}(X,Y) \rangle.$$

On the other hand, it follows from the equation of Gauss that

$$\tilde{K}(X,Y) = \langle \tilde{h}(X,X), \tilde{h}(Y,Y) \rangle - \langle \tilde{h}(X,Y), \tilde{h}(X,Y) \rangle.$$

Therefore we obtain the lemma. □

Lemma 9.4. *If M is an n-dimensional submanifold of $\mathbb{F}P^m$ imbedded in $H(m+1;\mathbb{F})$ by φ, then the squared mean curvature H^2 of M in $H(m+1;\mathbb{F})$ and H'^2 in $\mathbb{F}P^m$ are related by*

$$H^2 - H'^2 = \frac{4(n+2)}{3n} + \frac{2}{3n^2} \sum_{i \neq j} \tilde{K}(e_i, e_j), \quad (9.71)$$

where $e_1, \ldots, e_n$ is an orthonormal frame of $T(M)$.

Proof. Follows from (9.70) and Lemma 9.3. □

9.7 λ_1 of minimal submanifolds of projective spaces

Now, we apply finite type theory to obtain sharp estimate of λ_1 for minimal submanifolds of projective spaces.

We need the following.

Theorem 9.12. [Dimitric (2009)] *Let $\varphi : RP^m \to H(m+1;\mathbb{R})$ be the first standard imbedding of RP^m into $H(m+1;\mathbb{R})$. Then a submanifold M of RP^m is of 1-type in $H(m+1;\mathbb{R})$ if and only if it is a totally geodesic RP^n in RP^m with $n = \dim M$. In this case, M is order 1 in $H(m+1;\mathbb{R})$.*

Proof. Under the hypothesis, we shall identify RP^m with its image under the imbedding φ. For $A \in \varphi(RP)^m$, the second fundamental form $\tilde{h}$ and the shape operator $\tilde{A}$ of the imbedding φ satisfy

$$\langle \tilde{h}(X,Y), I \rangle = 0, \quad \langle \tilde{h}(X,Y), A \rangle = -\langle X, Y \rangle, \tag{9.72}$$

$$\langle \tilde{h}(X,Y), \tilde{h}(V,W) \rangle = 2 \langle X, Y \rangle \langle V, W \rangle \tag{9.73}$$
$$+ \langle X, V \rangle \langle Y, W \rangle + \langle X, W \rangle \langle Y, V \rangle,$$

$$\tilde{A}_{\tilde{h}(X,Y)} V = 2 \langle X, Y \rangle V + \langle X, V \rangle Y + \langle Y, V \rangle X. \tag{9.74}$$

It is easy to verify that one orthonormal basis of the normal bundle of $\varphi(RP^m)$ at A is formed by the following vectors

$$\left\{ \sqrt{2}A, \tfrac{1}{\sqrt{2}} [\tilde{h}(e_i, e_i) + 2A], \tilde{h}(e_i, e_j) : 1 \le i, j \le m, i < j \right\}. \tag{9.75}$$

Thus

$$I = (m+1)A + \frac{1}{2} \sum_{t=1}^{m} \tilde{h}(e_t, e_t). \tag{9.76}$$

Assume that $x : M \to RP^m$ is an isometric immersion of a Riemannian n-manifold M into RP^m. Denote by $\tilde{x} = \varphi \circ x : M \to H(m+1; \mathbb{R})$. From Beltrami's formula, we have

$$\Delta \tilde{x} = -n\tilde{H} = -nH - \sum_{i=1}^{n} \tilde{h}(e_i, e_i), \tag{9.77}$$

where $e_1, \ldots, e_n$ is an orthonormal frame of M and $\tilde{H}$ and H denote the mean curvature vector of M in $H(m+1; \mathbb{R})$ and in RP^m, respectively.

If $\tilde{x}$ is of 1-type, then we have $\tilde{x} = \tilde{x}_0 + \tilde{x}_1$, where $\tilde{x}_0$ is a constant vector and $\tilde{x}_1$ is a non-constant map which satisfies $\Delta \tilde{x}_1 = \lambda x_1$ for some real number λ. Then $\Delta \tilde{x} = \lambda(\tilde{x} - \tilde{x}_0)$, i.e., according to (9.77)

$$\lambda(\tilde{x} - \tilde{x}_0) + nH + \sum_{i=1}^{n} \tilde{h}(e_i, e_i) = 0. \tag{9.78}$$

Differentiate this formula with respect to a tangent vector X to get

$$\lambda X + n\tilde{h}(H, X) - nA_H X + nD_X H - \sum_i \tilde{A}_{\tilde{h}(e_i, e_i)} X$$
$$+ \sum_i \bar{D}_X \tilde{h}(e_i, e_i) = 0.$$

Using (9.74) and the parallelism of $\tilde{h}$ this is equivalent to

$$[\lambda - 2c(n+1)]X - nA_H X + nD_X H + n\tilde{h}(H, X) + 2 \sum_r \tilde{h}(e_r, A_r X) = 0.$$

Separating the part normal to RP^m yields $\sum_r \tilde{h}(B_r X, e_r) = 0$, where $B_r :=$ $(\operatorname{tr} A_r)I + 2A_r$. If ξ is an arbitrary normal vector field of M, then we find from (9.74) that $0 = \sum_r \bar{A}_{\tilde{h}(B_r X, e_r)}\xi = (\operatorname{Tr} A_\xi)X + 2A_\xi X$, i.e., $2A_\xi = -(\operatorname{tr} A_\xi)\,I$. Taking the trace of this gives $A_\xi = 0$. Hence the submanifold is totally geodesic. Therefore, M is then an open portion of a canonically embedded totally geodesic RP^n.

The converse can be verify by direct computation. $\qquad\square$

Remark 9.3. When M is a minimal submanifold of RP^m, Theorem 9.12 is due to [Ros (1983)].

We also need the following lemmas.

Lemma 9.5. *If M is an n-dimensional submanifold of $\mathbb{F}P^m$ imbedded in $H(m+1;\mathbb{F})$ by φ, then we have*

$$H^2 \geq H'^2 + \frac{2(n+1)}{n}, \tag{9.79}$$

with the equality holding if and only if M is totally real in $\mathbb{F}P^m$.

Proof. It is known that the sectional curvature $\tilde{K}(X,Y)$ of $\mathbb{F}P^m$ is ≥ 1, with the equality holding if and only if X, Y span a totally real plane. Thus we obtain this lemma from Lemma 9.4. $\qquad\square$

Definition 9.2. A submanifold M of CP^m is a *complex submanifold* if each tangent space of M is invariant under the action of the complex structure J of CP^m. Similarly, a submanifold M of QP^m is called *quaternionic* if $J_i(T_x(M)) = T_x(M)$, $x \in M$, for $i = 1, 2, 3$. Complex submanifolds and quaternionic submanifolds are also known as *invariant submanifolds*.

Lemma 9.6. *Let M be a minimal submanifold of $\mathbb{F}P^m$ ($\mathbb{F} = \mathbb{R}, \mathbb{C}$ or $\mathbb{Q}$) which is imbedded in $H(m+1;\mathbb{F})$ by φ. If $n = \dim M \geq d$, then we have*

$$H^2 \leq \frac{2(n+d)}{n}, \quad d = d(\mathbb{F}), \tag{9.80}$$

with the equality holding if and only if $n \equiv 0 \;(mod\,d)$ and M is an invariant submanifold of $\mathbb{F}P^m$

Proof. If $\mathbb{F} = \mathbb{R}$, this follows from $\tilde{K}(e_i, e_j) = 1$ for $1 \leq i \neq j \leq n$.

If $\mathbb{F} = \mathbb{C}$, then $\tilde{K}(e_i, e_j) = 1 + 3\langle e_i, Je_j\rangle^2$. Thus Lemma 9.4 gives

$$H^2 = \frac{2(n+1)}{n} + \frac{2}{n^2}\sum_{i \neq j} \tilde{K}(e_i, e_j). \tag{9.81}$$

Hence we have

$$H^2 = \frac{2(n+1)}{n} + \frac{2}{n^2}||P||^2, \tag{9.82}$$

where P is the endomorphism of $T(M)$ defined by $\langle PX, Y \rangle = \langle JX, Y \rangle$.

Since $||P||^2 \leq n$, with the equality holding if and only if n is even and $T(M)$ is invariant under J, (9.82) implies (9.80) with $d = 2$, with equality holding if and only if M is a complex submanifold.

Similar proof applies to minimal submanifolds in QP^m as well. □

Theorem 9.13. [Chen (1983c)] *If M is a minimal closed submanifold of $RP^m(1)$, we have*

$$\lambda_1 \leq 2(n+1), \quad n = \dim M. \tag{9.83}$$

Equality holding if and only if M is a totally geodesic $RP^n(1) \subset RP^m(1)$.

Proof. Under the hypothesis of the theorem, it follows from Lemma 9.6 that $H^2 \leq 2(n+1)/n$. Combining this with inequality (9.7) gives (9.83).

If the equality sign of (9.83) holds, Theorem 9.1 implies that M is of 1-type. Thus we conclude from Theorem 9.12 that M is a totally geodesic $RP^n(1)$ in $RP^m(1)$. □

Theorem 9.14. [Chen (1983c)] *If M is a closed n-dimensional minimal submanifold of $CP^m(4)$, then we have*

$$\lambda_1 \leq 2(n+2), \tag{9.84}$$

equality holding if and only if M is a totally geodesic $CP^{\frac{n}{2}}(4) \subset CP^m(4)$.

Proof. Let $CP^m(4)$ be holomorphically and isometrically imbedded in $H(m+1;\mathbb{C})$ by its first standard imbedding. If M is a closed n-dimensional minimal submanifold of $CP^m(4)$, then Lemma 9.6 gives

$$H^2 \leq \frac{2(n+2)}{n}, \tag{9.85}$$

with equality holding if and only if n is even and M is a complex submanifold of $CP^m(4)$. By combining this with inequality (9.7) we obtain inequality (9.84). If the equality sign of (9.85) holds, then the equality sign of (9.84) holds. Thus n is even and M is a complex submanifold. On the other hand, we also have $\int_M H^2 dV = (\lambda_1/n)\,\text{vol}(M)$. Thus we may conclude from Theorem 9.2 that M is of order 1 in $H(m+1;\mathbb{C})$. Hence, by Theorem 9.12, M is a totally geodesic $CP^{n/2}(4)$ in $CP^m(4)$.

The converse is trivial. □

Remark 9.4. If M is a closed complex submanifold of $CP^m(4)$, Theorem 9.14 is due to A. Ros [Ros (1984a)] and N. Ejiri (unpublished).

For minimal submanifolds of QP^m we have the following.

Theorem 9.15. [Chen (1983c)] *Let M be a closed minimal submanifold of $QP^m(4)$. Then we have $\lambda_1 \leq 2(n+4)$, with the equality holding if and only if M is a totally geodesic $QP^{n/4}(4)$.*

Proof. This can be done in the same way as Theorem 9.14.   □

Definition 9.3. A submanifold M of a Kähler manifold is called a *CR-submanifold* if there exist distributions $\mathcal{D}$ and $\mathcal{D}^\perp$ such that

$$T(M) = \mathcal{D} \oplus \mathcal{D}^\perp, \quad J\mathcal{D} = \mathcal{D}, \quad J\mathcal{D}^\perp \subset T^\perp(M).$$

For CR-submanifolds, A. Ros proved the following result.

Theorem 9.16. [Ros (1983)] *Let M be a closed minimal CR-submanifold of $CP^m(4)$. Then we have $\lambda_1 \leq 2(n^2 + n + 2r)/n$ with $r = \mathrm{rank}_\mathbb{C} \mathcal{D}$.*

9.8   Further applications to spectral geometry

In this section, we present further applications of finite type theory to spectral geometry of submanifolds involving λ_1, λ_2.

By applying the spectral decomposition of submanifolds, A. Ros proved the following theorems and corollaries.

Theorem 9.17. [Ros (1984a)] *If M is a closed Kähler submanifold of $CP^m(4)$, then we have*

$$\frac{n}{2}\{n + 1 + (n + 1 - \lambda_1)(n + 1 - \lambda_2)\}\mathrm{vol}(M) \geq \int_M \tau \, dV, \qquad (9.86)$$

where $n = \dim_\mathbb{C} M$ and τ is the scalar curvature of M.

If the equality sign of (9.86) holds, then M is Einstein-Kähler.

Theorem 9.18. [Ros (1984b)] *Let M is a closed n-dimensional minimal submanifold of $S^m(1)$. Then we have*

$$\{n[2(n + 1) - \lambda_1][2(n + 1) - \lambda_2] + 4n(n - 1)\}\mathrm{vol}(M) \geq 8 \int_M \tau \, dV. \quad (9.87)$$

If the equality sign of (9.87) holds, then M is Einstein.

Theorem 9.19. [Ros (1984b)] *Let M is a closed n-manifold which admits an order 1 immersion in a Euclidean space. Then we have*

$$n\{2(n+5)\lambda_1 - (n+2)\lambda_2\}\mathrm{vol}(M) \geq 8\int_M \tau\, dV.$$

Ros also gave applications of Theorem 9.18. E.g., he gave the following.

Corollary 9.6. [Ros (1984b)] *Let M be an n-dimensional closed Kähler manifold immersed in $CP^m(4)$. If $\lambda_1 = 2\left(\int_M \tau dV\right)/(n\,\mathrm{vol}(M))$ and if M is not totally geodesic, then $\lambda_2 \leq n+2$. The equality sign holds only when M is an Einstein submanifold with parallel second fundamental form.*

Corollary 9.7. [Ros (1984b)] *If a closed Riemannian n-manifold which admits a standard immersion of order 1, then $2m_1(n+2)\lambda_1 \geq n(m_1+1)\lambda_2$, where m_1 is the dimension of the λ_1-eigenspace of the Laplacian.*

Definition 9.4. Let M be a closed Riemannian manifold. Then it has a unique kernel of the heat equation: $K : M \times M \times \mathbb{R}_+^* \to \mathbb{R}$. If there is a function $\Psi : \mathbb{R}_+ \times \mathbb{R}_+^* \to \mathbb{R}$ such that $K(u,v,t) = \Psi(d(u,v),t)$ for $u,v \in M$ and $r \in \mathbb{R}_+^*$, then M is called *strongly harmonic*.

Since every strongly harmonic closed manifold admit order 1 immersion (cf. [Besse (1978)]), Corollary 9.7 implies the following.

Corollary 9.8. [Ros (1984b)] *Let M be a strongly harmonic n-manifold. Then $2m_1(n+2)\lambda_1 \geq n(m_1+1)\lambda_2$, where m_1 is the multiplicity of λ_1, with the equality sign holding if and only if M is a compact symmetric space of rank one.*

S. Montiel, A. Ros and F. Urbano proved the following.

Theorem 9.20. [Montiel et al. (1986)] *Let M be an n-dimensional minimal closed submanifold of $S^m(1)$ which is of order 1 in $\mathbb{E}^{m+1}$. Then*

$$n\{2(n+5)\lambda_1 - (n+2)\lambda_2\}\mathrm{vol}(M) \geq 8\int_M \tau\, dV$$

where τ is the scalar curvature. If the immersion is the standard one, then

$$\left(\frac{n+2}{n}\right)\left(\frac{2m_1}{m_1+1}\right)\lambda_1 \geq \lambda_2, \quad \int_M \tau dV \leq \frac{\lambda_2}{4}\left\{n-4+\frac{(n+2)^2}{m_1+1}\right\}\mathrm{vol}(M),$$

where m_1 is the dimension of the λ_1-eigenspace of Δ. Either equality above holds if and only if M is a rank one compact symmetric space.

For more applications of finite type theory to closed minimal surfaces of S^m, see [Barros and Urbano (1987); Udagawa (1986c)].

9.9   Application to variational principle: k-minimality

In this section, we present a new variational principle introduced in [Chen et al. (1995)]. This principle was motivated by the desire to understand via finite type theory two classical variational minimal principles in spirit of amongst others H. Hopf and H. Poincaré.

Let $\mathcal{E}$ be a class of variations acting on a closed submanifold M of a Riemannian manifold $\tilde{M}$. Then M is said to *satisfy the variational minimal principle in the class $\mathcal{E}$* if it is a critical point of the volume-functional in the class $\mathcal{E}$. It is well-known that a compact submanifold M in a Riemannian manifold is a minimal submanifold if and only if it satisfies the variational minimal principle under all normal variations of M.

A classical result of H. Hopf is the following: Let M be a imbedded closed hypersurface in $\mathbb{E}^{n+1}$ which encloses an $(n+1)$-dimensional volume. If we consider the class $\mathcal{E}_1$ of normal variations on M which preserve the enclosed volume, then M satisfies the variational minimal principle in the class $\mathcal{E}_1$ if and only if M is a CMC hypersurface. Thus, by Alexandrov's theorem, if and only if it is a hypersphere. Another well-known variational minimal principle is due to H. Poincaré: Let E be the equator of an ordinary 2-sphere S^2 and $\mathcal{E}_2$ the class of normal variations on E which bisects the area of S^2. Then the equator is stable under the action of $\mathcal{E}_2$; in particular, E satisfies the variational minimal principle in the class $\mathcal{E}_2$. We simply call the normal variations in the classes $\mathcal{E}_1$ and $\mathcal{E}_2$ the *Hopf and the Poincaré deformations*, respectively.

Inspired by the work of Poincaré, Y.-G. Oh introduced in [Oh (1990)] the notion of Hamiltonian deformations for Lagrangian submanifolds; also, the author and J.-M. Morvan introduced the notion of isotropic, exact and harmonic deformations for isotropic (or totally real) submanifolds in [Chen and Morvan (1994)]. By using harmonic minimality, the author and Morvan were able to solve an open problem on Maslov class of Lagrangian submanifolds proposed in [Le Khong Van and Fomenko (1988)]. Further, it was proved in [Chen et al. (1993a)] that submanifolds of finite type satisfy a spectral variational minimal principle.

In this section, following the ideas of [Chen et al. (1993a); Chen and Morvan (1994)] and using elementary spectral theory, we define the notion of k-deformations ($k \in \mathbb{N}$). From which we introduce the notions of k-minimality and k-stability. In terms of these notions, the "usual" minimality is nothing but 0-minimality and CMC hypersurfaces become 1-minimal hypersurfaces. Furthermore, both Hopf's and Poincaré's deformations are

1-deformations in this variational theory.

For each integer $q \geq 1$, let $\mathcal{F}_q$ denote the class of all normal deformations associated to functions $f \in \mathcal{D}_q := \sum_{i \geq q} V_i$, where V_i is the λ_i-eigenspace of the Laplacian of M. It is clear that $\mathcal{F}_0 \supset \mathcal{F}_1 \supset \mathcal{F}_2 \supset \cdots \supset \mathcal{F}_k \supset \cdots$.

Definition 9.5. Let M be an oriented closed hypersurface M of a Riemannian manifold. By a k-*deformation* of M we mean a deformation in the class $\mathcal{F}_k$. The hypersurface M is called k-*minimal* if it is a critical point of the volume functional $\mathcal{V}(t)$, i.e., $\mathcal{V}'(0) = 0$, for all variations in $\mathcal{F}_k$. For a k-minimal hypersurface M and for an integer $\ell \geq k$, M is called ℓ-*stable* if $\mathcal{V}''(0) \geq 0$ for all variations in $\mathcal{F}_\ell$. A k-minimal hypersurface is simply called *stable* if it is k-stable.

Analogous to Theorem 6.1 for finite type submanifolds, k-minimality can also be characterized in terms of the minimal polynomial.

Proposition 9.6. *Let M be an oriented closed hypersurface of $\mathbb{E}^{n+1}$. Then M is k-minimal for some k if and only if there is a nontrivial polynomial P such that the mean curvature function α of M in $\mathbb{E}^{n+1}$ satisfies $P(\Delta)(\alpha) = 0$, where Δ is the Laplacian of M.*

Proof. This can be done in the same as Theorem 6.1. $\square$

Let $\phi : M \to \tilde{M}$ be an isometric immersion from an oriented closed Riemannian n-manifold M into a Riemannian $(n + 1)$-manifold $\tilde{M}$ with a global unit normal vector field ξ. The normal variation

$$\phi_t(x) := \exp_{\phi(x)} tf(x)\xi(x), \quad x \in M, \quad t \in (-\epsilon, \epsilon), \tag{9.88}$$

is called an *exact deformation* if fdV is an exact form on M, where dV is the volume form of M.

The following proposition is classical.

Proposition 9.7. *Let M be an oriented closed hypersurface of a Riemannian manifold $\tilde{M}$, ϕ_t the variation of M given by (9.88), and α the mean curvature function. Then*

(a) $\mathcal{V}'(0) = -n \int_M \alpha f dV,$
(b) $\mathcal{V}''(0) = \int_M \left(||\nabla f||^2 + f^2(2\tau - 2\tilde{\tau} + \tilde{S}(\xi, \xi)) \right) dV,$

where τ denotes the scalar curvature of M and $\tilde{\tau}$ and $\tilde{S}$ denote the scalar curvature and Ricci tensor of $\tilde{M}$.

Corollary 9.9. *If M and $\tilde{M}$ are as in Proposition 9.7 and if the sectional curvatures $\tilde{K}$ of $\tilde{M}$ satisfy $\tilde{K} \leq c$, then*

$$\mathcal{V}''(0) \geq \int_M \left(\|\nabla f\|^2 + f^2(2\tau - n^2 c) \right) dV,$$

and if $\tilde{K} \equiv c$, then equality occurs.

Among the k-deformations, 1-deformations play a special role.

Proposition 9.8. [Chen et al. (1995)] *Let M be an oriented closed hypersurface of a Riemannian manifold $\tilde{M}$. Then (9.88) is 1-deformation if and only if it is an exact deformation.*

Proof. According to the Hodge decomposition theorem, every differential form can be written as the sum of an exact, a co-exact and a harmonic form. In particular, the n-form $f DA$ can be expressed as

$$f dV = d\eta + c dV, \tag{9.89}$$

where c is a constant. By integrating both sides of (9.89), we see that $c = 0$ if and only if $\int f dV = 0$. Hence $f dV$ is exact if and only if $\int f dV = 0$. On the other hand, by taking the spectral decomposition $f = f_0 + \sum_{i=1}^{\infty} f_i$ and integrating this, we have $\int f dV = 0$ if and only if $f_0 = 0$, i.e., $f \in \mathcal{D}_1$. Hence (9.88) is exact if and only if it is a 1-deformation. $\qquad\square$

In above, we consider normal variations of the form given by (9.88). These deformations are however rather special. For instance, it is in general not possible to find such variations that leave the enclosed volume fixed. A general normal variation is usually defined by

$$\phi_t(x) := \exp_{\phi(x)} t F(t, x) \xi(x). \tag{9.90}$$

If $f(x) = F(0, x)$, then (9.90) is regarded as a *linearization* of (9.88). A general normal deformation is called a k-*deformation* if its linearization is.

If we denote the volume under the deformation (9.88) by $\mathcal{V}(t)$ and the corresponding function under the deformation (9.90) by $\tilde{\mathcal{V}}(t)$, then $\mathcal{V}'(0) = \tilde{\mathcal{V}}'(0)$, if (9.88) is a linearization of (9.90). Moreover, if $\mathcal{V}'(0) = 0$, then $\mathcal{V}''(0) = \tilde{\mathcal{V}}''(0)$. Let M be an oriented closed hypersurface of $\tilde{M}$ immersed by ϕ. Suppose that M is the boundary of a compact manifold $\tilde{M}$ and ϕ can be extended to $\tilde{M}$. A sufficiently small variation of M can be extended to a variation of $\tilde{M}$. Let ϕ_t be a variation of M given by (9.88) and suppose ϕ_t can be extended to a deformation $\tilde{\phi}_t$ of $\tilde{M}$. Let $\hat{\mathcal{V}}(t)$ denote the volume of $\tilde{\phi}_t(\tilde{M})$. We call $\hat{\mathcal{V}}(t)$ the enclosed volume of M_t. The variation is called a *Hopf's deformation* if the enclosed volume $\hat{\mathcal{V}}(t)$ of M_t is constant.

Similarly, if we denote the enclosed volume under the deformation (9.88) by $\bar{\mathcal{V}}(t)$ and the corresponding function under (9.90) by $\tilde{\mathcal{V}}(t)$, then we have $\bar{\mathcal{V}}'(0) = \tilde{\mathcal{V}}'(0)$, whenever (9.88) is a linearization of (9.90).

The following variational formula for $\bar{\mathcal{V}}$ is well-known.

Proposition 9.9. *Let M, $\tilde{M}$ and $\tilde{M}$ be as above. Then $\bar{\mathcal{V}}'(0) = \int_M f dV$, where dV is the volume form of M.*

For Hopf deformations we have the following.

Proposition 9.10. [Chen et al. (1995)] *Hopf deformations are (general) 1-deformations.*

Proof. If a deformation with linearization given by (9.88) is a Hopf deformation, then it follows from Proposition 9.9 that $f_0 = 0$, i.e., $f \in \mathcal{D}_1$. Hence the deformation is a 1-deformation. $\square$

For Poincaré deformations we have

Proposition 9.11. [Chen et al. (1995)] *Let E be the equator of S^2. Then every Poincaré deformation on E is a 1-deformation.*

Proof. This follows from Proposition 9.9 and the fact that every Poincaré deformation on E preserves the area enclosed by E. $\square$

In the classical variational theory, minimal closed hypersurfaces are not always stable. On the contrary, we have the following stability theorem.

Theorem 9.21. [Chen et al. (1995)] *Every oriented k-minimal closed hypersurface is q-stable for some $q \geq k$.*

Proof. If M is an oriented closed hypersurface and if $f \in \mathcal{D}_k$. Then we have $(\nabla f, \nabla f) = (\Delta f, f) \geq \lambda_k(f, f)$. Thus Proposition 9.7 implies

$$\mathcal{V}''(0) \geq \int_M f^2(\lambda_k + 2\tau - 2\tilde{\tau} + \tilde{S}(\xi, \xi)) dV,$$

for $f \in \mathcal{D}_k$. Thus $\mathcal{V}''(0) \geq 0$ if q is chosen to be a sufficiently large. $\square$

The next result characterizes all k-minimal hypersurfaces.

Theorem 9.22. [Chen et al. (1995)] *Let M be an oriented closed hypersurface of a Riemannian manifold $\tilde{M}$ and α the mean curvature function of M. Then M is a k-minimal hypersurface if and only if $\alpha \in \sum_{i=0}^{k-1} V_i$.*

Theorem 9.22 implies the following.

Corollary 9.10. *An oriented closed hypersurface M of $\tilde{M}$ is*

(1) *0-minimal if and only if it is minimal in the usual sense,*
(2) *1-minimal if and only if it is a CMC hypersurface,*
(3) *2-minimal if and only if the mean curvature function α can be written as $\alpha = \alpha_0 + \alpha_1$, where α_0 is a constant and $\Delta\alpha_1 = \lambda\alpha_1$.*

Example 9.1. Consider the Clifford torus $T^2 = S^1(1/\sqrt{2}) \times S^1(1/\sqrt{2})$ in $S^3(1)$. Then T^2 is 0-minimal. Since $\lambda_1 = 2$ and $\lambda_2 = 4$, the Clifford torus is neither 0-stable nor 1-stable. But it is 2-stable in $S^3(1)$.

Example 9.2. Consider the flat torus $T^2_{a,b} = S^1(a) \times S^1(b)$ in $S^3(1)$ with $a^2 + b^2 = 1$, $a < b$. Then $1 < \lambda_1 \leq 2$ and $\lambda_3 \geq 4$, in general. When $a \leq \frac{1}{2}$, we have $\lambda_2 \geq 4$. Hence $T^2_{a,b}$ is 1-minimal which is 3-stable in general, and it is 2-stable whenever $a \leq \frac{1}{2}$.

Example 9.3. Let M be a closed hypersurface with scalar curvature τ in a Euclidean space. If M is CMC, then it is 1-minimal, and if it satisfies $2\tau + \lambda_1 \geq 0$, it is stable. In particular, if $\tau \geq 0$, it is always stable.

Example 9.4. A great circle of S^2 is 0-minimal and 1-stable, but it is not 0-stable. A small circle is 1-minimal and 1-stable. Conversely, a 0-minimal or 1-minimal curve on S^2 is a circle. Hence, if γ is an area-bisecting curve on S^2 and if we take a variation by area-bisecting curves, then the length of γ is minimal if and only if γ is a great circle (H. Poincaré).

An ordinary hypersphere S^n is 1-minimal. Conversely, we have:

Theorem 9.23. [Chen et al. (1995)] *The only k-minimal ellipsoids in a Euclidean space are ordinary hyperspheres.*

Consider a closed planar curve γ of length 2π. It follows from Proposition 9.7(b) that $\mathcal{V}''(0) \geq 0$. Thus each k-minimal curve in $\mathbb{E}^2$ is stable.

Let s be the arc length of γ. Then the Laplacian is given by $-d^2/ds^2$. The eigenvalues of the Laplacian are $\{\lambda_n = n^2 : n \in \mathbb{N}\}$ and a basis of the corresponding eigenspace is given by $\{\cos(ns), \sin(ns)\}$. Hence the spectral decomposition of a function is a Fourier series expansion.

Lemma 9.7. *A closed curve in the plane is k-minimal if and only if its curvature function does not contain any terms $\cos(ns)$ or $\sin(ns)$ for $n \geq k$.*

It follows from Lemma 9.7 that one can write the curvature function κ of a k-minimal curve γ as

$$\kappa(s) = a_0 + \sum_{n=1}^{k-1}\{a_n \cos(ns) + b_n \sin(ns)\}. \tag{9.91}$$

Integrating this over γ gives us $2\pi a_0 = 2\pi i_\gamma$, where i_γ is the rotation index of γ. Thus a_0 is an integer. Moreover, we can recover γ from κ by

$$\gamma(s) = \left(\int_0^s \cos\left(\int_0^u \kappa(t)dt\right) du, \int_0^s \sin\left(\int_0^u \kappa(t)dt\right) du \right). \tag{9.92}$$

However, not every curve (9.92) is a closed curve. A necessary and sufficient condition for γ, defined by (9.91) and (9.92), to be closed is

$$\int_0^{2\pi} \cos\left(\int_0^u \kappa(t)dt\right) du = \int_0^{2\pi} \sin\left(\int_0^u \kappa(t)dt\right) du = 0. \tag{9.93}$$

The original definition of *Bessel function* J_n of order n is given by

$$\begin{aligned}J_n(k) &= \frac{1}{2\pi} \int_0^{2\pi} \cos(nu - k\sin(u))du \\ &= (-1)^n \frac{1}{2\pi} \int_0^{2\pi} \cos(nu + k\sin(u))du.\end{aligned} \tag{9.94}$$

It is known that the positive zero's of J_n form an infinite discrete series $0 < j_{n,1} < \cdots < j_{n,m} < \cdots \nearrow \infty$, where $j_{n,m}$ is the m-th zero of J_n. The numbers $j_{n,m}$ are non-algebraic and mutually different. Hence we have the following classification of 2-minimal curves in terms of Bessel's functions.

Theorem 9.24. [Chen et al. (1995)] *For each root $j_{\ell,m}$ of the Bessel function J_ℓ of order ℓ, the curve $\gamma_{\ell,m}$ defined by*

$$\gamma_{\ell,m} = \left(\int_0^s \cos(\ell u + j_{\ell,m}\sin(u))du, \int_0^s \sin(\ell u + j_{\ell,m}\sin(u))du \right)$$

is a closed planar 2-minimal curve. Conversely, up to rigid motions, every 2-minimal curve can be obtained in this way.

Moreover, $\gamma_{\ell,m}$ and $\gamma_{\ell',m'}$ are congruent if and only if $\ell = \ell'$, $m = m'$.

Planar 2-minimal curves have some nice global geometric properties.

Theorem 9.25. [Chen et al. (1995)] *Every 2-minimal planar closed curve has a line of symmetry and a point of self-intersection.*

We have the following non-existence result for tubes.

Theorem 9.26. [Chen et al. (1995)] *There are no k-minimal closed tubes in Euclidean 3-space.*

Now we provide two examples of k-minimal planar closed curves.

Example 9.5. The following are 2-minimal and 5-minimal planar closed curves respectively whose curvature functions are given as indicated.

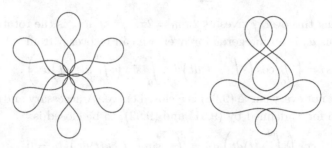

Figure 9.1: $\kappa(s) = 1 + 12\cos 6s$. Figure 9.2: $\kappa(s) = \cos s + 2\cos 2s$
$$+ (8.899655\cdots)\cos 4s.$$

Remark 9.5. For k-minimal closed curves in $\mathbb{E}^3$, see [Petrović et al. (2000)]. Also, for contact version of k-minimal curves for Legendre curves in Sasakian $\mathbf{R}^3(-3)$ see [Blair et al. (2010)]. For k-minimal closed curves in $\mathbb{E}^3$, see [Petrović et al. (2000)].

9.10 Applications to smooth maps

In this section, we present some applications of finite type theory to maps. Analogous to Theorem 6.1, we have the following.

Theorem 9.27. [Chen and Piccinni (1987)] *Let $\phi : M \to \mathbb{E}^m$ be a non-constant map from a closed Riemannian manifold M into $\mathbb{E}^m$. Then*

(a) *ϕ is of finite type if and only if there is a non-trivial polynomial Q such that $Q(\Delta)\tau_1 = 0$, where τ_1 is the tension field of the map ϕ;*
(b) *If ϕ is of finite type, there is a unique monic polynomial $P(t)$ such that $P(\Delta)\tau_1 = 0$;*
(c) *If ϕ is of finite type, then ϕ is of k-type if and only if $\deg P = k$.*

The same results holds if τ_1 is replaced by $\phi - \phi_0$, where ϕ_0 is the center of mass. The monic polynomial P is called the minimal polynomial of ϕ.

Proof. This can be proved in the same as Theorem 6.1. $\qquad\square$

Theorem 9.28. [Chen et al. (1986)] *Let $\phi : (M, g) \to \mathbb{E}^m$ be a non-constant map of a closed Riemannian manifold M into $\mathbb{E}^m$. We have*

$$2\lambda_p E(\phi) \leq \int_M ||\tau_1||^2 dV_g \leq \lambda_q E(\phi), \tag{9.95}$$

where $E(\phi)$ is the energy of ϕ and p and q are lower order p and upper order q of ϕ. Either equality sign in (9.95) holds if and only if ϕ is of 1-type.

Proof. Under the hypothesis, we have the spectral decomposition:

$$\phi = \phi_0 + \sum_{t=p}^{q} \phi_t. \tag{9.96}$$

Because $\Delta\phi = -\tau_1(\phi)$, (9.96) implies $-\tau_1 = \sum_{t=p}^{q} \lambda_t \phi_t$. Since Δ is self-adjoint, this together with (9.96) gives

$$2E(\phi) = (d\phi, d\phi) = (\phi, \Delta\phi) = \sum_{t=p}^{q} \lambda + t(\phi_t, \phi_t), \tag{9.97}$$

$$\int_M ||\tau_1||^2 dV = (\Delta\phi, \Delta\phi) = \sum_{t=p}^{q} \lambda_t^2 (\phi_t, \phi_t). \tag{9.98}$$

Hence

$$\int_M ||\tau_1||^2 dV - 2\lambda_p E(\phi) = \sum_{t=p}^{q} \lambda_t (\lambda_t - \lambda_p)(\phi_t, \phi_t) \geq 0,$$

with equality holding if and only if ϕ_p is the only nonzero component. The other inequality can be proved in the same way. $\qquad\square$

An immediate consequence of Theorem 9.28 is the following.

Corollary 9.11. [Chen et al. (1986)] *If $\phi : M \to \mathbb{E}^m$ is a non-constant map of a closed Riemannian manifold M into $\mathbb{E}^m$, then we have*

$$\int_M ||\tau_1||^2 dV \geq 2\lambda_1 E(x),$$

equality holding if and only if ϕ is of order 1.

We also have the following sharp estimate for total tension.

Theorem 9.29. [Chen (1987b)] *Let $\phi : M \to \mathbb{E}^m$ be a non-constant map of a closed Riemannian manifold M into $\mathbb{E}^m$. Then we have*

$$\int_M ||\tau_1||^2 dV \geq 2(\lambda_p + \lambda_{p+1})E(\phi) - \lambda_p \lambda_{p+1} \mathcal{M}(\phi), \tag{9.99}$$

where $\mathcal{M}(\phi)$ is the moment and p is the lower order of ϕ. With the equality holding if and only if ϕ is either of 1-type or of 2-type and of order $\{p, p+1\}$.

Proof. Under the hypothesis, the moment of ϕ is given by

$$\mathcal{M}(\phi) = \sum_{t=p}^{q} (\phi_t, \phi_t). \qquad (9.100)$$

Thus we find from (9.97), (9.98) and (9.100) that

$$\int_M \|\tau_1\|^2 dV - 2(\lambda_p + \lambda_q)E(\phi) - \lambda_p \lambda_q \mathcal{M}(\phi)$$

$$= \sum_{t=p}^{q} (\lambda_t - \lambda_p)(\lambda_t - \lambda_{p+1})(\phi_t, \phi_t) \geq 0, \qquad (9.101)$$

where implies (9.99). Clearly, the equality sign of (9.100) holds if and only if ϕ is either of 1-type or of 2-type with order $\{p, p+1\}$. $\qquad \square$

When $\phi : M \to \mathbb{E}^m$ is of finite type, we also have the following.

Theorem 9.30. [Chen (1987b)] *Let $\phi : M \to \mathbb{E}^m$ be a non-constant map of a closed Riemannian manifold M into $\mathbb{E}^m$. If ϕ is of finite type, we have*

$$\int_M \|\tau_1\|^2 dV \leq 2(\lambda_p + \lambda_q)E(\phi) - \lambda_p \lambda_q \mathcal{M}(\phi),$$

with the equality sign holding if and only if ϕ is of 1-type or of 2-type.

Analogous to Theorem 9.28, we have the next result.

Theorem 9.31. [Chen (1987b)] *For a non-constant map $\phi : M \to \mathbb{E}^m$ of a closed Riemannian manifold M into $\mathbb{E}^m$, we have*

$$\lambda_p \mathcal{M}(\phi) \leq 2E(\phi) \leq \lambda_q \mathcal{M}(\phi),$$

with either equality holding if and only if ϕ is of 1-type.

Theorem 9.31 implies immediately the following.

Corollary 9.12. *Let γ be a closed curve of length $2\pi r$ in $\mathbb{E}^m$. Then the moment of γ satisfies $\mathcal{M} \leq r^2$, with the equality holding if and only if γ is a planar circle of radius r.*

9.11 Application to Gauss map via topology

The main purpose of this section is to present some simple relations between finite type theory and topology of Gauss map of Euclidean submanifolds. Let V be a linear n-subspace of $\mathbb{E}^m$. If $e_{n+1}, \ldots, e_m$ is an oriented orthonormal basis of V, then $e_{n+1} \wedge \cdots \wedge e_m$ is a decomposable $(m-n)$-vector of norm one and $e_{n+1} \wedge \cdots \wedge e_m$ gives an orientation of V. Conversely, each decomposable $(m - n)$-vector of norm one determines a unique oriented linear n-subspace of $\mathbb{E}^m$. If $G(m, m-n)$ denotes the Grassmann manifold of oriented linear $(m-n)$-subspaces in $\mathbb{E}^m$, we may identify $G(m, m-n)$ with the set of decomposable $(m-n)$-vectors of norm one. Hence $G(m, m-n)$ can be regarded as an $n(m-n)$-dimensional submanifold of the unit hypersphere $S^{N_{m,n}-1}(1)$ centered at the origin in $\mathbb{E}_{m,n}^N = \wedge^{m-n}\mathbb{E}^m$ in a natural way. Therefore we have the following canonical inclusions:

$$G(m, m - n) \subset S^{N_{m,n}-1}(1) \subset \mathbb{E}^{N_{m,n}} = \wedge^{m-n}\mathbb{E}^m, \tag{9.102}$$

where $N_{m,n} = \binom{m}{n-m}$. The classical Gauss map was introduced by Gauss in his fundamental paper [Gauss (1827)] on surfaces. He used the classical Gauss map to define Gauss curvature. Since then the notion of Gauss map plays one of important roles in the theory of submanifolds.

For an n-dimensional submanifold M of $\mathbb{E}^m$ with arbitrary codimension, the *classical Gauss map* $\hat{\nu}$ can be defined to be the map

$$\hat{\nu} : M \to G(m, m - n) \tag{9.103}$$

which carries a point $x \in M$ to the linear $(m - n)$-subspace of $\mathbb{E}^m$ obtained from the normal space $T_x^\perp M$ via parallel displacement.

If $\{e_{n+1}, \ldots, e_m\}$ is an oriented orthonormal frame of $T^\perp M$, then the *Gauss map* of M in $\mathbb{E}^{N_{m,n}}$:

$$\nu : M \to G(m, m - n) \subset S^{N_{m,n}-1}(1) \subset \mathbb{E}^{N_{m,n}} = \wedge^{m-n}\mathbb{E}^m \tag{9.104}$$

is given by

$$\nu(x) = (e_{n+1} \wedge \cdots \wedge e_m)(x). \tag{9.105}$$

Remark 9.6. Using the canonical isomorphism $G(m, m - n) \cong G(m, n)$, one can also define the Gauss map to be the map

$$\bar{\nu} : M \to G(m, n) \subset \wedge^n \mathbb{E}^m \tag{9.106}$$

given by $\bar{\nu}(x) = (e_1 \wedge \cdots \wedge e_n)(x)$ as well.

Lemma 9.8. *Let M be an oriented closed submanifold of $\mathbb{E}^m$. Then the Gauss map ν given by (9.104) is mass-symmetric in $S^{N_{m,n}-1}(1)$.*

Proof. Let $\phi : M \to \mathbb{E}^m$ be an isometric immersion and $e_1, \ldots, e_n$ an oriented orthonormal local frame on M. Denote by $\omega^1, \ldots, \omega^n$ the dual frame. Then we have $d\phi = \sum_{i=1}^n e_i \omega^i$. It is easy to verify that

$$\overbrace{d\phi \wedge \cdots \wedge d\phi}^{n \text{ copies}} = n!(e_1 \wedge \cdots \wedge e_n)\omega^1 \wedge \cdots \wedge \omega^n = n!\, \bar{\nu}\, dV, \tag{9.107}$$

where $\bar{\nu} = e_1 \wedge \cdots \wedge e_n$. It follows from (9.107) that

$$n! \int_M \bar{\nu} dV = \int_M d\overbrace{d\phi \wedge \cdots \wedge d\phi}^{n \text{ copies}} = \int_M d(\phi \wedge \overbrace{d\phi \wedge \cdots \wedge d\phi}^{n-1 \text{ copies}}) = 0.$$

Therefore $\bar{\nu}$ is mass-symmetric. Consequently, the Gauss map ν is mass-symmetric in $S^{N-1}(1)$. $\qquad\square$

For a given isometric immersion $\phi : M \to \mathbb{E}^m$, we choose an oriented orthonormal local frame $e_1, \ldots, e_n, e_{n+1}, \ldots, e_m$ such that $e_1, \ldots, e_n$ is a local orthonormal tangent frame and $e_{n+1}, \ldots, e_m$ a local orthonormal normal frame of M. As before, we denote by h^r_{ij} the coefficients of the second fundamental form h, i.e., $h^r_{ij} = \langle h(e_i, e_j), e_r \rangle$, and by K^r_{sij} the coefficient of the normal curvature tensor R^D given by

$$K^r_{sij} = \langle R^D(e_i, e_j)e_r, e_s \rangle.$$

By a straightforward computation we obtain from (9.105) that

$$e_i \nu = -\sum h^r_{ij} e_{n+1} \wedge \cdots \wedge e^r_j \wedge \cdots \wedge e_m, \tag{9.108}$$

where e^r_j means to replace e_r by e_j. Since $\Delta \mu = -\sum_i e_i e_i \nu + \sum_i (\nabla_{e_i} e_i)\nu$, a direct computation shows that

$$\Delta \nu = \|h\|^2 \nu + n \sum_r e_{n+1} \wedge \cdots \wedge \nabla H_r \wedge \cdots \wedge e_m$$
$$- \sum_{r \neq s} \sum_{j < k} K^r_{sjk} e_{n+1} \wedge \cdots \wedge e^r_j \wedge \cdots \wedge e^s_k \wedge \cdots \wedge e_m, \tag{9.109}$$

where $H_r = \langle H, e_r \rangle$ and ∇H_r is the gradient of H_r.

Corollary 9.13. [Ruh and Vilms (1970)] *Let M be a submanifold of $\mathbb{E}^m$. Then the classical Gauss map $\hat{\nu} : M \to G(m, m-n)$ is harmonic if and only if $DH = 0$.*

Proof. Since the first term of the right-hand side of (9.109) is the only term which is tangent to $G(m, m-n)$ and the other terms are normal to $G(m, m-n)$, we obtain the corollary immediately from (9.109). $\qquad\square$

We have the following simple relation between type number of Gauss map ν and Euler class $e(T^\perp M)$ of the normal bundle.

Theorem 9.32. [Chen et al. (1985, 1986)] *Let M be an oriented closed submanifold of $\mathbb{E}^m$. If the Euler class $e(T^\perp M)$ of the normal bundle of M is non-trivial, then the Gauss map ν of M is of k-type with $k > m/2$.*

Proof. If $m - n$ is odd or $m > 2n$, then $e(T^\perp M)$ vanished automatically. So, we may assume that $m \le 2n$ and $m - n = 2\delta$ is an even integer. For each $\ell \le (m - n)/2$, let W_ℓ denote the subspace of $\wedge^{m-n}\mathbb{E}^m$ spanned by

$$\{e_{i_1} \wedge \cdots \wedge e_{i_{2\ell}} \wedge e_{r_1} \wedge \cdots \wedge e_{r_{m-n-2\ell}} : 1 \le i_1, \ldots, i_{2\ell} \le n \text{ and}$$

$$n + 1 \le r_1, \ldots, r_{m-n-2\ell} \le m\}.$$

Denote by $\pi_\ell : \wedge^{m-n}\mathbb{E}^m \to W_\ell$ the canonical projection. Then it follows from (9.109) that

$$\pi_\alpha(\nu) = 0, \ \alpha \ge 1; \ \pi_\alpha(\Delta\nu) = 0, \ \alpha \ge 2; \tag{9.110}$$

$$\pi_1(\Delta\nu) = -\sum K^r_{sjk}e_{n+1} \wedge \cdots \wedge e^r_j \wedge \cdots \wedge e^s_k \wedge \cdots \wedge e_m. \tag{9.111}$$

Assume ν is of k-type for some $k \le \delta$. It follows from Theorem 9.27 and Lemma 9.8 that there exists a monic polynomial P of degree k such that $P(\Delta)\nu = 0$. Since $k \le \delta$, $Q(t) = t^{\delta-k}P(t)$ is a monic polynomial of degree δ such that $Q(\Delta)\nu = 0$. In particular, we get $\pi_\delta(Q(\Delta)\nu) = 0$. By a direct computation we find $\pi_\delta(\Delta^\ell\nu) = 0$ for $\ell < \delta$. So, $\pi_\delta(\Delta^q\nu) = \pi_\delta(Q(\Delta)\nu) = 0$.

On the other hand, by a straightforward computation we have

$$\pi_q(\Delta^\delta\nu) = (-1)^\delta \sum K^{r_1}_{r_2j_1j_2} \cdots K^{r_{m-n-1}}_{r_{m-n}j_{m-n-1}j_{m-n}} e^{r_1}_{j_1} \wedge \cdots \wedge e^{r_{m-n}}_{j_{m-n}} = 0.$$

Thus we find $\sum K^{r_1}_{r_2j_1j_2} \cdots K^{r_{m-n-1}}_{r_{m-n}j_{m-n-1}j_{m-n}} e^{r_1}_{j_1} \wedge \cdots \wedge e^{r_{m-n}}_{j_{m-n}} = 0$, which is equivalent to

$$\sum \epsilon_{r_1\cdots r_{m-n}} K^{r_1}_{r_2j_1j_2} \cdots K^{r_{m-n-1}}_{r_{m-n}j_{m-n-1}j_{m-n}} \omega^{j_1} \wedge \cdots \wedge \omega^{j_{m-n}} = 0, \tag{9.112}$$

where $\epsilon_{r_1\ldots r_{m-n}}$ is 1 or -1 depending on $(r_1, \cdots, r_{m-n})$ is an even or odd permutation of $(n + 1, \ldots, m - n)$.

Let us put $\Omega^r_s = \frac{1}{2} \sum K^r_{sij}\omega^i \wedge \omega^j$. Then we derive from (9.112) that

$$\Psi = \frac{(-1)^\delta}{2^{2\delta}\pi^\delta\delta!} \sum \epsilon_{r_1\cdots r_{2\delta}} \Omega^{r_1}_{r_2} \wedge \cdots \wedge \Omega^{r_{2\delta-1}}_{r_{2\delta}} = 0.$$

Since Ψ represents the Euler class of $T^\perp M$, we obtain $e(T^\perp M) = 0$. $\quad\square$

When $\phi : M \to \mathbb{C}^n$ is Lagrangian, the tangent bundle is isomorphic to the normal bundle. Thus Theorem 9.32 implies immediately the following.

Corollary 9.14. [Chen et al. (1985, 1986)] *Let M be an oriented closed Lagrangian submanifold of $\mathbb{C}^n$. If the type number of the Gauss map of M is $\le n/2$, then M has zero Euler number.*

Theorem 9.32 also implies the following.

Theorem 9.33. [Chen et al. (1985, 1986)] *Let M be an n-dimensional closed submanifold of $\mathbb{E}^{2n}$. If M has nonzero self-intersection number, then the Gauss map of M is of k-type with $k > n/2$.*

Proof. This follows from Theorem 9.32 and the fact that, for an oriented closed n-dimensional submanifold M immersed in $\mathbb{E}^{2n}$, the Euler number of the normal bundle is equal to twice of the self-intersection number by a result of [Lashof and Smale (1958)]. $\qquad\square$

Corollary 9.15. [Chen et al. (1985, 1986)] *Let $\phi : M \to \mathbb{E}^{2n}$ be an oriented closed n-dimensional submanifold of $\mathbb{E}^{2n}$. If $e(T^{\perp}M)$ is non-trivial, then ϕ cannot be deformed regularly to an immersion with k-type Gauss map for any $k \leq n/2$.*

Proof. Follows from Theorem 9.32 and the fact that the self-intersection number is a regular homotopic invariant (cf. [Smale (1959)]). $\qquad\square$

Example 9.6. The *Whitney immersion* is defined as follows:

Let $\phi : \mathbb{E}^{2n+1} \to \mathbb{E}^{4n}$ be the map of $\mathbb{E}^{2n+1}$ into $\mathbb{E}^{4n}$ defined by

$$\phi(x_0, x_1, \ldots, x_{2n}) = (x_1, \ldots, x_{2n}, 2x_0x_1, \ldots, 2x_0x_{2n}).$$

The map ϕ induces a non-isometric immersion $w : S^{2n}(1) \to \mathbb{E}^{4n}$ from $S^{2n}(1)$ into $\mathbb{E}^{4n}$, known as the *Whitney's immersion*, which has a unique self-intersection point at $\phi(-1, 0, \ldots, 0) = \phi(1, 0, \ldots, 0)$.

Although the ordinary inclusion of $S^{2n}(1)$ in a totally geodesic $\mathbb{E}^{2n+1} \subset \mathbb{E}^{4n}$ has 1-type Gauss map, the Whitney immersion w of $S^{2n}(1)$ in $\mathbb{E}^{4n}$ cannot be deformed regularly into an immersion with k-type Gauss map for any $k \leq n/2$ according to Theorem 9.33, since the self-intersection number $I(w)$ of the Whitney immersion is one.

9.12 Linearly independence and orthogonal maps

In this section we provide some simple relations between linear algebras and differential geometry via finite type theory.

Let $\phi : M \to \mathbb{E}^m$ be a k-type map of a Riemannian manifold into $\mathbb{E}^m$ with spectral decomposition:

$$\phi = c + \phi_1 + \cdots + \phi_k, \quad \Delta\phi_i = \lambda_i\phi_i, \tag{9.113}$$

where c is a vector in $\mathbb{E}^m$, $\phi_1, \ldots, \phi_k$ are non-constant maps and $\lambda_1, \ldots, \lambda_k$ are distinct eigenvalues of the Laplacian Δ. We put

$$\mathcal{E}_i = \text{Span}\{\phi_i(x) : x \in M\} \subset \mathbb{E}^m,$$
$$V(\lambda_i) = \{f \in \mathcal{F}(M) : \Delta f = \lambda_i\}, \quad i = 1, \ldots, k.$$

The notions of linearly independent and orthogonal maps were introduced in [Chen (1991c)] as follows.

Definition 9.6. Let $\phi : M \to \mathbb{E}^m$ be a k-type map with spectral decomposition given by (9.113). Then ϕ is called *linearly independent* if $\mathcal{E}_1, \ldots, \mathcal{E}_k$ are linearly independent subspaces, i.e., if

$$\dim \left(\mathrm{Span}\{\mathcal{E}_1 \cup \ldots \cup \mathcal{E}_k\} \right) = \dim \mathcal{E}_1 + \cdots + \dim \mathcal{E}_k.$$

And ϕ is called *orthogonal* if $\mathcal{E}_1, \ldots, \mathcal{E}_k$ are mutually orthogonal subspaces.

Clearly, 1-type immersions are orthogonal and linearly independent.

Lemma 9.9. *Let* $\phi : M \to \mathbb{E}^m$ *be a k-type map of a Riemannian manifold M into $\mathbb{E}^m$. Then for each $i \in \{1, \ldots, k\}$ there exist linearly independent vectors $c_{ij} \in \mathbb{E}^m$ and linearly independent functions $f_{ij} \in V(\lambda_i)$ such that $\phi_i = \sum_{j=1}^{m_i} f_{ij} c_{ij}$, $i = 1, \ldots, k$.*

Proof. Since $\Delta\phi_i = \lambda_i \phi_i$, there exist vectors a_{ij} $(j = 1, \ldots, n_i)$ in $\mathbb{E}^m$ and functions φ_{ij} $(j = 1, \ldots, n_i)$ in $V(\lambda_i)$ such that $\phi_i = \sum_{j=1}^{m_i} a_{ij} \varphi_{ij}$.

Let $E_i = \mathrm{Span}\{a_{i1}, \ldots, a_{an_i}\}$ and $c_{i1}, \ldots, c_{im_i}$ be a basis of E_i. Since $V(\lambda_i)$ is a vector space, we have $\phi_i = \sum_{j=1}^{m_i} f_{ij} c_{ij}$ for some functions $f_{ij} \in V(\lambda_i)$, $j = 1, \ldots, m_j$. We claim that $f_{i1}, \ldots, f_{im_i}$ are linearly independent functions in $V(\lambda_i)$. This can be easily seen as follows. If not, one of $f_{i1}, \ldots, f_{im_i}$ is a linear combination of the others. Without loss of generality, we may assume that $f_{i1} = \sum_{j=2}^{m_i} b_j f_{ij}$, $b_j \in \mathbb{R}$. Hence we have $\phi_i = \sum_{j=2}^{m_i} (c_{ij} + b_j c_{i1}) f_{ij}$, which implies that $\dim E_i < m_i$. This is a contradiction. $\square$

Theorem 9.34. [Chen and Petrović (1991)] *A finite type immersion $\phi : M \to \mathbb{E}^m$ is linearly independent if and only if it satisfies $\Delta\phi = A\phi + B$ for some constant $m \times m$ matrix A and vector $B \in \mathbb{R}^m$.*

Proof. Let $\phi : M \to \mathbb{E}^m$ be an immersion of finite type with spectral decomposition given by (9.113). Without loss of generality, we may assume that ϕ is full.

($\Longleftarrow$) If ϕ satisfies $\Delta\phi = A\phi + B$ for $A \in \mathbb{R}^{m \times m}$ and $B \in \mathbb{R}^m$, then it follows from (9.113) that

$$Ac + B + (A\phi_1 - \lambda_1 \phi) + \cdots + (A\phi_k - \lambda_k \phi_k) = 0. \tag{9.114}$$

It follows from $\Delta(A\phi_i) = A(\Delta\phi_i) = \lambda_i A\phi_i$ and (9.114) that

$$\lambda_1^j (A\phi_1 - \lambda\phi) + \cdots + \lambda_k^j (A\phi_k - \lambda_k \phi_k) = 0, \quad j = 1, 2, \ldots. \tag{9.115}$$

Since $\lambda_1, \ldots, \lambda_k$ are mutually distinct, (9.115) implies that $A\phi_i = \lambda_i\phi_i$ for $i = 1, \ldots, k$. Combining this with Lemma 9.9 gives

$$\sum_{j=1}^{m_i}(Ac_{ij} - \lambda_i c_{ij})f_{ij} = 0, \quad i = 1, 2 \ldots, k. \tag{9.116}$$

From the linear independence of $f_{i1}, \ldots, f_{im_i}$ (see Lemma 9.9), we obtain $Ac_{ij} = \lambda_i c_{ij}$. Since eigenvectors belonging to distinct eigenspaces of A are linearly independent, the immersion ϕ is linearly independent.

($\Longrightarrow$) If ϕ is linearly independent, then by Lemma 9.9, we may assume each ϕ_i is expressed as $\phi_i = \sum_{j=1}^{m_i} f_{ij}c_{ij}$, where $c_{i1}, \ldots, c_{im_i}$ are independent vectors and $f_{i1}, \ldots, f_{im_i}$ are independent functions in $V(\lambda_i)$. By definition, $\{c_{ij} : i = 1, \ldots, k; j = 1, \ldots, m_i\}$ are linearly independent. We put

$$S = (c_{11}, \ldots, c_{1m_1}, \ldots, c_{k1}, \ldots, c_{km_k}). \tag{9.117}$$

Since ϕ is assumed to be full, we have $\sum_i m_i = m$ and S is a nonsingular. Let D denote the diagonal $m \times m$ matrix given by

$$D = \text{Diag}\,(\lambda_1, \ldots, \lambda_1, \ldots, \lambda_k, \ldots, \lambda_k), \tag{9.118}$$

where λ_i repeats m_i-times. We put $A = SDS^{-1}$ and $B = -Ac$. Then by direct computation we obtain $\Delta\phi = A\phi + B$. $\qquad\square$

Lemma 9.10. *Let $\phi : M \to \mathbb{E}^m$ be an isometric immersion.*

(a) *If M is a closed submanifold of $\mathbb{E}^m$ satisfying condition $\Delta\phi = A\phi + B$, then it is of k-type with $k \le m$. In particular, it is of finite type.*

(b) *If ϕ is a curve satisfying $\Delta\phi = A\phi + B$, then it is of finite type.*

Proof. Let M be a submanifold M of $\mathbb{E}^m$ satisfying $\Delta\phi = A\phi + B$. After applying a suitable translation we get $\Delta\phi = A\phi$. Thus $Q(\Delta)\phi = Q(A)\phi$ for any polynomial Q of one variable. If we choose P the characteristic polynomial of A, then Cayley-Hamilton's theorem gives $P(A) = 0$. Hence $P(\Delta)\phi = 0$. So M is of k-type with $k \le m$ according to Theorem 6.1. This gives (a). From $P(\Delta)\phi = 0$ and Proposition 6.3, we obtain (b). $\qquad\square$

Theorem 9.35. *[Chen and Petrović (1991)] A finite type immersion $\phi : M \to \mathbb{E}^m$ is orthogonal if and only if it satisfies $\Delta\phi = A\phi + B$ for some constant symmetric matrix $A \in \mathbb{R}^{m \times m}$ and vector $B \in \mathbb{R}^m$.*

Proof. Without loss of generality we may assume that ϕ is full. Suppose that there exist a symmetric matrix $A \in \mathbb{R}^{m \times m}$ and $B \in \mathbb{R}^m$ such that $A\phi = A\phi + B$. Let c_{ij}, $i = 1, \ldots, k; j = 1, \ldots, m_i$ be the vectors given in

Lemma 9.9. Then as in the proof of Theorem 9.34. we have $Ac_{ij} = \lambda_i c_{ij}$. Since A is symmetric, distinct eigenspaces of A are mutually orthogonal. Thus ϕ is an orthogonal immersion. Conversely, if ϕ is orthogonal, one may choose a Euclidean coordinate system such that

$$\phi_1 \in \mathrm{Span}\{\varepsilon_1, \ldots, \varepsilon_{m_1}\}, \ldots, \phi_k \in \mathrm{Span}\{\varepsilon_{m-m_k+1}, \ldots, \varepsilon_m\},$$

where $\{\varepsilon_1, \ldots, \varepsilon_m\}$ is the canonical orthonormal basis of $\mathbb{E}^m$. It is easy to see that with respect to this coordinate system, $\Delta\phi = D\phi - Dc$, where D is the diagonal matrix given by (9.118). Thus, with respect to the original coordinate system, we have $\Delta\phi = A\phi + B$ for some symmetric matrix A and vector B. ☐

Corollary 9.16. *Let $\phi : \mathbb{R} \supset I \to \mathbb{E}^m$ be a unit speed curve in $\mathbb{E}^m$. Then ϕ is a W-curve if and only if it satisfies $\Delta\phi = A\phi + B$ for some symmetric matrix $A \in \mathbb{R}^{m \times m}$ and vector $B \in \mathbb{R}^m$.*

Proof. Let $\phi : I \to \mathbb{E}^m$ be a unit speed curve. If $\Delta\phi = A\phi + B$ holds for some symmetric matrix $A \in \mathbb{R}^{m \times m}$ and vector $B \in \mathbb{R}^m$, then we have

$$\Delta H = AH, \tag{9.119}$$

where H is the mean curvature vector. If P denotes the characteristic polynomial of A, then by the Cayley-Hamilton theorem, we have $P(A) = 0$ and thus $P(\Delta)H = 0$. Hence ϕ is of finite type according to Proposition 6.3. So, after applying Theorem 9.35 we conclude that the spectral decomposition

$$\phi = c + \phi_1 + \cdots + \phi_k \tag{9.120}$$

is orthogonal. Thus ϕ can be expressed as the following form:

$$\phi = c + \sum_{i=1}^{k} (a_i \cos \lambda_i s + b_i \sin \lambda_i s),$$

where $a_i, b_i \in \mathbb{E}^m$ and $\lambda_1, \ldots, \lambda_k$ are mutually distinct non-negative real numbers. Since (9.120) is orthogonal, $E_i = \mathrm{Span}\{a_i, b_i\}$, $i = 1, \ldots, k$ are mutually orthogonal. Thus, by using the condition $\langle \phi'(s), \phi'(s) \rangle = 1$, we may conclude that either (a) $||a_i|| = ||b_i||$ and $a_i \perp b_i$ or (b) $\lambda_i = 0$ for each i. Consequently, ϕ is a W-curve. The converse is trivial. ☐

Definition 9.7. Let $\phi : M \to \mathbb{E}^m$ be a null finite type map whose spectral decomposition is given by $\phi = \phi_0 + \phi_1 + \cdots + \phi_k$ such that $\Delta\phi_0 = 0$ and $\Delta\phi_i = \lambda_i\phi_i$, where $\phi_0, \phi_1, \ldots, \phi_k$ are non-constant maps and $\lambda_1, \ldots, \lambda_k$ are distinct nonzero numbers. The map ϕ is called *weakly linearly independent* if $\mathcal{E}_1, \ldots, \mathcal{E}_k$ spanned by $\phi_1, \ldots, \phi_k$ are linearly independent. The map ϕ is called *weakly orthogonal* if $\mathcal{E}_1, \ldots, \mathcal{E}_k$ are mutually orthogonal subspaces.

The following is a characterization of weakly linearly independent immersions.

Theorem 9.36. [Chen (1994d)] *Let M be a null finite type submanifold of $\mathbb{E}^m$. Then M is weakly linearly independent if and only if it satisfies*

$$\Delta H = AH \tag{9.121}$$

for some matrix $A \in \mathbb{R}^{m \times m}$, where H is the mean curvature vector.

Remark 9.7. Beltrami's formula implies that every submanifold of $\mathbb{E}^m$ satisfies the condition $\Delta \phi = A\phi + B$ also satisfies $\Delta H = AH$. However, if M is a closed submanifold, condition $\Delta \phi = A\phi + B$ is the same as $\Delta H = AH$. On the other hand, when M is non-closed and ϕ is null, condition $\Delta H = AH$ is weaker than $\Delta \phi = A\phi + B$ (cf. [Chen (1994d)]).

9.13   Adjoint hyperquadrics and orthogonal immersions

Let $\phi : M \to \mathbb{E}^m$ be a linearly independent immersion of a Riemannian manifold M whose spectral decomposition is given by (9.113). If we choose c to be the origin, then we have

$$\phi = \phi_1 + \cdots + \phi_k, \quad \Delta \phi_i = \lambda_i \phi_i. \tag{9.122}$$

Then by Theorem 9.34, ϕ satisfies $\Delta \phi = A\phi$ for some matrix $A \in \mathbb{R}^{m \times m}$. Let us put $A = (a_{ij}), a_{ij} \in \mathbb{R}$.

Definition 9.8. [Chen (1991c)] Let $\phi : M \to \mathbb{E}^m$ be a non-minimal linearly independent immersion whose spectral decomposition is (9.122) and let $u = (u_1, \ldots, u_m)$ be a Euclidean coordinate system of $\mathbb{E}^m$. Then, for each fixed point $x \in M$,

$$\langle Au, u \rangle := \sum_{i,j}^{m} a_{ij} u_i u_j = c_x, \quad c_x = \langle A\phi, \phi \rangle(x),$$

defines a hyperquadric Q_x in $\mathbb{E}^m$, which is called the *adjoint hyperquadric* at x. If $\phi(M)$ lies in Q_x for some $x \in M$, all of the $\{Q_x : x \in M\}$ give the same adjoint hyperquadric Q, which is called the *adjoint hyperquadric* of ϕ.

Proposition 9.12. [Chen (1991c)] *Let $\phi : M \to \mathbb{E}^m$ be a linearly independent immersion of a Riemannian closed n-manifold M into $\mathbb{E}^m$. Then M lies in the adjoint hyperquadric Q via ϕ if and only if $\phi(M)$ lies in a hypersphere of $\mathbb{E}^m$ centered at the origin.*

Proof. Let $\phi : M \to \mathbb{E}^m$ be a linearly independent immersion of a Riemannian closed n-manifold M into $\mathbb{E}^m$. If $\phi(M)$ lies in a hypersphere $S^{m-1}(r)$ centered at the origin, then the mean curvature vectors $\hat{H}$ and H of M in $S^{m-1}(r)$ and $\mathbb{E}^m$ satisfy $H = \hat{H} - r^{-1}\phi$. Thus $\langle H, \phi \rangle = -r$. Hence, by Beltrami's formula, we find $\langle \Delta\phi, \phi \rangle = nr$. Therefore, using $\Delta\phi = A\phi$ we see that $\phi(M)$ lies in the adjoint hyperquadric Q given by $\langle Au, u \rangle = nr$.

Conversely, if $\phi(M)$ lies in an adjoint hyperquadric Q_x for some point x, then we have $\langle A\phi, \phi \rangle = c_x$, $c_x = \langle A\phi, \phi \rangle (x)$. Since $A\phi = \Delta\phi = -nH$, we have $\langle nH, x \rangle = -c_p$. Combining this with (9.1) yields $c_x = -n$. Hence we obtain

$$\Delta \langle \phi, \phi \rangle = 2 \langle \Delta\phi, \phi \rangle - 2n = -2n(\langle H, \phi \rangle + 1) = 0. \qquad (9.123)$$

Therefore, after applying Hopf's lemma to (9.123), we conclude that $\phi(M)$ lies in a hypersphere of $\mathbb{E}^m$ centered at the origin. $\qquad \square$

Lemma 9.11. *Let $\phi : M \to \mathbb{E}^m$ be a linearly independent immersion whose spectral decomposition is (9.122). Then we have $\tilde{\nabla}_X \phi_i = X_i$, $\forall X \in TM$, where $X = X_1 + \cdots + X_k$ with $X_i \in \mathcal{E}_i = \text{Span}\{\phi_i\}$ for $i = 1, \ldots, k$, and $\tilde{\nabla}$ is the Levi-Civita connection of $\mathbb{E}^m$.*

Proof. Follows from the linearly independence of ϕ and $\tilde{\nabla}_X \phi = X$. $\qquad \square$

Theorem 9.37. [Chen (1991c)] *If $\phi : M \to \mathbb{E}^m$ is a non-minimal linearly independent immersion, then M is immersed as a minimal submanifold in the adjoint hyperquadric Q of ϕ if and only if ϕ is an orthogonal immersion.*

Proof. Let $\phi : M \to \mathbb{E}^m$ be a non-minimal linearly independent immersion whose spectral decomposition is (9.122). Then there is an $m \times m$ matrix $A = (a_{ij})$ such that $\Delta\phi = A\phi$. Suppose that M is a minimal submanifold of the adjoint hyperquadric Q defined by $\sum_{i,j}^m a_{ij} u_i u_j = b$ with $b \in \mathbb{R}$.

If we put $f = \sum a_{ij} u_i u_j$, then Q is a level hypersurface of f and the gradient vector field of f is

$$\nabla f = (2a_{11}u_1 + (a_{12} + a_{21})u_2 + \cdots + (a_{1m} + a_{m1})u_m,$$
$$\ldots, (a_{1m} + a_{m1})u_1 + \cdots \qquad (9.124)$$
$$+ (a_{m-1\,1} + a_{1\,m-1})u_{m-1} + 2a_{mm}u_m).$$

On the other hand, it follows from $-nH = \Delta\phi = A\phi$ that

$$-nH = (a_{11}u_1 + \cdots + a_{1m}u_m, \ldots, a_{m1}u_1 + \cdots + a_{mm}u_m) \qquad (9.125)$$

at each $x \in M$. Since M lies in Q as a minimal submanifold, ∇f is parallel to the mean curvature vector H of M in $\mathbb{E}^m$ along $\phi(M)$. Therefore there is a function μ on M such that $\nabla f = \mu H$.

If ϕ is full, the coordinate functions $u_1, \ldots, u_m$ restricted to M are linearly independent, since the immersion is nonlinear. Thus, by using (9.124), (9.125) and $\nabla f = \mu H$, we conclude that $\mu = 2$ and A is a symmetric matrix. On the other hand, it follows from (9.122) and $\Delta \phi = A\phi$ that

$$(A\phi_1 - \lambda_1 \phi_1) + \cdots + (A\phi_k - \lambda_k \phi_k) = 0. \tag{9.126}$$

Because $\Delta(A\phi_i) = A(\Delta\phi_i) = \lambda_i A\phi_i$, equation (9.126) yields

$$\lambda_1^j (A\phi_1 - \lambda_1 \phi_1) + \cdots + \lambda_k^j (A\phi_k - \lambda_k \phi_k) = 0$$

for $j = 1, 2, \ldots, k$. Since $\lambda_1 < \cdots < \lambda_k$, we find $A\phi_i = \lambda_i \phi_i$, $i = 1, \ldots, k$. Due to the fact that A is symmetric and distinct eigenspaces of a symmetric matrix are mutually orthogonal, ϕ is an orthogonal immersion.

If ϕ is non-full, then ϕ gives rise a full linearly independent immersion $\bar{\phi} : M \to \mathbb{E}^\ell$ for linear ℓ-subspace $\mathbb{E}^\ell$ of $\mathbb{E}^m$. If $\bar{A}$ is the corresponding $\ell \times \ell$ matrix with $\Delta\bar{\phi} = \bar{A}\bar{\phi}$. Then A is obtained from $\bar{A}$ by putting zeros on the additional entries. Since M is minimal in the adjoint hyperquadric Q, it is immersed as a minimal submanifold in Q by $\bar{\phi}$. Thus, as previous case, $\bar{\phi}$ is an orthogonal immersion. Hence ϕ is orthogonal too.

Conversely, assume that ϕ is an orthogonal immersion. Put

$$f = \lambda_1 \langle \phi_1, \phi_1 \rangle + \cdots + \lambda_k \langle \phi_k, \phi_k \rangle .$$

Then Lemma 9.11 implies

$$Xf = 2\{\lambda_1 \langle \phi_1, X_1 \rangle + \cdots + \lambda_k \langle \phi_k, X_k \rangle\} = 2 \langle \Delta\phi, X \rangle = 0$$

for $X \in TM$. Hence M lies in an adjoint hyperquadric given by $\langle Au, u \rangle = b$ for some constant b and also the gradient of f is given by $-2nH$. Since the gradient of f is normal to the hyperquadric, M lies in the adjoint hyperquadric as a minimal submanifold. $\qquad \Box$

Theorem 9.38. [Chen (1991c)] *If $\phi : M \to \mathbb{E}^m$ is an equivariant isometric immersion from a compact connected homogeneous space M into $\mathbb{E}^m$, then ϕ is an orthogonal immersion. Moreover, M is immersed as a minimal submanifold of the adjoint hyperquadric via ϕ.*

Proof. Let $\phi : M \to \mathbb{E}^m$ be an equivariant isometric immersion of a compact connected homogeneous space M into $\mathbb{E}^m$. Without loss of generality we may assume that ϕ is full. Since ϕ is equivariant, ϕ is spherical. Thus $\phi(M)$ lies in a hypersphere S^{m-1} of $\mathbb{E}^m$. Without loss of generality, we may also assume that S^{m-1} is centered at the origin. Denote by $G = I_0(M)$ the identity component of the group of isometries of M. Then G is a compact Lie group acting transitively on M and there exists a Lie homomorphism $\psi : G \to SO(\mathbb{E}^m)$ such that $\phi(g(x)) = \psi(g)(\phi(x))$ for every $g \in G$ and $x \in M$. Because $(\psi, \mathbb{E}^m)$ is a representation of the compact Lie group G, $(\psi, \mathbb{E}^m)$ is the direct sum of irreducible subrepresentations $(\psi_1, \mathbb{E}_1), \ldots, (\psi_k, \mathbb{E}_k)$ such that $\mathbb{E}^m = \mathbb{E}_1 \oplus \cdots \oplus \mathbb{E}_k$. Let ϕ_i denote the $\mathbb{E}_i$-component of ϕ. Then, for each $i = 1, \ldots, k$, we have

$$\phi_i(g(x)) = \psi_i(g)(\phi_i(x)), \quad g \in G, \quad x \in M.$$

Because $\mathbb{E}_1, \ldots, \mathbb{E}_k$ are mutually orthogonal, the decomposition

$$\phi = \phi_1 + \cdots + \phi_k$$

is orthogonal.

Next, we claim that each ϕ_i is a 1-type map, i.e., $\Delta \phi_i = \lambda_i \phi_i$ for a real numbers λ_i. In order to do so, we choose a fixed point $o \in M$. Denote by K the isotropy subgroup of G at o. Then M can be identified with G/K in a natural way. Consider a bi-invariant Riemannian metric on G such that the projection $\pi : G \to M = G/K$ is a Riemannian submersion.

Let $e_1, \ldots, e_N$ be any orthonormal basis of the Lie algebra $\mathfrak{g} = T_e G$ of G, where e is the identity element of G. For each $i \in \{1, \ldots, k\}$, denote also by ψ_i the homomorphism $\mathfrak{g} \to \mathfrak{so}(\mathbb{E}_i)$ induced from $\psi_i : G \to \mathbb{E}_i$, where $\mathfrak{so}(\mathbb{E}_i)$ is the Lie algebra of $SO(\mathbb{E}_i)$. Then each $\psi_i(e_a)$ is a skew-symmetric linear transformation of $\mathbb{E}_i$ and $\sum_{a=1}^N \psi_i(e_a)^2$ is a symmetric linear transformation. Let $Ad : G \to GL(\mathfrak{g})$ denote the adjoint representation of G. Because $Ad(g)(h) = ghg^{-1}$ and ψ_i is a Lie homomorphism, we have

$$\psi_i(g)\psi_i(X)\psi_i(g^{-1}) = \psi_i(Ad(g)X)$$

for $X \in \mathfrak{g}$, $g \in G$ and $i \in \{1, \ldots, k\}$. Thus we obtain

$$\psi_i(g)\left(\sum_{a=1}^N \psi_i(e_a)^2\right)\psi_i(g^{-1}) = \sum_{a=1}^N \psi_i(Ad(g)e_a)^2 = \sum_{a=1}^N \psi_i(e_a)^2$$

for $g \in G$. Thus $\sum_{a=1}^N \psi_i(e_a)^2$ lies in the centralizer of $\psi_i(G)$. Since (ψ_i, E_i) is irreducible, Schur's lemma in representation theory gives

$$\sum_{a=1}^N \psi_i(e_a)^2 = -\lambda_i I_i \tag{9.127}$$

for some constants λ_i, where I_i is the identity transformation on $\mathbb{E}_i$.

On the other hand, the Laplacian of ϕ_i is given by

$$\Delta \phi_i(x) = -\sum_{a=1}^N \frac{d^2}{dt^2} \phi_i(\exp t e_a)\Big|_{t=0} = -\sum_{a=1}^N \psi_i(e_a)^2(\phi_i(x)). \tag{9.128}$$

So, we obtain the claim from (9.127) and (9.128). Thus ϕ is an orthogonal immersion. Consequently, M is immersed as a minimal submanifold of the adjoint hyperquadric according to Theorem 9.38. $\qquad\square$

9.14 Submanifolds satisfying $\Delta\phi = A\phi + B$

The class of 1-type immersions in $\mathbb{E}^m$ was classified in Corollary 6.3. In fact, a submanifold $\phi : M \to \mathbb{E}^m$ satisfies $\Delta\phi = \lambda\phi$ if and only if it is either a minimal submanifold of $\mathbb{E}^m$ or a minimal submanifolds of a hypersphere S^{m-1} in $\mathbb{E}^m$. As a generalization of $\Delta\phi = \lambda\phi$, O. J. Garay studies in [Garay (1990a)] immersions $\phi : M \to \mathbb{E}^m$ satisfying the condition:

$$\Delta\phi = A\phi \tag{9.129}$$

for some $m \times m$ diagonal matrix A. He called submanifolds via such an immersion *submanifolds of coordinate finite type*. Later, it was observed in [Dillen et al. (1990a)] that condition (9.129) is not coordinate-invariant. Consequently, submanifolds satisfying the condition:

$$\Delta\phi = A\phi + B, \quad A \in \mathbb{E}^{m \times m}, \quad B \in \mathbb{E}^m \tag{9.130}$$

were investigated in [Dillen et al. (1990a)]. This setting generalizes (9.129) in a way which is independent of the choice of coordinates.

Corollary 8.5 states that *every submanifold of $\mathbb{E}^m$ satisfying $\Delta\phi = B$ is always minimal.*

Lemma 9.12. *Let $\phi : M \to \mathbb{E}^m$ be a submanifold satisfying $\Delta\phi = A\phi + B$ form some matrix $A \in \mathbb{R}^{m \times m}$ and $B \in \mathbb{R}^m$. Then we have*

$$A_H X = -(AX)^T, \quad D_X H = (AX)^\perp, \tag{9.131}$$

for any $X \in TM$, where $(AX)^T$ and $(AX)^\perp$ denote the tangential and the normal components of AX.

Proof. Let $\phi : M \to \mathbb{E}^m$ be a submanifold satisfying $\Delta\phi = A\phi + B$. Then Beltrami's formula gives $-nH = A\phi + B$. By taking covariant derivative of this equation and using Weingarten's formula, we find

$$-A_H X + D_X H = AH, \quad X \in TM,$$

which implies the lemma. $\qquad\square$

Garay proved that if a hypersurface M in $\mathbb{E}^{n+1}$ satisfies (9.129), then it is either minimal in $\mathbb{E}^{n+1}$, or an open part of a hypersphere or of a spherical hypercylinder [Garay (1988c)]. Dillen, Pas and Verstraelen shown that a surface in $\mathbb{E}^3$ satisfies (9.130) if and only if it is an open part of a minimal surface, a sphere, or a circular cylinder [Dillen et al. (1990a)].

For hypersurfaces satisfying (9.130), we have the following classification result from [Chen and Petrović (1991); Hasanis-Vlachos (1992)].

Theorem 9.39. *A hypersurface of $\mathbb{E}^{n+1}$ satisfies condition (9.130) if and only if it is an open portion of one of the following hypersurfaces:*

(a) *a minimal hypersurface;*
(b) *a hypersphere S^n;*
(c) *a spherical hypercylinder $S^\ell \times \mathbb{E}^{n-\ell}$, $\ell \in \{1, 2, \ldots, n-1\}$.*

For a coordinate finite type submanifold M of $\mathbb{E}^m$ there are m numbers $\mu_1, \ldots, \mu_m$ such that each coordinate function x_i of M is an eigenfunction of Δ with eigenvalue μ_i. We denote such a submanifold by $M(\mu_1, \ldots, \mu_m)$.

For coordinate finite type submanifolds, we have the following.

Theorem 9.40. [Hasanis-Vlachos (1991c)] *Let $\phi : M \to \mathbb{E}^m$ be a submanifold of $\mathbb{E}^m$. Then it is a coordinate finite type submanifold $M(\mu_1, \ldots, \mu_m)$ if and only if it is one of the following:*

(1) *a minimal submanifold of $\mathbb{E}^m$;*
(2) *an open portion of a minimal submanifold of a quadratic hypersurface defined by equation $f(u_1, \ldots, u_m) = \sum_{i=1}^m \mu_i u_i^2 - c$ for some constant c satisfying $\nabla f = 2nH$, where $n = \dim M$.*

Theorem 9.41. [Hasanis-Vlachos (1991c)] *Let $\phi : M \to \mathbb{E}^{n+2}$ be a coordinate finite type closed submanifold of codimension 2. Then M is one of the following hypersurfaces:*

(1) *a minimal hypersurface of a hypersphere of $\mathbb{E}^{n+2}$;*
(2) *an ordinary hypersphere of a hyperplane of $\mathbb{E}^{n+2}$;*
(3) *the Riemannian product $S^k(a) \times S^{n-k}(b)$ such that $1 \le k \le n-1$ and $a^2/b^2 \ne k/(n-k)$.*

For non-closed surfaces in $\mathbb{E}^4$, O. J. Garay obtained the following.

Theorem 9.42. [Garay (1994b)] *The only coordinate finite type surfaces in $\mathbb{E}^4$ with constant mean curvature are the open parts of the following:*

(1) *a minimal surface in $\mathbb{E}^4$;*
(2) *a minimal surface in some hypersphere $S^3(r)$;*
(3) *a helical cylinder;*
(4) *a flat torus $S^1(a) \times S^1(b)$ in some hypersphere $S^3(r)$.*

For further results on submanifolds satisfying condition (9.129) or (9.130), see [Alías et. al. (1992, 1995b); Baikoussis et al. (1993c); Park (1994)]. For further results concerning linearly independent immersions and their corresponding hyperquadrics, see [Jang (1998)].

For surfaces satisfying $\Delta H = AH$, we have following results.

Theorem 9.43. [Chen (1994d)] *Minimal surfaces and circular cylinders are the only finite type surfaces in* $\mathbb{E}^3$ *satisfying the condition* $\Delta H = AH$ *for some* 3×3 *singular matrix* A.

Theorem 9.44. [Chen (1994d)] *A cylinder over a weakly linearly independent curve of finite type and part of a helicoid in a totally geodesic* $\mathbb{E}^3$ *are the only ruled surfaces in* $\mathbb{E}^m$ *satisfying the condition* $\Delta H = AH$.

Theorem 9.45. [Chen (1994d)] *Circular cylinders are the only tubes in* $\mathbb{E}^3$ *satisfying* $\Delta H = AH$ *for some* 3×3 *matrix* A.

For further results on submanifolds satisfying condition $\Delta H = AH$, see [Jung and Pak (1996); Pak and Yoon (2000); Kim at al. (2014c)].

9.15   Submanifolds of restricted type

Submanifolds of restricted type were introduced in [Chen et al. (1993b)].

Definition 9.9. A submanifold M of $\mathbb{E}^m$ is said to be *of restricted type* if the shape operator A_H at the mean curvature vector H is the restriction of a fixed endomorphism F of $\mathbb{E}^m$ to tangent space of M at each $x \in M$. In other words, M is of restricted type if there exists a fixed endomorphism F of $\mathbb{E}^m$ such that, for each $X \in T_x M$, $x \in M$, we have

$$A_H X = (FX)^T, \tag{9.132}$$

where $(FX)^T$ denotes the tangential component of FX.

Lemma 9.12 shows that all submanifolds in $\mathbb{E}^m$ satisfying the condition $\Delta \phi = A\phi + B$ are submanifolds of restricted type.

The class of submanifolds of restricted type is rather large. It includes 1-type submanifolds, pseudo-umbilical submanifolds with constant mean curvature, all submanifolds satisfying either Garay's or Dillen-Pas-Verstraelen's condition, all k-type curves lying fully in $\mathbb{E}^{2k}$, all null k-type curves lying fully in $\mathbb{E}^{2k-1}$, direct products of submanifolds of restricted type, diagonal immersions of restricted type submanifolds as well as all equivariant isometric immersions of compact homogeneous spaces.

On the other hand, Euclidean hypersurfaces of restricted type were completely classified in the next two theorems.

Theorem 9.46. [Chen et al. (1993b)] *A hypersurface M of $\mathbb{E}^{n+1}$ is of restricted type if and only if it is one of the following hypersurfaces:*

(a) *a minimal hypersurface;*
(b) *an open portion of a spherical hypercylinder $S^k \times \mathbb{E}^{n-k}$, $2 \leq k \leq n$;*
(c) *an open portion of a hypercylinder on a plane curve of restricted type.*

Proof. Assume that M is a hypersurface of restricted type in $\mathbb{E}^{n+1}$. Let ξ be a unit normal vector field, A_ξ the shape operator at ξ, and $\alpha = \frac{1}{n}\text{Tr}\, A_\xi$ the mean curvature. Denote by h_ξ the scalar-valued second fundamental form defined by $h_\xi(X,Y) = \langle h(X,Y), \xi \rangle$. Then for tangent vectors X, Y we have $\alpha \langle A_\xi X, Y \rangle = \alpha h_\xi \langle X, Y \rangle = \langle FX, Y \rangle$. We can assume that F is symmetric. Then we can write

$$FX = \alpha A_\xi(X) + \beta(X)\xi, \tag{9.133}$$

$$F\xi = Z + f\xi, \tag{9.134}$$

where β is a one-form on M, Z is a tangent vector field, f is a function and

$$\beta(X) = \langle Z, X \rangle. \tag{9.135}$$

Taking the derivative of (9.133) and using (9.134) and the formulas of Gauss and Weingarten, we obtain

$$h_\xi(X,Y)Z = (X\alpha)A_\xi Y + \alpha(\nabla_X A_\xi)Y - \beta(Y)A_\xi X, \tag{9.136}$$

$$fh_\xi(X,Y) = \alpha h_\xi(X, A_\xi Y) + (\nabla_X \beta)Y. \tag{9.137}$$

By interchanging X and Y in (9.136) and subtracting, we can obtain, by using the Codazzi equation $(\nabla_X A_\xi)Y = (\nabla_Y)X$ that

$$(d\alpha + \beta)(X)A_\xi Y = (d\alpha + \beta)(Y)A_\xi X. \tag{9.138}$$

The combination of (9.135) and (9.136) yields for tangent vector X, Y, W,

$$h_\xi(X,Y)\beta(W) = (d\alpha)(X)h_\xi(Y, W) + \alpha \langle (\nabla_X A_\xi)Y, W \rangle \\ - \beta(Y)h_\xi(X, W). \tag{9.139}$$

Since A_ξ is symmetric, it can be diagonalized at every point. Thus there exist continuous functions $\kappa_1, \ldots, \kappa_n$ such that $\kappa_1(x), \ldots, \kappa_n(x)$ are the eigenvalues of A_ξ at x for all x. Moreover, the functions $\kappa_1, \ldots, \kappa_n$ are differentiable on the open dense subset of points where the multiplicities of the eigenvalues of A_ξ do not change. Let x be such a point. Then there exists a local orthonormal frame $\{e_1, \ldots, e_n\}$ around x such that $A_\xi(e_i) = \kappa_i e_i$ for $i = 1, \ldots, n$. Then we have

$$(\nabla_{e_j} A_\xi)e_i = (e_j \kappa_i)e_i + (\kappa_i - A_\xi)(\nabla_{e_j} e_i).$$

This implies in particular that

$$\langle (\nabla_{e_j} A_\xi)e_i, e_i \rangle = e_j \kappa_i. \tag{9.140}$$

The equation of Codazzi can be written down as

$$(e_j \kappa_i)e_i + (\kappa_i - A_\xi)\nabla_{e_j} e_i = (e_i \kappa_j)e_j + (\kappa_j - A_\xi)\nabla_{e_i} e_j, \tag{9.141}$$

which in particular for $i \neq j$ gives

$$e_j \kappa_i = (\kappa_j - \kappa_i)\langle \nabla_{e_i} e_j, e_i \rangle. \tag{9.142}$$

Putting $X = Y = W = e_i$ in (9.139), we obtain from (9.140) that

$$2\kappa_i \beta(e_i) = (e_i \alpha)\kappa_i + \alpha e_i \kappa_i = e_i(\alpha \kappa_i), \tag{9.143}$$

and putting $X = W = e_i$ and $Y = e_j$ for $i \neq j$ in (9.139) we get

$$\alpha e_j \kappa_i = \beta(e_j)\kappa. \tag{9.144}$$

Now, put $V = \{x \in M : \alpha(x) \neq 0\}$. If V is empty, then M is minimal. So assume V is not empty. Let x be any point where the multiplicities of the principal curvatures do not change. We consider two cases

Case (1): Rank $(A_\xi)_x = 1$ *and* $x \in V$. Without loss of generality, we can assume $\kappa_1 \neq 0$ and $\kappa_2 = \cdots = \kappa_n = 0$. Then (9.138) gives for $i > 1$ that

$$(d\alpha + \beta)e_i = 0. \tag{9.145}$$

Putting $i = 1$ in (9.143), we get, using the fact that $\alpha = \kappa_1/n$ that

$$e_1 \kappa_1 = n\beta(e_1). \tag{9.146}$$

Let $i = 1$, $j > 1$ in (9.144),. Then (9.145) gives $\kappa_1 e_j \kappa_i = -\kappa_1 e_j \kappa_1$. Thus

$$e_j \kappa_1 = 0. \tag{9.147}$$

From (9.142) we then find, putting $i = 1$, $j > 1$, that $\langle \nabla_{e_1} e_1, e_j \rangle = 0$. Thus $\nabla_{e_1} e_1 = 0$. Putting $i = 1$, $j > 1$ in (9.141), we obtain from (9.147) that $\kappa_1 \nabla_{e_j} e_1 = -A_\xi(\nabla_{e_1} e_j)$, and so $\nabla_{e_j} e_1 = 0$. Hence e_1 is parallel. Now, a standard argument shows that a neighborhood of x is a cylinder on a plane curve. Since M is of restricted type, this curve is of restricted type.

Case (2): Rank $(A_\xi)_x > 1$, $x \in V$. In this case, (9.138) implies $d\alpha = -\beta$. It follows from (9.144) for that

$$e_j(\alpha \lambda_i) = 0, \quad i \neq j. \tag{9.148}$$

Summing (9.148) over all i different from j gives $e_j(\alpha(n\alpha - \kappa_j) = 0$. So by (9.143) we get that for all j

$$(n\alpha + \kappa_j)e_j\alpha = 0. \tag{9.149}$$

If there is a j such that $e_j\alpha \neq 0$, then (9.149) implies $n\alpha + \kappa_j = 0$. Putting this in (9.143) gives $e_j\kappa_j^2 = 0$. Consequently, we have $e_j\alpha = 0$, which is a contradiction. So for all j we have $e_j\alpha = 0$. Hence α is a nonzero constant. This implies that both β and Z are zero. So from (9.136) we see that $\nabla A_\xi = 0$. Therefore M has parallel second fundamental form around x, and hence must be an open part of a product $S^k \times \mathbb{E}^{n-k}$.

Now consider the following three sets, whose union is dense in M:

$$U_1 = \{x \in M : \alpha(x) \neq 0 \text{ and } \mathrm{Rank}\,(A_\xi)_x > 1\},$$
$$U_2 = \{x \in M : \alpha(x) = 0 \text{ and } \mathrm{Rank}\,(A_\xi)_x > 1\},$$
$$U_3 = \text{Interior of } \{x \in M : \mathrm{Rank}\,(A_\xi)_x \leq 1\}.$$

If U_1 is nonempty, then it follows from Case (2) and (9.133), (9.134) and (9.137) that $\mathrm{Rank}\,F > 2$. If U_2 is nonempty, then it follows from (9.133), (9.134) (9.137) and (9.138) that $F = 0$. If U_3 is nonempty, then it follows from Case (1) and (9.133), (9.134) that $\mathrm{Rank}\,F \leq 2$. Since F is constant, it immediately follows that either U_1 is empty or $U_1 = M$. In the latter case, M is an open part of $S^k \times \mathbb{E}^{n-k}$. So we assume that U_1 is empty. If U_2 is nonempty, then $F = 0$ and this implies that α is identically zero. So M is minimal. Hence it remains to consider the case where $M = U_3$. In this case that M is flat.

If M is flat, we can consider V defined above and the interior U of the complement of V. Then every each connected component of U is a part of a hyperplane, and from Case (1) we can easily obtain that each connected component of V is an open part of cylinder on a curve of restricted type.

The curves of restricted type are classified in the next theorem. From the explicit equations given there, it follows that it is impossible to paste a part of a plane to a cylinder on a curve of restricted type which is not a line. Then one immediately obtains that M itself is a cylinder on a curve of restricted type. $\qquad\square$

It follows from (9.132) that a unit speed curve $\gamma : I \to \mathbb{E}^m$ is of restricted type if and only if it satisfies

$$(\gamma'''(s))^T = (F\gamma'(s))^T, \tag{9.150}$$

for any $s \in I$, where $(\,\cdot\,)^T$ is the tangential component of $(\,\cdot\,)$ and F is a fixed endomorphism of $\mathbb{E}^m$.

By applying Frenet's formula, one can prove that a unit speed planar curve is of restricted type if and only if its curvature function κ satisfies

$$\kappa\kappa''' - \kappa'\kappa'' + 4\kappa^3\kappa' = 0. \tag{9.151}$$

Curves of restricted type are classified in the next theorem.

Theorem 9.47. [Chen et al. (1993b)] *A curve in* $\mathbb{E}^2$ *is of restricted type if and only if it is congruent to an open portion of one of the following curves:*

(1) *a circle;*
(2) *a line;*
(3) *a curve with equation:* $\cos(cu) = e^{-cv}$, *where* $c \neq 0$;
(4) *curve with equation:* $a\sin^2(\sqrt{c}\,u) + b\sinh^2(\sqrt{c}\,v) = c$, *where* $a > b > 0$ *and* $c = a - b$;
(5) *a curve with equation:* $a\sin^2(\sqrt{c}\,u) - b\cosh^2(\sqrt{c}\,v) = c$, *where* $a > 0 > b$ *and* $c = a - b$.

Proof. Let γ be a curve in $\mathbb{E}^2$ parametrized by its arclength s. If γ is of restricted type, then there exists a fixed endomorphism F such that

$$\langle FT, T \rangle = \kappa^2, \tag{9.152}$$

where $T = \gamma'$ and κ is the curvature of γ. Since F can be assumed to be symmetric, we can choose Euclidean coordinates on $\mathbb{E}^2$ such that F is in diagonal form, i.e., $A = \begin{pmatrix} a & 0 \\ 0 & b \end{pmatrix}$, and we can suppose $a \geq b$.

Put $T = (\cos\theta(s), \sin\theta(s))$. Then $\kappa^2(s) = \theta'^2$ so that (9.152) becomes

$$\left(\frac{d\theta}{ds}\right)^2 = a\cos^2\theta + b\sin^2\theta. \tag{9.153}$$

Case (1): $a = b = 0$. In this case, the curve is a portion of a straight line.

Case (2): $a = b \neq 0$. Here γ is a part of a circle.

Case (3): $a \neq b$. Certainly $a > 0$. We can divide γ into pieces such that on each piece, $d\theta/ds \neq 0$, so that we can compute the inverse function $s(\theta)$. Let $\beta(\theta) = \gamma(s(\theta))$ be the reparametrization of γ in terms of θ. Then

$$\frac{d\beta}{d\theta} = \pm\left(\frac{\cos\theta}{\sqrt{a\cos^2\theta + b\sin^2\theta}}, \frac{\sin\theta}{\sqrt{a\cos^2\theta + b\sin^2\theta}}\right). \tag{9.154}$$

We consider three subcases.

Case (3.1): $a > 0$, $b = 0$. We can integrate (9.154) immediately and obtain

$$\beta(\theta) = \pm\left(\frac{1}{\sqrt{a}}\theta + c_1, -\frac{1}{\sqrt{a}}\ln|\cos\theta| + c_2\right).$$

We can assume that $c_1 = c_2 = 0$. Then it is clear that γ is of the form (3).

Case (3.2): $a > b > 0$. Integrating (9.154) implies that, up to a translation such that γ is of the form (4).

Case (3.3): $a > 0 > b$. Integrating (9.154) implies that, up to a translation such that it is of the form (5). $\qquad\square$

Example 9.7. The following are examples of curves of restricted type.

Figure 9.3: $\cos x = e^{-y}$.   Figure 9.4: $2\sin^2 x + \sinh^2 y = 1$.

It follows from Theorem 9.47 that not every curve (or more generally, submanifold) of restricted type is of finite type.

Remark 9.8. A curve with equation (3) given in Theorem 9.47 consists of infinitely many disjoint congruent pieces which can all be obtained by translating over a distance $2\pi/c$ in the u-direction of the uv-plane. One such component, say the one that contains the origin, lies in the region bounded by vertical lines $u = -\pi/2c$ and $u = \pi/2c$. It is called a "*chain of equal resistance*", because it represents the equilibrium of an infinite wire with varying thickness, but such that the tension at all points is constant. It was studied amongst others by G. Coriolis (1972-1843) in 1836, and has some nice geometric properties. For instance, the projection of the curvature vector on a certain fixed line (the normal line at the origin) has constant length. Another nice property is that, if we let the curve roll on a line, then the center of curvature of the contact point describes a catenary. We would like to remark also that, if we cut the minimal surface of Scherk $e^w \cos u = \cos v$ with a plane $u = c$ or $v = c$ in the uvw-space, we get a chain of equal resistance.

Pseudo-Riemannian version of Theorem 9.46 is the following.

Theorem 9.48. [Baikoussis et al. (1996)] *A pseudo-Riemannian hypersurface of Minkowski space-time* $\mathbb{E}_1^{n+1}$ *is of restricted type if and only if it is one of the following:*

(1) *a minimal hypersurface;*

(2) *an open part of one of the hypersurfaces:*
$$S_1^n, \ H^n, \ S^k \times \mathbb{E}_1^{n-k}, \ S_1^k \times \mathbb{E}^{n-k}, \ H^k \times \mathbb{E}^{n-k}, \ 1 \leq k \leq n-1;$$

(3) *an open portion of a hypercylinder on a plane curve of restricted type.*

Curves of restricted type in $\mathbb{E}_1^2$ were completely classified as follows.

Theorem 9.49. [Dillen et al. (1993)] *Up to rigid motions of* $\mathbb{E}_1^2$, *a curve in Minkowski plane* $\mathbb{E}_1^2$ *is of restricted type if and only if it is an open portion of one of the following curves:*

(1) *a line;*
(2) *an orthogonal hyperbola;*
(3) *a curve with equation:* $\sinh(cv) = e^{cu}$, *where* $c > 0$;
(4) *a curve with equation:* $\cosh(cu) = e^{cv}$, *where* $c > 0$;
(5) *a curve with equation:* $b \cosh^2(\sqrt{c}\,v) - a \cosh^2(\sqrt{c}\,u) = c$ *with* $b > a > 0$ *and* $c = b - a$;
(6) *a curve with equation:* $a \sinh^2(\sqrt{c}\,u) + b \cosh^2(\sqrt{c}\,v) = c$, *with* $b > 0 > a$ *and* $c = b - a$;
(7) *a curve with equation:* $a \sinh^2(\sqrt{c}\,u) - b \sinh^2(\sqrt{c}\,v) = c$ *with* $0 > b > a$ *and* $c = b - a$;
(8) *a curve with equation:* $a \sin^2(\sqrt{c}\,u) - b \sin^2(\sqrt{c}\,v) = c$, *with* $0 > a > b$ *and* $c = a - b$;
(9) *a curve with equation:* $\big(\cos(\sqrt{2b}(u+v)) - d\big)\big(\cosh(\sqrt{2b}(u-v)) + d\big) = d^2$, *where* $b > 0, c = b/\sqrt{a^2+b^2}$ *and* $d = a/\sqrt{a^2+b^2}$;
(10) *a curve defined by* $\big(\cosh(\sqrt{2|b|}\,(u+v)) + d\big)\big(\cos(\sqrt{2|b|}\,(u-v)) - d\big) = d^2$, *where* $b < 0, c = b/\sqrt{a^2+b^2}$ *and* $d = a/\sqrt{a^2+b^2}$;
(11) *a curve with equation:* $\sqrt{4v}e^u = 1$;
(12) *a curve with equation:* $v = c\tanh u$, *where* $c < 0$;
(13) *a curve with equation:* $v = c\coth u$, *where* $c > 0$.

For spherical surfaces of restricted type, we have the following.

Theorem 9.50. [Decruyenaere et al. (1993)] *If* M *is a surface in* $S^3(r) \subset \mathbb{E}^4$ *which has constant mean curvature, then* M *is of restricted type if and only if it is one of the following:*

(1) *a minimal submanifold of* $S^3(r)$;
(2) *an open portion of a small sphere* $S^2(r')$ *of* $S^3(r)$;
(3) *an open portion of the product of two circles* $S^1(a) \times S^1(b)$, *for some real positive numbers* a, b *with* $a^2 + b^2 = r^2$.

Chapter 10

Additional Topics in Finite Type Theory

10.1 Pointwise finite type maps

Recall that a map $\phi : M \to \mathbb{E}^m$ of Riemannian manifold M into $\mathbb{E}^m$ is called k-type if it admits a spectral decomposition

$$\phi = c + \phi_1 + \cdots + \phi_k$$

for some $c \in \mathbb{E}^m$ and non-constant maps $\phi_1, \ldots, \phi_k$ such that $\Delta\phi_i = \lambda_i\phi_i$ for $i = 1, \ldots, k$, where $\lambda_1, \ldots, \lambda_k$ are distinct real numbers and Δ is the Laplacian on M.

Definition 10.1. A map $\phi : M \to \mathbb{E}^m$ of a Riemannian manifold M into $\mathbb{E}^m$ is said to be of *pointwise k-type* if it admits a finite decomposition: $\phi = c + \phi_1 + \cdots + \phi_k$, where c is constant vector and $\phi_1, \ldots, \phi_k$ are non-constant maps such that

$$\Delta\phi_i = f_i(\phi_i - c_i), \quad i = 1, \ldots, k, \tag{10.1}$$

for some functions $f_i \in \mathcal{F}(M)$ and vectors $c_i \in \mathbb{E}^m$. A pointwise k-type map is said to be *proper pointwise k-type* if at least one of $f_1, \ldots, f_k$ is non-constant.

Example 10.1. If $\gamma(s) = \big(u_1(s), \ldots, u_m(s)\big)$ is a unit speed curve in $\mathbb{E}^m$ with $m \geq 2$ such that $u_1, \ldots, u_m$ are nowhere zero, then γ is a pointwise finite type curve. This simple fact can be seen as follows: If we put $f_i = -u_i''/u_i$ for each i, then $\Delta u_i = f_i u_i$. Hence γ is pointwise finite type curve. Thus the product

$$\phi : \gamma \times \mathbb{E}^{n_1} \to \mathbb{E}^m \times \mathbb{E}^{n_1}$$

defines a pointwise k-type submanifold for some $k \geq 2$. If particular, if γ is chosen to be a proper pointwise finite type curve, then ϕ is also proper pointwise finite type.

Example 10.2. If $\gamma_1(s)$ and $\gamma_2(t)$ are two proper pointwise finite type curves, then $\phi(s,t) = (\gamma_1(s), \gamma_2(t))$ is a proper pointwise finite type immersion of a flat surface into a Euclidean space. Thus there are abundant examples of proper pointwise finite type surfaces in Euclidean spaces.

Examples 10.1 and 10.2 show that there exist ample examples of proper pointwise k-type submanifold with $k \geq 2$. In contrast, the next result shows that there are no proper pointwise 1-type submanifolds in any Euclidean space.

Theorem 10.1. *Let* $\phi : M \to \mathbb{E}^m$ *be an isometric immersion of a Riemannian* n-*manifold* M *into* $\mathbb{E}^m$ *with* $n \geq 1$. *If* ϕ *is pointwise 1-type, then it is 1-type. Hence it is not proper pointwise 1-type.*

Proof. Let $\phi : M \to \mathbb{E}^m$ be a pointwise 1-type isometric immersion. Then we have $\Delta\phi = f(\phi - c)$ for a function $f \in \mathcal{F}(M)$ and $c \in \mathbb{E}^m$. Without loss of generality, we may assume $c = 0$. Hence, by Beltrami's formula we get

$$f\phi = -nH. \tag{10.2}$$

If $f = 0$, then M is a minimal submanifold of $\mathbb{E}^m$, which is of 1-type.

Now, suppose that $f \neq 0$. Then $U = \{x \in M : f(x) \neq 0\}$ is a non-empty open subset of M. On each connected component of U, (10.2) implies that ϕ is normal to M. Hence $X\langle\phi, \phi\rangle = 2\langle X, \phi\rangle = 0$ for each tangent vector X. Thus each connected component of U lies in a hypersphere of $\mathbb{E}^m$ centered at the origin. Therefore f is nowhere zero by continuity. So we have $U = M$, which shows that $\phi(M)$ lies in a hypersphere S_0^{m-1} centered at the origin. By taking covariant derivative of (10.2) with respect to $X \in TM$ and by applying the formula of Weingarten, we find

$$(Xf)\phi + fX = nA_H X - nD_X H. \tag{10.3}$$

Now, by comparing the normal components of (10.3) we find

$$nD_X H = -(Xf)\phi. \tag{10.4}$$

Since ϕ is normal to S^{m-1} and $D_X H$ is tangent to S^{m-1}, we obtain from (10.4) that $Xf = 0$ for any $X \in TM$. Hence f is a constant. Consequently, $\phi : M \to \mathbb{E}^m$ is of 1-type by definition. $\square$

The next example shows that Theorem 10.1 is false if ϕ is non-isometric.

Example 10.3. Consider the catenoid M in $\mathbb{E}^3$ parametrized by

$$\psi(x,y) = \left(\sinh^{-1} x, \sqrt{1+x^2}\cos y, \sqrt{1+x^2}\sin y \right)$$

equipped with the induced metric $g = dx^2 + (1 + x^2)dy^2$. Define a map $\phi : (M, g) \to \mathbb{E}^3$ by

$$\phi(x, y) = \left(\frac{x}{\sqrt{1 + x^2}}, \frac{\cos y}{\sqrt{1 + x^2}}, \frac{\sin y}{\sqrt{1 + x^2}} \right).$$

Then a straightforward computation shows that ϕ satisfies

$$\Delta\phi = \frac{1}{(1 + x^2)^2}\phi.$$

Hence ϕ is a proper pointwise 1-type map.

10.2 Submanifolds with finite type Gauss map

In the framework of finite type theory, the author and P. Piccinni initiated the study of submanifolds with finite type Gauss map in [Chen and Piccinni (1987)].

For a closed curve γ in $\mathbb{E}^m$, the type number of γ and the type number of the Gauss map of γ coincide. More precisely, we have the following.

Proposition 10.1. [Chen and Piccinni (1987)] *Let* $\gamma : \mathbb{R} \supset I \to \mathbb{E}^m$ *be a unit speed curve in* $\mathbb{E}^m$. *Then* γ *is of finite-type if and only if its Gauss map is of finite type. In particular, if* γ *is a closed curve, then* γ *is of* k-*type if and only if the Gauss map of* γ *is of* k-*type.*

Proof. Let $\gamma : I \to \mathbb{E}^m$ be a unit speed curve with arclength function s. Then $e_1 = d\gamma/ds$ is a unit tangent vector field. The Gauss map ν of γ can be defined as $\nu(s) = e_1(s) \in S^{m-1}(1) \subset \mathbb{E}^m$.

If γ is of k-type, then γ admit a spectral decomposition of the form:

$$\gamma = c_0 + \gamma_1 + \cdots + \gamma_k, \quad \Delta\gamma_i = \lambda_i\gamma_i, \tag{10.5}$$

where c_0 is a vector, $\gamma_1, \ldots, \gamma_k$ are non-constant maps and $\lambda_1, \ldots, \lambda_k$ are distinct eigenvalues of Δ. Since $\Delta = -d^2/ds^2$, (10.5) implies

$$\nu(s) = \gamma_1'(s) + \cdots + \gamma_k'(s), \quad \Delta\gamma_i'(s) = \lambda_i\gamma_i'(s). \tag{10.6}$$

Therefore the Gauss map of γ is of at most k-type. And it is of k-type if γ is non-null.

Conversely, let us assume that the Gauss map ν of γ is of k-type with minimal polynomial P. Then $P(\Delta)\nu = P(\Delta)\gamma' = 0$. Hence $P(\Delta)H = 0$, which implies γ is of finite type according to Proposition 6.3. In particular, if γ is a closed curve, then Theorem 6.1 implies that γ is also of k-type. $\square$

In contrast, for a closed submanifold M of $\mathbb{E}^m$ with $\dim M \geq 2$, the type number of M and of its Gauss map are not necessary the same (see Corollary 10.3).

Proposition 10.2. *If $\phi : M \to \mathbb{E}^m$ is an equivariant isometric immersion of a compact homogeneous Riemannian n-manifold into $\mathbb{E}^m$, then the Gauss map ν of M is of finite type with type number $\leq \binom{m}{n}$.*

Proof. Let $\tau_1(\nu)$ be the tension field of the Gauss map ν. Then

$$\tau_1(\nu), \Delta\tau_1(\nu), \ldots, \Delta^{N_{m,n}}\tau_1(\nu),$$

with $N_{m,n} = \binom{m}{n}$, are linearly dependent at a given point $u \in M$. Thus there is a polynomial $Q(t)$ of degree $\leq N_{m,n}$ satisfying $Q(\Delta)\tau_1(\nu) = 0$ at u. Since ϕ is an equivariant immersion, the group of isometries of the Euclidean space acts transitively on M as well as on the tangent bundle of M. So it acts transitively on its Gauss map. Thus $Q(\Delta)\tau_1(\nu) = 0$ at each point in M. Hence, by Theorem 9.27, the Gauss map is of finite type. Since the degree of the minimal polynomial of Q is $\leq \binom{m}{n}$, the type number is at most $\binom{m}{n}$ according to Theorem 9.27. □

The next theorem characterizes submanifolds with 1-type Gauss map (cf. [Bleecker and Weiner (1976); Chen and Piccinni (1987)]).

Theorem 10.2. *Let M be a closed submanifold of $\mathbb{E}^m$. Then M has 1-type Gauss map if and only if it has parallel mean curvature vector, flat normal connection, and constant scalar curvature.*

Proof. It follows from Theorem 6.1 and (9.109) that the Gauss map ν of M is of 1-type if and only if $DH = R^D = 0$ and $||h||$ is a constant. From (4.29) of Proposition 4.4, we have

$$2\tau = n^2 H^2 - ||h||^2,$$

where τ is the scalar curvature. Since we have $DH = 0$, the squared mean curvature H^2 of M is constant. Hence M has constant scalar curvature as well. The converse is clear. □

Theorem 10.2 implies the following.

Corollary 10.1. *If M is an isoparametric hypersurface of a hypersphere $S_o^{n+1} \subset \mathbb{E}^{n+2}$, then $\phi : M \to S_o^{n+1} \subset \mathbb{E}^{n+2}$ has 1-type Gauss map.*

Proof. Follows from the fact that every isoparametric hypersurface of S_o^{n+1} has parallel mean curvature, flat normal connection and constant scalar curvature in S_o^{n+1} as well as in $\mathbb{E}^{n+2}$. □

Corollary 10.2. [Chen and Piccinni (1987)] *The only closed hypersurfaces of Euclidean space with 1-type Gauss map are hyperspheres.*

Proof. If M is a closed hypersurface of $\mathbb{E}^{n+1}$, then M has flat normal connection. Thus Theorem 10.2 implies that the Gauss map is of 1-type if and only if it has constant mean curvature and constant scalar curvature; consequently, if and only if M is a hypersphere. □

Similarly, because hypersurfaces of a hypersphere $S_o^{n+1} \subset \mathbb{E}^{n+2}$ have flat normal connection in $\mathbb{E}^{n+2}$, Theorem 10.2 also implies the following.

Corollary 10.3. [Chen and Piccinni (1987)] *A hypersurface of a hypersphere S_o^{n+1} of $\mathbb{E}^{n+2}$ has 1-type Gauss map if and only if it is one of the following hypersurfaces:*

(1) *a small hypersphere of S_o^{n+1};*
(2) *a hypersurface of S_o^{n+1} which is 2-type in $\mathbb{E}^{n+2}$;*
(3) *a minimal hypersurface of S_o^{n+1} with constant scalar curvature.*

The next result classifies closed surfaces of $\mathbb{E}^m$ with 1-type Gauss map.

Theorem 10.3. [Chen and Piccinni (1987)] *Let M be a closed surface of $\mathbb{E}^m$. Then M has 1-type Gauss map if and only if it is either*

(1) *an ordinary sphere $S^2 \subset \mathbb{E}^3 \subset \mathbb{E}^m$ or*
(2) *the product of two circles $S^1(a) \times S^1(b) \subset \mathbb{E}^4 \subset \mathbb{E}^m$.*

The next theorem provides examples of spherical surfaces with 2-type Gauss map.

Theorem 10.4. [Chen and Piccinni (1987)] *Let $\phi : S^2 \to S_o^{m-1}(1) \subset \mathbb{E}^m$ be a minimal isometric immersion of a 2-sphere into $S_o^{m-1}(1) \subset \mathbb{E}^m$. If ϕ is not totally geodesic, then it has 2-type Gauss map.*

To prove the following three theorems for spherical hypersurfaces with 2-type Gauss map, we need the following.

Lemma 10.1. [Chen and Li (1998)] *Let M be a closed hypersurface of a hypersphere $S_o^{n+1}(1) \subset \mathbb{E}^{n+2}$. If the Gauss map ν of M is of 2-type, then we have:*

(1) $A'(\nabla\beta) = -\frac{1}{n}\nabla\|h\|^2 + \frac{n}{2}\beta\nabla\beta;$

(2) $\Delta\|h\|^2 - n^2\beta\Delta\beta - \|h\|^4 + a\|h\|^2 + b = 0;$

(3) $|\Delta\beta|^2 + (\|h\|^2 - a)\|\nabla\beta\|^2 = 0,$

where $\{p,q\}$ is the order of ν, $a = \lambda_p + \lambda_q$, $b = -\lambda_p\lambda_q$, h is the second fundamental form of M in $\mathbb{E}^{n+2}$, and A' and β are the shape operator and the mean curvature of M in $S_o^{n+1}(1)$, respectively.

Proof. Let H and H' be the mean curvature vector of M in $\mathbb{E}^{n+2}$ and in $S_o^{n+1}(1)$, respectively. Hence we have $H = H' - \mathbf{x}$, where $\mathbf{x}$ is the position vector of M in $\mathbb{E}^{n+2}$. We choose an orthonormal local frame field $\{e_1, e_2, \ldots, e_n, e_{n+1}, e_{n+2}\}$ on M such that $e_1, \ldots, e_n$ are tangent to M, $e_{n+1} = \mathbf{x}$, and e_{n+2} is normal to M in S^{n+1} so that $H' = \beta e_{n+2}$. The Gauss map $\nu : M \to G(2, n+2) \subset \bigwedge^2 \mathbb{E}^{n+2} = \mathbb{E}^{N_{n+2,2}}$ is then given by

$$\nu = \mathbf{x} \wedge e_{n+2}. \tag{10.7}$$

By a straightforward long computation, we find

$$\Delta\nu = \|h\|^2\nu + n\mathbf{x} \wedge \nabla\beta, \tag{10.8}$$

and

$$\begin{aligned}
\Delta^2\nu = &-\big[n(\|h\|^2 + n - 1)\nabla\beta + 2A'(\nabla\|h\|^2) \\
&+ n\,\Delta^T(\nabla\beta) + nA'^2(\nabla\beta)\big] \wedge \mathbf{x} \\
&- \big\{2\nabla\|h\|^2 - n^2\beta\nabla\beta + 2nA'(\nabla\beta)\big\} \wedge e_{n+2} \\
&+ \Big(\Delta\|h\|^2 - 2n\sum_i \langle\nabla_{e_i}(A'(\nabla\beta)), e_i\rangle + \|h\|^4 + n^2\|\nabla\beta\|^2\Big)\nu,
\end{aligned} \tag{10.9}$$

where $\Delta^T(\nabla\beta)$ denotes the tangential component of $\Delta(\nabla\beta)$.

If M is a closed hypersurface of a hypersphere S_o^{n+1}. Then Lemma 9.8 implies that Gauss map ν of M is mass-symmetric. Thus, if ν is of 2-type with order $\{p, q\}$, it admits the following spectral decomposition:

$$\nu = \nu_p + \nu_q, \quad \Delta\nu_p = \lambda_p\nu_p, \quad \Delta\nu_q = \lambda_q\nu_q, \quad \lambda_p < \lambda_q. \tag{10.10}$$

Hence we have

$$\Delta^2\nu = a\Delta\nu + b\nu, \quad a = \lambda_p + \lambda_q, \quad b = -\lambda_p\lambda_q. \tag{10.11}$$

After applying (10.7), (10.8), (10.9) and (10.11) we obtain

$$\begin{aligned}
n\,\Delta^T(\nabla\beta) + 2A'(\nabla\|h\|^2) + n\,A'^2(\nabla\beta) \\
+ n(\|h\|^2 + n - 1)\nabla\beta = na\nabla\beta,
\end{aligned} \tag{10.12}$$

$$2nA'(\nabla\beta) + 2\nabla\|h\|^2 - n^2\beta\nabla\beta = 0, \tag{10.13}$$

$$\begin{aligned}
\Delta\|h\|^2 - 2n\sum_i \langle\nabla_{e_i}(A'(\nabla\beta)), e_i\rangle + \|h\|^4 + n^2\|\nabla\beta\|^2 \\
= a\|h\|^2 + b.
\end{aligned} \tag{10.14}$$

Equality (10.13) is equivalent to

$$A'(\nabla\beta) = -\frac{1}{n}\nabla\|h\|^2 + \frac{n}{2}\beta\nabla\beta,\tag{10.15}$$

which gives formula (1).

Substituting (10.15) into into (10.14) gives

$$-2n\sum_i \langle\nabla_{e_i}(A'(\nabla\beta)), e_i\rangle = \sum_i \Big\{2\langle\nabla_{e_i}(\nabla\|h\|^2), e_i\rangle$$
$$-n^2\langle\nabla_{e_i}(\beta\nabla\beta), e_i\rangle\Big\} = -2\Delta\|h\|^2 + \frac{n^2}{2}\Delta\beta^2.\tag{10.16}$$

Hence (10.14) becomes

$$\Delta\|h\|^2 - \frac{n^2}{2}\Delta\beta^2 - n^2|\nabla\beta|^2 - (\|h\|^4 - a\|h\|^2 - b) = 0.\tag{10.17}$$

By using the identity

$$\Delta\beta^2 = 2\beta\Delta\beta - 2|\nabla\beta|^2,\tag{10.18}$$

equation (10.17) reduces to

$$\Delta\|h\|^2 - n^2\beta\Delta\beta - (\|h\|^4 - a\|h\|^2 - b) = 0.\tag{10.19}$$

This gives formula (2). Similarly, by applying (10.12) and (10.15), we have

$$\Delta^T(\nabla\beta) + \frac{1}{n}A'(\nabla\|h\|^2) - \frac{1}{2}\beta\nabla\|h\|^2$$
$$+ \left(\frac{n^2}{4}\beta^2 + \|h\|^2 + n - 1 - a\right)\nabla\beta = 0.\tag{10.20}$$

By taking the inner product with $\nabla\beta$, the first term of (10.20) gives

$$\langle\Delta^T(\nabla\beta), \nabla\beta\rangle = \frac{1}{2}\Delta(\|\nabla\beta\|^2) + \|H^\beta\|^2,\tag{10.21}$$

where H^β denotes the Hessian of β. For a given function $f \in \mathcal{F}(M)$ we have the following formula of Bochner-Lichnerowicz:

$$-\frac{1}{2}\Delta(\|\nabla f\|^2) = \|H^f\|^2 - |\Delta f|^2 + Ric(\nabla f, \nabla f).\tag{10.22}$$

Make use of (10.21) in (10.20), we obtain

$$\langle\Delta^T(\nabla\beta), \nabla\beta\rangle = |\Delta\beta|^2 - Ric(\nabla\beta, \nabla\beta).\tag{10.23}$$

On the other hand, by applying Gauss' equation, we have

$$Ric(\nabla\beta, \nabla\beta) = -\sum_i \|h(\nabla\beta, e_i)\|^2 + n\langle H, h(\nabla\beta, \nabla\beta)\rangle$$
$$= -\|A'\nabla\beta\|^2 - \|\nabla\beta\|^2 + n\langle A'(\nabla\beta), \beta\nabla\beta\rangle + n\|\nabla\beta\|^2$$
$$= (n-1)\|\nabla\beta\|^2 - \|A'(\nabla\beta)\|^2 + n\langle A'(\nabla\beta), \beta\nabla\beta\rangle.\tag{10.24}$$

Consequently, we have

$$\langle \Delta^T(\nabla\beta), \nabla\beta \rangle = |\Delta\beta|^2 - (n-1)||\nabla\beta||^2 + ||A'(\nabla\beta)||^2$$
$$- n \langle A'(\nabla\beta), \beta\nabla\beta \rangle. \tag{10.25}$$

For the other terms of (10.20), with the help of (10.15) we have

$$\frac{1}{n} \langle A'(\nabla||h||^2), \nabla\beta \rangle = \frac{n}{2} \langle A'(\nabla\beta), \beta\nabla\beta \rangle - ||A'(\nabla\beta)||^2, \tag{10.26}$$

$$\frac{1}{2} \langle \beta\nabla||h||^2, \nabla\beta \rangle = \frac{n^2}{4}\beta^2||\nabla\beta||^2 - \frac{n}{2} \langle A'(\nabla\beta), \beta\nabla\beta \rangle. \tag{10.27}$$

Now, after combining (10.20) and (10.25)-(10.27), we obtain formula (3) by taking the inner product of (10.20) with $\nabla\beta$. □

Theorem 10.5. [Chen and Li (1998)] *Let M be a closed hypersurface of $S_o^{n+1}(1) \subset \mathbb{E}^{n+2}$. If M has 2-type Gauss map, then M has non-constant mean curvature and non-constant length of second fundamental form.*

Proof. Suppose that M has constant mean curvature. Then β is constant. Therefore, by applying Lemma 10.19(1), we obtain $\nabla||h|| = 0$ identically. Thus M has constant length of second fundamental form. Since the scalar curvature τ of M satisfies $2\tau = n^2H^2 - ||h||^2$ by Gauss' equation, M has constant scalar curvature as well. But this is impossible by Theorem 10.2. Consequently, M has non-constant mean curvature.

Next, suppose that M has constant length $||h||$ of second fundamental form. Since the Gauss map of M is of 2-type, M must have non-constant mean curvature β in $S_o^{n+1}(1)$. Thus β^2 has a positive maximum at some point $y \in M$. So $\nabla\beta = 0$ at y. Thus, it follows from Lemma 10.19(3) that $\Delta\beta = 0$ at y. Since $\Delta\beta^2 = 2\beta\Delta\beta - 2||\nabla\beta||^2$, these yield $(\beta\Delta\beta)(y) = 0$. Substituting these into Lemma 10.19(2) yields

$$||h||^4 - a||h||^2 - b = \left(||h||^2 - \lambda_q\right)\left(||h||^2 - \lambda_p\right) = 0 \tag{10.28}$$

at y. Because $||h||$ is constant, (10.28) must holds at every point on M. Consequently, we find from (10.28) that

$$||h||^2 = \lambda_q \quad \text{or} \quad ||h||^2 = \lambda_p. \tag{10.29}$$

Now, from Lemma 10.19(2) and (10.28), we obtain $\Delta\beta = 0$ identically by continuity. Thus $\left(||h||^2 - \lambda_p - \lambda_q\right)|\nabla\beta|^2 = 0$ according to Lemma 10.19(3). Therefore β is constant according to (10.29), which is a contradiction. □

Theorem 10.6. [Chen and Li (1998)] *Every closed Einstein n-manifold M does not admit an isometric immersion into $S_o^{n+1}(1) \subset \mathbb{E}^{n+2}$ with 2-type Gauss map.*

Proof. Let M be a closed Einstein n-manifold with constant scalar curvature τ. If M has 2-type Gauss map, then, by Theorem 10.5, M has non-constant mean curvature. Let $U = \{u \in M : \nabla\beta^2 \neq 0 \text{ at } u\}$. From $2\tau = n^2 H^2 - \|h\|^2$ we obtain $\nabla\|h\|^2 = 2n^2\beta\nabla\beta$. Substituting this into Lemma 10.19(1) gives $A'(\nabla\beta) = -(3n/2)\beta\nabla\beta$. Thus $\nabla\beta$ is an eigenvector of A' with eigenvalue $-(3n/2)\beta$ on U. Without loss of generality, we may choose e_1 in the direction of $\nabla\beta$ on U. Denote by $\kappa_2, \ldots, \kappa_n$ the other eigenvalue of A'. Then we have $\kappa_2 + \cdots + \kappa_n = 5n\beta/2$ on U. By combining these with Gauss' equation, we obtain

$$Ric(e_1) = n - 1 - \frac{15n^2}{4}\beta^2. \tag{10.30}$$

Since M is Einsteinian with constant scalar curvature, (10.30) implies that M has constant mean curvature by continuity, which is impossible. $\square$

Theorem 10.7. [Chen and Li (1998)] *Let M be an oriented closed Riemannian n-manifold of class C^ω with constant scalar curvature. Then M admits no isometric immersion in $S_o^{n+1} \subset \mathbb{E}^{n+2}$ with 2-type Gauss map unless M is homeomorphic to an n-sphere.*

Proof. Let M be an oriented closed Riemannian n-manifold of class C^ω with constant scalar curvature. It follows from $2\tau = n^2 H^2 - \|h\|^2$ that

$$\Delta\|h\|^2 = n^2\Delta\beta^2. \tag{10.31}$$

Substituting (10.18) and (10.31) into Lemma 10.19(3) gives

$$n^2\Delta\beta^2 - 2n^2|\nabla\beta|^2 = 2\left(\|h\|^2 - \lambda_q\right)\left(\|h\|^2 - \lambda_p\right). \tag{10.32}$$

Assume that $\|h\|^2$ attends its absolute maximum at a point $u_1 \in M$. Then $\nabla\|h\|^2(u_1) = 0$. Since $\nabla\|h\|^2 = 2n^2\beta\nabla\beta$, Lemma 10.19(2) implies $\Delta\beta(u_1) = 0$. So, by (10.18) and (10.31) we find $\Delta\beta^2(u_1) = \Delta\|h\|^2(u_1) = 0$. Substituting these into (10.32), we have $\max\|h\|^2 = \lambda_q$ or $\max\|h\|^2 = \lambda_p$.

Similarly, if $\|h\|^2$ has its absolute minimum at $u_2 \in M$, then by (10.32), we find $\min\|h\|^2 = \lambda_q$ or $\min\|h\|^2 = \lambda_p$. Since M has 2-type Gauss map, $\|h\|^2$ is non-constant according to Theorem 10.5. Thus we must have

$$\max\|h\|^2 = \|h\|^2(u_1) = \lambda_q \text{ and } \min\|h\|^2 = \|h\|^2(u_2) = \lambda_p. \tag{10.33}$$

Because M is of class C^ω and $\|h\|^2$ is non-constant, the maximum and the minimum points of $(\|h\|^2)$ are isolated maximum and minimum points.

Now, we claim that the function $\|h\|^2$ has no other critical points on M. This can be seem as follows. If u_3 is another critical point of $\|h\|^2$, then $\nabla\beta(u_3) = \Delta\beta^2(u_3) = 0$ according to Lemma 10.19(3) and (10.18).

This implies that $\|h\|^2(u_3) = \lambda_q$ or $\|h\|^2(u_3) = \lambda_p$ by (10.32). But this is impossible, since the maximum and the minimum points of $\|h\|^2$ are isolated. Consequently, M admits a functions with only two critical points. Hence M is homeomorphic to an n-sphere by Reeb's theorem. □

The above results support the following conjecture posed in [Chen and Li (1998)].

Conjecture 10.1. *The only hypersurface in a hypersphere $S^{n+1} \subset \mathbb{E}^{n+2}$ with finite type Gauss map are hypersurfaces with 1-type Gauss map.*

The next theorem completely classifies minimal closed surfaces of S_o^{m-1} with 2-type Gauss map.

Theorem 10.8. [Chen and Piccinni (1987)] *Let M be a closed minimal surface of $S_o^{m-1}(1) \subset \mathbb{E}^m$. Then M has 2-type Gauss map ν if and only if it is one of the following surfaces:*

(1) *A sphere of radius $r_k = \sqrt{k(k+1)/2}$ for some integer $k \geq 2$ and the immersion is given by the k-th standard isometric immersion of $S^2(r_k) : \psi_k : S^2(r_k) \to S_o^{2k}(1) \subset \mathbb{E}^{2k+1}$ via an orthonormal basis of spherical harmonic functions of degree k;*

(2) *A flat torus $T_{(n,k,h)}^2 := \mathbb{E}^2/\Lambda$, immersed in a totally geodesic $S^5(1) \subset S_o^{m-1}(1)$, where the lattice Λ is generated by the vectors:*

$$\left(0, 2\sqrt{2/3}\, n\pi\right), \quad \left(\sqrt{2}k\pi, \sqrt{2/3}(2h-k)\pi\right), \quad n, k, h \in \mathbb{Z}; \; n, k > 0$$

and the immersion $\phi : T_{(n,k,h)}^2 \to S^5(1) \subset \mathbb{E}^6$ is induced from the isometric immersion $\psi : \mathbb{E}^2 \to \mathbb{C}^3 \cong \mathbb{E}^6$ defined by

$$\psi(s,t) = \frac{1}{\sqrt{3}}\left(e^{\frac{i}{\sqrt{2}}(s+\sqrt{3}t)}, e^{\frac{i}{\sqrt{2}}(-s+\sqrt{3}t)}, e^{\sqrt{2}si}\right).$$

The classification of all surfaces of revolution and ruled surfaces with finite type Gauss map was archived in the next two theorems.

Theorem 10.9. [Baikoussis et al. (1993a)] *A tube in $\mathbb{E}^3$ has finite type Gauss map if and only if it is a circular cylinder.*

Theorem 10.10. [Baikoussis et al. (1993a)] *A ruled surface M of $\mathbb{E}^m$ has finite type Gauss map if and only if it is a cylinder over a curve of finite type.*

Theorem 10.10 has been generalized by C. Baikoussis in [Baikoussis (1994)] to ruled submanifolds with finite type Gauss map.

For Gauss map of cyclides of Dupin we have the following result.

Theorem 10.11. [Baikoussis et al. (1993b)] *The Gauss map of cyclides of Dupin, closed or non-closed, in* $\mathbb{E}^3$ *are of infinite type.*

Theorem 10.12. [Kim and Kim (1995a)] *We have:*

(1) *The hyperplanes are the only hypercones shaped on spherical submanifolds whose Gauss map is of finite type.*
(2) *The hyperplanes, the hyperspheres and the cylinders over round spheres are the only quadric hypersurfaces whose Gauss map is of finite type.*

The following results study Euclidean submanifolds whose Gauss map satisfying the condition: $\Delta \nu = A\nu$.

Theorem 10.13. [Dillen et al. (1990b)] *Open portions of planes, spheres and circular cylinders are the only surfaces of revolution in* $\mathbb{E}^3$ *whose Gauss map satisfies* $\Delta \nu = A\nu$ *for some* $A \in \mathbb{R}^{3 \times 3}$.

Theorem 10.14. [Baikoussis and Blair (1992)] *Open portions of planes and circular cylinders are the only ruled surfaces in* $\mathbb{E}^3$ *whose Gauss map satisfies* $\Delta \nu = A\nu$ *for some* $A \in \mathbb{R}^{3 \times 3}$.

Theorem 10.15. [Baikoussis and Verstraelen (1995)] *Open portion of planes are the only spiral surfaces in* $\mathbb{E}^3$ *whose Gauss map satisfies* $\Delta \nu = A\nu$ *for some fixed endomorphism A of Euclidean 3-space.*

Theorem 10.16. [Baikoussis and Verstraelen (1993)] *Open portions of planes, spheres and circular cylinders are the only helicoidal surfaces in* $\mathbb{E}^3$ *whose Gauss map satisfies* $\Delta \nu = A\nu$ *for some* $A \in \mathbb{R}^{3 \times 3}$.

Let M be a spacelike (resp., timelike) surface in $\mathbb{E}^3_1$ and ξ be a unit vector field normal to M. Then ξ can be regarded as a point in $H^2(-1)$ (resp., in S^2_1) by parallel translation. The map $\nu : M \to H^2_1$ (resp., $\nu : M \to S^2_1$) is called the Gauss map of M in $\mathbb{E}^3_1$.

For Gauss map of surfaces in $\mathbb{E}^3_1$, we have the following.

Theorem 10.17. [Choi (1995)] *The only spacelike ruled surfaces in* $\mathbb{E}^3_1$ *whose Gauss map satisfies* $\Delta \nu = A\nu$ *for some fixed endomorphism A of* $\mathbb{E}^3_1$ *are open portions of* $\mathbb{E}^2_1$, $S^1_1 \times \mathbb{R}$ *and* $\mathbb{E}^1_1 \times S^1$.

Theorem 10.18. [Choi (1995)] *The only timelike ruled surfaces in* $\mathbb{E}^3_1$ *whose Gauss map satisfies* $\Delta\nu = A\nu$ *for some fixed endomorphism* A *of* $\mathbb{E}^3_1$ *are open portions of* $\mathbb{E}^2$ *and* $H^1 \times \mathbb{R}$.

Theorem 10.19. [Liu and Liu (1994)] *Let* M *be a spacelike surface of revolution in* $\mathbb{E}^3_1$. *If the Gauss map of* M *satisfies* $\Delta\nu = \lambda\nu$ *for some* $\lambda \in \mathbb{R}$, *then* M *is congruent to an open part of one of the following surfaces:*

(1) *a hyperbolic cylinder defined by* $x_3^2 - x_1^2 = 1/\lambda$, *with* $\lambda < 0$;
(2) *a hyperbolic space defined by* $-x_1^2 + x_2^2 + x_3^2 = 2/\lambda$ *with* $\lambda < 0$ *and* $x_3 > 0$, *where* x_1, x_2, x_3 *denote the standard coordinates of* $\mathbb{E}^3_1$.

Theorem 10.20. [Liu and Liu (1994)] *Let* M *be a timelike surface of revolution in* $\mathbb{E}^3_1$. *If the Gauss map of* M *satisfies* $\Delta\nu = \lambda\nu$ *for some* $\lambda \in \mathbb{R}$, *then* M *is congruent to an open part of one of the following surfaces:*

(1) *a right circular cylinder defined by* $x_2^2 + x_3^2 = 1/\lambda$, *with* $\lambda > 0$,
(2) *a hyperbolic cylinder defined by* $x_3^2 - x_1^2 = 1/\lambda$ *with* $\lambda > 0$;
(3) *a pseudo-Riemannian sphere defined by* $-x_1^2 + x_2^2 + x_3^2 = 2/\lambda$ *with* $\lambda > 0$.

For further results on submanifolds with finite type Gauss map, see [Kim at al. (2014b); Kim and Yoon (2005)].

10.3   Submanifolds with pointwise 1-type Gauss map

If a submanifold M of a Euclidean (or more generally, pseudo-Euclidean) space has 1-type Gauss map ν, then we have

$$\Delta\nu = \lambda(\nu + c) \tag{10.34}$$

for some $\lambda \in \mathbb{R}$ and some constant vector c. However, the Laplacian $\Delta\nu$ of the Gauss map of several important surfaces, such as helicoid, catenoid and right cones, take a more general form; namely,

$$\Delta\nu = f(\nu + c) \tag{10.35}$$

for some non-constant function f. Thus it is natural to study Euclidean submanifolds with *pointwise 1-type Gauss map*.

Definition 10.2. [Chen et al. (2005)] A Euclidean submanifold M with pointwise 1-type Gauss map is said to be *of the first kind* if the vector c in (10.35) is the zero vector. Otherwise, it is said to have pointwise Gauss map *of the second kind*.

Example 10.4. Consider a catenoid parameterized by

$$\phi(u, v) = \left(\sqrt{1 + v^2} \cos u, \sqrt{1 + v^2} \sin u, \sinh^{-1} v \right).$$

Its Gauss map ν is given by

$$\nu = \left(\frac{\cos u}{\sqrt{1 + v^2}}, \frac{\sin u}{\sqrt{1 + v^2}}, \frac{-v}{\sqrt{1 + v^2}} \right).$$

It is direct to show that the Laplacian $\Delta \nu$ of ν satisfies

$$\Delta \nu = \frac{2\nu}{(1 + v^2)^2}.$$

Hence the catenoid is a surface of revolution with proper pointwise 1-type Gauss map of the first kind.

Example 10.5. Consider the right cone C_a parametrized by

$$\phi(u, v) = \left(v \cos u, v \sin u, av \right), \quad a > 0.$$

Then the Gauss map ν is given by

$$\nu = \left(\frac{a \cos u}{\sqrt{1 + a^2}}, \frac{a \sin u}{\sqrt{1 + a^2}}, \frac{-1}{\sqrt{1 + a^2}} \right).$$

Moreover, the Laplacian $\Delta \nu$ of the Gauss map satisfies

$$\Delta \nu = \frac{1}{v^2} \left\{ \nu + \left(0, 0, \frac{1}{\sqrt{1 + a^2}} \right) \right\}.$$

Hence the right cone C_a has proper pointwise 1-type Gauss map of the second kind.

The next result is an easy consequence of formula (9.109).

Theorem 10.21. *We have:*

(a) *A submanifold of $\mathbb{E}^m$ (or a submanifold of $S^{m-1} \subset \mathbb{E}^m$) has pointwise 1-type Gauss map of first kind if and only if it has parallel mean curvature vector and flat normal connection.*
(b) *If a submanifold of $\mathbb{E}^m$ has harmonic Gauss map, i.e., $\Delta \nu = 0$, then it is totally geodesic.*

An immediate consequence of Theorem 10.21(a) is the following.

Corollary 10.4. *A hypersurface of $\mathbb{E}^{n+1}$ (or of $S^{n+1} \subset \mathbb{E}^{n+2}$) has pointwise 1-type Gauss map of the first kind if and only if it has constant mean curvature.*

The notion of surfaces of revolution of polynomial or rational kinds was introduced in [Chen and Ishikawa (1993)] as follows.

Definition 10.3. Let M be a surface of revolution in $\mathbb{E}^3$ parametrized by

$$x(\theta, t) = (t\cos\theta, t\sin\theta, g(t)) \qquad (10.36)$$

for some smooth function $g(t)$. Then M is said to be of *polynomial kind* if $g(t)$ in (10.36) is a polynomial in t; and it is of *rational kind* if $g(t)$ is a rational function. A surface of revolution of rational kind is simply called a *rational surface of revolution*.

For surfaces in $\mathbb{E}^3$, we have the following.

Theorem 10.22. [Chen et al. (2005)] *There do not exist rational surfaces of revolution in $\mathbb{E}^3$, except the polynomial kind, with pointwise 1-type Gauss map of the second kind.*

Theorem 10.23. [Chen et al. (2005)] *A surface of revolution of polynomial kind in $\mathbb{E}^3$ has pointwise 1-type Gauss map of the second kind if and only if it is a right cone.*

Theorem 10.24. [Chen et al. (2005)] *A rational surface of revolution in $\mathbb{E}^3$ has pointwise 1-type Gauss map if and only if it is an open part of a plane, a circular cylinder, or a right cone.*

Theorem 10.25. [Choi and Kim (2001)] *Open parts of the plane and the circular cylinder are the only cylindrical ruled surfaces in $\mathbb{E}^3$ with pointwise 1-type Gauss map of first kind.*

Theorem 10.26. [Choi and Kim (2001)] *The only ruled surfaces in $\mathbb{E}^3$ with pointwise 1-type Gauss map of the first kind are the open portions of the plane, the circular cylinder and the minimal helicoid.*

Theorem 10.27. [Ki at al. (2009)] *Let M be a cylindrical ruled surface in $\mathbb{E}^3$. If the Gauss map of M is of pointwise 1-type of the second kind, then M is an open part of a plane or a cylinder of an infinite type.*

Theorem 10.28. [Choi et al. (2009)] *A rational helicoidal surfaces in $\mathbb{E}^3$ with pointwise 1-type Gauss map if and only if M is part of a circular cylinder, a right cone or an ordinary helicoid.*

For flat surfaces with pointwise 1-type Gauss map we have

Theorem 10.29. [Dursun (2010a,b)] *Let M be an oriented flat regular surface in $\mathbb{E}^3$. Then M has pointwise 1-type Gauss map of the second kind* *if and only if it is an open portion of a right circular cone, a plane, or a cylinder congruent to*

$$\phi(s,t) = \left(\pm \frac{a^2}{b} \mu(s) - \frac{s}{b}, \frac{a}{2b\kappa^2(s)}, t \right),$$

where a, b are nonzero real numbers and the function $\mu(s)$ and the curvature function $\kappa(s)$ of the base curve are related by

$$\mu(s) = \int \frac{d\kappa}{\kappa^3 \sqrt{(b^2 - 1)\kappa^2 + 2a\kappa - a^2}}$$

and $\kappa(s)$ satisfies the differential equation $a^2 \kappa'^2 = \kappa^4 [(b^2 - 1)\kappa^2 + 2a\kappa - a^2]$.

Surfaces in $\mathbb{E}^4$ with pointwise 1-type Gauss map were studied in [Arslan et. al. (2011); Arsan et al. (2008)].

Theorem 10.30. [Dursun and Arsan (2011)] *An oriented non-minimal surface M in $\mathbb{E}^4$ has pointwise 1-type Gauss map of the first kind if and only if M has parallel mean curvature vector in $\mathbb{E}^4$.*

Theorem 10.31. [Dursun and Arsan (2011)] *A non-planar minimal oriented surface M in $\mathbb{E}^4$ has pointwise 1-type Gauss map of the second kind if and only if, with respect to some suitable local orthonormal frame on M, the shape operators of M are given by*

$$A_3 = \text{Diag}(\rho, -\rho), \quad A_4 = \text{anti-Diag}(\pm\rho, \pm\rho),$$

where ρ is a nonzero function on and anti-Diag(a, b) means a 2×2 anti-diagonal matrix.

Definition 10.4. *Vranceanu surfaces in $\mathbb{E}^4$ are the surfaces defined by*

$$\phi(s,t) = \left(r(s) \cos s \cos t, r(s) \cos s \sin t, r(s) \sin x \cos t, r(s) \sin s \sin t \right),$$

for some nowhere zero function $r(s)$.

A necessary and sufficient condition for Vranceanu surfaces to have pointwise 1-type Gauss map of the second kind was obtained in [Arslan et. al. (2011)]. For results on submanifolds with pointwise 1-type Gauss map, see [Arslan et. al. (2014); Choi et al. (2010b, 2011); Dursun (2009a); Ki at al. (2009); Kim and Kim (2012); Kim at al. (2014a); Kim and Yoon (2000); Kim at al. (2007); Miloushcva (2013); Niang (2004); Turgay (2014)].

Definition 10.5. A submanifold M of $\mathbb{E}^m$ is said to have *biharmonic Gauss map* if its Gauss map ν, defined by (9.104), satisfies $\Delta^2 \nu = 0$ identically.

Proposition 10.3. *Every hypersurface of $\mathbb{E}^{n+1}$ with constant mean curvature has pointwise 1-type Gauss map. Moreover, it has biharmonic Gauss map ν if and only if the shape operator A and the second fundamental form h of M satisfy*

$$\Delta||h||^2 + ||h||^4 = 0, \tag{10.37}$$

$$A(\nabla||h||^2) = 0. \tag{10.38}$$

Proof. The first statement follows from Corollary 10.4. For the second statement, let M be a hypersurface with constant mean curvature in $\mathbb{E}^{n+1}$. Choose an orthonormal local frame $e_1, \ldots, e_n, e_{n+1}$ such that $e_1, \ldots, e_n$ are tangent to M. Then the Gauss map ν can be defined as $\nu(p) = e_{n+1}(p)$. Then formula (9.109) yields $\Delta\nu = ||h||^2\nu$. Thus we find

$$\Delta^2\nu = \left(\Delta||h||^2 + ||h||^4\right)\nu - 2A(\nabla||h||^2), \tag{10.39}$$

which shows that ν is biharmonic if and only if (10.37) and (10.38) hold. $\quad\square$

The next three results follow easily from Proposition 10.3.

Corollary 10.5. *Let M be a surface in $\mathbb{E}^3$ with constant mean curvature. If M has biharmonic Gauss map ν, then it is totally geodesic.*

Proof. Under the hypothesis, we have (10.37) and (10.38) by Proposition 10.3. If $||h||^2$ is constant, (10.37) implies $h = 0$. If $||h||^2$ is non-constant on M, then $\nabla||h||^2$ is nowhere zero on a non-empty open subset $U \subset M$. Let us choose an orthonormal frame e_1, e_2 on U such that e_1 is in the direction of $\nabla||h||^2$. Then it follows from (10.38) of Proposition 10.3 that $Ae_1 = 0$. Thus one of the two principal curvatures is zero. Hence the other principal curvature is also constant since M has constant mean curvature. Consequently, $||h||^2$ is constant on U which is a contradiction. $\quad\square$

Corollary 10.6. *Let M be a hypersurface in $\mathbb{E}^{n+1}$ with constant mean curvature and with biharmonic Gauss map ν. If $||h||^2$ attends a maximum at some point $u \in M$, then M is totally geodesic.*

Proof. Under the hypothesis, if $||h||^2$ has a maximum at a point $u \in M$, then $\nabla||h||^2 = 0$ and the Hessian of $||h||^2$ is semi-negative definite at u. Thus $(\Delta||h||^2)(u) \geq 0$. Hence (10.37) gives $||h||^4(u) = -(\Delta||h||^2)(u) \leq 0$. Consequently, we obtain $h = 0$. $\quad\square$

Theorem 10.32. *Let M be a complete hypersurface with constant mean curvature in $\mathbb{E}^{n+1}$. If M has biharmonic Gauss map, it is totally geodesic.*

Proof. Under the hypothesis, if we put $u = \|h\|^2$, then u is a smooth non-negative function on M. It follows from Proposition 10.3 that u satisfies $\Delta u = -u^2$ on M. Hence, after applying Theorem 2.4 we obtain $h = 0$. $\square$

For curves with biharmonic Gauss map, we have the following.

Proposition 10.4. *The only curves in $\mathbb{E}^m$ with biharmonic Gauss map are open portions of lines.*

Proof. Let $\gamma : I \to \mathbb{E}^m$ be a unit speed curve. Then the Gauss map of γ can be defined to be $\nu(s) = \gamma'(s)$. If ν is biharmonic, then $\gamma^{(v)} = 0$. Hence γ is a tri-harmonic curve in $\mathbb{E}^m$. Therefore, by Corollary 7.22, γ is an open portion of a line. $\square$

In views of the above results, the author proposed the following.

Conjecture 10.2. *Every hypersurface of $\mathbb{E}^m$ with biharmonic Gauss map is totally geodesic.*

For spherical hypersurfaces with biharmonic Gauss map we have the following.

Theorem 10.33. [Chen and Li (1998)] *There do not exist hypersurfaces of $S_o^{n+1}(1) \subset \mathbb{E}^{n+2}$ with biharmonic Gauss map. In particular, there do not exist hypersurfaces in $S_o^{n+1} \subset \mathbb{E}^{n+2}$ with harmonic Gauss map.*

Proof. Assume that M is a hypersurface of $S_o^{n+1}(1) \subset \mathbb{E}^{n+2}$ with biharmonic Gauss map, then $\Delta^2 \nu = 0$. By applying arguments similar to the proof of Lemma 10.19, we obtain

$$A'(\nabla\beta) = -\frac{1}{n}\nabla\|h\|^2 + \frac{n}{2}\beta\nabla\beta, \qquad (10.40)$$

$$\Delta\|h\|^2 - n^2\beta\Delta\beta - \|h\|^4 = 0, \qquad (10.41)$$

$$|\Delta\beta|^2 + \|h\|^2|\nabla\beta|^2 = 0. \qquad (10.42)$$

Since $\|h\|^2 = \|A'\|^2 + n \neq 0$, (10.42) implies that β is a constant. Hence $\|h\|^2$ is also a constant by (10.40). Therefore we obtain from (10.41) that $h = 0$ which is a contradiction, since $\|h\|^2 \geq n$. $\square$

Remark 10.1. For the biharmonicity of the classical Gauss map $\hat{\nu} : M^n \to G(m, n - n)$ of a submanifold $M^n \subset \mathbb{E}^m$, see [Balmuş et al. (2010c)].

10.4   Submanifolds with finite type spherical Gauss map

Let $S_o^{m-1}(1)$ denote the unit hypersphere of $\mathbb{E}^m$ centered at the origin o. Assume that $\phi : M \to S_o^{m-1}(1) \subset \mathbb{E}^m$ is an isometric immersion of an oriented Riemannian n-manifold M into $S_o^{m-1}(1)$. For each $x \in M$, let $e_1, \ldots, e_n$ be an oriented orthonormal basis of $T_x M$. Then $\phi(x), e_1, \ldots, e_n$ determine an oriented $(n+1)$-subspace L_x of $\mathbb{E}^m$. Thus, for a spherical immersion $\phi : M \to S_o^{m-1}(1)$, there exists a natural map

$$\hat{\eta} : M \to G(m, n+1) \tag{10.43}$$

which carries $x \in M$ into a linear $(n+1)$-subspace in $G(m, n+1)$ obtained by L_x via parallel translation. In a natural way, $G(m, n+1)$ can be considered as a submanifold of $S_o^{N-1}(1) \subset \wedge^{n+1}\mathbb{E}^m = \mathbb{E}^N$ with $N = \binom{m}{n+1}$. Hence, for an immersion $\phi : M \to S_o^{m-1}(1)$ of an n-manifold M, there exists a map

$$\eta_\phi : M \to G(m, n+1) \subset S_o^{N-1}(1) \subset \mathbb{E}^N = \wedge^{n+1}\mathbb{E}^m$$

given by

$$\eta_\phi(x) = \phi(x) \wedge e_1 \wedge \cdots \wedge e_n \in \mathbb{E}^N = \wedge^{n+1}\mathbb{E}^m, \tag{10.44}$$

which is called the *spherical Gauss map* of ϕ. The corresponding map $\tilde{\eta}_\phi : M \to S_o^{N-1}(1)$ is called the *associated spherical Gauss map* of ϕ.

Proposition 10.5. [Chen and Lue (2007)] *If $\phi : M \to S_o^{m-1}(1) \subset \mathbb{E}^m$ is an equivariant isometric immersion of a compact homogeneous Riemannian n-manifold into $S_o^{m-1}(1)$, then its spherical Gauss map η_ϕ is of finite type. Moreover, the type number of $\tilde{\nu}$ is at most $\binom{m}{n+1}$.*

Proof. This can be proved in the same as Proposition 10.2.   □

By differentiating η_ϕ in (10.44) we find

$$e_j \eta_\phi = \sum_{r,k} h_{jk}^r \, \phi \wedge e_1 \wedge \cdots \wedge e_{k-1} \wedge e_r \wedge e_{k+1} \wedge \cdots \wedge e_n. \tag{10.45}$$

Since the Laplacian of η_ϕ is defined by $\Delta\eta_\phi = -\sum_i e_i e_i \eta_\phi + (\nabla_{e_i} e_i)\eta_\phi$, we obtain from (10.45) after a direct computation that

$$\Delta\eta_\phi = \|\hat{h}\|^2 \eta_\phi + n\hat{H} \wedge e_1 \wedge \cdots \wedge e_n$$
$$- n \sum_k \phi \wedge e_1 \wedge \cdots \wedge e_{k-1} \wedge D_{e_k}\hat{H} \wedge e_{k+1} \wedge \cdots \wedge e_n$$
$$+ \sum_{r,s,j,k} R_{sjk}^r \, \phi \wedge e_1 \wedge \cdots \wedge \overset{k-th}{\widehat{e_s}} \wedge \cdots \wedge \overset{j-th}{\widehat{e_r}} \wedge \cdots \wedge e_n, \tag{10.46}$$

where $\hat{h}$ and $\hat{H}$ denote the second fundamental form and the mean curvature vector of M in $S_o^{m-1}(1)$.

Proposition 10.6. [Chen and Lue (2007)] *Let* $\phi : M \to S_o^{m-1}(1)$ *be an isometric immersion. Then the spherical Gauss map η_ϕ is harmonic if and only if $\phi : M \to S_o^{m-1}(1)$ is totally geodesic.*

Proof. If the spherical Gauss map is harmonic, we have $\Delta\eta_\phi = 0$. Since the first term in the right-hand side of (10.46) is perpendicular to other terms, we obtain $\hat{h} = 0$. Hence ϕ is totally geodesic. $\square$

Similarly, we have the following.

Proposition 10.7. [Chen and Lue (2007)] *Let* $\phi : M \to S_o^{m-1}(1)$ *be an isometric immersion. Then the associated spherical Gauss map $\tilde{\eta}_\phi$ of ϕ is a harmonic map if and only if $\phi : M \to S_o^{m-1}(1)$ is a minimal immersion with flat normal connection.*

Theorem 10.34. [Chen and Lue (2007)] *Let* $\phi : M \to S_o^{m-1}(1)$ *be an isometric immersion. Then has mass-symmetric 1-type spherical Gauss map if and only if M is a minimal submanifold of $S_o^{m-1}(1)$ with constant scalar curvature and flat normal connection.*

Proof. If ϕ has mass-symmetric 1-type spherical Gauss map η_ϕ, then $\Delta\eta_\phi$ and η_ϕ are proportional. Since the three terms on the right hand side of (10.46) are perpendicular to η_ϕ, we conclude from (10.46) that η_ϕ is mass-symmetric 1-type if and only if $\hat{H} = R^r_{sjk} = 0$ and $||\hat{h}||^2$ is constant. Hence we obtain the theorem. $\square$

Theorem 10.35. [Chen and Lue (2007)] *A non-totally geodesic surface in $S_o^{m-1}(1)$ has mass-symmetric 1-type spherical Gauss map if and only if it is an open portion of the Clifford minimal torus lying fully in a totally geodesic 3-sphere $S_o^3(1) \subset S_o^{m-1}(1)$.*

Proof. It is easy to verify that the spherical Gauss map η_ϕ of the Clifford minimal torus satisfies $\Delta\eta_\phi = 2\eta_\phi$. Thus it has mass-symmetric 1-type spherical Gauss map. The converse follows from Theorem 10.34 and the fact that the only minimal surfaces of $S_o^{m-1}(1)$ with constant Gauss curvature and flat normal connection are open portions of a totally geodesic 2-sphere or of the Clifford minimal torus (cf. [Chen (1972c)]). $\square$

Theorem 10.36. [Bektas and Dursun (2014)] *An n-dimensional submanifold of $S_o^{m-1}(1)$ has non-mass-symmetric 1-type spherical Gauss map if and only if it is an open portion of small hypersphere in a totally geodesic $(n+1)$-sphere $S^{n+1} \subset S_o^{m-1}(1)$.*

An immediate consequence of Theorem 10.36 is the following.

Corollary 10.7. [Chen and Lue (2007); Bektas and Dursun (2014)] *Every isoparametric minimal hypersurface of $S_o^{n+1}(1)$ has mass-symmetric 1-type spherical Gauss map.*

The next two results classify minimal surfaces of $S_o^{m-1}(1)$ with 2-type spherical Gauss map.

Theorem 10.37. [Chen and Lue (2007)] *A minimal surface M of $S_o^{m-1}(1)$ is an open portion of the Veronese surface lying fully in a totally geodesic $S_o^4(1) \subset S_o^{m-1}(1)$ if and only if it has mass-symmetric 2-type spherical Gauss map.*

Theorem 10.38. [Chen and Lue (2007)] *A minimal surface of $S_o^{m-1}(1)$ is an open portion of the equilateral minimal torus lying fully in a totally geodesic $S_o^5(1) \subset S_o^{m-1}(1)$ if and only if it has non-mass-symmetric 2-type spherical Gauss map.*

10.5 Finite type submanifolds in Sasakian manifolds

A $(2n+1)$-dimensional manifold M is called an *almost contact manifold* if the structure group $GL(2n+1; \mathbf{R})$ of its linear frame bundle is reducible to $U(n) \times \{1\}$. This is equivalent to the existence of a tensor field ϕ of type $(1,1)$, a vector field ξ and 1-form η satisfying

$$\phi^2 = -I + \eta \otimes \xi, \quad \eta(\xi) = 1. \tag{10.47}$$

It follows that $\eta \circ \phi = 0$ and $\phi \xi = 0$. Further, since $U(n) \times \{1\} \subset O(2n+1)$, there exists a Riemannian metric g which satisfies

$$g(\phi X, \phi Y) = g(X, Y) - \eta(X)\eta(Y),$$
$$g(\xi, X) = \eta(X), \tag{10.48}$$

for all $X, Y \in TM$. The structure (ϕ, ξ, η, g) is called an *almost contact metric structure* and the manifold M^{2n+1} with an almost contact metric structure is called an *almost contact metric manifold*. If an almost contact metric manifold satisfies

$$d\eta(X, Y) = g(X, \phi Y), \tag{10.49}$$

for all $X, Y \in TM$, then M is called a *contact metric manifold*. On a contact metric manifold, the vector field ϕ is called the *characteristic vector field*.

A contact metric manifold is called a *Sasakian manifold* if it satisfies $[\phi, \phi] + 2d\eta \otimes \phi = 0$ on M^{2n+1}, where $[\phi, \phi]$ is the Nijenhuis torsion of ϕ

Example 10.6. Consider the unit hypersphere $S^{2n+1}(1) \subset \mathbb{C}^{n+1}$. Denote by $\mathbf{x}$ the position vector field of $S^{2n+1}(1)$ in $\mathbb{C}^{n+1}$ and by g the induced metric. Let $\xi = -J\mathbf{x}$, where J is the complex structure of $\mathbb{C}^{n+1}$. Let η be the dual 1-form given by $\eta(X) = g(\xi, X)$ and ϕ the tensor field of type $(1, 1)$ defined by $\phi = \pi \circ J$, where π is the orthogonal projection from $T_x\mathbb{C}^{n+1}$ onto $T_x S^{2n+1}(1)$, $x \in S^{2n+1}(1)$. Then $(S^{2n+1}(1), \phi, \xi, \eta, g)$ is a Sasakian manifold of constant ϕ-sectional curvature one.

Example 10.7. We consider $\mathbb{R}^{2n+1}$ with Cartesian coordinates (x_i, y_i, z) $(i = 1, \ldots, n)$ and its usual contact form

$$\eta = \frac{1}{2}\left(dz - \sum_{i=1}^{n} y_i dx_i\right). \tag{10.50}$$

The characteristic vector field ξ is given by $2\partial/\partial z$ and its Sasakian metric and the tensor field ϕ are given by

$$g = \eta \otimes \eta + \frac{1}{4}\sum_{i=1}^{n}\{(dx_i)^2 + (dy_i)^2\}, \tag{10.51}$$

$$\phi = \begin{pmatrix} 0 & \delta_{ij} & 0 \\ -\delta_{ij} & 0 & 0 \\ 0 & y_i & 0 \end{pmatrix}, \quad i = 1, \ldots, n. \tag{10.52}$$

Its curvature tensor $\bar{R}$ is given by

$$\bar{R}(X, Y)Z = -\eta(X)\eta(Z)Y + \eta(Y)\eta(Z)X - g(X, Z)\eta(Y)\xi$$
$$+ g(Y, Z)\eta(X)\xi - g(Z, \phi Y)\phi X + g(Z, \phi X)\phi Y - 2g(X, \phi Y)\phi Z.$$

We denote $\mathbb{R}^{2n+1}$ equipped this Sasakian structure by $\mathbb{R}^{2n+1}(-3)$.

For $\mathbb{R}^{2n+1}(-3)$, we can easily see that

$$\phi\xi = 0, \quad \eta \circ \phi = 0, \quad \bar{\nabla}_X\xi = -\phi X,$$
$$g(\phi X, \phi Y) = g(X, Y) - \eta(X)\eta(Y), \tag{10.53}$$
$$(\bar{\nabla}_X\phi)Y = g(X, Y)\xi - \eta(Y)X,$$

where $\bar{\nabla}$ is the Levi-Civita connection of $\mathbb{R}^{2n+1}(-3)$. The vector fields

$$e_i = 2\frac{\partial}{\partial y_i}, \quad \phi e_i = 2\Big(\frac{\partial}{\partial x_i} + y_i\frac{\partial}{\partial z}\Big),$$

$$\xi = 2\frac{\partial}{\partial z}, \quad i = 1, \ldots, n,$$

(10.54)

form an orthonormal basis of $\mathbb{R}^{2n+1}(-3)$ which is called a ϕ-*basis*. We find

$$\bar{\nabla}_{e_i}\phi e_j = \delta_{ij} = -\bar{\nabla}_{\phi e_i}e_j, \quad \bar{\nabla}_\xi e_i = -\phi e_i = \bar{\nabla}_{e_i}\xi,$$

$$\bar{\nabla}_\xi \phi e_i = e_i = \bar{\nabla}_{\phi e_i}\xi, \quad \bar{\nabla}_{e_i}e_j = \bar{\nabla}_{\phi e_i}\phi e_j = \bar{\nabla}_\xi \xi = 0.$$

Let $x : M \to \mathbb{R}^{2n+1}(-3)$ be an isometric immersion. We put

$$e_{n+i} = \phi e_i,$$

(10.55)

$$x = \sum_{i=1}^{n}(\bar{x}_i e_i + \bar{x}_{n+i}e_{n+i}) + \bar{x}_{2n+1}\xi,$$

(10.56)

$$H = \sum_{i=1}^{n}(\bar{H}_i e_i + \bar{H}_{n+i}e_{n+i}) + \bar{H}_{2n+1}\xi.$$

(10.57)

The ϕ-*position vector* x_ϕ and the ϕ-*mean curvature vector* H_ϕ are given respectively by $(\bar{x}_1, \ldots, \bar{x}_{2n})$ and $(\bar{H}_1, \ldots, \bar{H}_{2n})$. If x_ϕ satisfies $\Delta^2 x_\phi = 0$ which is equivalent to $\Delta H_\phi = 0$, we say that x_ϕ is ϕ-*biharmonic*.

Lemma 10.2. [Baikoussis and Blair (1991b); Sasahara (2002)] *Let $x : M \to \mathbb{R}^{2n+1}(-3)$ be an isometric immersion of an r-dimensional Riemannian manifold into $\mathbb{R}^{2n+1}(-3)$ which is either tangent or normal to ξ. Then the ϕ-position vector and ϕ-mean curvature vector are related by*

$$\Delta x_\phi = -r H_\phi.$$

(10.58)

Proof. Since $X\langle x, e_A\rangle = \langle X, e_A\rangle$ for any vector fields X, we have

$$-XX\langle x, e_A\rangle + \langle \nabla_X X, e_A\rangle = \langle h(X, X), e_A\rangle + 2\sum_{A=1}^{2n}\langle \phi X, e_A\rangle\langle \xi, X\rangle,$$

which implies (10.58). $\qquad\qquad\square$

Since we have the spectral decomposition for each coordinate function $\bar{x}^A$, we have the spectral decomposition for the ϕ-position vector field:

$$x_\phi = (x_\phi)_0 + \sum_{t+p}^{q}(x_\phi)_t.$$

(10.59)

Analogous to what we did in Chapter 6, an immersion of x of M in $\mathbb{R}^{2n+1}(-3)$ is said to be *of finite type* if q is finite; and it is *of k-type* if there are exactly k nonzero terms in the decomposition (10.59). Consequently, one can study finite type submanifolds of $\mathbb{R}^{2n+1}(-3)$ just in the same way as for finite type Euclidean submanifolds.

Definition 10.6. A Riemannian manifold M isometrically immersed in $\mathbb{R}^{2n+1}(-3)$ is called an *integral submanifold* if η restricted to M vanishes identically. In particular if $\dim M = n$, it is called a *Legendre submanifold*. Let M be a submanifold tangent to the characteristic vector field. Then M is called a *contact CR submanifold* if there exists a differential ϕ-invariant distribution $\mathcal{D}$ normal to ξ such that the complementary orthogonal distribution $\mathcal{D}^{\perp}$ normal to ξ is anti-invariant to ϕ, i.e., $\phi \mathcal{D}_x^{\perp} \subset T_x M^{\perp}$ for each $x \in M$. If $\dim \mathcal{D} = 0$, then M is called an *anti-invariant submanifold*.

Remark 10.2. Consider the canonical fibration $\pi : \mathbb{R}^{2n+1}(-3) \to \mathbb{C}^n$ given by

$$\pi(x_1, \ldots, x_n, y_1, \ldots, y_n, z) = \frac{1}{2}(y_1, \ldots, y_n, x_1, \ldots, x_n). \qquad (10.60)$$

Let $\varphi : M \to \mathbb{C}^n$ be a Lagrangian immersion of a Riemannian n-manifold M into $\mathbb{C}^n$. Then there is a covering map $\tau : \hat{M} \to M$ and a horizontal immersion $\hat{\varphi} : \hat{M} \to \mathbb{R}^{2n+1}(-3)$ such that $\varphi \circ \tau = \pi \circ \hat{\varphi}$. Thus, each Lagrangian submanifold of $\mathbb{C}^n$ can be lifted (locally) (or globally if M is simply-connected) to a Legendre immersion of the same manifold.

Conversely, let $\psi : \hat{M} \to \mathbb{R}^{2n+1}(-3)$ be a Legendre immersion. Then $\varphi = \pi \circ \psi : M \to \mathbb{C}^n$ is a Lagrangian immersion. Under this correspondence, we have a one-to-one correspondence between Lagrangian submanifold $\hat{M}$ of $\mathbb{C}^n$ and Legendre submanifolds M of $\mathbb{R}^{2n+1}(-3)$.

The existence and uniqueness theorems for Legendre submanifolds of $\mathbb{R}^{2n+1}(-3)$ are the following.

Theorem 10.39. *Let (M, g) be a simply connected Riemannian n-manifold and let σ be a symmetric bilinear TM-valued form on M satisfying*

(1) $\langle \sigma(X, Y), Z \rangle$ *is totally symmetric,*
(2) $(\nabla \sigma)(X, Y, Z) = \nabla_X \sigma(Y, Z) - \sigma(\nabla_X Y, Z) - \alpha(Y, \nabla_X Z)$ *is totally symmetric,*
(3) $R(X, Y)Z = \sigma(\sigma(Y, Z), X) - \sigma(\sigma(X, Z), Y).$

Then there exists a Legendre immersion $x : (M, g) \to \mathbb{R}^{2n+1}(-3)$ such that the second fundamental form h satisfies $h(X, Y) = \phi \sigma(X, Y)$.

Theorem 10.40. *Let* $x^1, x^2 : M \to \mathbb{R}^{2n+1}(-3)$ *be two Legendre isometric immersions of a connected Riemannian n-manifold into the Sasakian manifold* $\mathbb{R}^{2n+1}(-3)$ *with second fundamental forms* h^1, h^2, *respectively. If*

$$\langle h^1(X, Y), \phi(x^1_* Z) \rangle = \langle h^2(X, Y), \phi(x^2_* Z) \rangle$$

for all vector fields X, Y, Z *tangent to* M, *there exists an isometry* Φ *of* $\mathbb{R}^{2n+1}(-3)$ *such that* $x^1 = \Phi \circ x^2$.

Define the cylinder $N^{2n}(c) \subset \mathbb{R}^{2n+1}(-3)$ defined by

$$N^{2n}(c) = \left\{ x \in \mathbb{R}^{2n+1}(-3) : \sum_{i=1}^{2n} (x^i - x^i_0)^2 = 4c^2 \right\}, \tag{10.61}$$

where c is a constant and $\phi^2 x_0 \in \mathbb{R}^{2n+1}(-3)$.

We have the following lemma from [Baikoussis and Blair (1991b)].

Lemma 10.3. *An integral submanifold* M *of* $\mathbb{R}^{2n+1}(-3)$ *lies in* $N^{2n}(c)$ *if and only if the vector* $x - x_0$ *is perpendicular to* M.

C. Baikoussis and D. E. Blair studied finite type integral submanifolds of $\mathbb{R}^{2n+1}(-3)$ and obtained the following results.

Theorem 10.41. [Baikoussis and Blair (1991b)] *Let* $x : M \to \mathbb{R}^{2n+1}(-3)$ *be an r-dimensional integral closed submanifold whose* ϕ*-position vector field satisfies* (10.59). *Then* M *is of 1-type with order* p *if and only if it is a minimal submanifold of the cylinder* $N^{2n}(c)$ *with* $c^2 = r/\lambda_p$.

Theorem 10.42. [Baikoussis and Blair (1991b)] *Every Legendre curve of* $\mathbb{R}^3(-3)$ *lying in* $N^2(c)$ *defined by* (10.61) *is of 1-type and it is of the form*

$$x(t) = x_0 + \left(2c \cos t, 2c \sin t, -c^2(2t - \sin 2t) + 2cc_2 \cos t + c_0 \right)$$

with $x_0 = (c_1, c_2, 0)$.

Theorem 10.43. [Baikoussis and Blair (1991b)] *Let* M *be a Legendre closed submanifold of* $\mathbb{R}^{2n+1}(-3)$ *lying in* $N^{2n}(c)$. *If* M *is of 2-type, then the squared mean curvature* $\hat{H}^2$ *of* M *in* $N^{2n}(c)$ *is constant given by*

$$\hat{H}^2 = \frac{c^2}{n^2} \left(\frac{n}{c^2} - \lambda_p \right) \left(\lambda_q - \frac{n}{c^2} \right),$$

where p *and* q *are the lower order and the upper order of* M, *respectively.*

Theorem 10.44. [Baikoussis and Blair (1991b)] *Let* M *be a 2-type Legendre surface of* $\mathbb{R}^5(-3)$ *lying in* $N^4(c)$. *Then* M *is locally the Riemannian product of two curves given by*

(1) *two helices of rank 3, or*
(2) *a helix of rank 3 and a helix of rank 4, or*
(3) *a geodesic of* $\mathbb{R}^5(-3)$ *and a helix of rank 3, or*
(4) *a circle and a helix of rank 3.*

C. Baikoussis and D. E. Blair also proved the following.

Theorem 10.45. [Baikoussis and Blair (1991a)] *If M is a mass-symmetric 2-type Legendre surface of the Sasakian $S^5(1)$, then M is locally the product of a circle and a helix of rank 4 or the product of two circles.*

Theorem 10.46. [Baikoussis and Blair (1991a)] *If M is a mass-symmetric, 2-type Legendre surface of the Sasakian $S^5(1)$, then M is not stationary.*

Theorem 10.47. [Baikoussis and Blair (1995)] *Let $x : M \to S^7(1) \subset \mathbb{E}^8$ be a flat mass-symmetric 2-type Legendre submanifold in $S^7(1)$. Then M lies fully in $S^7(1) \subset \mathbb{E}^8 \cong \mathbb{C}^4$ and the position vector field of M in $\mathbb{E}^8$ is given by*

$$
\begin{aligned}
x = {} & \frac{\lambda}{\sqrt{\lambda^2+1}} \cos\left(\frac{u}{\lambda}\right) e_1 + \frac{1}{\sqrt{\sigma_2(\sigma_1+\sigma_2)}} \sin(\lambda u - \sigma_2 v) e_2 \\
& + \frac{\sin(\lambda u + \sigma_1 v + \rho_1 w)}{\sqrt{\rho_1(\rho_1+\rho_2)}} e_3 + \frac{\sin(\lambda u + \sigma_1 v - \rho_2 w)}{\sqrt{\rho_2(\rho_1+\rho_2)}} e_4 \\
& + \frac{\lambda}{\sqrt{\lambda^2+1}} \sin\left(\frac{u}{\lambda}\right) e_5 + \frac{\cos(\lambda u - \sigma_2 v)}{\sqrt{\sigma_2(\sigma_1+\sigma_2)}} e_6 \\
& + \frac{\cos(\lambda u + \sigma_1 v + \rho_1 w)}{\sqrt{\rho_1(\rho_1+\rho_2)}} e_7 + \frac{\cos(\lambda u + \sigma_1 v - \rho_2 w)}{\sqrt{\rho_2(\rho_1+\rho_2)}} e_8
\end{aligned}
\tag{10.62}
$$

with

$$
\begin{aligned}
\rho_1 &= \frac{\sqrt{4c(2c-a)+d^2}+d}{2}, \\
\rho_2 &= \frac{\sqrt{4c(2c-a)+d^2}-d}{2}, \\
\sigma_1 &= c, \quad \sigma_2 = c-a,
\end{aligned}
\tag{10.63}
$$

where $\{e_1, e_2, e_3, e_4, e_5 = -Je_1, e_6 = -Je_2, e_7 = -Je_3, e_8 = -Je_4\}$ is an orthonormal basis of $\mathbb{C}^4$ and a, c, d, λ are constants satisfying

$$
-1 \le \lambda < 0, \quad 1 + \lambda^2 + ac - c^2 = 0, \quad a \ge 0, \quad a^2 \ge d^2.
$$

10.6   Legendre submanifolds satisfying $\Delta H_\phi = \lambda H_\phi$

On the Sasakian $\mathbb{R}^{2n+1}(-3)$, we consider the canonical ϕ-basis

$$\{e_1, \ldots, e_n, e_{n+1} = \phi e_1, \ldots, e_{2n} = \phi e_n, \xi\}$$

defined by (10.54).

For integral submanifolds of $\mathbb{R}^{2n+1}(-3)$, we have the following.

Lemma 10.4. [Baikoussis and Blair (1995)] *Let M be an integral submanifold in $\mathbb{R}^{2n+1}(-3)$. Then*

$$\Delta \langle H, e_A \rangle = \langle \operatorname{Tr}(\bar{\nabla} A_H) + \Delta^D H + (\operatorname{Tr} A^2_{E_{k+1}})H$$
$$+ \mathfrak{a}(H) - \sum_{i=1}^{k} \eta(D_{E_i}H)\phi E_i, e_A \rangle. \tag{10.64}$$

For a k-dimensional submanifold M tangent to the characteristic vector field ξ, we choose an orthonormal frame $\{E_1, \ldots, E_{2n+1}\}$ on M so that $E_1, \ldots, E_{k-1}, E_k = \xi$ are tangent to M and H is parallel to E_{k+1}.

The next formula is analogous to formulas (4.46)-(4.47) for Euclidean submanifolds and also to formula (10.64) for integral submanifolds.

Lemma 10.5. [Sasahara (2002)] *Let M be a k-dimensional submanifold tangent to ξ in $\mathbb{R}^{2n+1}(-3)$. Then we have*

$$\Delta \langle H, e_A \rangle = \Big\langle \operatorname{Tr}(\bar{\nabla} A_H) + \Delta^D H + (\operatorname{Tr} A^2_{E_{k+1}} + 1)H + \mathfrak{a}(H)$$
$$+ \sum_{i=1}^{k-1} \eta(A_H E_i)\phi E_i + 2\phi A_H \xi - 2\phi D_\xi H, e_A \Big\rangle$$

for $A = 1, \ldots, 2n$, where $\mathfrak{a}(H) = \sum_{r=k+2}^{n} \operatorname{Tr}(A_H A_{E_r})E_r$.

Analogously to biharmonic submanifolds in Euclidean space, we have the following result for biharmonic Legendre submanifolds in the Sasakian $\mathbb{R}^{2n+1}(-3)$.

Proposition 10.8. [Sasahara (2002)] *Let M be a Legendre submanifold of the Sasakian $\mathbb{R}^{2n+1}(-3)$ with constant mean curvature. If $\Delta H_\phi = 0$, then M is minimal.*

Proof. Let us choose an orthonormal frame $\{E_1, \ldots, E_n\}$ on M such that

$H = \alpha\phi E_1$, where α is a constant. Then by using (10.53) we obtain

$$- D_{E_i} D_{E_i} H = \alpha\Big\{ \delta_{1i}\phi E_i - \sum_{j=1}^{n} E_i \langle \nabla_{E_i} E_1, E_j \rangle \phi E_j$$

$$- \sum_{j=1}^{n} \langle \nabla_{E_i} E_1, E_j \rangle \Big(\delta_{ij}\xi + \sum_{k=1}^{n} \langle \nabla_{E_i} E_j, E_k \rangle \phi E_k \Big) \Big\},$$

$$D_{\nabla_{E_i} E_i} H = \alpha\Big\{ \langle E_1, \nabla_{E_i} E_i \rangle \xi + \phi \sum_{k=1}^{n} \langle \nabla_{E_i} E_i, E_k \rangle \nabla_{E_k} E_1 \Big\}.$$

Combining this with $\Delta H_\phi = 0$ yields

$$\alpha\Big\{ \sum \langle \nabla_{E_i} E_j, E_1 \rangle^2 + \mathrm{tr} A_{\phi E_1}^2 \Big\} = 0$$

due to Lemma 10.5. Hence M is minimal. $\qquad\square$

The follows are examples of Legendre surfaces in $\mathbb{R}^5(-3)$ satisfying the condition $\Delta H_\phi = \lambda H_\phi$.

Example 10.8. Consider the immersion of a flat surface into $\mathbb{R}^5(-3)$ defined by

$$x(u, v) = \left(r\cos u, 0, r\sin u, v, \frac{1}{4}(\sin 2u - 2u)r^2 \right), \qquad (10.65)$$

where r is a nonzero constant. Then (10.65) defines a Legendre surface, and

$$\left\{ E_1 = \frac{2}{r}\frac{\partial}{\partial u}, E_2 = 2\frac{\partial}{\partial v} \right\}$$

is an orthonormal frame for (10.65). The shape operator of (10.65) takes the following forms,

$$A_1 = \begin{pmatrix} -\frac{2}{r} & 0 \\ 0 & 0 \end{pmatrix}, \quad A_2 = \begin{pmatrix} 0 & 0 \\ 0 & 0 \end{pmatrix}, \qquad (10.66)$$

where $A_i = A_{\phi E_i}$. This immersion satisfies $\Delta H_\phi = 4r^{-2} H_\phi$.

Example 10.9. Consider the immersion of a flat surface into $\mathbb{R}^5(-3)$:

$$x(u, v) = \Big(r\cos u, r\cos v, r\sin u, r\sin v,$$

$$\frac{r^2}{4}(\sin 2u + \sin 2v - 2u - 2v) \Big), \qquad (10.67)$$

where r is a nonzero constant. Then (10.67) defines a Legendre surface and

$$\{E_1 = 2r^{-1}\partial/\partial u, E_2 = 2r^{-1}\partial/\partial v\}$$

is an orthonormal frame for (10.67). The shape operator of (10.67) takes the following forms,

$$A_1 = \begin{pmatrix} -\frac{2}{r} & 0 \\ 0 & 0 \end{pmatrix}, \quad A_2 = \begin{pmatrix} 0 & 0 \\ 0 & -\frac{2}{r} \end{pmatrix}, \tag{10.68}$$

where $A_i = A_{\phi E_i}$. This immersions satisfies $\Delta H_\phi = 4r^{-2}H_\phi$.

It is known that both surfaces defined by (10.65) and (10.67) are Chen surfaces. The next theorem classifies Legendre Chen surfaces of $\mathbb{R}^5(-3)$ satisfying $\Delta H_\phi = \lambda H_\phi$.

Theorem 10.48. [Sasahara (2002)] *Let M be a Legendre Chen surface of $\mathbb{R}^5(-3)$ satisfying $\Delta H_\phi = \lambda H_\phi$. If the mean curvature function is nowhere zero, then there exists a coordinate system $\{x, y\}$ defined in a neighborhood $U \subset I \times \mathbb{R}$ of $p \in M$ and functions $a, b : I \to \mathbb{R} : x \to a(x), b(x)$ such that $a + b$ is nowhere zero and they satisfy*

$$ab - b^2 = -p' - p^2,$$
$$\frac{3a + b}{2}a' + \frac{3a(a - b)}{2(a - 2b)}b' = 0, \tag{10.69}$$
$$(a + b)'' + p(a + b)' - (a^2 + b^2 + p^2 - \lambda)(a + b) = 0,$$

where $p = b'/(a - 2b)$. Moreover the metric tensor of M is given by

$$g = dx^2 + G^2 dy^2, \quad G = q(y)\exp\left(\int^x p dx\right), \tag{10.70}$$

for some function $q(y)$ and the second fundamental form is given by

$$h(E_1, E_1) = a\phi E_1, \quad h(E_2, E_2) = b\phi E_1, \quad h(E_1, E_2) = b\phi E_2, \tag{10.71}$$

where $E_1 = \partial/\partial x$ and $E_2 = G^{-1}\partial/\partial y$.

Conversely, suppose that $a(x), b(x)$ are functions defined on an open interval I such that $a + b$ is nowhere zero and they satisfy (10.69). Let g be the metric tensor on a simply-connected domain $U \subset I \times \mathbb{R}$ defined by (10.70). Then, up to rigid motions of $\mathbb{R}^5(-3)$, there exists a unique Legendre Chen immersion of (U, g) into $\mathbb{R}^5(-3)$ whose second fundamental form is given by (10.71). Moreover such surface satisfies $\Delta H_\phi = \lambda H_\phi$.

Proof. Let $x : M \to \mathbb{R}^5(-3)$ be a Legendre immersion. Suppose that M is not minimal. Let U be a neighborhood of $p \in M$ such that $H \neq 0$. If M is a Chen surface, there exist a local orthonormal frame $\{E_1, E_2\}$ on U such that the shape operators take the following forms,

$$A_1 = \begin{pmatrix} a & 0 \\ 0 & b \end{pmatrix}, \quad A_2 = \begin{pmatrix} 0 & b \\ b & 0 \end{pmatrix}. \tag{10.72}$$

The Codazzi equation becomes

$$(\nabla_{E_1} A_i) E_2 - A_{D_{E_1} \phi E_i} E_2 - (\nabla_{E_2} A_i) E_1 + A_{D_{E_2} \phi E_i} E_1 = 0,$$

for $i = 1, 2$. Hence we have

$$E_1 b - (a - 2b) \omega_1^2(E_2) = 0, \tag{10.73}$$

$$E_2 a - (a - 2b) \omega_1^2(E_1) = 0, \quad E_2 b - 3b \omega_1^2(E_1) = 0, \tag{10.74}$$

where $\{\omega_i^j\}$ are connection forms. Assume that M satisfies $\Delta H_\phi = \lambda H_\phi$. Then since $\mathrm{Tr}(\bar{\nabla} A_H) = 0$ by Lemma 10.5, we obtain

$$aE_1(a + b) E_1 + b E_2(a + b) E_2 + \frac{1}{2}(a + b)\{b \omega_1^2(E_1) E_2$$
$$+ b \omega_1^2(E_2) E_1 + (E_1 a) E_1 + (a - b) \omega_1^2(E_1) E_2$$
$$+ (E_2 b) E_2 + (a - b) \omega_1^2(E_2) E_1\} = 0.$$

This yields

$$aE_1(a + b) + \frac{1}{2}(a + b)(a \omega_1^2(E_2) + E_1 a) = 0, \tag{10.75}$$

$$bE_2(a + b) + \frac{1}{2}(a + b)(a \omega_1^2(E_1) + E_2 b) = 0. \tag{10.76}$$

We put $W = \{p \in U : \omega_1^2(E_1) \neq 0\}$. Suppose that W is not empty. Then from (10.74) and (10.76) we obtain $(a^2 + 6ab + 5b^2)\omega_1^2(E_1) = 0$. It follows that

$$2aE_2 a + 6(E_2 a)b + 6(E_2 b)a + 10(E_2 b)b = 0.$$

This together with (10.74) give $a^2 + 10ab + 9b^2 = 0$. Thus W is totally geodesic, which is impossible since $H \neq 0$ on W. Hence $\omega_1^2(E_1) = 0$ and $E_2 a = E_2 b = 0$ on U.

Lemma 10.6. *Let M be a non-minimal Legendre Chen surface of $\mathbb{R}^5(-3)$ satisfying $\Delta H_\phi = \lambda H_\phi$. Then $a \neq 2b$.*

Proof. Assume that $a = 2b$. It follows from (10.73) and (10.74) that a and b are constants. Thus (10.75) gives $a \omega_1^2(E_2) = 0$. Since M is non-minimal, we find $\omega_1^2(E_2) = 0$ and so $\omega_1^2 = 0$. Hence, by the Gauss equation, we have $ab - b^2 = 0$, which is a contradiction. $\qquad \square$

Now, it follows from $\omega_1^2(E_1) = 0$ that there exists a local coordinate system $\{x, y\}$ such that the metric tensor takes the form $g = dx^2 + G^2 dy^2$ and $E_1 = \frac{\partial}{\partial x}$, $E_2 = G^{-1}\frac{\partial}{\partial y}$. We find

$$G = q(y)\exp \int^x (b'/(a - 2b))dx$$

from $\omega_1^2(E_2) = G_x/G$. The equation of Gauss implies the first equation of (10.69). Combining (10.73) and (10.75) yields the second equation of (10.69). In the same as [Baikoussis and Blair (1991b)] we have $\sum_{i=1}^2 \eta(D_{E_i}H)\phi E_i = H$. Also, from Lemma 10.5, we have

$$\langle \Delta^D H + (\mathrm{Tr}A_{\phi E_1}^2)H - H, \phi E_i \rangle = \langle \lambda H, \phi E_i \rangle , \quad i = 1, 2. \tag{10.77}$$

Let us put $H = \alpha\phi E_1$. Then by a direct computation we find that (10.77) is equivalent to the following relations:

$$- E_1 E_1 \alpha + \alpha\omega_1^2(E_2)^2 - \omega_1^2(E_2)(E_1\alpha) + (a^2 + b^2)\alpha = \lambda\alpha, \tag{10.78}$$

$$\alpha E_2(\omega_1^2(E_2)) + \alpha\omega_1^2(E_2)\omega_1^2(E_1) = 0. \tag{10.79}$$

Since $\omega_1^2(E_1) = 0$ and $E_2 a = E_2 b = 0$, relation (10.79) holds automatically. Hence (10.77) is equivalent to the last equation of (10.69).

Conversely, suppose that $a(x), b(x)$ are functions defined on an open interval I such that $a + b$ is nowhere zero and they satisfy (10.69). Let g be the metric tensor on $U \subset I \times \mathbb{R}$ defined by (10.70). We define a symmetric bilinear form σ on (U, g) by

$$\sigma(E_1, E_1) = aE_1, \quad \sigma(E_2, E_2) = bE_1, \quad \sigma(E_1, E_2) = bE_2,$$

where $E_1 = \partial/\partial x$ and $E_2 = G^{-1}\partial/\partial y$. We can easily obtain that $((U, g), \sigma)$ satisfies (1)-(3) of Theorem 10.39. Moreover we find that such immersion satisfies $\Delta H_\phi = \lambda H_\phi$ by using (10.69). $\qquad \square$

Corollary 10.8. [Sasahara (2002)] *Let M be a Legendre Chen surface of $\mathbb{R}^5(-3)$ satisfying $\Delta H_\phi = \lambda H_\phi$. Suppose that the mean curvature is nowhere zero. If a or b is constant, then M is an open part of the surface defined by (10.65) or (10.67).*

Proof. If $a = 0$, then from the first equation of (10.69) we have

$$4b^4 + 2bb'' - 3b'^2 = 0. \tag{10.80}$$

By applying $db'^2/db = 2b''$, we see that (10.80) is equivalent to the relation $bdb'^2/db - 3b'^2 + 4b^4 = 0$. Hence we get

$$b'^2 = Cb^3 - 4b^4 \tag{10.81}$$

for some constant C. Also, the last equation of (10.69) implies

$$\frac{1}{2}\frac{db'^2}{db} - \frac{b'^2}{2b} - \left(b^2 + \frac{b'^2}{4b^2} - \lambda\right)b = 0. \tag{10.82}$$

By combining (10.81) and (10.82) we conclude that b is constant and hence $b = 0$ by (10.80), which is a contradiction.

If a is a nonzero constant, we have $(a-b)b' = 0$ from the second equation of (10.69). It follows from (10.73) and (10.74) that $\omega_1^2 = 0$ and $ab - b^2 = 0$.

If b is a constant, we obtain $ab - b^2 = 0$ from the first equation of (10.69). In case $b = 0$, then M is an open part of (10.65) by applying the uniqueness theorem. In case $a = b$, the shape operators take the following form after rotating the frame $\{E_1, E_2\}$ through a constant angle:

$$A_1 = \begin{pmatrix} \alpha & 0 \\ 0 & 0 \end{pmatrix}, \quad A_2 = \begin{pmatrix} 0 & 0 \\ 0 & \alpha \end{pmatrix}, \quad \alpha \in \mathbb{R}.$$

Hence M is an open part of the surface defined by (10.67). $\qquad \square$

For further results, see [Sasahara (2003a,b, 2005b,c,d)].

10.7 Geometry of tensor product immersions

Let V and W be two vector spaces over the field of real numbers $\mathbb{R}$. We have the notion of the tensor product $V \otimes W$. In fact, if V and W are vector spaces equipped with inner products $\langle \ , \ \rangle_V$ and $\langle \ , \ \rangle_W$, respectively, $V \otimes W$ is a vector space equipped with the inner product given by

$$\langle x \otimes v, y \otimes w \rangle = \langle x, y \rangle_V \langle v, w \rangle_W, \tag{10.83}$$

for $x, y \in V$, $v, w \in W$. Obviously, $\mathbb{E}^m \otimes \mathbb{E}^{m'}$ is isometric to $\mathbb{E}^{mm'}$ in a natural way.

Definition 10.7. [Chen (1993a)] Let $\phi : M \to \mathbb{E}^m$ and $\psi : M \to \mathbb{E}^{m'}$ be two immersions of a manifold M, then the *tensor product immersion* $\phi \otimes \psi$ of ϕ and ψ is given by

$$(\phi \otimes \psi)(u) = \phi(u) \otimes \psi(u) \in \mathbb{E}^m \otimes \mathbb{E}^{m'}, \quad u \in M. \tag{10.84}$$

Definition 10.8. An immersion $\phi : M \to \mathbb{E}^m$ is called *transversal* at a point $u \in M$ if its position vector $\phi(u)$ at u is not tangent to M at u. An immersion ϕ is called *transversal* if it is transversal at every point on M. For a transversal immersion ϕ of M and an integer $k \geq 2$, $\phi^k := \phi \otimes \cdots \otimes \phi$ (k-times) is also transversal. The immersions ϕ^2 and ϕ^3 are called the *quadric* and *cubic* representation of M, respectively.

Let $\mathcal{R}(M)$ denote the set of all transversal immersions of a manifold M into Euclidean spaces. In the following, an immersion $\phi : M \to \mathbb{E}^m$ is called *spherical* if the image $\phi(M)$ lies in a hypersphere of $\mathbb{E}^m$ centered at the origin. Clearly, every spherical immersion is transversal. For each spherical immersion ϕ, ϕ^k is spherical for each integer $k \geq 2$.

The main question on tensor product immersions is the following.

Question 10.1. *To what extent do the tensor product immersions ϕ^k, $k \geq 2$, determine the immersion $\phi : M \to \mathbb{E}^m$? In particular, to what extent do the quadric and cubic representations ϕ^2 and ϕ^3 of ϕ determine ϕ?*

Remark 10.3. Let $\phi : M \to S_o^m(1) \subset \mathbb{E}^{m+1}$ be an isometric immersion of a Riemannian n-manifold M into the unit hypersphere $S_o^m(1)$ centered at the origin o in $\mathbb{E}^{m+1}$. Let

$$\epsilon_1 = (1, 0, \ldots, 0), \ldots, \epsilon_{m+1} = (0, \ldots, 0, 1)$$

be the standard orthonormal basis of $\mathbb{E}^{m+1}$. Then ϕ can be expressed as

$$\phi(u) = \sum_{a=1}^{m+1} x_a(u)\epsilon_a, \quad u \in M. \tag{10.85}$$

The quadric representation $\phi^2 : M \to \mathbb{E}^{m+1} \otimes \mathbb{E}^{m+1}$ of ϕ is then given by

$$\phi^2(u) = \sum_{a,b=1}^{m+1} x_a(u)x_b(u)\epsilon_a \otimes \epsilon_b. \tag{10.86}$$

If we put $u = (x_1, \ldots, x_{m+1})^T$, then the quadric representation $\phi^2 : M \to \mathbb{E}^{m+1} \otimes \mathbb{E}^{m+1}$ can be identified in a natural way with the map

$$\tilde{x} = xx^T : M \to H(m+1; \mathbb{R}); \quad u \mapsto x(u)x^T(u)$$

of M into the space $H(m+1; \mathbb{R})$ of real $(m+1) \times (m+1)$ symmetric matrices. Consequently, studying submanifolds with finite type quadric representation amounts to studying spectral behavior of the product $x_i \cdot x_j$ of coordinate functions.

Proposition 10.9. [Chen (1994b)] *Let $\phi : M \to \mathbb{E}^m$ be a transversal isometric immersion of a Riemannian n-manifold M into $\mathbb{E}^m$ with $n \geq 2$. Then, for $k \geq 2$, ϕ^k is a conformal map if and only if ϕ is spherical.*

Proof. For any tangent vector $X \in T_u M$, we have

$$(\phi^k)_*(X) = \phi_* X \otimes \phi^{k-1}(u) + \phi(u) \otimes \phi_* X \otimes \phi^{k-2}(u)$$
$$+ \cdots + \phi^{k-1}(u) \otimes \phi_* X. \tag{10.87}$$

Thus, for any $X, Y \in T_x M$, we have

$$
\begin{aligned}
\langle (\phi^k)_*(X), (\phi^k)_*(Y) \rangle = {} & k f(u)^{k-1} \langle \phi_* X, \phi_* Y \rangle \\
& + k(k-1) f(u)^{k-2} \langle \phi_* X, \phi(u) \rangle \langle \phi_* Y, \phi(u) \rangle
\end{aligned}
\tag{10.88}
$$

with $f(u) = \langle \phi(u), \phi(u) \rangle$. For any $X \in T_u M$, we choose $Y \in T_u M$ which is perpendicular to X. If ϕ^k is conformal, then (10.88) yields

$$
0 = \langle (\phi^k)_*(X), (\phi^k)_*(Y) \rangle = k(k-1) f(u)^{k-2} \langle \phi_* X, \phi(u) \rangle \langle \phi_* Y, \phi(u) \rangle,
$$

from which we get $\langle \phi_* X, \phi(u) \rangle \langle \phi_* Y, \phi(u) \rangle = 0$ since $\phi \in \mathcal{R}(M)$. Because this is true for any orthogonal X, Y, the linearity and the continuity of ϕ_* implies that $\langle \phi_* X, \phi(u) \rangle = 0$ for any $X \in T_u M, u \in M$. From this we conclude that $\phi(M)$ is spherical.

Conversely, if ϕ is spherical, then (10.88) implies

$$
\langle (\phi^k)_*(X), (\phi^k)_*(Y) \rangle = k f(u)^{k-1} \langle \phi_* X, \phi_* Y \rangle
$$

for any $X, Y \in T_u M$. This implies that ϕ^k is conformal. □

Remark 10.4. For $k = 2$, Proposition 10.9 is due to [Dimitric (1992b)].

The next result classifies 1-type tensor product immersions $\phi \otimes \psi$.

Theorem 10.49. [Chen (1993a)] *Let $\phi : M \to S_o^{m-1}(a) \subset \mathbb{E}^m$ and $\psi : M \to S_o^{m'-1}(b) \subset \mathbb{E}^{m'}$ be two spherical isometric immersions of a Ricmannian n-manifold M with $n \geq 2$. Then $\phi \otimes \psi$ is of 1-type if and only if M is an open portion of an n-sphere $S^n(r)$ and both ϕ and ψ are the standard imbedding of $S^n(r)$ into a totally geodesic $\mathbb{E}^{n+1} \subset \mathbb{E}^m$.*

Proof. Under the hypothesis, denote by $\hat{\phi} : M \to S_o^{m-1}(a)$ the immersion induced from $\phi : M \to \mathbb{E}^m$ and by $H_{\hat{\phi}}$ and H_ϕ the mean curvature vectors of $\hat{\phi}$ and of ϕ, respectively. Similar notations apply to $\psi : M \to S_o^{m'-1}(b) \subset \mathbb{E}^{m'}$. If $\phi \otimes \psi$ is of 1-type, then there is a constant map $c : M \to \mathbb{E}^m \otimes \mathbb{E}^{m'}$ and a real number λ such that

$$
\Delta(\phi \otimes \psi) = \lambda \phi \otimes \psi - c.
\tag{10.89}
$$

Since ϕ and ψ are isometric, Beltrami's formula and (10.89) imply that

$$
n H_\phi \otimes \psi + n \phi \otimes H_\psi + 2 \chi_{\phi \otimes \psi} = c - \lambda \phi \otimes \psi,
\tag{10.90}
$$

where $\chi_{\phi \otimes \psi}$ is the well-defined *tensor product field* given by

$$
\chi_{\phi \otimes \psi} = \sum_{i=1}^{n} \phi_*(e_i) \otimes \psi_*(e_i),
\tag{10.91}
$$

and $e_1, \ldots, e_n$ is an orthonormal frame of M. By taking the covariant derivative of (10.91) with respect to a tangent vector X, we find

$$n(D_X H_\phi - A_{H_\phi} X) \otimes \psi + n\phi \otimes (D_X H_\psi - A_{H_\psi} X)$$
$$+ n(H_{\hat\phi} - a^{-2}\phi) \otimes \psi_* X + n\phi_* X \otimes (H_{\hat\psi} - b^{-2}\psi)$$
$$+ 2\sum_i \left\{ h_{\hat\phi}(X, e_i) \otimes \psi_* e_i + \phi_* e_i \otimes h_{\hat\psi}(X, e_i) \right\} \qquad (10.92)$$
$$- a^{-2}\phi \otimes \psi_* X - b^{-2}\phi_* X \otimes \psi$$
$$= -\lambda\{\phi \otimes \psi_* X + \phi_* X \otimes \psi\},$$

where h and D denote the second fundamental form and normal connection. Since ϕ and ψ are spherical and isometric, we have both $DH_\phi = DH_{\hat\phi}$ and $DH_\psi = DH_{\hat\psi}$.

It is easy to verify that the tangent space of $\phi \otimes \psi$ at a point $u \in M$ is spanned by

$$\{\phi_* Y \otimes \psi(u) + \phi(u) \otimes \psi_* Y : Y \in T_u M\}.$$

Thus we find from (10.92) that

$$nH_{\hat\phi} \otimes \psi_* X + 2\sum_i h_{\hat\phi}(X, e_i) \otimes \phi_* e_i = 0 \qquad (10.93)$$

for any vector X tangent to M. If we choose $e_1, \ldots, e_n$ such that $X = e_1$, then we find from (10.93) that $nH_{\hat\phi} + 2h_{\hat\phi}(X, X) = 0$. Since this holds for all tangent vectors X, we obtain $H_{\hat\phi} = 0$. Combining this with (10.93) yields $\sum_i h_{\hat\phi}(X, e_i) \otimes \phi_* e_i = 0$ which implies that $\hat\phi$ is totally geodesic. Thus M is an open portion of an n-sphere. Same argument applies to $\hat\psi$. Consequently, we have $a = b$.

The converse can be verify by direct computation. $\qquad\qquad\qquad \square$

Theorem 10.49 implies immediately the following

Corollary 10.9. [Dimitric (1992b); Chen (1993a)] *If $\phi : M \to S_o^{m-1} \subset \mathbb{E}^m$ is an isometric immersion of a Riemannian n-manifold M into $\mathbb{E}^m$ with $n \geq 2$, then the quadric representation ϕ^2 is of 1-type if and only if ϕ is an open part of an ordinary hypersphere lying in a totally geodesic $\mathbb{E}^{n+1} \subset \mathbb{E}^m$.*

The next result provides further relations between type number and tensor product immersions.

Theorem 10.50. [Chen (1993a)] *Let $\phi : I \to \mathbb{E}^m$ and $\psi : I \to \mathbb{E}^{m'}$ be two finite type unit speed curves. Then we have:*

(1) *If ϕ and ψ are non-null curves, then $\phi \otimes \psi$ is of finite type.*
(2) *If at least one of ϕ and ψ is null, then $\phi \otimes \psi$ is of infinite type.*
(3) *If ϕ is null, then ϕ^k is of infinite type for any $k \geq 2$.*

Proof. Under the hypothesis, ϕ and ψ admit spectral decompositions:

$$\phi(s) = a_0 + b_0 s + \sum_{r=1}^{k_1}(a_r \cos p_r s + b_r \sin p_r s), \qquad (10.94)$$

$$\psi(s) = c_0 + d_0 s + \sum_{t=1}^{k_2}(c_t \cos q_r s + d_t \sin q_t s), \qquad (10.95)$$

for some constant vectors $a_0, b_0, c_0, d_0, a_r, b_r, c_t, d_t$ and positive numbers p_r, q_t. If ϕ and ψ are non-null finite type curves, then $b_0 = d_0 = 0$. Since the products of two sine and/or cosine functions are sine or cosine function, the tensor product $\phi \otimes \psi$ is of finite type.

If ϕ is null, the $b_0 \neq 0$. Thus there exists coordinate function of $\phi \otimes \psi$ which contain nonzero terms of the form: $s(c_t \cos q_r s + d_t \sin q_t s)$. Clearly, such functions are of infinite type. Hence $\phi \otimes \psi$ is of infinite type. This proves (2). Similar proof applies to (3). $\qquad \square$

The next two results provide sharp estimates of type number for tensor product immersions.

Theorem 10.51. [Chen (1993c)] *Let $\phi_i : M \to S_o^{m_i-1}(r_i) \subset \mathbb{E}^{m_i}$, $i = 1, \ldots, t$, be spherical isometric immersions of a Riemannian manifold M. Then $f_1 \otimes \cdots \otimes f_t$ is of k-type with $k \geq [\frac{t+1}{2}]$, where $[a]$ denotes the greatest integer $\leq a$.*

Proof. Under the hypothesis, we find from direct computation that

$$\Delta(\phi_1 \otimes \cdots \otimes \phi_t) = \overset{\text{sym}}{\sum} \Delta\phi_1 \otimes \phi_2 \otimes \cdots \otimes \phi_t$$
$$- 2 \overset{\text{sym}}{\sum} \phi_{1*}e_i \otimes \phi_{2*}e_i \otimes \phi_3 \otimes \cdots \otimes \phi_t, \qquad (10.96)$$

where $\sum^{\text{sym}}$ denotes the symmetric sum and $e_1, \ldots, e_n$ is an orthonormal tangent frame of M. In particular, we have

$$\overset{\text{sym}}{\sum} \Delta\phi_1 \otimes \phi_2 \otimes \cdots \otimes \phi_t = \sum_i \phi_1 \otimes \cdots \otimes \phi_{i-1} \otimes \Delta\phi_i \otimes \phi_{i+1} \otimes \cdots \otimes \phi_t,$$

$$\overset{\text{sym}}{\sum} \phi_{1*}e_i \otimes \phi_{2*}e_i \otimes \phi_3 \cdots \otimes \phi_t = \sum_{j<k}\sum_i \phi_1 \otimes \cdots \otimes \phi_{j-1} \otimes \phi_{j*}e_i \otimes$$
$$\otimes \phi_{j+1} \otimes \cdots \otimes \phi_{k-1} \otimes \phi_{k*}e_i \otimes \phi_{k+1} \otimes \cdots \otimes \phi_t.$$

Assume $t \geq 4$. It follows from (10.96) and a direct computation that
$$\Delta^2(\phi_1 \otimes \cdots \otimes \phi_t)$$

$$= 8 \sum^{\text{sym}} \phi_{1*}e_{i_1} \otimes \phi_{2*}e_{i_1} \otimes \phi_{3*}e_{i_2} \otimes \phi_{4*}e_{i_2} \otimes \phi_5 \otimes \cdots \otimes \phi_t$$

$$- 4 \sum^{\text{sym}} \Delta\phi_1 \otimes \phi_{2*}e_i \otimes \phi_{3*}e_i \otimes \phi_4 \otimes \cdots \otimes \phi_t \qquad (10.97)$$

$$+ 8 \sum^{\text{sym}} \sigma_{\phi_1}(e_{i_1}, e_{i_2}) \otimes \phi_{2*}e_{i_1} \otimes \phi_{3*}e_{i_2} \otimes \phi_4 \otimes \cdots \otimes \phi_t$$

$$+ \text{ terms with at least } (t-2) \ \phi_i\text{'s}.$$

Moreover, if $t \geq 2k$, induction and a long computation give
$$\Delta^k(\phi_1 \otimes \cdots \otimes \phi_t) =$$

$$(-1)^k a_k \sum^{\text{sym}} \phi_{1*}e_{i_1} \otimes \phi_{2*}e_{i_1} \otimes \cdots \otimes \phi_{2k-1*}e_{i_k} \otimes \phi_{2k*}e_{i_k} \otimes \phi_{2k+1} \otimes \cdots \otimes \phi_t$$

$$+ (-1)^{k-1} b_k \sum^{\text{sym}} \Delta\phi_1 \otimes \phi_{2*}e_{i_2} \otimes \phi_{3*}e_{i_2} \otimes \cdots$$

$$\cdots \otimes \phi_{2k-2*}e_{i_k} \otimes \phi_{2k-1*}e_{i_k} \otimes \phi_{2k} \otimes \cdots \otimes \phi_t \qquad (10.98)$$

$$+ (-1)^k c_k \sum^{\text{sym}} \sigma_{\phi_1}(e_{i_1}, e_{i_2}) \otimes \phi_{2*}e_{i_1} \otimes \phi_{3*}e_{i_2} \otimes \phi_{4*}e_{i_3} \otimes \phi_{5*}e_{i_3} \otimes$$

$$\cdots \otimes \phi_{2k-2*}e_{i_k} \otimes \phi_{2k-1*}e_{i_k} \otimes \phi_{2k} \otimes \cdots \otimes \phi_t$$

$$+ \text{ terms with at least } (t+2-2k) \ \phi_i\text{'s},$$

where
$$a_k = 2^k \cdot k!, \ \ b_1 = 1,$$
$$b_k = 2(k-1)b_{k-1} + a_{k-1}, \ \ k \geq 2,$$
$$c_1 = 0, \ \ c_2 = 8, \ \ c_k = 4a_{k-1} + 2(k-2)c_{k-1}, \ \ k \geq 3.$$

If $\phi_1 \otimes \cdots \otimes \phi_t$ is of k-type, then there exist real numbers $\alpha_1, \ldots, \alpha_k$ and a constant vector $c \in \mathbb{E}^{m_1} \otimes \cdots \otimes \mathbb{E}^{m_t}$ such that
$$\Delta^k \phi_1 \otimes \cdots \otimes \phi_t + \alpha_1 \Delta^{k-1} \phi_1 \otimes \cdots \otimes \phi_t$$
$$+ \cdots + \alpha_k(\phi_1 \otimes \cdots \otimes \phi_t - c) = 0. \qquad (10.99)$$

Assume $k < [\frac{t+1}{2}]$. By taking the covariant derivative of (10.99) with respect to a tangent vector X of M and using (10.96), (10.97) and (10.98), we obtain
$$a_k \sum^{\text{sym}} \phi_{1*}e_{i_1} \otimes \phi_{2*}e_{i_1} \otimes \cdots \otimes \phi_{2k-1*}e_{i_k} \otimes \phi_{2k*}e_{i_k}$$

$$\otimes \phi_{2k+1*}X \otimes \phi_{2k+2} \otimes \cdots \otimes \phi_t \qquad (10.100)$$

$$+ \text{ terms with at least } (t-2k) \ \phi_i\text{'s} = 0.$$

Since ϕ_i is a spherical immersion in $\mathbb{E}^{m_i}$, each ϕ_i can be regarded as a normal vector field along M. So, by taking the inner product of

$$\phi_{1*}e_1 \otimes \phi_{2*}e_1 \otimes \cdots \otimes \phi_{2k-1*}e_1 \otimes \phi_{2k*}e_1 \otimes \phi_{2k+1*}X \otimes \phi_{2k+2} \otimes \cdots \otimes \phi_t$$

with (10.100), we get $X = 0$ which is a contradiction, because this holds for any arbitrary tangent vector of M. $\qquad\square$

Theorem 10.52. [Chen (1993c)] *Let* $\phi_i : M \to S_o^{m_i-1}(r_i) \subset \mathbb{E}^{m_i}, i = 1, \ldots, 2k$, *be spherical isometric immersions of a Riemannian n-manifold. Then we have:*

(a) *The type number of* $\phi_1 \otimes \cdots \otimes \phi_{2k}$ *is at least k.*
(b) $\phi_1 \otimes \cdots \otimes \phi_{2k}$ *is of k-type if and only if M is an open portion of an n-sphere and each* $\hat{\phi}_i : M \to S^{m_i-1}(r_i)$ *is totally geodesic.*

If $\phi_1 = \cdots = \phi_{2k}$, Theorem 10.52 implies

Corollary 10.10. *Let* $\phi : M \to S_o^{m-1}(r) \subset \mathbb{E}^m$ *be a spherical isometric immersion of a Riemannian n-manifold. Then*

(1) *The type number of ϕ^{2k} is $\geq k$.*
(2) ϕ^{2k} *is of k-type if and only if ϕ is an open portion of an ordinary hypersphere lying in a totally geodesic $\mathbb{E}^{n+1}$.*

The next six results from [Chen (1994c)] determine tensor product immersions $\phi_1 \otimes \cdots \otimes \phi_{2k}$ with type number $k + 1$.

Proposition 10.10. *Let* $\phi : S^n(r) \to \mathbb{E}^{n+1}$ *be the standard imbedding and let* $\psi : S^n(r) \to S_o^{n(n+3)/2-1}(a) \subset \mathbb{E}^{n(n+3)/2}$ *be the second standard immersion. Then $\phi \otimes \psi$ is mass-symmetric and of 2-type.*

Theorem 10.53. *Let* $\phi_i : M \to S_o^{m_i-1}(r_i) \subset \mathbb{E}^{m_i}, i = 1, \ldots, 2k, k \geq 2$, *be spherical isometric immersions. If* $\phi_1 \otimes \cdots \otimes \phi_{2k}$ *is mass-symmetric and of $(k+1)$-type, then each ϕ_i is of 1-type and it has parallel mean curvature vector and parallel second fundamental form.*

Theorem 10.54. *Let* $\phi_i : M \to S_o^{m_i-1}(r_i) \subset \mathbb{E}^{m_i}, i = 1, \ldots, 2k$, *with $k \geq 2$ be spherical isometric immersions with $\phi_1 = \phi_2$. If* $\phi_1 \otimes \cdots \otimes \phi_{2k}$ *is mass-symmetric and of $(k+1)$-type, then*

(1) *M is an open portion of an n-sphere.*
(2) *Exactly $2k - 1$ of $\phi_1, \ldots, \phi_{2k}$ are the ordinary hypersphere.*
(3) *the remaining one is the second standard isometric immersion.*

Theorem 10.55. *Let* $\phi : M \to S_o^{m-1}(r) \subset \mathbb{E}^m$ *be a spherical isometric immersion of a Riemannian n-manifold M into* $\mathbb{E}^m$. *Assume the quadric representation of* ϕ *is of 2-type.*

(1) *If M is Einstein, then the quadric representation is mass-symmetric and* ϕ *is of 1-type.*

(2) *If the quadric representation of* ϕ *is mass-symmetric, then* $\phi : M \to S_o^{m-1}(r)$ *has parallel second fundamental form, i.e., M is a parallel submanifold of* $S_o^{m-1}(r)$.

Theorem 10.56. *Let* $\phi_i : M \to S_o^{m_i-1}(r_i) \subset \mathbb{E}^{m_i}$ $(i = 1, \ldots, 2k, \; k \geq 2)$ *be spherical isometric immersions of a Riemannian 2-manifold. Then the immersion* $\phi_1 \otimes \cdots \otimes \phi_{2k}$ *is of* $(k + 1)$-*type if and only if*

(1) *M is an open portion of a 2-sphere* S^2 *and*

(2) *exactly one of* $\phi_1, \ldots, \phi_{2k}$ *is the second standard isometric immersion and the others are the first standard imbedding of* S^2.

Remark 10.5. Since parallel submanifolds of S^{m-1} have been classified, Theorem 10.53 yields complete classification of spherical submanifolds whose quadric representations are mass-symmetric and of 2-type.

Remark 10.6. For tensor product immersions $\phi_1 \otimes \cdots \otimes \phi_{2k+1}$ given by odd number of spherical immersions, see [Chen (1993c)].

Remark 10.7. The notion of tensor product immersions was extended to tensor product $\phi_1 \otimes \phi_2$ of immersions $\phi_1 : M_1 \to \mathbb{E}^{m_1}$ and $\phi_2 : M_2 \to \mathbb{E}^{m_2}$ of different manifolds in [Decruyenaere et al. (1994)].

10.8   Finite type quadric and cubic representations

If M is a minimal submanifold of $S_o^{m-1}(1)$, then the Euclidean coordinate functions restricted to M are eigenfunctions of the Laplacian Δ on M with the same eigenvalue n. Spectral behavior of a spherical submanifold can be also nicely related to the second standard immersion of $S_o^{m-1}(1)$. Thus, in frame work of the finite type theory, if one wants to study spectral geometry of minimal submanifolds in the sphere, it is natural to look for the spectral behavior of the products of coordinate functions $x_i \cdot x_j$ and then to deal with a very special case of the following problem (see Remark 10.3):

What is the eigenvalue behavior of the products of eigenfunctions?

In this case one can organize the product of coordinate functions to give a new isometric immersion in the Euclidean space of symmetric matrices over $\mathbb{R}$, this is nothing but the composition of the isometric immersion in a sphere with the second standard immersion of the sphere in the Euclidean space according to the description in [Sakamoto (1977)] and then one can study its type number. This idea was first used in [Ros (1984a,b)] to give a characterization for minimal submanifolds in the sphere for which the spectral behaviors of $x_i \cdot x_j$ involve exactly two different eigenvalues.

In this second edition, we present numerous results on quadric and cubic representations obtained by many mathematicians via this method.

Theorem 10.57. [Ros (1984a)] *Let $\phi : M \to S_o^{m-1}(1) \subset \mathbb{E}^m$ be a minimal immersion of a closed Riemannian n-manifold such that the immersion is full. Then the quadric representation of ϕ is of 2-type if and only if M is an Einstein manifold and it satisfies $T(\xi, \eta) = c \langle \xi, \eta \rangle$ for normal vectors ξ, η of M in $S_o^{m-1}(1)$, where c is a real number, $T(\xi, \eta) = \operatorname{Tr} A_\xi A_\eta$, and A is the shape operator of M in $S_o^{m-1}(1)$.*

Theorem 10.58. [Barros and Chen (1987b)] *Let $\pi : M \to S_o^{m-1} \subset \mathbb{E}^m$ be an isometric immersion of a closed Riemannian n-manifold such that the immersion is full. If the quadric representation of ϕ is mass-symmetric and of 2-type, then either*

(1) ϕ *is of 1-type and so M is a minimal submanifold of a hypersphere of $\mathbb{E}^m$ or*

(2) ϕ *is of 2-type and it is mass-symmetric in $S_o^{m-1} \subset \mathbb{E}^m$.*

Theorem 10.59. [Barros and Chen (1987b)] *Let $\phi : M \to S_o^{m-1} \subset \mathbb{E}^m$ be an isometric immersion of a closed Einstein manifold such that the immersion is full. If the quadric representation of ϕ is mass-symmetric and of 2-type, then either*

(1) M *is a minimal submanifold of S_o^{m-1} or*

(2) M *is a minimal submanifold of a small hypersphere of S_o^{m-1}.*

In both cases, ϕ is of 1-type.

Theorem 10.60. [Barros and Chen (1987b)] *Let $\phi : M \to S_o^{m-1} \subset \mathbb{E}^m$ be an isometric isometric full immersion of a closed Riemannian n-manifold. If the quadric representation ϕ is mass-symmetric and of 2-type, then $m \leq 1 + n(n+3)/2$. In particular, if $m = 1 + n(n+3)/2$, it is immersed as a minimal submanifold of a small hypersphere of S_o^{m-1} via ϕ.*

Recall that a submanifold M of a Riemannian manifold is called *constant isotropic* if $||h(v,v)|| = c$ is a constant, which is independent of the choice of the unit tangent vector $v \in T_u M$ and the choice of $u \in M$.

Theorem 10.61. [Barros and Chen (1987b)] *Let* $\phi : M \to S_o^{n(n+3)/2} \subset \mathbb{E}^{1+n(n+3)/2}$ *be an isometric full immersion of a closed Riemannian n-manifold. If its quadric representation is mass-symmetric and of 2-type, then M is a real-space-form immersed fully in a small hypersphere of $S_o^{n(n+3)/2}$ as a constant isotropic minimal submanifold.*

For spherical minimal immersions, we also have the following.

Theorem 10.62. [Barros and Chen (1987b)] *Let* $\phi : M \to S_o^{m-1}$ *be an isometric minimal full immersion of a closed Riemannian n-manifold. If its quadric representation is mass-symmetric and of 2-type, then we have $m \leq n(n+3)/2$. In particular, if $m = n(n+3)/2$, then M is a real-space-form immersed as an constant isotropic submanifold of $S_o^{n(n+3)/2}$.*

Theorem 10.63. [Barros and Chen (1987b)] *Let* $\phi : M \to S_o^{m-1}(1) \subset \mathbb{E}^m$ *be a full immersion of a closed Riemannian 2-manifold. Then its quadric representation is mass-symmetric and of 2-type if and only if one of the following four statements holds:*

(1) $m = 3$ *and M is immersed as a small hypersphere $S^2(r)$ with $r = \sqrt{3}/2$.*
(2) $m = 3$ *and M is immersed as a Clifford torus in $S_o^3(1)$.*
(3) $m = 4$ *and M is immersed as a Veronese surface in $S_o^4(1)$.*
(4) $m = 5$ *and M is immersed as a Veronese surface lying in the small hypersphere $S^4(\sqrt{5/6})$ of $S_o^5(1)$.*

For spherical hypersurfaces, we have the following.

Theorem 10.64. [Barros and Chen (1987b)] *Let* $\phi : M \to S_o^{n+1}(1) \subset \mathbb{E}^{n+2}$ *be an isometric immersion of a closed Riemannian n-manifold. Then the quadric representation of ϕ is mass-symmetric and of 2-type if and only if*

(1) M *is a small hypersphere of $S_o^{n+1}(1)$ or*
(2) M *is the Riemannian product of two spheres $S^k(a) \times S^{n-k}(b)$ such that $a^2 = (k+1)/((n-2)$ and $b^2 = (n-k+1)/(n+2)$,*

where the immersions of M in $S_o^{n+1}(1)$ in (1) and (2) are given in a natural way.

By applying Theorem 10.55 and a result from [Itoh and Ogiue (1993)], Theorems 10.61 and 10.62 can be improved to the next two theorems.

Theorem 10.65. *Let $\phi : M \to S_o^{n(n+3)/2}(1) \subset \mathbb{E}^{1+n(n+3)/2}$ be an isometric full immersion of a closed Riemannian n-manifold. If its quadric representation is mass-symmetric and of 2-type, then M is of constant curvature $c < n/2(n+1)$ immersed fully in a small hypersphere of $S_o^{n(n+3)/2}$ by its second standard immersion.*

Theorem 10.66. *Let $\phi : M \to S_o^{m-1}(1)$ be an isometric minimal full immersion of a closed Riemannian n-manifold. If its quadric representation is mass-symmetric and of 2-type, then $m \leq n(n+3)/2$. In particular, if $m = n(n+3)/2$, then M is of constant curvature $n/2(n+1)$, immersed as an open portion of a Veronese minimal submanifold of $S_o^{n(n+3)/2}(1)$.*

I. Dimitric extended Theorem 10.64 to the following.

Theorem 10.67. *[Dimitric (1990, 2009)] Let $\phi : M \to S_o^{n+1}(1) \subset \mathbb{E}^{n+2}$ be an isometric immersion of a complete Riemannian n-manifold. Then the quadric representation of ϕ is of 2-type if and only if either*

(1) *M is a geodesic hypersphere of any radius $r \in (0, \pi/2)$, or*
(2) *M is the Riemannian product of two spheres $S^k(a) \times S^{n-k}(b)$ with the following three possibilities for the radii:*

$$a^2 = \frac{k}{n+2}, \quad a^2 = \frac{k+1}{n+2}, \quad a^2 = \frac{k+2}{n+2},$$
$$\text{and } b^2 = 1 - a^2.$$

All of the immersions of M mentioned above are given in a natural way.

M. Barros and O. J. Garay studied closed minimal surfaces of S^3 whose images under their quadric representations lie in some hyperquadric as minimal surfaces. They obtained the following two results.

Theorem 10.68. *[Barros and Garay (1994a)] Let $\phi : M \to S_o^3(1) \subset \mathbb{E}^4$ be a closed minimal surface of $S^3(1)$. If its quadric representation has center of mass at ϕ_0, then the quadric representation is mass-symmetric in some hypersphere and minimal in some hyperquadric centered at ϕ_0 if and only if either $\phi : M \to S_o^3(1)$ is totally geodesic or ϕ is a Clifford torus.*

Theorem 10.69. *[Barros and Garay (1994b)] Let $\phi : M \to S_o^3(1) \subset \mathbb{E}^4$ be an isometric minimal immersion of closed surface in $S_o^3(1)$. Then its* *quadric representation is minimal in some canonical hyperquadric centered at ϕ_0 if and only if $\phi : M \to S_o^3(1)$ is either totally geodesic or a Clifford torus in $S_o^3(1)$.*

T. Hasanis and T. Vlachos extended Theorem 10.68 to the following.

Theorem 10.70. [Hasanis-Vlachos (1994b)] *Let $\phi : M \to S_o^{n+1}(1) \subset \mathbb{E}^{n+2}$ be a closed minimal hypersurface of $S_o^{n+1}(1)$. If the quadric representation of ϕ has center of mass at ϕ_0, then the quadric representation of ϕ is mass-symmetric in some hypersphere and minimal in some hyperquadric of centered at ϕ_0 if and only if either*

(1) *$\phi : M \to S_o^{n+1}(1)$ is totally geodesic or*
(2) *M is the product $S^k(\sqrt{k/n}) \times S^{n-k}(\sqrt{(n-k)/n})$ immersed in $S^{n+1}(1)$ in a standard way for some k, $1 \leq k < n$.*

Consider the isometric immersion $y : \mathbb{E}^2 \to S_o^5(1) \subset \mathbb{E}^6$ by

$$y(u,v) = \frac{1}{\sqrt{3}} \left(\cos u, \sin u, \cos v, \sin v, \cos(u+v), \sin(u+v) \right). \quad (10.101)$$

Then y induces an isometric immersion of the flat torus $T^2 = \mathbb{E}^2/\Lambda$ into $S_o^5(1)$, where Λ is the lattice generated by $\{(1/\sqrt{2}, 1/\sqrt{6}), (0, \sqrt{6}/3)\}$. This isometric immersion of $\mathbb{E}^2/\Lambda$ in $\mathbb{E}^6$ is known as the *equilateral torus*.

Theorem 10.71. [Barros and Urbano (1987)] *Let $\phi : M \to S_o^{m-1}(1) \subset \mathbb{E}^m$ be a minimal full isometric immersion of a surface into $S_o^{m-1}(1)$. Then the quadric representation of ϕ is of 3-type if and only if either*

(1) *M has constant Gauss curvature $K = 1/6$ and ϕ is the third standard immersion of M in $S_o^6(1)$ or*
(2) *M is flat and ϕ is the equilateral torus in $S_o^5(1)$.*

É. Cartan investigated the hypersurface in $S_o^4(1) \subset \mathbb{E}^5$ defined by

$$2x_5^3 + 3(x_1^2 + x_2^2)x_5 - 6(x_3^2 + x_4^2)x_5 + 3\sqrt{3}(x_1^2 - x_2^2)x_4$$
$$+ 6\sqrt{3}x_1 x_2 x_3 = 0. \quad (10.102)$$

He proved that this hypersurface in $S_o^4(1)$ is isometric to the homogeneous Riemannian manifold $SO(3)/(\mathbb{Z}_2 \times \mathbb{Z}_2)$ and that the principal curvatures of the hypersurface are equal to $\sqrt{3}, 0, -\sqrt{3}$. This isoparametric hypersurface of $S_o^4(1)$ is well-known as the *Cartan hypersurface*.

In his doctoral program at Michigan State University, I. Dimitric also undertook a project to study spherical hypersurfaces with 3-type quadric representation. He proved the next four results.

Theorem 10.72. [Dimitric (1989)] *Let $\phi : M \to S_o^{n+1} \subset \mathbb{E}^{n+2}$ be a minimal isometric immersion of a closed Riemannian n-manifold into S_o^{n+1}. If the quadric representation of ϕ is mass-symmetric and of 3-type, then*

(1) $\operatorname{Tr} A = \operatorname{Tr} A^3 = 0$,
(2) *Both* $\operatorname{Tr} A^2$ *and* $\operatorname{Tr} A^4$ *are constant,*
(3) $\operatorname{Tr}(\nabla_X A)^2 = \langle A^2 X, A^2 X \rangle + p \langle AX, AX \rangle + q \langle X, X \rangle$ *for every tangent vector X of M, where p and q are constants.*

Conversely, if conditions (1)-(3) hold, then M is mass-symmetric and the quadric representation of ϕ is either of 1-type, of 2-type, or of 3-type.

By applying Theorem 10.72, I. Dimitric obtained the following.

Theorem 10.73. [Dimitric (1989)] *Let $\phi : M \to S_o^{n+1}(1) \subset \mathbb{E}^{n+2}$ be a minimal isoparametric closed hypersurface of $S_o^{n+1}(1)$ with mass-symmetric 3-type quadric representation. Then M can possibly have only 3, 4 or 6 distinct principal curvatures of the same multiplicity.*

Theorem 10.74. [Dimitric (1989)] *Let $\phi : M \to S_o^{n+1}(1) \subset \mathbb{E}^{n+2}$ be a minimal isoparametric closed hypersurface of $S_o^{n+1}(1)$ with three distinct principal curvatures. Then the quadric representation of ϕ is of 3-type.*

The class of minimal isoparametric closed hypersurface of $S_o^{n+1}(1)$ with exactly three distinct principal curvatures includes

$$SO(3)/\mathbb{Z}_2 \times \mathbb{Z}_2, \quad SU(3)/T^2, \quad Sp(3)/Sp(1), \quad \text{and} \quad F_4/Spin(8),$$

and no other such examples are known.

Theorem 10.75. [Dimitric (1989, 2009)] *Let $\phi : M \subset S_o^{n+1}(1) \subset \mathbb{E}^{n+2}$ be a complete CMC hypersurface of $S_o^{n+1}(1)$. If $2 \leq n \leq 5$, then the quadric representation of ϕ is mass-symmetric and of 3-type if and only if $n = 3$ and $M = SO(3)/\mathbb{Z}_2 \times \mathbb{Z}_2$ is immersed as the Cartan hypersurface in $S_o^4(1)$.*

For closed minimal spherical submanifolds with 3-type quadric representation we have the following.

Theorem 10.76. [Zhang and Ma (1993)] *Let $\phi : M \to S_o^m(1) \subset \mathbb{E}^{m+1}$ be a closed minimal submanifold of S_o^m with 3-type quadric representation. Then M has constant scalar curvature.*

If M is a minimal closed hypersurface of $S_o^{n+1}(1)$, Theorem 10.76 was also obtained in [Lu (1993)]. J.-T. Lu applied this theorem to show that there do not exist closed minimal surfaces of $S_o^3(1)$ with 3-type quadric representation; a result which is a consequence of Theorem 10.70.

Next, we consider the following simple geometric question.

Problem 10.1. *To what extent do the cubic representation of a submanifold determine the submanifold?*

First we mention the following.

Theorem 10.77. *[Chen (1993a)] Let $\phi : M \to S_o^{m-1} \subset \mathbb{E}^m$ be a spherical immersion of a Riemannian n-manifold M. Then*

(1) *The type number of the cubic representation ϕ^3 of ϕ is at least two.*
(2) *The cubic representation of ϕ is of 2-type if and only if M is an open portion of an n-sphere S^n and ϕ is the first standard imbedding of S^n.*

Theorem 10.77 determines all spherical submanifolds with 2-type cubic representation. Spherical submanifolds with 3-type cubic representations were investigated by the author and W. E. Kuan.

Proposition 10.11. *[Chen and Kuan (1994)] Let $\phi : M \to S_o^{m-1}(1) \subset \mathbb{E}^m$ be an isometric minimal immersion of M into $S_o^{m-1}(1)$. If the cubic representation of ϕ is mass-symmetric and of 3-type, then we have:*

(a) *M has constant scalar curvature.*
(b) *The second fundamental form $\tilde{\sigma}$ of $\tilde{\phi} : M \to S_o^{m-1}$ satisfies*

$$\sum_{i,j=1}^{n} \{(\bar{\nabla}_v \tilde{\sigma})(e_i, e_j) \otimes \tilde{\sigma}(e_i, e_j) + \tilde{\sigma}(e_i, e_j) \otimes (\bar{\nabla}_v \tilde{\sigma})(e_i, e_j)\} = 0$$

for any vector v tangent to M.

Theorem 10.78. *[Chen and Kuan (1994)] Let $\phi : M \to S_o^{m-1}(1) \subset \mathbb{E}^m$ be a spherical immersion of a Riemannian 2-manifold M. Then the cubic representation of ϕ is mass-symmetric and of 3-type if and only if M is an open part of the flat torus $S^1(1/\sqrt{2}) \times S^1(1/\sqrt{2})$ and $\phi : M \to S_o^{m-1}(1)$ is the Clifford immersion.*

For Cartan's hypersurface we have the following.

Theorem 10.79. *[Chen and Kuan (1994)] The cubic representation of the Cartan hypersurface in $S_o^4(1) \subset \mathbb{E}^5$ is mass-symmetric and of 5-type.*

By applying Theorem 10.79, we have the following.

Theorem 10.80. [Chen and Kuan (1994)] *There exists no 3-dimensional minimal hypersurface in $S_o^4(1) \subset \mathbb{E}^5$ whose cubic representation is mass-symmetric and of 3-type.*

The next result classifies all minimal surfaces of spheres with 3-type cubic representation.

Theorem 10.81. [Chen and Kuan (1994)] *Let $\phi : M \to S_o^{m-1}(1) \subset \mathbb{E}^m$ be a minimal immersion of a Riemannian 2-manifold M. Then its cubic representation of ϕ is of 3-type if and only if either*

(a) *M is an open portion of the flat torus $S^1(1/\sqrt{2}) \times S^1(1/\sqrt{2})$ and $\phi : M \to S_o^{m-1}(1)$ is the Clifford immersion or*
(b) *M has constant Gauss curvature $1/3$ and ϕ is the Veronese immersion.*

Remark 10.8. I. Dimitric derived in [Dimitric (2009)], a necessary and sufficient condition for a submanifold M in $S_o^{m-1}(1)$ to be mass-symmetric and of 2-type via quadric representation. He also obtained a necessary and sufficient conditions for a spherical submanifold M in $S_o^{m-1}(1)$ with parallel mean curvature vector to be of 2-type via its quadric representation.

10.9 Finite type submanifolds of complex projective space

Denote by $CP^m(4)$ the complex projective m-space endowed with Fubini-Study metric of constant holomorphic sectional curvature 4. Let

$$\psi_1 : CP^m(4) \to \mathbb{E}^N = H(m+1, \mathbb{C})$$

be the first standard isometric imbedding of $CP^m(4)$.

If $\phi : M \to CP^m(4)$ is an isometric immersion of a Riemannian manifold, we call the composition $\psi_1 \circ \phi$ the *quadratic representation* of ϕ. Submanifolds of $CP^m(4)$ whose quadratic representations being of 1-type or 2-type were investigated in [Ros (1983)]. In particular, A. Ros classified CR-minimal submanifolds of $CP^m(4)$ whose quadratic representations are of 1-type in the next theorem.

Theorem 10.82. [Ros (1983)] *Let M be an n-dimensional closed minimal CR submanifold of $CP^m(4)$. Then the quadratic representation of M is of 1-type if and only if either*

(a) *n is even and M is a totally geodesic $CP^{n/2}(4) \subset CP^m(4)$, or*

(b) M *is a totally real minimal submanifold of a totally geodesic* $CP^n(4) \subset$ $CP^m(4)$.

Real hypersurfaces of $CP^m(4)$ with 1-type quadratic representation via $\overset{\bullet}{\psi_1}$ were classified in [Martínez and Ros (1984)]. The complete classification of all 1-type submanifolds of $CP^m(4)$ via ψ_1 was achieved by I. Dimitric in the following.

Theorem 10.83. [Dimitric (1991, 1997)] *Let* $\phi : M \to CP^m(4)$ *be an isometric immersion of a Riemannian n-manifold M into $CP^m(4)$. Then the quadratic representation ϕ is of 1-type if and only if one of the following three cases occurs:*

(a) *n is even and M is an open portion of $CP^{n/2}(4)$ immersed in $CP^m(4)$ as a totally geodesic complex submanifold.*

(b) *M is a totally real minimal submanifold of a totally geodesic $\mathbb{C}P^n(4) \subset$ $CP^m(4)$.*

(c) *n is odd and M is an open portion of the geodesic hypersphere*
$$\pi\big(S^1(1/\sqrt{n+3}) \times S^n(\sqrt{n+2}/\sqrt{n+3})\big)$$
of $CP^{(n+1)/2}$ immersed as a totally geodesic submanifold of $CP^m(4)$, where π is the Hopf's fibration.

By a *Kähler immersion* we mean a holomorphic isometric immersion into a Kähler manifold. The next result characterizes all 2-type Kähler submanifolds of $CP^m(4)$.

Theorem 10.84. [Ros (1984a)] *Let* $\phi : M \to \mathbb{C}P^m(4)$ *be a full Kähler immersion of a closed Kähler manifold. Then the quadratic representation of ϕ is of 2-type if and only if*

(1) *M is an Einstein manifold and*
(2) *$\mathrm{Tr}(A_\xi A_\eta) = c \langle \xi, \eta \rangle$ for normal vectors ξ, η of M, where c is a constant.*

An immersion ϕ of a manifold M into $CP^m(4)$ is called *full* if $\phi(M)$ is not contained in any proper linear subvariety of $CP^m(4)$.

S. Udagawa noticed in [Udagawa (1986a)] that condition (2) in Theorem 10.84 is equivalent to the existence of Einstein-Kähler metric on the normal bundle of ϕ. Subsequently, he characterized all 2-type Kähler submanifolds as non-totally geodesic Einstein parallel submanifolds.

Theorem 10.85. [Udagawa (1986a)] *Let* $\phi : M \to CP^m(4)$ *be a full, non-totally geodesic, Kähler immersion of a closed Kähler submanifold M into $CP^m(4)$. Then the following five conditions are mutually equivalent:*

(1) *The quadratic representation of ϕ is of 2-type.*
(2) *M is an Einstein parallel submanifold.*
(3) $\text{Tr}(A_\xi A_\eta) = c \langle \xi, \eta \rangle$ *for a constant c and any normal vectors ξ, η of ϕ.*
(4) *The quadratic representation is of order $\{1, 2\}$.*
(5) *M is an Einstein submanifold whose normal bundle admits an Einstein-Kähler metric.*

By applying Theorem 10.85, S. Udagawa proved the following.

Theorem 10.86. [Udagawa (1986a,b)] *Let $\phi : M \to CP^m(4)$ be a full non-totally geodesic Kähler immersion of a closed Kähler manifold M into $CP^m(4)$. If the quadratic representation of ϕ is of 2-type, then it is one of the following six Kähler submanifolds:*

(1) *$CP^n(\frac{1}{2})$ with complex codimension $n(n+1)/2$;*
(2) *Complex quadric Q^n with complex codimension one;*
(3) *$CP^n(4) \times CP^n(4)$ with complex codimension n^2;*
(4) *$U(s+2)/U(s) \times U(2)$, $s \geq 3$, with complex codimension $s(s-1)/2$;*
(5) *$SO(10)/U(5)$ with complex codimension 5;*
(6) *$E_6/Spin(10) \times T$ with complex codimension 10.*

Theorem 10.87. [Udagawa (1986b)] *Compact irreducible Hermitian symmetric spaces of rank 3 in $CP^m(4)$ are of 3 type and of order $\{1, 2, 3\}$ via their quadratic representation.*

For a Kähler submanifold M of $CP^m(4)$, M is said to be of *degree k* if the pure part of the $(k-2)$-nd covariant derivative of the second fundamental form σ is not zero and the pure part of the $(k-1)$-st covariant derivative of σ is zero (see [Takagi and Takeuch (1977); Udagawa (1986b)] for details).

For irreducible Hermitian symmetric closed submanifolds of degree 2 and of degree 3 of $CP^m(4)$, S. Udagawa proved the following.

Theorem 10.88. [Udagawa (1986b)] *We have*

(1) *If $\phi : M \to CP^m(4)$ is a closed Einstein-Kähler submanifolds of degree 2, then the quadratic representation of ϕ is of 2-type and of order $\{1, 2\}$.*
(2) *Let M be a closed Einstein Kähler submanifold immersed in $CP^m(4)$, and let M be one of the compact Einstein Hermitian symmetric submanifolds of degree 3. If M and M' have the same spectrum, then M is congruent to M'.*

Definition 10.9. Let M be a surface in $CP^n(4)$ and let $z = x + iy$ be an isothermal coordinate in M. We call the angle θ between $J\partial/\partial x$ and $\partial/\partial y$

the *Kähler angle*, where J is the complex structure of $CP^n(4)$. The surface M is called a *holomorphic* (resp. an *anti-holomorphic*) surface if and only if θ is equal to 0 (resp. θ is equal to π).

Y.-B. Shen studied 2-type minimal surfaces of complex projective space and obtained the following

Theorem 10.89. [Shen (1995)] *Let $\phi : M \to CP^m(4)$ be a full minimal surface in $CP^m(4)$ which is neither holomorphic nor anti-holomorphic and not of 1-type. If the quadratic representation of ϕ is of 2-type, then either M is a totally real Veronese surface in CP^4 or it is a totally real flat superminimal surface in CP^3 or in CP^4. Moreover, such immersions are locally unique.*

S. J. Li and Q.-B. Chen studied minimal surfaces of $CP^m(4)$ whose quadratic representation satisfying the condition $\Delta H = \lambda H$. They proved the next two results.

Theorem 10.90. [Li and Chen (1995)] *Let $\phi : M \to CP^m(4)$ be a minimal surface in $CP^m(4)$. Then the mean curvature vector of the quadratic representation satisfies $\Delta H = \lambda H$ for some nonzero constant λ if and only if either M is an open portion of $CP^1(4)$ immersed as a totally geodesic complex surface in $CP^m(4)$ or it is a totally real isotropic minimal surface in a totally geodesic $CP^2(4) \subset CP^m(4)$.*

Theorem 10.91. [Li and Chen (1995)] *Let $\phi : M \to CP^m(4)$ be a full minimal surface in $CP^m(4)$ which is neither holomorphic nor anti-holomorphic. If M has constant Gauss curvature, then the mean curvature vector of the quadratic representation is an eigenvector of the Laplacian of M if and only if $m = 2$ and M is an open portion of one of the following surfaces:*

(1) *The real projective plane $RP^2(1)$ isometrically immersed as a totally real totally geodesic surface of $CP^2(4) \subset CP^m(4)$;*
(2) *M is a totally real flat minimal surface in $CP^2(4)$ whose homogeneous coordinates satisfy $\{[z_0, z_1, z_2] \in CP^2(4) : z_0\bar{z}_0 = z_1\bar{z}_1 = z_2\bar{z}_2\}$.*

Definition 10.10. A real hypersurface M of $CP^m(4)$ (or of $CH^m(4)$) is called a *Hopf hypersurface* if $J\xi$ is a principal direction of M, i.e., $J\xi$ is an eigenvector of the shape operator A, where ξ is a unit normal vector of M and J is the complex structure of the ambient space.

All 2-type Hopf hypersurfaces in $CP^m(4)$ were classified by I. Dimitric as follows.

Theorem 10.92. [Dimitric (2011)] *Let M be a Hopf hypersurface of $CP^m(4)\,(m \geq 2)$. Then the quadratic representation of M is of 2-type in $H(m+1;\mathbb{C})$ if and only if it is an open portion of one of the following:*

(1) *A geodesic hypersphere of any radius $r \in (0, \pi/2)$ except for $r = \cot^{-1}\left(1/\sqrt{2m+1}\right)$.*
(2) *A tube of radius $r = \cot^{-1}\left(\sqrt{(k+1)/(m-k)}\right)$ about a totally geodesic $CP^k(4) \subset CP^m(4)$, for any $k = 1, \ldots, m-2$.*
(3) *The tube of radius $r = \cot^{-1}\left(\sqrt{(2k+1)/(2(m-k)+1)}\right)$ about a totally geodesic $CP^k(4) \subset CP^m(4)$, for any $k = 1, \ldots, m-2$.*
(4) *The tube of radius $r = \cot^{-1}\left(\sqrt{m} + \sqrt{m+1}\right)$ about a complex quadric $Q^{m-1} \subset CP^m(4)$.*
(5) *The tube of radius $r = \cot^{-1}\left(\sqrt{\sqrt{2m^2-1} + \sqrt{2m^2-2}}\right)$ about a complex quadric $Q^{m-1} \subset CP^m(4)$.*

Remark 10.9. Theorem 10.92 also holds for CMC real hypersurfaces of $CP^m(4)$, since for 2-type real hypersurfaces of a complex projective m-space (or complex hyperbolic m-space), being a Hopf hypersurface is equivalent to having constant mean curvature (cf. [Dimitric (2011)]).

Remark 10.10. Hopf hypersurfaces of $CP^m(4)$ and $CH^m(-4)$ for $m \geq 2$ with constant principal curvatures are homogeneous and they are known. By a result of [Takagi (1975)] there are six types or six classes of Hopf hypersurfaces with constant principal curvatures in $CP^m(4)$:

(A1) A geodesic hypersphere of radius $r \in (0, \pi/2)$;
(A2) A tube of any radius $r \in (0, \pi/2)$ around a canonically embedded totally geodesic CP^k for some $k = 1, \ldots, m-2$;
(B) A tube of any radius $r \in (0, \pi/4)$ around a canonically embedded complex quadric $Q(m-1 = SO(m+1)/SO(2) \times SO(m-1)$;
(C) A tube of radius $r \in (0, \pi/4)$ around the Segre embedding of $CP^1 \times CP^k$ in CP^m, $m = 2k+1$;
(D) A tube of radius $r \in (0, \pi/4)$ of dimension 17 in CP^9 around Plücker's imbedding of the complex Grassmannian of 2-planes $G^{\mathbb{C}}(5,2)$;
(E) A tube of radius $r \in (0, \pi/4)$ of dimension 29 in CP^{15} around the canonical imbedding of the Hermitian symmetric space $SO(10)/U(5)$.

According to [Berndt (1989)], a Hopf hypersurface in $CH^m(-4)$ with constant principal curvatures is congruent to an open part of

(i) a tube over totally geodesic $CH^k, (0 \leq k \leq m-1)$;
(ii) a tube over totally geodesic H^m, or
(iii) a horosphere.

For 3-type real hypersurfaces in CP^2, I. Dimitric proved the following.

Theorem 10.93. [Dimitric (2011)] *Let M be a Hopf hypersurface of $CP^2(4)$ with constant mean curvature. Then M is mass-symmetric and of 3-type in $H(3;\mathbb{C})$ via its quadratic representation if and only if M is an open portion of a tube of any radius $r \in (0, \pi/4)$ about the complex quadric Q^1, except when $\cot r = \sqrt{2} + \sqrt{3}$ and $\cot r = \sqrt{\sqrt{6} + \sqrt{7}}$.*

One of the most typical examples of irreducible submanifolds in $CP^n(4)$ is a 2-sphere. Let $S^2(c)$ be the 2-sphere of constant curvature $c > 0$. In [Bando and Ohnita (1987)], the family $\{\varphi_{n,k}\}$ of all full isometric minimal immersions of $S^2(c)$ into $CP^n(4)$ are constructed, using irreducible unitary representations of $SU(2)$. Independently, the same family $\{\varphi_{n.k}\}$ was also obtained in [Bolton et al (1988)], using the method of harmonic sequence. They called this family the *Veronese sequence*.

S. Bando and Y. Ohnita constructed the family $\{\varphi_{n,k}\}$ of full minimal immersions of $S^2(c)$ into $CP^n(4)$ as follows: $SU(2)$ is defined by

$$SU(2) = \left\{ \begin{pmatrix} a & b \\ -\bar{b} & \bar{a} \end{pmatrix} : a, b \in \mathbb{C}, |a|^2 + |b|^2 = 1 \right\}.$$

The Lie algebra $\mathfrak{su}(2)$ of $SU(2)$ is given by

$$\mathfrak{su}(2) = \left\{ \begin{pmatrix} ix & y \\ -\bar{y} & -ix \end{pmatrix} : y = y' + iy'', x, y', y'' \in \mathbb{R} \right\}.$$

Define a basis $\{\varepsilon_0, \varepsilon_1, \varepsilon_2\}$ of $\mathfrak{su}(2)$ by

$$\varepsilon_0 = \begin{pmatrix} i & 0 \\ 0 & -i \end{pmatrix}, \quad \varepsilon_1 = \begin{pmatrix} 0 & 1 \\ -1 & 0 \end{pmatrix}, \quad \varepsilon_3 = \begin{pmatrix} 0 & i \\ i & 0 \end{pmatrix},$$

which satisfy

$$[\varepsilon_0, \varepsilon_1] = 2\varepsilon_2, \quad [\varepsilon_1, \varepsilon_2] = 2\varepsilon_0, \quad [\varepsilon_2, \varepsilon_0] = 2\varepsilon_1.$$

Let V_n be an $(n+1)$-dimensional complex vector space of all complex homogeneous polynomials of degree n with respect to z_0, z_1. We define a Hermitian inner product $\langle \ , \ \rangle$ of V_n in such a way that

$$\left\{ u_k^{(n)} = \frac{z_0^k z_1^{n-k}}{\sqrt{k!(n-k)!}} : 0 \leq k \leq n \right\}$$

is a unitary basis for V_n. We define a real inner product by $(\ ,\ ) = \mathrm{Re}\langle\ ,\ \rangle$. A unitary representation ρ_n of $SU(2)$ on V_n is defined by

$$\rho_n(g)f(z_0, z_1) = f((z_0, z_1)g) = f(az_0 - \bar{b}z_1, bz_0 + \bar{a}z_1)$$

for $g \in SU(2)$ and $f \in V_n$. We also denote by ρ_n the action of $\mathfrak{su}(2)$ on V_n so that

$$\begin{aligned}
\rho_n(X)(u_k^{(n)}) = {}& (k - (n-k))i x u_k^{(n)} \\
& - \sqrt{k(n-k+1)}\, \bar{y} u_{k-1}^{(n)} + \sqrt{(k+1)(n-k)}\, y u_{k+1}^{(n)}
\end{aligned} \tag{10.103}$$

for $0 \leq k \leq n$; $X \in \mathfrak{su}(2)$. It is known that $\{(\rho_n, V_n) : n = 0, 1, 2, \ldots\}$ is the set of all inequivariant irreducible unitary representations of $SU(2)$. Put

$$T = \{\exp(t\varepsilon_0) \in \mathfrak{su}(2) : t \in \mathbb{R}\}$$

and we have $S^2 = CP^1 = SU(2)/T$. We identify the tangent space of S^2 at $o = \{T\} \in S^2 = SU(2)/T$ with a subspace $\mathfrak{m} = \mathrm{Span}\{\varepsilon_1, \varepsilon_2\}$ of $\mathfrak{su}(2)$. We fix a complex structure on S^2 so that $\varepsilon_1 - i\varepsilon_2$ is a vector of type $(1,0)$.

Let g_c be an $SU(2)$-invariant Riemannian metric on S^2 defined by

$$g_c(X, Y) = -\frac{2}{c}\mathrm{Tr}\,(XY)$$

for $X, Y \in \mathfrak{m}$ and c is a positive constant. It is the restriction of $SU(2)$-invariant inner product on $\mathfrak{su}(2)$. Clearly, $\{(\sqrt{c}/2)\varepsilon_1, (\sqrt{c}/2)\varepsilon_2\}$ forms an orthonormal basis of $\mathfrak{m} \cong T_o S^2$ and (S^2, g_c) has the constant curvature c, denoted by $S^2(c)$. The spectrum of the Laplacian Δ of $S^2(c)$ is given by $Spec(S^2(c)) = \{\lambda_\ell = c\ell(\ell+1)\,|\,\ell \geq 0\}$.

Put $S^{2n+1} = \{v \in V_n\,|\,\langle v, v\rangle = 4/\tilde{c}\}$, where $\tilde{c}$ is a positive constant. Let $\pi : S^{2n+1} \to CP^n(\tilde{c})$ be the Hopf fibration so that the action of $\rho_n(SU(2))$ on S^{2n+1} induces the action on $CP^n(\tilde{c})$ through π. For integers $n \geq 0$ and k with $0 \leq k \leq n$, we denote by $\varphi_{n,k}$ the $SU(2)$-equivariant mapping of a Riemann sphere $S^2(c)$ into $CP^n(\tilde{c})$ defined by

$$\varphi_{n,k} : S^2(c) = SU(2)/T \in gT \mapsto \pi\left(\rho_n(g)\frac{2u_k^{(n)}}{\sqrt{\tilde{c}}}\right) \in CP^n(\tilde{c}). \tag{10.104}$$

S. Bando and Y. Ohnita proved the following results.

Theorem 10.94. [Bando and Ohnita (1987)]

(1) $\varphi_{n,k}$ is a full isometric immersion.
(2) c is equal to $\tilde{c}/(2k(n-k)+n)$.
(3) $\varphi_{n,k}$ is a minimal immersion.

(4.a) If $k = 0$ (resp. $k = n$), then $\varphi_{n,k}$ is holomorphic (resp. anti-holomorphic).

(4.b) If n is even and $k = n/2$, then $\varphi_{2k,k}$ is totally real and $\varphi_{2k,k}(S^2(c))$ lies in a totally geodesic totally real $RP^{2k}(\tilde{c}/4) \subset CP^{2k}(\tilde{c})$.

(4.c) Otherwise, $\varphi_{n,k}$ is not holomorphic, anti-holomorphic or totally real.

(5) $\varphi_{n,k}(S^2(c)) = \varphi_{n,n-k}(S^2(c))$.

Moreover, they proved the following rigidity theorem.

Theorem 10.95. [Bando and Ohnita (1987)] *Let $\varphi : S^2(c) \to CP^n(\tilde{c})$ be a full isometric minimal immersion. Then there exists an integer k with $0 \leq k \leq n$ such that $c = \tilde{c}/(2k(n-k)+n)$ and φ is congruent to $\varphi_{n,k}$ up to a holomorphic isometry of $CP^n(\tilde{c})$.*

Y. Miyata proved the following results.

Theorem 10.96. [Miyata (1994)]

(1) $\varphi_{n,k}$ *is mass-symmetric and of at most n-type. For integers n, k, ℓ with $n \geq 1, 0 \leq k, l \leq n$, define*

$$q_\ell^k = \frac{1}{\ell!} \sum_{m=0}^{\ell} (-1)^m \binom{\ell}{m} \prod_{j=1}^{\ell} (k+j-m)(n-k-j+m+1).$$

Then the order of $\varphi_{n,k}$ is $\{\ell : 1 \leq \ell \leq n, q_\ell^k \neq 0\}$.

(2) *A holomorphic imbedding $\varphi_{n,0}$ and its antipodal $\varphi_{n,n}$ are of n-type and of order $\{1, 2, 3, \ldots, n\}$.*

(3) *If n is even, then a totally real minimal immersion $\varphi_{n,n/2}$ is of $n/2$-type and of order $\{2, 4, 6, \ldots, n\}$.*

Theorem 10.97. [Miyata (1994)] *Let $\phi : S^2(c) \to CP^n(4)$ be a minimal immersion of $S^2(c)$ into $CP^n(4)$. If the quadratic representation of ϕ is mass-symmetric and of k-type, then n satisfies $n \leq 2k$.*

Theorem 10.98. [Miyata (1994)] *Let $\phi : S^2 \to CP^n(4)$ be a minimal immersion of a 2-sphere into $CP^n(4)$. If the quadratic representation of ϕ is of at most 2-type, then S^2 is of constant curvature and the immersion is congruent to either $\varphi_{1,0}, \varphi_{1,1}, \varphi_{2,0}, \varphi_{2,1}, \varphi_{2,2}$ or $\varphi_{4,2}$.*

Theorem 10.99. [Miyata (1994)] *Let $\phi : S^2 \to CP^n(4)$ be a minimal immersion of a 2-sphere into $CP^n(4)$ with constant Kähler angle. If the quadratic representation of ϕ is mass-symmetric and with at most 3-type,* *then S^2 is of constant curvature and the immersion is congruent to either* $\varphi_{n,k}, (n = 1, 2, 3; 0 \le k \le n), \varphi_{4.2}$ *or* $\varphi_{6,3}$.

Remark 10.11. Without the assumption of 3-type, it is shown in [Bolton et al (1988)] that Theorem 10.99 remains true if $n \le 4$ and the immersion is neither holomorphic, anti-holomorphic nor totally real.

For further results on submanifolds of $CP^m(4)$ with finite type quadratic representations, see [Chen (1996b); Chen and Maeda (1996)] and [Miyata (1994)].

10.10 Finite type submanifolds of complex hyperbolic space

Let Ψ be the standard Hermitian form on $\mathbb{C}^{m+1}$ given by

$$\Psi(z, w) = -\bar{z}_1 w_1 + \bar{z}_2 w_2 + \cdots + \bar{z}_{m+1} w_{m+1}$$

and let $g_1 = \text{Re}(\Psi)$ denote the real part of Ψ. Then g_1 defines a pseudo-Euclidean metric on $\mathbb{R}^{2m+2} \cong \mathbb{C}^{m+1}$. Let $\mathbb{C}_1^{m+1}$ denote the pair $(\mathbb{C}^{m+1}, g_1)$ and denote the inner product on $\mathbb{C}_1^{m+1}$ by $\langle\ ,\ \rangle$.

Put

$$H_1^{2m+1}(\mathbb{C}; -1) = \{z = (z_1, z_2, \dots, z_{m+1}) \in \mathbb{C}_1^{m+1} : \langle z, z \rangle = -1\},$$

which is an anti-de Sitter space in $\mathbb{C}_1^{m+1}$. Let

$$T_z' = \{z \in \mathbb{C}^{m+1} : \text{Re}\,\langle u, z \rangle = \text{Re}\,\langle u, iz \rangle = 0\}$$
$$H_1^1 = \{\lambda \in \mathbb{C} : \lambda\bar{\lambda} = 1\}.$$

Then we have an H_1^1-action on $H_1^{2m+1}(-1)$ defined by $z \mapsto \lambda z$. At each $z \in H_1^{2m+1}(\mathbb{C}; -1)$, iz is tangent to the flow of the action. Since g_1 is Hermitian, we have $\text{Real}\,g_1(iz, iz) = -1$. Note that the orbit is given by

$$x_t = (\cos t + i \sin t)z \quad \text{and} \quad \frac{dx_t}{dt} = iz_t.$$

Thus the orbit lies in the negative definite plane spanned by z and iz. The quotient space $H_1^{2m+1}(\mathbb{C}; -1)/\sim$, under the identification induced from the action, is the complex hyperbolic space CH^m with constant holomorphic sectional curvature -4, equipped with the complex structure J induced from the canonical complex structure on $\mathbb{C}_1^{m+1}$ via the totally geodesic Riemannian submersion:

$$\pi : H_1^{2m+1}(\mathbb{C}; -1) \to CH^m. \tag{10.105}$$

The standard embedding $\varphi_{\mathbb{C}}$ of $CH^m(-4)$ into the set of Ψ-Hermitian matrices $H^{(1)}(m+1;\mathbb{C})$ is achieved by identifying a point, i.e., a time-like complex line, with the projection operator onto it. Then one gets the following matrix representation of $\varphi_{\mathbb{C}}$ at a point $p = [z]$, where $z = (z_j) \in H_1^{2m+1}(\mathbb{C};-1) \subset \mathbb{C}_1^{m+1}$

$$
\varphi_{\mathbb{C}}([z]) = \begin{pmatrix} |z_1|^2 & -z_1\bar{z}_2 & \cdots & -z_1\bar{z}_{m+1} \\ z_2\bar{z}_1 & -|z_2|^2 & \cdots & -z_2\bar{z}_{m+1} \\ \vdots & \vdots & \ddots & \vdots \\ z_{m+1}\bar{z}_1 & -z_{m+1}\bar{z}_2 & \cdots & -|z_{m+1}|^2 \end{pmatrix}. \tag{10.106}
$$

The second fundamental form h of this embedding is parallel and the image $\varphi_{\mathbb{C}}(CH^m(-4))$ is contained in the hyperquadric of $H^{(1)}(m+1;\mathbb{C})$ centered at $I/(m+1)$ and defined by the equation:

$$
\left\langle P - \frac{I}{m+1}, P - \frac{I}{m+1} \right\rangle = -\frac{m}{2(m+1)},
$$

where I denotes the $(m+1) \times (m+1)$ identity matrix.

By applying this imbedding $\varphi_{\mathbb{C}}$ given in (10.106), O. J. Garay and A. Romero studied 1-type real hypersurfaces of $CH^m(-4)$ and obtained the following.

Theorem 10.100. [Garay and Romero (1990)] *There exist no real hypersurfaces of $CH^m(-4)$ for any $m \geq 2$ which are of 1-type via $\varphi_{\mathbb{C}}$.*

In [Garay and Romero (1990)], O. J. Garay and A. Romero also studied real hypersurfaces of $CH^m(-4)$ via $\varphi_{\mathbb{C}}$ whose mean curvature vector H in $H^{(1)}(m+1;\mathbb{C})$ satisfies $\Delta H = Q$ for some non-zero constant matrix Q of $H^{(1)}(m+1;\mathbb{C})$ and showed that such a real hypersurface is locally congruent to the horosphere defined by $\pi\big(\{z \in H_1^{2m+1}(\mathbb{C};-1) : |z_0 - z_1|^2 = 1\}\big)$, where π is given by (10.105).

The complete classification of 1-type submanifolds of $CH^m(-4)$ was achieved by I. Dimitric in the following.

Theorem 10.101. [Dimitric (1997)] *Let $\phi : M \to CH^m(-4)$ be an isometric immersion of a Riemannian n-manifold. Then M is of 1-type via $\varphi_{\mathbb{C}}$ if and only if one of the following two cases occurs:*

(1) *$n = 2d$ is even and M is a complex-space-form of constant holomorphic sectional curvature -4, immersed as a totally geodesic $CH^d(-4) \subset CH^m(-4)$.*

(2) M is a totally real minimal submanifold of a complex totally geodesic $CH^n(-4) \subset CH^m(-4)$.

For hypersurfaces of $CH^m(-4)$, I. Dimitric proved the following

Theorem 10.102. [Dimitric (2011)] *Let M be a real hypersurface of $CH^m(-4), m \geq 2$, for which we assume that it is a Hopf hypersurface or has constant mean curvature. Then M is of 2-type in $H^{(1)}(m+1; \mathbb{C})$ via $\varphi_{\mathbb{C}}$ if and only if it is an open portion of one of the following:*

(a) *a geodesic hypersphere of arbitrary radius $r > 0$;*
(b) *a tube of arbitrary radius $r > 0$ about a canonically embedded totally geodesic complex hyperbolic hyperplane $CH^{m-1}(-4)$.*

Theorem 10.103. [Dimitric (2011)] *Let M be a Hopf hypersurface of $CH^2(-4)$ with constant mean curvature and $(\mathrm{Tr}\, A)^2 \neq 4$. Then M is mass-symmetric and of 3-type in $H^{(1)}(3; \mathbb{C})$ via $\varphi_{\mathbb{C}}$ if and only if it is an open portion of a tube of any radius $r > 0$ about a canonically embedded, totally real totally geodesic $H^2(-1) \subset CH^2(-4)$.*

10.11 Finite type submanifolds of real hyperbolic space

Let

$$g(x, y) = -x_0 y_0 + \sum_{j=1}^{m} x_j y_j$$

denote the Lorentzian metric on $\mathbb{E}_1^{m+1}$, where $x = (x_0, \ldots, x_m)$ and $y = (y_0, \ldots, y_m)$. We define the hyperquadric $Q^m(-1)$ in $\mathbb{E}_1^{m+1}$ by

$$Q^m(-1) = \{x \in \mathbb{E}_1^{m+1} : g(x, x) = -1\}.$$

Then $Q^m(-1)$ consists of two copies of the hyperbolic space. The metric g is $\mathbb{Z}_2$-invariant, where $\mathbb{Z}_2 = \{\pm I\}$. The quotient space defines the real hyperbolic space as

$$H^m(-1) = Q^m(-1)/\mathbb{Z}_2.$$

Equivalently, the hyperbolic space is the set of time-like lines in $\mathbb{E}_1^{m+1}$ through the origin.

We define the embedding Φ of the real hyperbolic m-space $H^m(-1)$ into a suitable pseudo-Euclidean space by identifying a line $L = [x]$ with $g(x, x) = -1$ with the operator of the orthogonal projection P_L with respect to g onto that line. We get the embedding $L \to P_L$ of $H^m(-1)$ given by $P_L(v) = Mv$, where

$$M = \begin{pmatrix} x_0^2 & -x_0 x_1 & \cdots & -x_0 x_m \\ -x_1 x_0 & -x_1^2 & \cdots & -x_1 x_m \\ \vdots & \vdots & \ddots & \vdots \\ -x_m x_0 & -x_m x_1 & \cdots & -x_m^2 \end{pmatrix}.$$

Let $M_{m+1}(\mathbb{R})$ denote the space of all $(m+1) \times (m+1)$ matrices over $\mathbb{R}$. Equipped with the trace metric $\langle A, B \rangle = -\frac{1}{2}\text{Tr}(AB)$, $M_{m+1}(\mathbb{R})$ becomes a pseudo-Euclidean space. The image of $H^m(-1)$ under this embedding is

$$\Phi(H^m(-1)) = \{P \in M_{m+1}(\mathbb{R}) : P^T = P, \ P^2 = P \text{ and } \text{Tr}\,P = 1\},$$

which lies fully in the hyperplane as a space-like submanifold with a time-like normal bundle in the hyperbolic space. If we put

$$H^{(1)}(m+1; \mathbb{R}) = \{P \in M_{m+1}(\mathbb{R}) : P^T = P\},$$

then we obtain the following corresponding standard imbedding:

$$\varphi_{\mathbb{R}} : H^m(-1) \to H^{(1)}(m+1; \mathbb{R}).$$

For finite type submanifolds of $H^m(-1)$ via this standard imbedding, I. Dimitric proved the following results.

Theorem 10.104. [Dimitric (2009)] *Let M be a submanifold of the hyperbolic m-space $H^m(-1)$. Then M is of 1-type in $H^{(1)}(m+1; \mathbb{R})$ via $\varphi_{\mathbb{R}}$ if and only if M is an open portion of a totally geodesic $H^n(-1) \subset H^m(-1)$.*

Theorem 10.105. [Dimitric (2009)] *Let $\phi : M \to H^m(-1)$ be a full minimal immersion of a Riemannian manifold M in $H^m(-1)$. Then M is of 2-type in $H^{(1)}(m+1; \mathbb{R})$ via $\varphi_{\mathbb{R}}$ if and only if M is an Einstein manifold such that $\text{Tr}\,(A_\xi, A_\eta) = \rho g(\xi, \eta)$ for some constant ρ and for all normal vectors ξ, η.*

Theorem 10.106. [Dimitric (2009)] *A complete hypersurface M of $H^{n+1}(-1)$ is of 2-type via $\varphi_{\mathbb{R}}$ if and only if M is a geodesic sphere of arbitrary radius or an equidistant hypersurface to a totally geodesic hyperbolic space $H^n(-1) \subset H^{n+1}(-1)$ with an arbitrary nonzero distance to it.*

Theorem 10.107. [Dimitric (2009)] *There are no minimal 3-type hypersurfaces in $H^{n+1}(-1)$ via $\varphi_{\mathbb{R}}$ for $2 \leq n \leq 5$.*

Theorem 10.108. [Dimitric (2009)] *Assuming* $\operatorname{Tr} A \neq 2$, *the only CMC hypersurfaces in* $H^{n+1}(-1)$ *which are mass-symmetric and of 3-type via* $\varphi_{\mathbb{R}}$ *are the product hypersurfaces* $H^k\left(-\frac{k-\ell}{k}\right) \times S^\ell\left(\frac{k-\ell}{\ell}\right)$, $k + \ell = n$, $k > \ell$, *with the sectional curvatures of* H^k *and* S^ℓ *as indicated.*

Remark 10.12. In [Dimitric (2009)], Dimitric also derived a necessary and sufficient condition for a submanifold M in $H^m(-1)$ to be mass-symmetric and of 2-type via $\varphi_{\mathbb{R}}$. Furthermore, he obtained a necessary and sufficient conditions for a submanifold M in $H^m(-1)$ with parallel mean curvature vector to be of 2-type via $\varphi_{\mathbb{R}}$.

10.12 L_r finite type hypersurfaces

Let M be a hypersurface of a real space form $R^{n+1}(c)$ of constant curvature c. Then the shape operator S of M in $R^{n+1}(c)$ is a self-adjoint linear operator on each tangent plane $T_x M$, and its eigenvalues $\kappa_1(x), \ldots, \kappa_n(x)$ are the principal curvatures of the hypersurface. Associated to the shape operator there are n algebraic invariants given by

$$s_r(x) = \sigma_r(\kappa_1(x), \ldots, \kappa_n(x)), \quad 1 \leq r \leq n,$$

where $\sigma_r : \mathbb{R}^n \to \mathbb{R}$ is the elementary symmetric function in $\mathbb{R}^n$ given by

$$\sigma_r(x_1, \ldots, x_n) = \sum_{i_1 < \cdots < i_r} x_{i_1} \ldots x_{i_r}.$$

The characteristic polynomial of S can be written in terms of the s_r's as

$$Q_S(t) = \det(tI - S) = \sum_{r=0}^{n} (-1)^r s_r t^{n-r}, \tag{10.107}$$

where $s_0 = 1$. The r-*th mean curvature* H_r of the hypersurface is

$$s_r = \binom{n}{r} H_r, \quad 0 \leq r \leq n.$$

In fact, H_1 is the mean curvature of M, H_r is intrinsic for even r and it is extrinsic for odd r. The *Newton transformations* $P_r : \mathfrak{X}(M) \to \mathfrak{X}(M)$ for $r = 1, \ldots, n$, are defined inductively from the shape operator S by

$$P_0 = I,$$

$$P_r = s_r I - S \circ P_{r-1} = \binom{n}{r} H_r I - S \circ P_{r-1}, \tag{10.108}$$

where I is the identity map on $\mathfrak{X}(M)$. Equivalently,

$$P_r = \sum_{j=0}^{r}(-1)^j s_{r-j}S^j = \sum_{j=0}^{r}(-1)^j \binom{n}{r-j}H_{r-j}S^j. \tag{10.109}$$

By the Cayley-Hamilton theorem, we find $P_n = 0$ from (10.107). When r is even, the definition of P_r does not depend on the chosen orientation, but when r is odd there is a change of sign in the definition of P_r.

Each $P_r(x)$ is a self-adjoint linear operator on each tangent plane T_xM which commutes with $S(x)$. Indeed, $S(x)$ and $P_r(x)$ can be simultaneously diagonalized: If $\{e_1, \ldots, e_n\}$ are the eigenvectors of $S(x)$ corresponding to the eigenvalues $\kappa_1(x), \ldots, \kappa_n(x)$, respectively, then they are the eigenvectors of $P_r(x)$ with corresponding eigenvalues $\mu_{1,r}, \ldots, \mu_{n,r}$ given by

$$\mu_{i,r} = \sum_{i_1 < \cdots < i_r, i_j \neq i} \kappa_{i_1}(x) \cdots \kappa_{i_r}(x), \quad 1 \leq i \leq n. \tag{10.110}$$

From here it can be easily seen that

$$\mathrm{Tr}(P_r) = (n-r)s_r = c_r H_r, \tag{10.111}$$

$$\mathrm{Tr}(S \circ P_r) = (r+1)s_{r+1} = c_r H_{r+1}, \tag{10.112}$$

and

$$\mathrm{Tr}(S^2 \circ P_r) = (s_1 s_{r+1} - (r+2)s_{r+2})$$
$$= \binom{n}{r+1}(nH_1 H_{r+1} - (n-r-1)H_{r+2}), \tag{10.113}$$

where

$$c_r = (n-r)\binom{n}{r} = (r+1)\binom{n}{r+1}. \tag{10.114}$$

These properties were derived in [Reilly (1973)].

On the other hand, by local computation, we also have (cf. [Alías and Gürbüz (2006)]).

$$\mathrm{Tr}(P_k \circ \nabla_X S) = \langle \nabla s_{k+1}, X \rangle = \binom{n}{k+1}\langle \nabla H_{k+1}, X \rangle \tag{10.115}$$

for $X \in \mathfrak{X}(M)$. Associated to each Newton transformation P_r, consider the second order linear differential operator $L_r : \mathcal{F}(M) \to \mathcal{F}(M)$ given by

$$L_r(f) = -\mathrm{Tr}(P_r \circ \nabla^2 f),$$

where $\nabla^2 f : \mathfrak{X}(M) \to \mathfrak{X}(M)$ is the self-adjoint linear operator metrically equivalent to the hessian of f and given by

$$\langle \nabla^2 f(X), Y \rangle = \langle \nabla_X(\nabla f), Y \rangle, \quad X, Y \in \mathfrak{X}(M).$$

Consider a local orthonormal frame $\{E_1, \ldots, E_n\}$ on M and observe that

$$\text{div}(P_r(\nabla f)) = \sum_{i=1}^{n} \langle (\nabla_{E_i} P_r)(\nabla f), E_i \rangle + \sum_{i=1}^{n} \langle P_r(\nabla_{E_i} \nabla f), E_i \rangle$$

$$= \langle \text{div} P_r, \nabla f \rangle + L_r(f),$$

where div is the divergence on M so that

$$\text{div} P_r = \text{Tr}(\nabla P_r) = \sum_{i=1}^{n} (\nabla_{E_i} P_r)(E_i).$$

Obviously, div $P_0 = \text{div } I = 0$. Now, Codazzi equation jointly with (10.115) imply that div $P_r = 0$ also for every $r \geq 1$. To see it observe that, from the inductive definition of P_r, we have

$$(\nabla_{E_i} P_r)(E_i)$$

$$= \binom{n}{r} \langle \nabla H_r, E_i \rangle E_i - (\nabla_{E_i} S \circ P_{r-1}) E_i - (S \circ \nabla_{E_i} P_{r-1}) E_i,$$

so that

$$\text{div} P_r = \binom{n}{r} \nabla H_r - \sum_{i=1}^{n} (\nabla_{E_i} S)(P_{r-1} E_i) - S(\text{div} P_{r-1}).$$

It follows from Codazzi's equation that ∇S is symmetric. Thus we find

$$\sum_{i=1}^{n} \langle (\nabla_{E_i} S, (P_{r-1} E_i) X \rangle = \sum_{i=1}^{n} \langle P_{r-1} E_i, (\nabla_{E_i} S) X \rangle \sum_{i=1}^{n} \langle P_{r-1} E_i, (\nabla_X S) E_i \rangle$$

$$= \text{Tr}(P_{r-1} \circ \nabla_X S) = \binom{n}{k} \langle \nabla H_r, X \rangle,$$

for any $X \in \mathfrak{X}(M)$. In other words,

$$\sum_{i=1}^{n} (\nabla_{E_i} S)(P_{r-1} E_i) = \binom{n}{r} \nabla H_r,$$

and the div $P_r = -S(\text{div} P_{r-1})$. Since div $P_0 = 0$, this gives div $P_r = 0$ for every r. As a consequence, one has (cf. [Alías and Gürbüz (2006)])

$$L_r(f) = -\text{div}(P_r(\nabla f)), \tag{10.116}$$

which is a divergence form differential operator on M. Equivalently, in terms of a local coordinate system on M, L_r is given by

$$L_r(f) = -\frac{1}{\sqrt{\mathfrak{g}}} \sum_{i,j,\ell} \partial_i \big(\sqrt{\mathfrak{g}}^{j\ell} P_{ri,\ell} \partial_j f \big), \tag{10.117}$$

where $P_r \partial_i = \sum_\ell P_{ri,\ell} \partial_\ell$. Clearly, we may extend the action of L_r from scalar functions to $\mathbb{E}^{n+1}$-valued functions.

It is well known that the Laplace operator Δ of a hypersurface M in $\mathbb{E}^{n+1}$ is an intrinsic second-order linear differential operator which arises naturally as the linearized operator of the first variation of the mean curvature for normal variations of the hypersurface. From this point of view, the Laplace operator can be regard as the first one of a sequence of n operators $L_0 = \Delta, L_1, \ldots, L_{n-1}$, where L_r stands for the linearized operator of the first variation of the $(r+1)$-th mean curvature arising from normal variations of the hypersurface (cf. [Reilly (1973)]).

In contrast to Δ which is elliptic, the operators L_r, $r = 1, \ldots, n$, are not elliptic in general, but they still share some nice properties with the Laplacian of M. For instance, by using (10.116) or (10.117), we have the following formula similar to Beltrami's formula.

Proposition 10.12. [Reilly (1973)] *Let $\phi : M \to \mathbb{E}^{n+1}$ be a hypersurface. Then we have*

$$L_r \phi = -c_r H_{r+1} \xi, \tag{10.118}$$

where c_r is defined by (10.114) and ξ is a unit normal vector field of ϕ.

In theory, one may compute $L_r^2 \phi, L_r^3 \phi$, etc. by using (10.116). Thus it is natural to extend the notion of finite type hypersurfaces developed in Chapter 6, replacing Δ by L_r and study the corresponding properties for such hypersurfaces.

Definition 10.11. A hypersurface $\phi : M \to \mathbb{E}^{n+1}$ is said to be of L_r *finite type* if ϕ admits a finite L_r-spectral decomposition:

$$\phi = \phi_0 + \sum_{j=1}^{k} \phi_j, \quad L_r \phi_j = \mu_j \phi_j, \tag{10.119}$$

for some positive integer k, where ϕ_0 is a constant vector, $\phi_1, \ldots, \phi_k$ are non-constant maps, and $\mu_1, \ldots, \mu_k$ are real L_r-eigenvalues. If all of the L_r-eigenvalues $\mu_1, \ldots, \mu_k$ are mutually different, M is said to be of L_r k-*type*. Analogous to what we did for finite type maps, an L_r k-type hypersurface is called *null* if one of $\mu_1, \ldots, \mu_k$ is zero.

Remark 10.13. For an L_r k-type hypersurface M, $P(t) = \prod_{j=1}^{k} (t - \mu_j)$ is called the L_r *minimal polynomial* as we did in Theorem 6.1. Analogous to Theorem 6.3, if a hypersurface $\phi : M \to \mathbb{E}^{n+1}$ is of L_r finite type, then $P(L_r)(\phi - \phi_0) = 0$ holds identically.

Definition 10.12. A hypersurface $\phi : M \to \mathbb{E}^{n+1}$ is called L_r k-*harmonic* if $L_r^k \phi = 0$. More generally, it is called L_r (k, ℓ, λ)-*harmonic* if it satisfies $L_r^k \phi + \lambda L_r^\ell \phi = 0$ for some integers $k > \ell \geq 1$ and real number λ.

First we mention the following two propositions.

Proposition 10.13. *If a hypersurface $\phi : M \to \mathbb{E}^{n+1}$ satisfies $L_r^k \phi = 0$ for some $r \in \{1, \ldots, n-1\}$ and $k \in \{1, 2, \ldots\}$, then either ϕ is null L_r 1-type (i.e., $H_{r+1} = 0$), or ϕ is of L_r infinite type.*

Proposition 10.14. *If a hypersurface $\phi : M \to \mathbb{E}^{n+1}$ is L_r (k, ℓ, λ)-harmonic, then it is of L_r 1-type, or L_r null 2-type, or L_r infinite type.*

Proposition 10.13 and Proposition 10.14 can be proved exactly in the same way as Proposition 7.11 and Proposition 8.1, respectively.

For L_r finite type hypersurfaces, we have the following.

Theorem 10.109. [Mohammadpouri and Kashani (2012)] *If M is an L_r 1-type hypersurface of $\mathbb{E}^{n+1}$, then it is an open part of a hypersphere or a hypersurface with $H_{r+1} = 0$.*

Proof. If $r = 0$, this is nothing but Corollary 6.3. Thus we may assume $r \geq 1$. If $\phi : M \to \mathbb{E}^{n+1}$ is of L_r 1-type, then ϕ admits the spectral decomposition:

$$\phi = \phi_0 + \phi_1, \quad L_r \phi_1 = \kappa_1 \phi_1, \quad \phi_0 \in \mathbb{E}^{n+1}, \quad \kappa_1 \in \mathbf{R}. \tag{10.120}$$

If $\kappa_1 = 0$, we have $L_r \phi = 0$. Therefore Theorem 10.12 implies that M is r-minimal in $\mathbb{E}^{n+1}$, i.e., $H_{r+1} = 0$.

Next, assume that $H_{r+1} \neq 0$. From (10.120) we find $L_r \phi = \kappa_1(\phi - \phi_0)$. Combining this with (10.118) implies that $\phi - \phi_0$ is always normal to the hypersurface M. Thus $\langle \phi - \phi_0, \phi - \phi_0 \rangle$ is a positive constant. Consequently, M is an open part of a hypersphere of $\mathbb{E}^{n+1}$. $\square$

Remark 10.14. This result also follows from Theorem 10.117.

Theorem 10.110. [Mohammadpouri and Kashani (2012)] *If $\phi : M \to \mathbb{E}^{n+1}$ is a hypersurface of L_r 2-type for some $0 \leq r \leq n-1$ and if H_{r+1} is a nonzero constant, then it is of L_r null-2-type.*

Proof. If $r = 0$, this result is an immediate consequence of Corollary 8.6. Thus we only consider the case $r \geq 1$. Assume that $\phi : M \to \mathbb{E}^{n+1}$ is a hypersurface of L_r 2-type for some $0 \leq r \leq n-1$, then we have the following spectral decomposition:

$$\phi = \phi_0 + \phi_1 + \phi_2, \quad L_r\phi_1 = \kappa_1\phi_1, \quad L_r\phi_2 = \kappa_2\phi_2 \tag{10.121}$$

for some vector $\phi_0 \in \mathbb{E}^{n+1}$ and non-constant maps ϕ_1, ϕ_2 and distinct constants κ_1, κ_2. Thus we have

$$L_r^2\phi = (\kappa_1 + \kappa_2)L_r\phi - \kappa_1\kappa_2(\phi - \phi_0). \tag{10.122}$$

On the other hand, if H_{r+1} is constant, we also have

$$L_r\phi = -c_r H_{r+1}\xi, \tag{10.123}$$

$$L_r^2\phi = c_r\binom{n}{r+1} H_{r+1}\{nH_1 H_{r+1} - (n-r-1)H_{r+2}\}\xi. \tag{10.124}$$

From (10.122), (10.123) and (10.124) we obtain

$$c_r\binom{n}{r+1} H_{r+1}\{nH_1 H_{r+1} - (n-r-1)H_{r+2}\}\xi$$
$$= -c_r(\kappa_1 + \kappa_2)H_{r+1}\xi - \kappa_1\kappa_2(\phi - \phi_0). \tag{10.125}$$

It follows from (10.125) that we have either $\kappa_1\kappa_2 = 0$ or $\phi - \phi_0$ is parallel to ξ, i.e., it is normal to M. If $\kappa_1\kappa_2 = 0$, then M is of L_r null-2-type. If $\phi - \phi_0$ is parallel to ξ, then M is an open part of a hypersphere, which is not of L_r 2-type. Thus the second case cannot occur. $\qquad\square$

Theorem 10.111. [Mohammadpouri and Kashani (2012)] *There does not exist L_{n-1} null 2-type hypersurface with at most two distinct principal curvatures in $\mathbb{E}^{n+1}$.*

In particular, this implies the following.

Corollary 10.11. [Mohammadpouri and Kashani (2012)] *There does not exist L_1 null 2-type surface in $\mathbb{E}^3$.*

Theorem 10.112. [Mohammadpouri and Kashani (2012)] *Let $\phi : M \to \mathbb{E}^{n+1}$ be a hypersurface with at most two distinct principal curvatures and multiplicities are greater than one. Then M is of L_r null 2-type $(r \neq n-1)$, if and only if M is isoparametric, so locally isometric to $S^s \times \mathbb{E}^{n-s}$, $s \geq r+1$.*

The following is a corollary of Theorem 10.112.

Corollary 10.12. *Let M be a conformally flat hypersurface of $\mathbb{E}^{n+1}$ with $n > 3$. Then M is of L_r null 2-type if and only if M is locally isometric to the cylinder $\mathbb{R} \times S^{n-1}$ with $1 \leq r < n-1$.*

Proof. Follows immediately from Theorem 10.112 and the fact that every conformally flat hypersurface of $\mathbb{E}^{n+1}$ with $n > 3$ has at most two distinct principal curvatures according to a result of Cartan and Schouten (cf. page 154 of [Chen (1973b)]). $\qquad\square$

Theorem 10.113. [Mohammadpouri and Kashani (2012)] *Let $\phi : M \to \mathbb{E}^{n+1}$ be a hypersurface with at most two distinct principal curvatures, one of them is simple. Then M is of L_r null 2-type $(r \neq n-1)$ if and only if M is isoparametric, so locally isometric to $\mathbb{R} \times S^{n-1}$ for $0 \leq r < n-1$ or $\mathbb{E}^{n-1} \times S^1$ for $r = 0$.*

Theorem 10.114. [Mohammadpouri and Kashani (2012)] *There does not exist L_1 3-type surface with constant Gauss curvature in $\mathbb{E}^3$.*

Theorem 10.115. [Mohammadpouri and Kashani (2012)] *There does not exist hypersurface of L_r null 3-type in $\mathbb{E}^{n+1}$ with constant H_{r+1} and with at most two distinct principal curvatures.*

Analogous to Theorem 6.21 we have the following.

Theorem 10.116. [Mohammadpouri and Kashani (2013)] *A quadric hypersurface M in $\mathbb{E}^{n+1}$ is of L_r finite type if and only if it is one of the following hypersurfaces:*

(a) *A hypersphere;*
(b) *One of the r-minimal algebraic conic hypersurfaces of degree 2;*
(c) *The product of a linear $(n-\ell)$-subspace $\mathbb{E}^{n-\ell}$ and a hypersphere of $\mathbb{E}^{\ell+1}$, $\ell \geq r+1$;*
(d) *The product of a linear $(n-\ell)$-subspace $\mathbb{E}^{n-\ell}$ and one of the r-minimal algebraic conic hypersurfaces of degree 2 of $\mathbb{E}^{\ell+1}$, $\ell \geq r+1$;*
(e) *The product of a linear $(n-\ell)$-subspace $\mathbb{E}^{n-\ell}$ and one of the quadric hypersurfaces of $\mathbb{E}^{\ell+1}$, $0 < \ell < r+1$.*

L. J. Alías and N. Gürbüz studied hypersurfaces of $\mathbb{E}^{n+1}$ satisfying the condition $L_r\phi = A\phi + b$ for some constant $(n+1) \times (n+1)$-matrix A and constant vector $b \in \mathbb{E}^{n+1}$. They obtained the next two results.

Theorem 10.117. [Alías and Gürbüz (2006)] *Let $r \in \{1, \ldots, n-1\}$ and let $\phi : M \to \mathbb{E}^{n+1}$ be an orientable hypersurface of $\mathbb{E}^{n+1}$. Then the immersion satisfies the condition $L_r\phi = A\phi + b$ for some constant $(n+1) \times (n+1)$-matrix A and some vector $b \in \mathbb{E}^{n+1}$ if and only if it is one of the following hypersurfaces in $\mathbb{E}^{n+1}$:*

(1) *A hypersurface with zero $(r + 1)$-th mean curvature;*
(2) *An open piece of an ordinary hypersphere;*
(3) *An open piece of a generalized right spherical cylinder $S^s \times \mathbb{E}^{n-s}$, with $k + 1 \leq s \leq n - 1$.*

Corollary 10.13. [Alías and Gürbüz (2006)] *Hyperspheres are the only orientable closed hypersurfaces in $\mathbb{E}^{n+1}$ satisfying $L_r \phi = A\phi + b$ for some constant $(n + 1) \times (n + 1)$-matrix A and vector $b \in \mathbb{E}^{n+1}$.*

Analogous to Conjecture 6.1 on finite type hypersurfaces, the following conjecture was made in [Mohammadpouri et al. (2013)]. They named it a *Generalized Chen's (finite type) Conjecture.*

Conjecture 10.3. *The only L_1 finite type closed surfaces in $\mathbb{E}^3$ are the ordinary spheres.*

All of results given above support this conjecture. The following four results further support this conjecture.

Theorem 10.118. [Mohammadpouri et al. (2013)] *Let M be a surface of revolution in $\mathbb{E}^3$ such that M has constant mean curvature. If M is of L_1 finite type, then it is an open portion of a sphere, a plane, or a circular cylinder.*

Theorem 10.119. [Mohammadpouri et al. (2013)] *Let M be a surface of revolution in $\mathbb{E}^3$ whose generating curve is a finite type curve. If M is of L_1 finite type, then it must be a plane, a circular cylinder, a circular cone or a sphere.*

Theorem 10.120. [Mohammadpouri et al. (2013)] *There is no L_1 finite type spiral closed surface in $\mathbb{E}^3$.*

Theorem 10.121. [Mohammadpouri et al. (2013)] *A tube in $\mathbb{E}^3$ is of L_1 finite type if and only if it is a circular cylinder.*

Remark 10.15. See [Lucas and Ramirez-Ospina (2013a)] for the pseudo-Euclidean version of Theorem 10.13.

For hypersurfaces in non-flat real space forms satisfying $L_r \phi = A\phi + b$, see [Alías and Kashani (2010); Lucas and Ramirez-Ospina (2012, 2013b)] and [Pashaie and Kashani (2013)].

Remark 10.16. For further results on L_r finite type hypersurfaces, see [Kashani (2009); Kim and Turgay (2013a,b); Lucas and Ramirez-Ospina (2011)] and [Lucas and Ramirez-Ospina (2014a,b)].

Bibliography

Adem, J. (1953). Relations on iterated reduced powers, *Proc. Nat. Acad. Sci. U.S.A.* **39**, 636–638.

Ahn, S. S., Kim D. S, and Kim, Y.-H. (1996). Totally umbilical Lorentzian submanifolds, *J. Korean Math. Soc.* **33**, 507–512.

Akutagawa , K. and Maeta, S. (2013). Biharmonic properly immersed submanifolds in Euclidean spaces, *Geom. Dedicata* **164**, 351–355.

Alías, L. J., García-Martínez, S. C. and Rigoli, M. (2013). Biharmonic hypersurfaces in complete Riemannian manifolds, *Pacific J. Math.* **263**, 1–12.

Alías, L. J., Ferrández, A. and Lucas, P. (1992). Submanifolds in pseudo-Euclidean spaces satisfying the condition $\Delta x = Ax + B$, *Geom. Dedicata* **42**, 345–354.

Alías, L. J., Ferrández, A. and Lucas, P. (1994). 2-type surfaces in S_1^3 and H_1^3, *Tokyo J. Math.* **17**, 447–454.

Alías, L. J., Ferrández, A. and Lucas, P. (1995a). Hypersurfaces in the non-flat Lorentzian space forms with a characteristic eigenvector field, *J. Geom.* **52**, 10–24.

Alías, L. J., Ferrández, A. and Lucas, P. (1995b). Hypersurfaces in space forms satisfying the condition $\Delta x = Ax + B$, *Trans. Amer. Math. Soc.* **347**, 1793–1801.

Alías, L. J. and Gürbüz, N. (2006). An extension of Takahashi theorem for the linearized operators of the higher order mean curvatures, *Geom. Dedicata*, **121**, 113–127.

Alías, L. J. and Kashani, S. M. B. (2010). Hypersurfaces in space forms satisfying the condition $L_k x = Ax + b$, *Taiwanese J. Math.* **14**, 1957–1977.

Almgren, F. J. (1966). Some interior regularity theorems for minimal surfaces and an extension of Bernstein's theorem, *Ann. Math.* **84**, 277–292.

Anciaux, H. (2011). *Minimal submanifolds in pseudo-Riemannian geometry*, World Scientific (Hackensack, NJ).

Arroyo, J., Barros, M. and Garay, O. J. (1997). A characterization of helices and Cornu spirals in real space forms, *Bull. Austral. Math. Soc.* **56**, 37–49.

Arroyo, J., Garay, O. J. and Mencia, J. J. (1998). On a family of surfaces of revolution of finite Chen-type, *Kodai Math. J.* **21**, 73–80.

Arslan, K., Aydin, Y., Öztürk, G. and Ugail, H. (2009). Biminimal curves in Euclidean spaces, *Int. Electron. J. Geom.* **2**, no. 2, 46–52.

Arslan, K., Bayram, B. K.; Bulca, B., Kim, Y.-H.; Murathan, C.; Öztürk, G. (2011). Vranceanu surface in $\mathbb{E}^4$ with pointwise 1-type Gauss map, *Indian J. Pure Appl. Math.* **42**, 41–51.

Arslan, K., Bulca, B., Kilic, B., Kim, Y.-H., Murathan, C. and Öztürk, G. (2011). Tensor product surfaces with pointwise 1-type Gauss map, *Bull. Korean Math. Soc.* **48**, 601–609.

Arslan, K., Bulca, B. and Milousheva, V. (2014). Meridian surfaces in $\mathbb{E}^4$ with pointwise 1-type Gauss map, *Bull. Korean Math. Soc.* **51**, no. 3, 911–922.

Arslan, K., Ezentas, R., Murathan, C. and Sasahara, T. (2005). Biharmonic submanifolds in 3-dimensional (κ, μ)-manifolds, *Int. J. Math. Math. Sci.* **2005**, no. 22, 3575–3586.

Arslan, K., Ezentas, R., Murathan, C. and Sasahara, T. (2007). Biharmonic anti-invariant submanifolds in Sasakian space forms, *Beiträge Algebra Geom.* **48**, 191–207.

Arslan, K., Kilic, B., Bengü; B., Kim, Y.-H., Murathan, C. and Öztürk, G. (2011a). Rotational embeddings in E^4 with pointwise 1-type Gauss map, *Turkish J. Math.* **35**, 493–499.

Arsan, G. G., Canfes, E. O. and Dursun, U. (2008). On null 2-type submanifolds of pseudo Euclidean space E_t^5, *Int. Math. Forum*, **3**, no. 13, 609–622.

Arvanitoyeorgos, A., Defever, F. and Kaimakamis, G. (2007a). Hypersurfaces of $\mathbb{E}_s^4$ with proper mean curvature vector, *J. Math. Soc. Japan*, **59**, 797–809.

Arvanitoyeorgos, A., Defever, F., Kaimakamis, G. and Papantoniou, V. (2009a). Biharmonic Lorentz hypersurfaces in $\mathbb{E}_1^4$, *Pacific J. Math.* **229**, 293–305.

Arvanitoyeorgos, A. and Kaimakamis, G. (2013). Hypersurfaces of type M_2^3 in E_2^4 with proper mean curvature vector, *J. Geom. Phys.* **63**, 99–106.

Arvanitoyeorgos, A., Kaimakamis, G. and Magid, M. (2009b). Lorentz hypersurfaces in E_1^4 satisfying $\Delta H = \alpha H$, *Illinois J. Math.* **53**, 581–590.

Arvanitoyeorgos, A., Kaimakamis, G. and Palamourdas, D. (2007b). Chen's conjecture and generalizations, *Symp. Diff. Geom. Submanifolds*, 109–112.

Asperti, A. C. (1980). Immersions of surfaces into 4-dimensional spaces with nonzero normal curvature, *Ann. Mat. Pura Appl.* **125**, 313–328.

Asperti, A. C., Ferus, D. and Rodríguez, L. (1982). Surfaces with nonzero normal curvature tensor, *Atti Accad. Naz. Lincei Rend.* **73**, no. 5, 109–115.

Asserda, S. and Kassi, M. (2013). Immersions biharmoniques dans une variété de Cartan-Hadamard, *C. R. Math. Acad. Sci. Paris*, **351**, 627–630.

Baikoussis, C. (1994). Ruled submanifolds with finite type Gauss map, *J. Geom.* **49**, 42–45.

Baikoussis, C. and Blair, D. E. (1991a). 2-type integral surfaces in S^5, *Tokyo J. Math.* **14**, 345–356.

Baikoussis, C. and Blair, D. E. (1991b). Finite type integral submanifolds of the contact manifold $R^{2n+1}(-3)$, *Bull. Inst. Math. Acad. Sinica*, **19**, 327–350.

Baikoussis, C. and Blair, D. E. (1992). On the Gauss map of ruled surfaces, *Glasgow Math. J.* **34**, 355–359.

Baikoussis, C. and Blair, D. E. (1994). On Legendre curves in contact 3-manifolds, *Geom. Dedicata*, **49**, 135–142.

Baikoussis, C. and Blair, D. E. (1995). 2-type flat integral submanifolds in $S^7(1)$, *Hokkaido Math. J.* **24**, 473–490.

Baikoussis, C., Blair, D. E., Chen, B. Y. and Defever, F. (1996). Hypersurfaces of restricted type in Minkowski space, *Geom. Dedicata* **62**, 319–332.

Baikoussis, C., Chen, B. Y. and Verstraelen, L. (1991). Surfaces with finite type Gauss maps, *Geometry and Topology of Submanifolds* **IV**, 214–216.

Baikoussis, C., Chen, B. Y. and Verstraelen, L. (1993a). Ruled surfaces and tubes with finite type Gauss maps, *Tokyo J. Math.* **16**, 341–349.

Baikoussis, C., Defever, F., Embrechts, P. and Verstraelen, L. (1993b). On the Gauss map of the cyclides of Dupin, *Soochow J. Math.* **19**, 417–428.

Baikoussis, C., Defever, F., Koufogiorgos, T. and Verstraelen, L. (1995). Finite type immersions of flat tori into Euclidean spaces, *Proc. Edinburgh Math. Soc.* **38**, 413–420.

Baikoussis, C., Koufogiorgos, T. and Defever, F. (1993c). Coordinate finite type integral surfaces in S^5, *Algebras Groups Geom.* **10**, 227–239.

Baikoussis, C. and Verstraelen, L. (1993). On the Gauss map of helicoidal surfaces, *Rend. Sem. Mat. Messina Ser. II* **2(16)**, 31–42.

Baikoussis, C. and Verstraelen, L. (1995). The Chen-type of the spiral surfaces, *Results Math.* **28**, 214–223.

Baird, P. and Eells, J. (1981). A conservation law for harmonic maps. *Geometry Symposium*, Utrecht 1980, *Lecture Notes in Math.*, **894**, 1–25.

Baird, P., Fardoun, A. and Ouakkas, S. (2008). Conformal and semi-conformal biharmonic maps, *Ann. Global Anal. Geom.* **34**, 403–414.

Baird, P., Fardoun, A. and Ouakkas, S. (2010). Liouville-type theorems for biharmonic maps between Riemannian manifolds, *Adv. Calc. Var.* **3**, 49–68.

Baird, P., Loubeau, E. and Oniciuc, C. (2011). Harmonic and biharmonic maps from surfaces, *Contemp. Math.* **542**, 223–230.

Balmuş, A. (2004). Biharmonic properties and conformal changes, *An. Stiint. Univ. Al. I. Cuza Iasi. Mat.* (N.S.) **50**, 361–372.

Balmuş, A., Montaldo, S. and Oniciuc, C. (2007). Biharmonic maps between warped product manifolds, *J. Geom. Phys.* **57**, 449–466.

Balmuş, A., Montaldo, S. and Oniciuc, C. (2008a). Classification results for biharmonic submanifolds in spheres, *Israel J. Math.* **168**, 201–220.

Balmuş, A., Montaldo, S. and Oniciuc, C. (2008b). Classification results and new examples of proper biharmonic submanifolds in spheres, *Note Mat.* **1**, suppl. no. 1, 49–61.

Balmuş, A., Montaldo, S. and Oniciuc, C. (2010a). Properties of biharmonic submanifolds in spheres, *J. Geom. Symmetry Phys.* **17**, 87–102.

Balmuş, A., Montaldo, S. and Oniciuc, C. (2010b). Biharmonic hypersurfaces in 4-dimensional space forms, *Math. Nachr.* **283**, 1696–1705.

Balmuş, A., Montaldo, S. and Oniciuc, C. (2012). New results toward the classification of biharmonic submanifolds in S^n, *An. St. Univ. Ovidius Constanta*, **20**, no. 2, 89–114.

Balmuş, A., Montaldo, S. and Oniciuc, C. (2013). Biharmonic PNMC submanifolds in spheres, *Ark. Mat.* **51**, 197–221.

Balmuş, A. and Oniciuc, C. (2009). Biharmonic surfaces of S^4, *Kyushu J. Math.* **63**, 339–345.

Balmuş, A. and Oniciuc, C. (2012). Biharmonic submanifolds with parallel mean curvature vector field in spheres, *J. Math. Anal. Appl.* **386**, 619–630.

Balmuş, A., Oniciuc, C., and Montaldo, S. (2010c). Submanifolds with biharmonic Gauss map, *Internat. J. Math.* **21**, 1585–1603.

Bando, S. and Ohnita, Y. (1987). Minimal 2-spheres with constant curvature in $P_n(C)$, *J. Math. Soc. Japan*, **39**, 477–487.

Barros, M. (1992). There exist no 2-type surfaces in E^3 which are images under stereographic projection of minimal surfaces in S^3, *Ann. Global Anal. Geom.* **10**, 219–226.

Barros, M. (1999). Superelasticity and conformal total tension in unified theories. Implications for the Willmore-Chen variational problem in particle physics, *Rev. Acad. Canaria Cienc.* **11**, 193–221.

Barros, M. (2000). Willmore-Chen branes and Hopf T-duality, *Classical Quantum Gravity*, **17**, 1979–1988.

Barros, M. (2001). Critical points of Willmore-Chen tension functionals, *Differential Geometry*, Valencia, 72–83.

Barros, M. and Chen. B. Y. (1985a). Finite type spherical submanifolds, *Lecture Notes in Math.* **1209**, 73–93.

Barros, M. and Chen. B. Y. (1985b). Classification of stationary 2-type surfaces of hyperspheres, *C. R. Math. Rep. Acad. Sci. Canada*, **7**, 309–314.

Barros, M. and Chen. B. Y. (1987a). Stationary 2-type surfaces in a hypersphere, *J. Math. Soc. Japan*, **39**, 627–648.

Barros, M. and Chen. B. Y. (1987b). Spherical submanifolds which are of 2-type via the second standard immersion of the sphere, *Nagoya Math. J.* **108**, 77–91.

Barros, M., Ferrández, A., Lucas, P. and Merono, M. A. (2000). A criterion for reduction of variables in the Willmore-Chen variational problem and its applications, *Trans. Amer. Math. Soc.* **352**, 3015–3027.

Barros, M. and Garay, O. J. (1987). 2-type surfaces of S^3, *Geom. Dedicata*, **24**, 329–336.

Barros, M. and Garay, O. J. (1998). Hopf submanifolds in S^7 which are Willmore-Chen submanifolds, *Math. Z.* **228**, 121–129.

Barros, M. and Garay, O. J. (1993). Minimal submanifolds of the sphere with finite type quadratic representation, *Geometry and Topology of Submanifolds* **V**, 10–25.

Barros, M. and Garay, O. J. (1994a). A new characterization of the Clifford torus in S^3 via the quadric representation, *Japan. J. Math.* **20**, 213–224.

Barros, M. and Garay, O. J. (1994b). Spherical minimal surfaces with minimal quadric representation in some hyperquadric, *Tokyo J. Math.* **17**, 479–493.

Barros, M. and Garay, O. J. (1994c). Quadric representation of minimal surfaces of S^3, *C. R. Acad. Sci. Paris Sér. I Math.* **319**, 175–179.

Barros, M. and Garay, O. J. (1995). On submanifolds with harmonic mean curvature, *Proc. Amer. Math. Soc.* **123**, 2545–2549.

Barros, M. and Ros, A. (1984). Spectral geometry of submanifolds, *Note Mat.* **4**, 1–56.

Barros, M. and Urbano, F. (1987). Spectral geometry of minimal surfaces in the sphere, *Tohoku Math. J.* **39**, 575–588.

Bauer, M. and Kuwert, E. (2003). Existence of minimizing Willmore surfaces of prescribed genus, Int. Math. Res. Not. **2003**, 553–576.

Bejancu, A. (1986). *Geometry of CR-Submanifolds* (D. Reidel, Dordrecht).

Bektas, B. and Dursun, U. (2014). On spherical submanifolds with finite type spherical Gauss map, (preprint).

Beltrami, E. (1864). Ricerche di analisi applicata alla geometria, *Giornale di Math.* **II**, 150–162.

Berger, M., Gauduchon, P. and Mazet, E. (1971). *Le Spectre d'une Variété Riemannienne*, Lecture Notes in Mathematics, Vol. **194**.

Berndt, J. (1989). Real hypersurfaces with constant principal curvatures in complex hyperbolic space, *J. Reine Angew. Math.* **395**, 132–141.

Besse, A. (1978). *Einstein Manifolds*, (Springer-Verlag, New York-Berlin).

Bishop, R. L. and O'Neill, B. (1969). Manifolds of negative curvature, *Trans. Amer. Math. Soc.* **145**, 1–49.

Black, F. and Scholes, M. (1973). The pricing of options and corporate liabilities, *J. Political Econ.* **81**, no. 3, 637–654.

Blair, D. E. (1995). A classification of 3-type curves, *Soochow J. Math.* **21**, 145–158.

Blair, D. E. (2010). *Riemannian Geometry of Contact and Symplectic Manifolds*, 2nd edition (Birkhäuser Boston, Inc., MA).

Blair, D. E. and Chen, B. Y. (1979). On CR-submanifolds of Hermitian manifold, *Israel J. Math.* **34**, 353–363.

Blair, D. E., Dillen, F., Verstraelen, L. and Vrancken, L. (1995). Deformations of Legendre curves, *Note Mat.* **15**, 99–110.

Blaschke, W. (1929). *Vorlesungen über Differentialgeometrie III*, Springer, Berlin.

Bleecker, D. D. and Weiner, J. (1976). Extrinsic bounds on λ_1 of Δ on a compact manifold, *Comment Math. Helv.* **51**, 601–609.

Bolton, J., Jensen, G. R., Rigoli, M. and Wooward, L. M.W(1988). On conformal minimal immersions of S^2 in CP^n, *Math. Ann.* **279**, 599–620.

Bonnet, O. (1867). Mémoire sur la theéorie des surfaces applicables sur une surface donnée, *École Polytech.* **42**, 31–151.

Borrelli, V., Chen, B. Y. and Morvan, J. M. (1995). Une caractérisation géométrique de la sphère de Whitney, *C.R. Acad. Sc. Paris,* **321**, 1485–1490.

Borsuk, K. (1947). Sur la courbure totale des courbes, *Ann. Soc. Math. Polon.* **20**, 251–256.

Brada, D. and Niglio, L. (1992). Connected compact minimal Chen–type–1 submanifolds of Grassmannian manifold, *Bull. Soc. Math. Belg.* **44**, 299–310.

Brockett, R. W. and Park, P. C. Kinematic dexterity of robatic mechanisms, *Intern. J. Robotics Res.* **13**, 1–15.

Bryant, R. L. (1984). A duality theorem for Willmore surfaces, *J. Differential Geom.* **20**, 23–53.

Bryant, R. L. (1985). Minimal surfaces of constant curvature in S^n, *Trans. Amer.*

Math. Soc. **290**, 259–271.

Burago, Yu. D. and Zalgaller, V. A. (1988). *Geometric Inequalities*, Springer.

Burstin, C. (1931). Ein Betrag zum Problem der Einbettung der Riemannschen Räume in eukildische Räume, *Math Sb.* **38**, 74–85.

Cabrerizo, J. L. (2001). A bridge between Willmore-Chen submanifolds and elastic curves, *Nonlinear Anal.* **47**, 5145–5156.

Caddeo, R., Montaldo, S. and Oniciuc, C. (2001a). Biharmonic submanifolds of S^3, *Internat. J. Math.* **12**, 867–876.

Caddeo, R., Montaldo, S. and Oniciuc, C. (2002). Biharmonic submanifolds in spheres, *Israel J. Math.* **130**, 109–123.

Caddeo, R., Montaldo, S. and Piu, P. (2001b). On biharmonic maps, *Contemp. Math.* **288**, 286–290.

Caddeo, R., Montaldo, S., Oniciuc, C. and Piu, P. (2014). Surfaces in three-dimensional space forms with divergence-free stress-bienergy tensor, *Ann. Mat. Pura Appl.* **193**, 529–550.

Calabi, E. (1967). Minimal immersions of surfaces in Euclidean spheres, *J. Differential Geom.* **1**, 111–125.

Camci, C. and Hacisalihoglu, H. H. (2010). Finite type curve in 3-dimensional Sasakian manifold, *Bull. Korean Math. Soc.* **47**, 1163–1170.

Cao, S. (2011). Biminimal hypersurfaces in a sphere, *J. Geom. Phys.* **61**, 2378–2383.

Cartan, É. (1927). Sur la possibilité de plonger un espace riemannien donné dans un espace euclidien, *Ann. Soc. Math. Polon.* **6**, 1–17.

Cartan, É. (1939). Sur les familles remarquables d'hypersurfaces isoparamétriques dans les espaces sphériques, *Math. Z.* **45**, 335–367.

Castro, I. and Chen, B. Y. (2006). Lagrangian surfaces in complex Euclidean plane via spherical and hyperbolic curves, *Tohoku Math. J.* **58**, 565–579.

Catalan, E. C. (1842). Sur les surfaces reglees dont l'aire est un minimum, *J. Math. Pures Appl.* **7**, 203–211.

Cecil, T. E. and Ryan, P. J. (1978). Focal sets, taut embeddings and the cyclides of Dupin, *Math. Ann.* **236**, 177–190.

Cecil, T. E. and Ryan, P. J. (1985). *Tight and Taut Immersions of Manifolds*, Research Notes in Mathematics, **107**, (Pitman, Boston, MA).

Chang, S. (1993). A closed hypersurface with constant scalar and mean curvatures in S^4 is isoparametric, *Comm. Anal. Geom.* **1**, 71–100.

Chen, B. Y. (1971). On total curvature of immersed manifolds, I, *Amer. J. Math.* **93**, 148–162.

Chen, B. Y. (1972a). On total curvature of immersed manifolds, II, *Amer. J. Math.* **94**, 899–907.

Chen, B. Y. (1972b). On a variational problem of hypersurfaces, *J. London Math. Soc.* **6**, 321–325.

Chen, B. Y. (1972c). Minimal surfaces with constant Gauss curvature, *Proc. Amer. Math. Soc.* **34**, 504–508.

Chen, B. Y. (1973a). On total curvature of immersed manifolds, III, *Amer. J. Math.* **95**, 636–642.

Chen, B. Y. (1973b). *Geometry of Submanifolds*, (M. Dekker, New York, NY).

Chen, B. Y. (1973c). An invariant of conformal mappings, *Proc. Amer. Math. Soc.* **40**, 563–564.

Chen, B. Y. (1973d). Mean curvature vector of a submanifold, *Proc. Symp. Pure Math.* **27**, Part I, 119–123.

Chen, B. Y. (1974). Some conformal invariants of submanifolds and their applications, *Boll. Un. Mat. Ital.* **10**, 380–385.

Chen, B. Y. (1975). Mean curvature vector of a submanifold, *Proc. Symp. Pure Math.* Part I, **27**, 119–123.

Chen, B. Y. (1976). Some relations between differential geometric invariants and topological invariants of submanifolds, *Nagoya Math. J.* **60**, 1–6.

Chen, B. Y. (1979a). Conformal mappings and first eigenvalue of Laplacian on surfaces, *Bull. Inst. Math. Acad. Sinica*, **7**, 395–400.

Chen, B. Y. (1979b). On the total curvature of immersed manifolds, IV: Spectrum and total mean curvature, *Bull. Inst. Math. Acad. Sinica*, **7**, 301–311.

Chen, B. Y. (1979c). On submanifolds of finite type, *Soochow J. Math.* **9**, 65–81.

Chen, B. Y. (1980). Surfaces with parallel normalized mean curvature vector, *Monatsh. Math.* **90**, 185–194.

Chen, B. Y. (1981a). *Geometry of Submanifolds and Its Applications* (Science University of Tokyo, Tokyo, Japan).

Chen, B. Y. (1981b). *CR*-submanifolds of a Kähler manifold I, II, *J. Differential Geom.* **16**, 305–322.; *ibid*, **16**, 493–509.

Chen, B. Y. (1981c). Cohomology of *CR*-submanifolds, *Ann. Fac. Sc. Toulouse Math.* **3**, 167–172.

Chen, B. Y. (1981d). Differential geometry of real submanifolds in a Kähler manifold, *Monatsh. Math.* **91**, 257–274.

Chen, B. Y. (1981e). On the total curvature of immersed manifolds, V, *Bull. Inst. Math. Acad. Sinica*, **9**, 509–516.

Chen, B. Y. (1983a). On the total curvature of immersed manifolds, VI, *Bull. Inst. Math. Acad. Sinica*, **11**, 309–328.

Chen, B. Y. (1983b). On submanifolds of finite type, *Soochow J. Math.* **9**, 65–81.

Chen, B. Y. (1983c). On the first eigenvalue of Laplacian of compact minimal submanifolds of rank one symmetric spaces, *Chinese J. Math.* **11**, 259–273.

Chen, B. Y. (1984). *Total Mean Curvature and Submanifolds of Finite Type* (World Scientific Publ., New Jersey).

Chen, B. Y. (1985a). *Finite Type Submanifolds and Generalizations*, University of Rome, (Rome, Italy).

Chen, B. Y. (1985b). Finite type submanifolds in pseudo-Euclidean spaces and applications, *Kodai Math. J.* **8**, 358–374.

Chen, B. Y. (1986a). 2-type submanifolds and their application, *Chinese J. Math.* **14**, 1–14.

Chen, B. Y. (1986b). Finite type pseudo-Riemannian submanifolds, *Tamkang J. Math.*, **17**, 137–151.

Chen, B. Y. (1987a). Surfaces of finite type in Euclidean 3-space, *Bull. Soc. Math. Belg. Ser. B*, **39**, 243–254.

Chen, B. Y. (1987b). Some estimates of total tension and their applications, *Kodai Math. J.*, **10**, 93–101.

Chen, B. Y. (1987c). *A new approach to compact symmetries spaces and applications*, Katholieke Universiteit te Leuven (Leuven, Belgium).

Chen, B. Y. (1988a). Null 2-type surfaces in E^3 are circular cylinders, *Kodai Math. J.* **11**, 295–299.

Chen, B. Y. (1988b). Null 2-type surfaces in Euclidean space, *Algebra, Analysis and Geometry*, 1–18 (World Sci. Publ., River Edge, NJ).

Chen, B. Y. (1990a). *Geometry of Slant Submanifolds*, KU Leuven, Belgium.

Chen, B. Y. (1990b). 3-type surfaces in S^3, *Bull. Soc. Math. Belg. Sér. B*, **42**, 379–381.

Chen, B. Y. (1991a). Local rigidity theorems of 2-type hypersurfaces in a hypersphere, *Nagoya Math. J.* **122**, 139–148.

Chen, B. Y. (1991b). Some open problems and conjectures on submanifolds of finite type, *Soochow J. Math.* **17**, 169–188.

Chen, B. Y. (1991c). Linearly independent, orthogonal, and equivariant immersions, *Kodai Math. J.* **14**, 341–349.

Chen, B. Y. (1992). Submanifolds of finite type in hyperbolic spaces, *Chinese J. Math.* **20**, 5–21.

Chen, B. Y. (1993a). Differential geometry of semiring of immersions, I: General theory, *Bull. Inst. Math. Acad. Sinica* **21**, 1–34.

Chen, B. Y. (1993b). Some pinching and classification theorems for minimal submanifolds, *Arch. Math.* **60**, 568–578.

Chen, B. Y. (1993c). Differential geometry of tensor product immersions, *Ann. Global Anal. Geom.* **11**, 345–359.

Chen, B. Y. (1994a). Some classification theorems for submanifolds in Minkowski space-time, *Arch. Math.* **62**, 177–182.

Chen, B. Y. (1994b). Classification of tensor product immersions which are of 1–type, *Glasgow Math. J.* **36**, 255–264.

Chen, B. Y. (1994c). Differential geometry of tensor product immersions II, *Ann. Global Anal. Geom.* **12**, 87–96.

Chen, B. Y. (1994d). Submanifolds Euclidean spaces satisfying $\Delta H = AH$, *Tamkang J. Math.* **25**, 71–81.

Chen, B. Y. (1995). Submanifolds in de Sitter space-time satisfying $\Delta H = \lambda H$, *Israel J. Math.* **91**, 373–391.

Chen, B. Y. (1996a). Mean curvature and shape operator of isometric immersions in real-space-form, *Glasgow Math. J.* **38**, 87–97.

Chen, B. Y. (1996b). A report of submanifolds of finite type, *Soochow J. Math.* **22**, 117–337.

Chen, B. Y. (1997a). Complex extensors and Lagrangian submanifolds in complex Euclidean spaces, *Tohoku Math. J.* **49**, 277–297.

Chen, B. Y. (1997b). Interaction of Legendre curves and Lagrangian submanifolds, *Israel J. Math.* **99**, 69–108.

Chen, B. Y. (1998). Strings of Riemannian invariants, inequalities, ideal immersions and their applications, The Third Pacific Rim Geometry Conference (Seoul, 1996), 7–60, *Monogr. Geom. Topology*, **25** (Intern. Press).

Chen, B. Y. (1999). Relations between Ricci curvature and shape operator for submanifolds with arbitrary codimension, *Glasgow Math. J.* **41**, 33–41.

Chen, B. Y. (2000a). Some new obstructions to minimal and Lagrangian isometric immersions, *Japan. J. Math.* **26**, 105–127.

Chen, B. Y. (2000b). Riemannian Submanifolds, *Handbook of Differential Geometry*, vol. I, 187–418 (eds. F. Dillen and L. Verstraelen).

Chen, B. Y. (2001a). Geometry of warped product CR-submanifolds in Kähler manifolds, I, II, *Monatsh. Math.* **133**, 177–195; *ibid* **134**, 103–119.

Chen, B. Y. (2001b). Riemannian geometry of Lagrangian submanifolds, *Taiwanese J. Math.* **5**, 681–723.

Chen, B. Y. (2001c). Helgason spheres of compact symmetric spaces and immersions of finite type, *Bull. Austral. Math. Soc.* **63**, 243–255.

Chen, B. Y. (2002). On isometric minimal immersions from warped products into real space forms, *Proc. Edinburgh Math. Soc.* **45**, 579–587.

Chen, B. Y. (2003a). Another general inequality for CR-warped products in complex space forms, *Hokkaido Math. J.* **32**, 415–444.

Chen, B. Y. (2003b). What can we do with Nash's embedding theorem?, *Soochow J. Math.* **30**, 303–338.

Chen, B. Y. (2004). CR-warped products in complex projective spaces with compact holomorphic factor, *Monatsh. Math.* **141**, 177–186.

Chen, B. Y. (2005). Riemannian submersions, minimal immersions and cohomology class, *Proc. Japan Acad. Ser. A Math. Sci.* **81**, 162–167.

Chen, B. Y. (2007). Tension field, iterated Laplacian, type number and Gauss maps, *Houston J. Math.* **33**, 461–481.

Chen, B. Y. (2008). Classification of marginally trapped Lorentzian flat surfaces in $\mathbb{E}_2^4$ and its application to biharmonic surfaces, *J. Math. Anal. Appl.* **340**, 861–875.

Chen, B. Y. (2009a). Dependence of the Gauss-Codazzi equations and the Ricci equation of Lorentz surfaces *Publ. Math. Debrecen* **74**, 341–349.

Chen, B. Y. (2009b). Classification of spatial surfaces with parallel mean curvature vector in pseudo-Euclidean spaces with arbitrary codimension, *J. Math. Phys.* **50**, 043503, 14 pages.

Chen, B. Y. (2009c). Complete classification of spatial surfaces with parallel mean curvature vector in arbitrary non-flat pseudo-Riemannian space forms, *Cent. Eur. J. Math.* **7**, 400–428.

Chen, B. Y. (2009d). Classification of spatial surfaces with parallel mean curvature vector in pseudo-Euclidean spaces with arbitrary codimension, *J. Math. Phys.* **50** (2009), 043503, 14 pages.

Chen, B. Y. (2010a). Complete classification of parallel spatial surfaces in pseudo-Riemannian space forms with arbitrary index and dimension, *J. Geom. Phys.* **60**, 260–280.

Chen, B. Y. (2010b). Complete explicit classification of parallel Lorentz surfaces in arbitrary pseudo-Euclidean spaces, *J. Geom. Phys.* **60**, 1333–1351.

Chen, B. Y. (2010c). Complete classification of Lorentz surfaces with parallel mean curvature vector in arbitrary pseudo-Euclidean space, *Kyushu J. Math.* **64**, 261–279.

Chen, B. Y. (2010d). Complete classification of parallel Lorentz surfaces in neutral pseudo hyperbolic 4-space, *Cent. Eur. J. Math.* **8**, 706–734.

Chen, B. Y. (2010e). Complete classification of parallel Lorentz surfaces in 4D neutral pseudo-sphere, *J. Math. Phys.* **51**, no. 8, 083518, 22 pages.

Chen, B. Y. (2010f). Explicit classification of parallel Lorentz surfaces in 4D indefinite space forms with index 3, *Bull. Inst. Math. Acad. Sinica* (N.S.) **5**, 311–345.

Chen, B. Y. (2010g). Submanifolds with parallel mean curvature vector in Riemannian and indefinite space forms, *Arab J. Math. Sci.* **16**, 1–45.

Chen, B. Y. (2011a). Classification of minimal Lorentz surfaces in indefinite space forms with arbitrary codimension and arbitrary index, *Publ. Math. Debrecen*, **78**, 485–503.

Chen, B. Y. (2011b). *Pseudo-Riemannian Geometry, δ-invariants and Applications*, (World Scientific, Hackensack, NJ).

Chen, B. Y. (2013a). A tour through δ-invariants: From Nash embedding theorem to ideal immersions, best ways of living and beyond, *Publ. Inst. Math.* (Beograd) (N.S.), **94(108)** (2013), 67–80.

Chen, B. Y. (2013b). Recent developments of biharmonic conjectures and modified biharmonic conjectures, Pure and Applied Differential Geometry – PADGE 2012, pp. 81–90, Shaker Verlag, Aachen.

Chen, B. Y. (2013c). On ideal hypersurfaces of Euclidean 4-space, *Arab J. Math. Sci.* **19**, 129–144.

Chen, B. Y. (2013d). A tour through δ-invariants: From Nash embedding theorem to ideal immersions, best ways of living and beyond, *Publ. Inst. Math.* (Beograd) (N.S.), **94(108)**, 67–80.

Chen, B. Y. (2013e). The 2-ranks of connected compact Lie groups, *Taiwanese J. Math.* **17**, 815–831.

Chen, B. Y. (2014). Some open problems and conjectures on submanifolds of finite type: recent development, *Tamkang J. Math.* **45**, 87–108.

Chen, B. Y., Barros, M. and Garay, O. J. (1987). Spherical finite type hypersurfaces, *Algebras, Groups and Geom.* **4**, 58–72.

Chen, B. Y., Choi, M. and Kim, Y.-H. (2005). Surfaces of revolution with pointwise 1-type Gauss map, *J. Korean Math. Soc.* **42**, 447–455.

Chen, B. Y., Decu, S. and Verstraelen, L. (2014). Notes on isotropic geometry of production models, *Kragujevac J. Math.* **38**, 23–33.

Chen, B. Y., Deprez, J., Dillen, F., Verstraelen, L. and Vrancken, L. (1990a). Finite type curves, *Geometry and Topology of Submanifolds* **II**, 76–110.

Chen, B. Y., Deprez, J. and Verheyen, P. (1987). Immersions, dans un espace euclidien, d'un espace symétrique compact de rang un à géodésiques simples, *C. R. Acad. Sc. Paris, Math.* **304**, 567–570.

Chen, B. Y., Deprez, J. and Verheyen, P. (1992a). Immersions with geodesics of 2–type, *Geometry and Topology of Submanifolds* **IV**, 87–110.

Chen, B. Y., Deprez, J. and Verheyen, P. (1992b). Immersions with many circular geodesics, *Geometry and Topology of Submanifolds* **IV**, 111–132.

Chen, B. Y., Deprez, J. and Verheyen, P. (1992c). A note on the centroid set of compact symmetric spaces, *Geometry and Topology of Submanifolds* **IV**, 3–10.

Chen, B. Y. and Dillen, F. (1990a). Surfaces of finite type and constant curvature in the 3-sphere, *C. R. Math. Rep. Acad. Sci. Canada*, **12**, 47-49.

Chen, B. Y. and Dillen, F. (1990b). Quadrics of finite type, *J. Geom.* **38**, 16–22.

Chen, B. Y. and Dillen, F. (2008). Optimal inequalities for multiply warped product, *Int. Electron. J. Geom.* **1**, 1–11. Erratum, *ibid,* **4** (2011), 138.

Chen, B. Y. and Dillen, F. (2011). Optimal general inequalities for Lagrangian submanifolds in complex space forms, *J. Math. Anal. Appl.* **379**, 229–239.

Chen, B. Y., Dillen, F. and Song, H.-Z. (1992). Quadric hypersurfaces of finite type, *Colloq. Math.* **63**, 145–152.

Chen, B. Y., Dillen, F. and Van der Veken, J. (2010). Complete classification of parallel Lorentzian surfaces in Lorentzian complex space forms, *Intern. J. Math.* **21**, 665–686.

Chen, B. Y., Dillen, F. and Verstraelen, L. (1986). Finite type space curves, *Soochow J. Math.* **12**, 1–10.

Chen, B. Y., Dillen, F., Verstraelen, L. and Vrancken, L. (1990b). Ruled surfaces of finite type, *Bull. Austral. Math. Soc.* **42**, 447–453.

Chen, B. Y., Dillen, F., Verstraelen, L. and Vrancken, L. (1990c). Curves of finite type, *Geometry and Topology of Submanifolds* **II**, 76–110.

Chen, B. Y., Dillen, F., Verstraelen, L. and Vrancken, L. (1993a). A variational minimal principle characterizes submanifolds of finite type, *C. R. Acad. Sc. Paris*, **317**, 961–965

Chen, B. Y., Dillen, F., Verstraelen, L. and Vrancken, L. (1993b). Submanifolds of restricted type, *J. Geom.* **46**, 20–32.

Chen, B. Y., Dillen, F., Verstraelen, L. and Vrancken, L. (1995). Compact hypersurfaces determined by a spectral variational principle, *Kyushu J. Math.* **49**, 103–121.

Chen, B. Y. and Garay, O. J. (2009). Complete classification of quasi-minimal surfaces with parallel mean curvature vector in neutral pseudo-Euclidean 4-space $\mathbb{E}_2^4$, *Results. Math.* **55**, 23–38.

Chen, B. Y. and Garay, O. J. (2012). $\delta(2)$-ideal null 2-type hypersurfaces of Euclidean space are spherical cylinders, *Kodai Math. J.* **35**, 382–391.

Chen, B. Y. and Houh, C. S. (1975). On stable submanifolds with parallel mean curvature, *Quart. J. Math.* (Oxford), **26**, 229–236.

Chen, B. Y. and Ishikawa, S. (1991). Biharmonic surfaces in pseudo-Euclidean spaces, *Memoirs Fac. Sci. Kyushu Univ. Ser. A, Math.* **45**, 323–347.

Chen, B. Y. and Ishikawa, S. (1993). On classification of some surfaces of revolution of finite type, *Tsukuba J. Math.* **17**, 287–298.

Chen, B. Y. and Ishikawa, S. (1998). Biharmonic pseudo-Riemannian submanifolds in pseudo-Euclidean spaces, *Kyushu J. Math.* **52**, 1–18.

Chen, B. Y. and Jiang, S. (1995). Inequalities between volume, center of mass, circumscribed radius, order and mean curvature, *Bull. Belg. Math. Soc.* **2**, 75–85.

Chen, B. Y. and Kuan, W. E. (1981). The Segre imbedding and its converse, *Ann. Fac. Sc. Toulouse Math.* **7**, 1–28.

Chen, B. Y. and Kuan, W. E. (1994). The cubic representation of a submanifold, *Beiträge Algebra Geom.* **35**, 55–66.

Chen, B. Y. and Li, S. J. (1991). 3-type hypersurfaces in a hypersphere, *Bull.*

Soc. Math. Belg. Sér. B, **43**, 135–141.

Chen, B. Y. and Li, S. J. (1998). Spherical hypersurfaces with 2-type Gauss map, *Beiträge Algebra Geom.* **39**, 169–179.

Chen, B. Y. and Lue, H. S. (1988). Some 2-type submanifolds and applications, *Ann. Fac. Sc. Toulouse Math. Ser. V*, **9**, 121–131.

Chen, B. Y. and Lue, H. S. (2007). Spherical submanifolds with finite type spherical Gauss map, *J. Korean Math. Soc.* **44**, 407–442.

Chen, B. Y. and Maeda, S. (1996). Extrinsic characterizations of circles in a complex projective space imbedded in a Euclidean space, *Tokyo J. Math.* **19**, 169–185.

Chen, B. Y. and Morvan, J.-M. (1994). Deformations of isotropic submanifolds in Kaehler manifolds, *J. Geom. Phys.* **13**, 79–104.

Chen, B. Y., Morvan, J.-M. and Nore, T. (1985). Énergie, tension et order des applications à valeurs dans un espace euclidien, *C. R. Acad. Sc. Paris*, **301**, 123–126.

Chen, B. Y., Morvan, J.-M. and Nore, T. (1986). Energy, tension and finite type maps, *Kodai Math. J.* **9**, 406–418.

Chen, B. Y. and Munteanu, M. I. (2013). Biharmonic ideal hypersurfaces of Euclidean spaces, *Differential Geom. Appl.* **31**, 1–16.

Chen, B. Y. and Nagano, T. (1977). Totally geodesic submanifolds of symmetric spaces, *Duke Math. J.* **44**, 745–755.

Chen, B. Y. and Nagano, T. (1978). Totally geodesic submanifolds of symmetric spaces, II, *Duke Math. J.* **45**, 405–425.

Chen, B. Y. and Nagano, T. (1984). Harmonic metrics, harmonic tensors, and Gauss maps, *J. Math. Soc. Japan*, **36**, 295–313.

Chen, B. Y. and Nagano, T. (1988). A Riemannian geometric invariant and its applications to a problem of Borel and Serre, *Trans. Amer. Math. Soc.* **308**, 273–297.

Chen, B. Y. and Ogiue, K. (1974a). On totally real submanifolds, *Trans. Amer. Math. Soc.* **193**, 257–266.

Chen, B. Y. and Ogiue, K. (1974b). Two theorems on Kähler manifolds, *Michigan Math. J.* **21**, 257–266.

Chen, B. Y. and Petrovic, M. (1991). On spectral decomposition of immersions of finite type, *Bull. Austral. Math. Soc.* **44**, 117–129.

Chen, B. Y. and Petrovic, M. (1992). Spectral decomposition of submanifolds, *Geometry and Topology of Submanifolds* **IV**, 207–213.

Chen, B. Y. and Piccinni, P. (1987). Submanifolds with finite type Gauss map, *Bull. Austral. Math. Soc.* **44**, 161–186.

Chen, B. Y. and Song, H.-Z. (1989a). Null 2–type surfaces in Minkowski space-time, *Algebras, Groups and Geom.* **6**, 333–352.

Chen, B. Y. and Song, H.-Z. (1989b). Null 2-type surfaces in Minkowski space-time. II, *Atti Accad. Pelor. Peri. Cl. Sci. Fis. Mat. Natur.* **67**, 421–432.

Chen, B. Y. and Tazawa, Y. (2000). Slant submanifolds of complex projective and complex hyperbolic spaces, *Glasgow Math. J.* **42**, 439–454.

Chen, B. Y. and Van der Veken, J. (2009). Complete classification of parallel surfaces in 4-dimensional Lorentzian space forms, *Tohoku Math. J.* **61**, 1–40.

Chen, B. Y. and Vanhecke, L. (1981). Differential geometry of geodesic spheres, *J. Reine Angew. Math.* **325**, 28–67.

Chen, B. Y. and Verheyen, P. (1981). Sous-variétés dont les sections normales sont des géodésiques, *C. R. Acad. Sci. Paris Sér. I Math.* **293**, (1981), 611–613.

Chen, B. Y. and Verheyen, P. (1984). Submanifolds with geodesic normal sections, *Math. Ann.* **269**, 417–429.

Chen, B. Y. and Verstraelen, L. (1995). *Laplace Transformation of Submanifolds*, PADGE, vol. **I** (Brussel-Leuven, Belgium).

Chen, B. Y. and Yano, K. (1971). Minimal submanifolds of a higher dimensional sphere, *Tensor* (N.S.) **22**, 369–393.

Chen, J. H. (1993). Compact 2-harmonic hypersurfaces in $S^{n+1}(1)$, *Acta Math. Sinica* **36**, 341–347.

Cheng, S.-Y. and Yau, S.-T. (1976). Maximal space-like hypersurfaces in the LorentzMinkowski spaces, *Ann. Math.* **104**, 407–419.

Chern, S. S. (1968). *Minimal Submanifolds in a Riemannian Manifold,* (University of Kansas).

Chern, S. S. and Lashof, R. K. (1957). On the total curvature of immersed manifold, *Amer. J. Math.* **79**, 306–318.

Chern, S. S. and Lashof, R. K. (1958). On the total curvature of immersed manifold, II, *Michigan Math. J.* **5**, 5–12.

Cho, J. T. Inoguchi, J.-I. Lee, J. E. (2007). Biharmonic curves in 3-dimensional Sasakian space forms, *Ann. Mat. Pura Appl.* **186**, 685–701.

Cho, J. T. Inoguchi, J.-I. Lee, J. E. (2009). Affine biharmonic submanifolds in 3-dimensional pseudo-Hermitian geometry, *Abh. Math. Semin. Univ. Hambg.* **79**, 113–133.

Choi, S. M. (1995). On the Gauss map of ruled surfaces in a 3-dimensional Minkowski space, *Tsukuba J. Math.* **19**, 285–304.

Choi, M., Kim, D.-S. and Kim, Y.-H. (2009). Helicoidal surfaces with pointwise 1-type Gauss map, *J. Korean Math. Soc.* **46**, 215–223.

Choi, M., Kim, D.-S. and Kim, Y.-H. (2010a). Classification of ruled surfaces with pointwise 1-type Gauss map, *Taiwanese J. Math.* **14**, 1297–1308.

Choi, M. and Kim, Y.-H. (2001). Characterization of the helicoid as ruled surfaces with pointwise 1-type Gauss map, *Bull. Korean Math. Soc.* **38**, 753–761.

Choi, M., Kim, Y.-H. and Yoon, D. W. (2011). Classification of ruled surfaces with pointwise 1-type Gauss map in Minkowski 3-space, *Taiwanese J. Math.* **15**, 1141–1161.

Choi, M., Kim, Y.-H., Liu, H. L. and Yoon, D. W. (2010b). Helicoidal surfaces and their Gauss map in Minkowski 3-space, *Bull. Korean Math. Soc.* **47**, 859–881.

Christoffel, E. B. (1869), ÜUber die Transformation der homogenen Differentialausdrücke zweiten Grades, *Jour. Reine Angew. Math.* **70**, 46–70.

Clarke, C. J. S. (1970). On the global isometric embedding of pseudo-Riemannian manifolds, *Proc. Roy. Soc. London, A*, **314**, 417–428.

Clebsch, A. (1866). Über die simultane Integration linearer partieller Differentialgleichungen, *J. Reine. Angew. Math.* **65**, 257–268.

Codazzi, D. (1868). Sulle coordinate curvilinee, *Ann. Mat. Pura Appl.* **2**, 1101–

1119.

Chung, H. S., Kim, D. S. and Sohn, K. H. (1995). Finite type curves in the Lorentz Minkowski plane, *Honam Math. J.* **17**, 41–47.

Dajczer, M. (1980). *Reducao da Codimensao de Imersoes Isometricas Regulares*, Doctoral thesis, IMPA.

Dajczer, M. and Nomizu, K. (1981). On flat surfaces in S_1^3 and H_1^3, *Manifolds and Lie Groups*, Birkhäuser, 71–108.

Deahna, F. (1840). Über die Bedingungen der Integrabilitat, *J. Reine Angew. Math.* **20**, 340–350.

Decruyenaere, F., Dillen, F. and Verstraelen, L. (1993). Spherical surfaces of restricted type, *Geometry and Topology of Submanifolds* **V**, 103–108.

Decruyenaere, F., Dillen, F., Verstraelen, L. and Vrancken, L. (1994). The semiring of immersions of manifolds, *Beiträge Algebra Geom.* **34**, 209–215.

Defever, F. (1997). Hypersurfaces of $\mathbb{E}^4$ satisfying $\Delta H = \lambda H$, *Michigan Math. J.* **44**, 355–363.

Defever, F. (1998). Hypersurfaces of $\mathbb{E}^4$ with harmonic mean curvature vector, *Math. Nachr.* **196**, 61–69.

Defever, F., Deszcz, R. and Verstraelen, L. (1993). The compact cylcides of Dupin and a conjecture of B.-Y. Chen, *J. Geom.* **46**, 33–38.

Defever, F., Deszcz, R. and Verstraelen, L. (1994). The Chen-type of noncompact cylcides of Dupin, *Glasgow Math. J.* **36**, 71–75.

Defever, F., Kaimakamis, G. and Papantoniou, V. (2006). Biharmonic hypersurfaces of the 4-dimensional semi-Euclidean space $\mathbb{E}_s^4$, *J. Math. Anal. Appl.* **315**, 276–286.

Deng, S. (2009). An improved Chen-Ricci inequality, *Int. Electron. J. Geom.* **2**, no. 2, 39–45.

Deprez, J. (1988). *Immersions of finite type of compact homogeneous Riemannian manifolds*, Ph.D. Thesis (Katholieke Universiteit Leuven).

Deprez, J. (1990). Immersions of finite type, *Geometry and Topology of Submanifolds* **II**, 111–133.

Deprez, J., Dillen, F. and Vrancken, L. (1990). Finite type curves on quadrics, *Chinese J. Math.* **18**, 95–121.

de Rham, G. (1931). Sur l'analysis situs des variétés à n dimensions, *J. Math. Pures Appl.*, **10**, 115–200.

Dillen, F. (1992). Ruled submanifolds of finite type, *Proc. Amer. Math. Soc.* **114**,795–798.

Dillen, F., Pas, J. and Verstraelen, L. (1990a). On surfaces of finite type in Euclidean 3-space, *Kodai Math. J.* **13**, 10–21.

Dillen, F., Pas, J. and Verstraelen, L. (1990b). On Gauss map of surfaces of revolution, *Bull. Inst. Math. Acad. Sinica*, **18**, 239–246.

Dillen, F., Petrovic, M., Verstraelen, L. and Vrancken, L. (1991). Classification of curves of Chen type 2, *Differential Geometry, in honor of Rosca*, KU Leuven, 101–106 .

Dillen, F., Van de Woestijne, I., Verstraelen, L. and Walrave, J. (1993). Curves of restricted type in Minkowski space, *Geometry and Topology of Submanifolds* **V**, 161–168.

Dillen, F., Van de Woestijne, I., Verstraelen, L. and Walrave, J. (1995a). Curves and ruled surfaces of finite type in Minkowski space, *Geometry and Topology of Submanifolds* **VII**, 124–127.

Dillen, F., Van de Woestijne, I., Verstraelen, L. and Walrave, J. (1995b). Ruled surfaces of finite type in 3-dimensional Minkowski space, *Results Math.* **27**, 250–255.

Dillen, F., Verstraelen, L., Vrancken, L. and Zafindratafa, G. (1995c). Classification of polynomial translation hypersurfaces of finite type, *Results Math.* **27**, 244–249.

Dimitric, I. (1989). *Quadric representation and submanifold of finite type*, Doctoral Thesis (Michigan State University).

Dimitric, I. (1990). Spherical submanifolds with low type quadric representation, *Tokyo J. Math.* **13**, 469–492.

Dimitric, I. (1991). 1-type submanifolds of the complex projective space, *Kodai Math. J.* **14**, 281–295.

Dimitric, I. (1992a). Submanifolds of $\mathbb{E}^m$ with harmonic mean curvature vector, *Bull. Inst. Math. Acad. Sinica*, **20**, 53–65.

Dimitric, I. (1992b). Quadric representation of a submanifold, *Proc. Amer. Math. Soc.* **114**, 201–210 .

Dimitric, I. (1993). Quadric representation of a submanifold and spectral geometry, *Proc. Symp. Pure Math.* **54**, Part 3, 155–168.

Dimitric, I. (1997). 1-type submanifolds of non-Euclidean complex space forms, *Bull. Belg. Math. Soc. Simon Stevin*, **4**, 673–684.

Dimitric, I. (2000). CR-submanifolds of HP^m and hypersurfaces of the Cayley plane whose Chen-type is 1, *Kyungpook Math. J.* **40**, 407–429.

Dimitric, I. (2009). Low-type submanifolds of real space forms via the immersions by projectors, *Differential Geom. Appl.* **27**, 507–526.

Dimitric, I. (2011). Hopf hypersurfaces of low type in non-flat complex space forms, *Kodai Math. J.* **34**, 202–243.

Dupin, C. (1822). *Applications de Géometrie et de Méchanique*, (Paris, France).

Dursun, U. (2005). Null 2-type submanifolds of the Euclidean space $\mathbb{E}^5$ with parallel normalized mean curvature vector, *Kodai Math. J.* **28**, 191–198.

Dursun, U. (2006a). Null 2-type submanifolds of the Euclidean space $\mathbb{E}^5$ with non-parallel mean curvature vector, *J. Geom.* **86**, 73–80.

Dursun, U. (2006b). Null 2-type space-like submanifolds of E_t^5 with normalized parallel mean curvature vector, *Balkan J. Geom. Appl.* **11**, 61–72.

Dursun, U. (2009a). Hypersurfaces with pointwise 1-type Gauss map in Lorentz-Minkowski space, *Proc. Est. Acad. Sci.* **58**, 146–161.

Dursun, U. (2009b). On null 2-type submanifolds of Euclidean spaces, *Int. Electron. J. Geom.* **2**, no. 2, 20–26.

Dursun, U. (2010a). Flat surfaces in the Euclidean space E^3 with pointwise 1-Type Gauss map, *Bull. Malay. Math. Sci. Soc.* **33**, 469–478.

Dursun, U. (2010b). On space-like surfaces in Minkowski 4-space with pointwise 1-type Gauss map of the second kind, *Balkan J. Geom. Appl.* **17(2)**, 34–45.

Dursun, U. and Arsan, G. G. (2011). Surface in the Euclidean space $\mathbb{E}^4$ with pointwise 1-type Gauss map, *Hacettepe J. Math. Stat.* **40**, 617–625.

Eells, J. and Lemaire, L. (1978). A report on harmonic maps, *Bull. London Math. Soc.* **10**, 1–68.

Eells, J. and Lemaire, L. (1983). *Selected topics in harmonic maps*, CBMS **50**, Amer. Math. Soc.

Eells, J. and Lemaire, L. (1988). Another report on harmonic maps, *Bull. London Math. Soc.* **20**, 385–524.

Eells, J. and Sampson, J. H. (1964). Harmonic mappings of Riemannian manifolds, *Amer. J. Math.* **86**, 109–160.

Ejiri, N. (1981). Totally real submanifolds in a 6-sphere, *Proc. Amer. Math. Soc.* **83**, 759–763.

Ejiri, N. (1982a). A counter example for Weiner's open problem, *Indiana Univ. Math. J.* **31**, 209–211.

Erbacher, J. (1971). Reduction of the codimension of an isometric immersion, *J. Differential Geom.* **5**, 333–340.

Eschenburg, J.-H. and Tribuzy, R. (1993). Existence and uniqueness of maps into affine homogeneous spaces, *Rend. Sem. Mat. Univ. Padova*, **89**, 11–18.

Fary, I. (1949). Sur la courbure totale d'une courbe gauche faisant un noeud, *Bull. Soc. Math. France* **77**, 128–138.

Fenchel, W. (1929). Über die Krümmung und Windung geschlossenen Raumkurven, *Math. Ann.* **101**, 238–252.

Ferrández, A., Garay, O. J. and Lucas, P. (1991a). Finite type ruled manifolds shaped on spherrical submanifolds, *Arch. fur Math.* **57**, 97–104.

Ferrández, A., Garay, O. J. and Lucas, P. (1991b). On a certain class of conformally flat Euclidean hypersurfaces, *Lecture Notes in Math.* **1481**, 48–54.

Ferrández, A. and Lucas, P. (1991). Null finite type hypersurfaces in space forms, *Kodai Math. J.* **14**, 406–419.

Ferrández, A. and Lucas, P. (1992a). Null 2-type hypersurfaces in a Lorentz space, *Canad. Math. Bull.* **35**, 354–360.

Ferrández, A. and Lucas, P. (1992b). On surfaces in the 3-dimensional Lorentz Minkowski space, *Pacific J. Math.* **152**, 93–100.

Ferrández, A., Lucas, P. and Meroño, M. A. (1996). Semi-Riemannian constant mean curvature surfaces via its quadric representation, *Houston J. Math.* **22**, 533–546.

Ferrández, A., Lucas, P. and Meroño, M. A. (1998). Biharmonic Hopf cylinders, *Rocky Mountain J. Math.* **28**, 957–975.

Ferus, D. (1974). Immersions with parallel second fundamental form, *Math. Z.* **140**, 87–93.

Fetcu, D. (2008). Biharmonic Legendre curves in Sasakian space forms, *J. Korean Math. Soc.* **45**, 393–404.

Fetcu, D., Loubeau, E., Montaldo, S. and Oniciuc, C. (2010). Biharmonic submanifolds of CP^n, *Math. Z.* **266**, 505–531.

Fetcu, D. and Oniciuc, C. (2005). Some remarks on the biharmonic submanifolds of S^3 and their stability, *An. Sti. Univ. Al. I. Cuza Iasi. Mat.* **51**, 171–190.

Fetcu, D. and Oniciuc, C. (2007). Explicit formulas for biharmonic submanifolds in non-Euclidean 3-spheres, *Abh. Math. Sem. Univ. Hambg.* **77**, 179–190.

Fetcu, D. and Oniciuc, C. (2009a). Biharmonic hypersurfaces in Sasakian space forms, *Differential Geom. Appl.* **27**, 713–722.

Fetcu, D. and Oniciuc, C. (2009b). Explicit formulas for biharmonic submanifolds in Sasakian space forms, *Pacific J. Math.* **240**, 85–107.

Fetcu, D. and Oniciuc, C. (2009c). Explicit formulas for biharmonic submanifolds in non-Euclidean 3-spheres, *Abh. Math. Sem. Univ. Hambg.* **77**, 179–190.

Fetcu, D. and Oniciuc, C. (2009d). On the geometry of biharmonic submanifolds in Sasakian space forms, *J. Geom. Symmetry Phys.* **14**, 21–34.

Fetcu, D. and Oniciuc, C. (2012). Biharmonic integral C-parallel submanifolds in 7-dimensional Sasakian space forms., *Tohoku Math. J.* **64**, 195–222.

Frobenius, G. (1877). Über das Pfaffsche probleme, *J. Reine Agnew. Math.* **82**, 230–315.

Fu, Y. (2013a). Biharmonic and quasi-biharmonic slant surfaces in Lorentzian complex space forms, *Abstr. Appl. Anal.* Art. ID 412709, 7 pp.

Fu, Y. (2013b). Biharmonic submanifolds with parallel mean curvature vector in pseudo-Euclidean spaces, *Math. Phys. Anal. Geom.* **16**, 331–344.

Fu, Y. (2013c). On bi-conservative surfaces in Minkowski 3-space, *J. Geom. Phys.* **66**, 71–79.

Fu, Y. (2014a). Biharmonic hypersurfaces with three distinct principal curvatures in Euclidean 5-space, *J. Geom. Phys.* **75**, 113–119.

Fu, Y. (2014b). Biharmonic hypersurfaces with three distinct principal curvatures in Euclidean space, *Tohoku Math. J.* (to appear).

Fu, Y. (2014c). Biharmonic hypersurfaces with three distinct principal curvatures in spheres, (preprint).

Fu, Y. (2014d). Explicit classification of biconservative surfaces in Lorentz 3-space forms, *Ann. Mat. Pura Appl.* DOI 10.1007/s10231-014-0399-1.

Fu, Y. (2014e). Null 2-type hypersurfaces with at most three distinct principal curvatures in Euclidean space, *Taiwanese J. Math* (to appear).

Fu, Y. and Hou, Z. H. (2010). Classification of Lorentzian surfaces with parallel mean curvature vector in pseudo-Euclidean spaces, *J. Math. Anal. Appl.* **371**, 25–40.

Fu, Y. and Li, L. (2013). A class of Weingarten surfaces in Euclidean 3-space, *Abstr. Appl. Anal.* **2013**, Art. ID 398158, 6 pp.

Garay, O. J. (1988a). Spherical Chen surfaces which are mass-symmetric and of 2-type, *J. Geom.* **33**, 39–52.

Garay, O. J. (1988b). Finite type cones shaped on spherical submanifolds, *Proc. Amer. Math. Soc.* **104**, 868–870.

Garay, O. J. (1988c). On a certain class of finite type surfaces of revolution, *Kodai Math. J.* **11**, 25–31.

Garay, O. J. (1990a). An extension of Takahashi's theorem, *Geom. Dedicata*, **34**, 105–112.

Garay, O. J. (1990b). Pseudo-umbilical 2-type surfaces in spheres, *Canad. Math. Bull.* **33**, 65–68.

Garay, O. J. (1994a). A classification of certain 3-dimensional conformally flat Euclidean hypersurfaces, *Pacific J. Math.* **162**, 13–25.

Garay, O. J. (1994b). Orthogonal surfaces with constant mean curvature in the Euclidean 4-space, *Ann. Global Anal. Geom.* **12**, 79–86.

Garay, O. J. and Romero, A. (1990). An isometric embedding of the complex hyperbolic space in a pseudo-Euclidean space and its application to the study of real hypersurfaces, *Tsukuba J. Math.* **14**, 293–313.

Garay, O. J. and Verstraelen, L. (1992). On submanifolds of finite Chen type and some related topics, *Preprint Series, Dept. Math. KU Leuven*, **4**, 5–28.

Gastel, A. and Zorn, F. (2012). Biharmonic maps of cohomogeneity one between spheres, *J. Math. Anal. Appl.* **387**, 384–399.

Gauss, C. F. (1827). *Disquisitiones generales circa superficies curvas*, Comment. Soc. Sci. Gotting. Recent. Classis Math. **6**

Germain, S. (1831). Mémoire sur la coubure des surfaces, *J. Reine Angrew. Math.* **7**, 1–29.

Germain, S. (1921). *Recherches sur la theories des surfaces élastiques*, Paris.

Gheysens, L., Verheyen, P. and Verstraelen, L. (1983). Characterization and examples of Chen submanifolds, *J. Geom.* **20**, 47–62.

Grassmann, H. (1844). *Die Ausdehnungslehre*, (Wigand, Leipzig).

Graves, L. K. (1979a). On codimension one isometric immersions between indefinite space forms, *Tsukuba J. Math.* **3**, 17–29.

Graves, L. K. (1979b). Codimension one isometric immersions between Lorentz spaces, *Trans. Amer. Math. Soc.* **252**, 367–392.

Greene, R. E. (1970). Isometric embeddings of Riemannian and pseudo-Riemannian manifolds, *Mem. Amer. Math. Soc.* **97**, 1–63.

Gromov, M. L. and Rokhlin, V. A. (1970). Embeddings and immersions in Riemannian geometry, *Russian Math. Surveys*, **25**, 1–57.

Guadalupe, I. V. and Rodríguez, L. (1983). Normal curvature of surfaces in space form, *Pacific J. Math.* **106**, 95–103.

Hasanis, T. and Vlachos, T. (1991a). A local classification of 2-type surfaces in S^3, *Proc. Amer. Math. Soc.* **122**, 533–538.

Hasanis, T. and Vlachos, T. (1991b). Spherical 2-type hypersurfaces, *J. Geom.* **40**, 82–94.

Hasanis, T. and Vlachos, T. (1991c). Coordinate finite-type submanifolds, *Geometria Dedicata* **37**, 155–165.

Hasanis, T. and Vlachos, T. (1992). Hypersurfaces of E^{n+1} satisfying $\Delta x = Ax + B$, *J. Austral. Math. Soc. A.* **53**, 377–384.

Hasanis, T. and Vlachos, T. (1993). Surfaces of finite type with constant mean curvature, *Kodai Math. J.* **16**, 244–352.

Hasanis, T. and Vlachos, T. (1994a). A classification of ruled surfaces of finite type in S^3, *J. Geom.*, **50**, 84–94.

Hasanis, T. and Vlachos, T. (1994b). Quadric representation and Clifford minimal hypersurfaces, *Bull. Belg. Math. Soc.* **1**, 559–568.

Hasanis, T. and Vlachos, T. (1995a). Hypersurfaces in E^4 with harmonic mean curvature vector field, *Math. Nachr.* **172**, 145–169.

Hasanis, T. and Vlachos, T. (1995b). Hypersurfaces with constant scalar curvature and constant mean curvature, *Ann. Global Anal. Geom.* **13**, 69–77.

Hasanis, T. and Vlachos, T. (1996a). 2-type surfaces in a hypersurfaces, *Kodai Math. J.* **19**, 26–38.

Hasanis, T. and Vlachos, T. (1996b). Spherical 2-type surfaces, *Arch. Math.* **67**,

430–440.

Helfrich, W. (1973). Elastic properties of lipid bilayers: Theory and possible experiments, *Z. Naturforsch.* **28**, 693–703.

Helgason, S. (1978). *Differential Geometry, Lie groups and Symmetric Spaces,* (Academic Press, NY).

Hiepko, S. (1979). Eine innere Kennzeichnung der verzerrten Produkte, *Math. Ann.* **241**, 209–215.

Hilbert, D. (1915). Die Grundlagen der Physik, *Konigl. Gesell. d. Wiss. Gttingen, Nachr. Math.-Phys. Kl.* 395–407.

Hilbert, D. (1924). Die grundlagen der physik, *Math. Ann.* **92**, 1–32.

Houh, C. S. (1988a). Null 2-type surfaces in $\mathbb{R}^3_1$ and in S^3_1, *Algebra, Analysis and Geom.* 19–37.

Houh, C. S. (1988b). Some low type spacelike surfaces in a Minkowski space-time, *Soochow J. Math.* **14**, 167–178.

Houh, C. S. (1988c). On Chen surfaces in a Minkowski space time, *J. Geom.* **32**, 40–50.

Houh, C. S. (1990). Rotation surfaces of finite type, *Algebras, Groups Geom.* **7**, 199–209.

Houh, C. S. (1993). Some 2-type surface in E^4, *Bull. Soc. Math. Belg. Sér. B,* **45**, 105–110.

Huang, A. M., Li, Z. Q. and Ouyang, C. Z. (2003). Rotational surfaces of finite type in $H^{n+2}(c)$, *J. Math. Wuhan,* **23**, 95–101.

Ichiyama, T., Inoguchi, J.-I. and Urakawa, H. (2009). Bi-harmonic maps and bi-Yang-Mills fields, *Note Mat.* **28**, suppl. 1, 233–275.

Impera, D. and Montaldo, S. (2009). Totally biharmonic submanifolds, *Differential geometry, World Scientific,* 237–246.

Inoguchi, J.-I. (2003). Biharmonic curves in Minkowski 3-space, *Int. J. Math. Math. Sci.* **2003**, no. 21, 1365–1368.

Inoguchi, J.-I. (2004). Submanifolds with harmonic mean curvature vector field in contact 3-manifolds, *Colloq. Math.* **100**, 163–179.

Inoguchi, J.-I. (2006). Biharmonic curves in Minkowski 3-space, II, *Int. J. Math. Math. Sci.* **2006**, Art. ID 92349, 4 pp.

Inoguchi, J.-I. (2007). Biminimal submanifolds in contact 3-manifolds, *Balkan J. Geom. Appl.* **12**, 56–67.

Inoguchi, J.-I. and Lee. J.-E. (2012a). Submanifolds with harmonic mean curvature in pseudo-Hermitian geometry, *Arch. Math.* (Brno), **48**, no. 1, 15–26.

Inoguchi, J.-I. and Lee. J.-E. (2012b). Biminimal curves in 2-dimensional space forms, *Commun. Korean Math. Soc.* **27**, 771–780.

Ishikawa, S. (1992a). *Classification problems of finite type submanifolds and biharmonic submanifolds,* Doctoral Thesis, Kyushu University (Kyushu, Japan).

Ishikawa, S. (1992b). Biharmonic W-surfaces in 4-dimensional pseudo-Euclidean space, *Memoirs Fac. Sci. Kyushu Univ. Ser. A,* **46**, 269–286.

Ishikawa, S. and Miyasato, S. (1992). On finite type curves in hyperbolic space, *Soochow J. Math.* **19**, 339–356.

Itoh, T. and Ogiue, K. (1973). Isotropic immersions, *J. Diff. Geom.* **8**, 305–316.

Janet, M. (1926). Sur la possibilité de plonger un espace riemannien donné dans un espace euclidien, *Ann. Soc. Math. Polon.* **5**, 38–43.

Jang, C. (1996). Some linearly independent immersions into their adjoint hyper quadrics, *J. Korean Math. Soc.* **33**, 169–181.

Jang, K.-O. and Kim, Y.-H. (1997). 2-type surfaces with 1-type Gauss map, Commun, *Korean Math. Soc.* **12**, 79–86.

Jiang, G. Y. (1986a). 2-harmonic isometric immersions between Riemannian manifolds, *Chinese Ann. Math. Ser. A*, **7**, 130–144.

Jiang, G. Y. (1986b). 2-harmonic maps and their first and second variational formulas, *Chinese Ann. Math. Ser. A*, **7**, 389–402.

Jiang, G. Y. (1987). Some nonexistence theorems on 2-harmonic and isometric immersions in Euclidean space, *Chinese Ann. Math. Ser. A*, **8**, 377–383.

Jung, S. D. (2013). Variation formulas for transversally harmonic and biharmonic maps, *J. Geom. Phys.* **70**, 9–20.

Jung, S. D. and Pak, J. S. (1996). Classification of cylindrical ruled surfaces satisfying $\Delta H = AH$ in a 3-dimensional Minkowski space, *Bull. Korean Math. Soc.* **33**, 97–106.

Kashani, S. M. B. (2009). On some L_1-finite type (hyper)surfaces in $\mathbb{R}^{n+1}$, *Bull. Korean Math. Soc.* **46**, 35–43.

Katz, V. J. (1979). The history of Stokes' theorem, *Math. Mag.* **52**, 146–156.

Ki, U. H., Kim, D.-S., Kim, Y.-H. and Roh. Y. M. (2009) Surfaces of revolution with pointwise 1-type Gauss map in Minkowski 3-space, *Taiwanese J. Math.* **13**, 317–338.

Kim, D.-S. (2010). Ruled submanifolds of finite type in Lorentzian space-times, *Honam Math. J.* **32**, 261–269.

Kim, D.-S. and Chung, H. S. (1994). Space curves satisfying $\Delta H = AH$, *Bull. Korean Math. Soc.* **31**, 193–200.

Kim, D. S. and Kim, S.-B. (1995a). Quadric hypersurfaces with finite type Gauss maps, *Kyungpook Math. J.* **35**, 377–385.

Kim, D.-S. and Kim, Y.-H. (1996a). Null 2-type surfaces in Minkowski 4-space, *Houston J. Math.* **22**, 179–296.

Kim, D.-S. and Kim, Y.-H. (1996b). Spherical submanifolds of null 2-type, *Kyungpook Math. J.* **36**, 361–369.

Kim, D.-S. and Kim, Y.-H. (1996c). Null 2-type surfaces in pseudo-Euclidean 4-space with null mean curvature vector, *Kyungpook Math. J.* **35**, 563–570.

Kim, D.-S. and Kim, Y.-H. (2008). Finite type ruled hypersurfaces in Lorentz-Minkowski space, *Honam Math. J.* **30**, 743–748.

Kim, D.-S. and Kim, Y.-H. (2012). Some classification results on finite-type ruled submanifolds in a Lorentz-Minkowski space, *Taiwanese J. Math.* **16**, 1475–1488.

Kim, D.-S., Kim, Y.-H. and Jung, S. M. (2014a). Ruled submanifolds with harmonic Gauss map, *Taiwanese J. Math.* **18**, 53–76.

Kim, D.-S., Kim, Y.-H. and Jung, S. M. (2014b). Some classification of ruled submanifolds in Minkwoski space and their Gauss map, *Taiwanese J. Math.* **18**, no. 4, 1021–1040.

Kim, D.-S., Kim, Y.-H. and Jung, S. M. (2014c). Some classification of ruled submanifolds, *Bull. Korean Math. Soc.* **51**, 823–829.

Kim, D.-S., Kim, Y.-H. and Yoon, D. W. (2003). Characterization of generalized B-scrolls and cylinders over finite type curves, *Indian J. Pure Appl. Math.* **34**, 1523–1532.

Kim, D.-S., Kim, Y.-H. and Yoon, D. W. (2007). Finite type ruled surfaces in Lorentz-Minkowski space, *Taiwanese J. Math.* **11**, 1–13.

Kim, D. W. (2009). Finite type surfaces of revolution in Minkowski 3-spaces, *Indian J. Math.* **51**, 401–418.

Kim, Y,-H. (1997). Low type pseudo-Riemannian submanifolds, *J. Korean Math. Soc.* **34**, 437–452.

Kim, Y,-H. (1998). Null 2-type surfaces with constant mean curvature, *Kyungpook Math. J.* **38**, 459–472.

Kim, Y.-H. and Kim, Y. W. (1995b). Pseudo-umbilical surfaces in a pseudo-Riemannian sphere or a pseudo-hyperbolic space, *J. Korean Math. Soc.* **32**, 151–160.

Kim, Y, H. and Turgay, N. C. (2013a). Surfaces in $\mathbb{E}^3$ with L_1-pointwise 1-type Gauss map, *Bull. Korean Math. Soc.* **50**, 935–949.

Kim, Y, H. and Turgay, N. C. (2013b). Classifications of helicoidal surfaces with L_1-pointwise 1-type Gauss map, *Bull. Korean Math. Soc.* **50**, 1345–1356.

Kim, Y.-H. and Yoon, D. W. (2000). Ruled surfaces with pointwise 1-type Gauss map, *J. Geom. Phys.* **34**, 191–205.

Kim, Y, H. and Yoon, D. W. (2000). Ruled surfaces with finite type Gauss map in Minkowski spaces, *Soochow J. Math.* **26**, 85–96; *ibid* **31** (2005), 1–3.

Kim, Y, H. and Yoon, D. W. (2005). On the Gauss map of ruled surfaces in Minkowski space, *Rocky Mountain J. Math.* **35**, 1555–1581.

Kobayashi, S. and Nomizu, K. (1963). *Foundations of Differential Geometry*, I, (J. Wiley, New York).

Kobayashi, S. and Nomizu, K. (1968). *Foundations of Differential Geometry*, II, (J. Wiley, New York).

Kocayigit, H. and Hacsalihoglu H. H. (2012). Biharmonic curves in contact geometry, *Commun. Fac. Sci. Univ. Ank. Sér. A1 Math. Stat.* **61**, no. 2, 35–43.

Kotani, M. (1990). A decomposition theorem of 2-type immersions, *Nagoya Math. J.* **118**, 55–64.

Kuiper, N. H. (1958). Immersions with minimal total absolute curvature, *Colloq. Géom. Diff. Globale*, Bruxelles, 75–88.

Kuiper, N. H. and Meeks, W. H. (1984). Total curvature for knotted surfaces, *Invent. Math.* **77**, 25–69.

Kulkarni, R. S. (1979). The values of sectional curvature in indefinite metrics, *Comm. Math. Helv.* **54**, 173–176.

Kusner, R. (1989). Comparison surfaces for the Willmore problem, *Pacific J. Math.* **138**, 317–345.

Langer, J. and Singer, D. (1984). Curves in the hyperbolic plane and mean curvature of tori in 3-space, *Bull. London Math. Soc.* **16**, 531–534.

Langevin, R. and Rosenberg, H. (1976). On curvature integrals and knots, *Topology* **15**, 405–416.

Larsen, J. C. (1996). Complex analysis, maximal immersions and metric singularities, *Monatsh. Math.* **122**, 105–156.

Lashof, R. K. and Smale, S. (1958). On the immersions of manifolds in Euclidean spaces, *Ann. Math.* **68**, 562–583.

Lawson, H. B. (1969). Local rigidity theorems for minimal hypersurfaces, *Ann. Math.* **89**, 187–197.

Lawson, H. B. (1970). Complete minimal surfaces in S^3, *Ann. Math.* **92**, 335–374.

Le Khong Van and Fomenko, A. T. (1988). Lagrangian manifolds and the Maslov index in the theory of minimal surfaces, *Soviet Math. Dokl.* **37**, 330–333.

Li, S. J. (1995). Null 2-type Chen surfaces, *Glasgow Math. J.* **37**, 233–242.

Li, S. J. (1991a). Finite type pseudo-umbilical submanifolds in a hypersphere, *Bull. Austral. Math. Soc.* **44**, 391–396.

Li, S. J. (1991b). 2-type pseudo-umbilical submanifolds, *Bull. Soc. Math. Belg. Sér. B*, **43**, 69–74.

Li, S. J. (1994). Null 2-type surfaces in E^m with parallel normalized mean curvature vector, *Math. J. Toyama Univ.* **17**, 23–30.

Li, S. J. and Chen, Q.-B. (1995). Minimal surfaces in a complex projective space whose mean curvature vectors are eigenvectors, *Beiträge Algbra Geom.* **36**, 135–143.

Li, S. J. and Huang, J. N. (1995). Quadric representation of minimal surfaces in an m-dimensional sphere, *Kyushu J. Math.* **50**, 207–219.

Liang, T. and Ou, Y.-L. (2013). Biharmonic hypersurfaces in a conformally flat space, *Results Math.* **64**, 91–104.

Little, J. (1969). On singularities of submanifolds of higher dimensional Euclidean spaces, *Ann. Mat. Pura Appl.* **88**, 261–336.

Little, J. (1976). Manifolds with planar geodesics, *J. Differen. Geom.* **11**, 265–285.

Liu, H. L. (1993). Rotation surfaces of finite type in pseudo-Euclidean space, *Algebras Groups Geom.* **10**, 253–261.

Liu, H. L. and Liu, G. L. (1994). On the Gauss map of rotation surfaces in 3-dimensional Minkowski space, *Kyushu J. Math.* **48**, 347–356.

Liu, J. and Yang, C. (2014). Hypersurfaces in $\mathbb{E}_s^{n+1}$ satisfying $\Delta H = \lambda H$ with at most three distinct principal curvatures, *J. Math. Anal. Appl.* **419**, 562–573.

Liu, M. and Song, W. D. (2012). Totally real 2-harmonic submanifolds in the complex projective space, *J. Math. Wuhan*, **32**, 129–134.

Loubeau, E. and Montaldo. S. (2005). Examples of biminimal surfaces of Thurston's three-dimensional geometries, *Math. Contemp.* **29**, 1–12.

Loubeau, E. and Montaldo. S. (2008). Binimimal immersions, *Proc. Edin. Math. Soc.* **51**, 421–437.

Loubeau, E., Montaldo, S. and Oniciuc, C. (2008). The stress-energy tensor for biharmonic maps. *Math. Z.* **259**, 503–524.

Loubeau, E. and Oniciuc, C. (2014). CMC proper-biharmonic surfaces of constant Gaussian curvature in spheres, airXiv:1403.1703v1, 2014.

Loubeau, E. and Ou, Y.-L. (2010). Biharmonic maps and morphisms from conformal mappings, *Tohoku Math. J.* **62**, 55–73.

Lu, J. T. (1994). Hypersurfaces of a sphere with 3-type quadric representation, *Kodai Math. J.* **17**, 290–298.

Lucas, P. and Ramirez-Ospina, H. F. (2011). Hypersurfaces in the Lorentz-Minkowski space satisfying $L_k \psi = A\psi + v$, *Geom. Dedicata* **153**, 151–175.

Lucas, P. and Ramirez-Ospina, H. F. (2012). Hypersurfaces in non-flat Lorentzian space forms satisfying $L_k \psi = A\psi + v$, *Taiwanese J. Math.* **16**, 1173–1203.

Lucas, P. and Ramirez-Ospina, H. F. (2013a). Hypersurfaces in pseudo-Euclidean spaces satisfying a linear condition on the linearized operator of a higher order mean curvature, *Differential Geom. Appl.* **31**, 175–189.

Lucas, P. and Ramirez-Ospina, H. F. (2013b). Hypersurfaces in non-flat pseudo-Riemannian space forms satisfying a linear condition in the linearized operator of a higher order mean curvature, *Taiwanese J. Math.* **17**, 15–45.

Lucas, P. and Ramirez-Ospina, H. F. (2014a). L_k-2-type hypersurfaces in hyperbolic spaces, *Taiwanese J. Math.* (to appear).

Lucas, P. and Ramirez-Ospina, H. F. (2014b). L_k-2-type hypersurfaces in S^4, (preprint).

Luo, Y (2013a). Weakly convex biharmonic hypersurfaces in nonpositive curvature space forms are minimal, *Results Math.* **65**, 49–56.

Luo, Y. (2013b). On biharmonic submanifolds in non-positively curved manifolds, airXiv:1306.6069v2.

Luo, L. (2014). On biminimal submanifolds in nonpositively curved manifolds, *Differn. Geom. Appl.* **35**, 1–8.

Maeta, S. (2012a). k-harmonic maps into a Riemannian manifold with constant sectional curvature, *Proc. Amer. Math. Soc.* **140**, 1635–1847.

Maeta, S. (2012b). Biminimal properly immersed submanifolds in the Euclidean spaces, *J. Geom. Phys.* **62**, 2288–2293.

Maeta, S. (2014a). Biharmonic maps from a complete Riemannian manifold into a non-positively curved manifold, *Ann. Glob. Anal. Geom.* **46**, 75–85.

Maeta, S. (2014b). Polyharmonic maps of order k with finite L^p k-energy into Euclidean spaces, *Proc. Amer. Math. Soc.* (to appear).

Maeta, S. (2014c). Properly immersed submanifolds in complete Riemannian manifolds, *Adv. Math.* **253**, 139–151.

Maeta, S., Nakauchi, N. and Urakawa, H. (2013). Triharmonic isometric immersions into a manifold of non-positively constant curvature, (preprint).

Maeta, S. and Urakawa, H. (2013). Biharmonic Lagrangian submanifolds in Kähler manifolds, *Glasgow Math. J.* **55**, 465–480.

Magid, M. A. (1984). Isometric immersions of Lorentz space with parallel second fundamental forms. *Tsukuba J. Math.* **8**, 31–54.

Mainardi, G. (1856). Sulla teria generale delle superficie, *Giornale dell' Istituteo Lombardo di Sci. Lett. Art.* **9**, 385–398.

Markellos, M. and Papantoniou, V. J. (2011). Biharmonic submanifolds in non-Sasakian contact metric 3-manifolds, *Kodai Math. J.* **34**, 144–167.

Marques, F. C. and Neves, A. (2014). Min-Max theory and the Willmore conjecture, *Ann. Math.* **179**, 683–782.

Martínez, A. and Ros, A. (1984). On real hypersurfaces of finite type of $\mathbb{C}P^m$, *Kodai Math. J.* **7**, 304–316.

McKean, H. P. and Singer, I. M. (1967). Curvature and the eigenvalues of the Laplacian, *J. Differential Geom.* **1**, 43–69.

Miyata, Y. (1988). 2-type surfaces of constant curvature in S^n, *Tokyo J. Math.* **11**, 157–204.

Miyata, Y. (1994). Finite type minimal 2-spheres in a complex projective space, *Tokyo J. Math.* **17**, 77–100.

Milnor, J. W. (1950). On the total curvature of knots, *Ann. Math.* **52**, 248–257.

Milnor, J. W. (1956). On manifolds homeomorphic to the 7-sphere, *Ann. Math.* **69**, 399–405.

Milnor, J. W. (1963). *Morse theory*, Based on lecture notes by M. Spivak and R. Wells, Annals of Mathematics Studies, **51** (Princeton, NJ).

Milnor, J. W. (1964). Eigenvalues of the Laplace operator on certain manifolds, *Proc. Nat. Acad. Sci. U.S.A.* **51**, p. 542.

Milousheva, V. (2013). Marginally trapped surfaces with pointwise 1-type Gauss in Minkowski 4-space, *Intern. J. Geom.* **2**, 34–43.

Minakshisundaram, S. and Pleijel, A. (1949). Some properties of the eigenfunctions of the Laplace-operator on Riemannian manifolds, *Canad. J. Math.* **1**, 242–256.

Mohammadpouri, A. and Kashani, S. M. B. (2012). On some L_k-finite-type Euclidean hypersurfaces, *ISRN Geom.* **2012**, Article ID 591296, 23 pages.

Mohammadpouri, A. and Kashani, S. M. B. (2013). Quadric hypersurfaces of L_r-finite type, *Beitr. Algebra Geom.* **54**, 625–641.

Mohammadpouri, A., Kashani, S. M. B. and Pashaie, F. (2013). On some L_1-finite type Euclidean surfaces, *Acta Math. Vietnam*, **38**, 303–316.

Montaldo, S., Oniciuc, C. and Ratto, A. (2013). Proper biconservative immersion into Euclidean space, airXiv:1312.3053v1.

Montiel, S., Ros, A. and Urbano, F. (1986). Curvature pinching and eigenvalue rigidity for minimal submanifolds, *Math. Z.* **191**, 537–548.

Moore, J. D. (1971). Isometric immersions of riemannian products, *J. Differential Geom.* **5**, 159–168.

Nagano, T. (1961). On the miniumum eigenvalues of the Laplacian in Riemannian manifolds, *Sci. Papers College Gen. Edu. Univ. Tokyo*, **11**, 177–182.

Nagatomo, Y. (1991). Finite type hypersurfaces of a sphere, *Tokyo J. Math.* **14**, 85–92.

Nakauchi, N. and Urakawa, H. (2011). Biharmonic hypersurfaces in a Riemannian manifold with non-positive Ricci curvature, *Ann. Glob. Anal. Geom.* **40**, 125–131.

Nakauchi, N. and Urakawa, H. (2013a). Biharmonic submanifolds in a Riemannian manifold with non-positive curvature, *Results Math.* **63**, 467–471.

Nakauchi, N. and Urakawa, H. (2013b). Polyharmonic maps into the Euclidean space, ArXiv: 1307.5089v2.

Nakauchi, N. and Urakawa, H. (2013c). Polyharmonic maps into the Euclidean space, ArXiv: 1307.5089v2.

Nakauchi, N., Urakawa, H. and Gudmundsson, S. (2014). Biharmonic maps into a Riemannian manifold of non-positive curvature, *Geom. Ded.* **169**, 263–272.

Nash, J. F. (1956). The imbedding problem for Riemannian manifolds, *Ann. Math.* **63**, 20–63.

Niang, A. (2004). On rotation surfaces in Minkowski 3-dimensional space with pointwise 1-type Gauss map, *J. Korean Math. Soc.* **41**, 1007–1021.

Nölker, S. (1996). Isometric immersions of warped products, *Differential Geom.*

Appl. **6**, 1–30.

Ogiue, K. (1974). Differential geometry of Kähler submanifolds, *Adv. Math.* **13**, 73–114.

Oh, Y.-G. (1990). Second variation and stability of minimal Lagrangian submanifolds, *Invent. Math.* **101**, 501–519.

Omori, H. (1967). Isometric immersions of Riemannian manifolds, *J. Math. Soc. Japan*, **19**, 205–214.

O'Neill, B. (1966). The fundamental equations of a submersion, *Michigan Math. J.* **13**, 459–469.

O'Neill, B. (1983). *Semi-Riemannian Geometry with Applications to Relativity*, (Academic Press, New York).

Oniciuc, C. (2002). Biharmonic maps between Riemannian manifolds, *An. Stii, Al. Univ. "Al. I. Cuza" Iasi*, **68**, 237–248.

Osserman, R. (1990). Curvature in the eighties, *Amer. Math. Mon.* **97**, 731–756.

Ou, Y.-L. (2006). *p*-harmonic morphisms, biharmonic morphisms, and nonharmonic biharmonic maps, *J. Geom. Phys.* **56**, 358–374.

Ou, Y.-L. (2009). On conformal biharmonic immersions, *Ann. Global Anal. Geom.* **36**, 133–142.

Ou, Y.-L. (2010). Biharmonic hypersurfaces in Riemannian manifolds, *Pacific J. Math.* **248**, 217–232.

Ou, Y.-L. (2012). Some constructions of biharmonic maps and Chen's conjecture on biharmonic hypersurfaces, *J. Geom. Phys.* **62**, 751–762.

Ou, Y.-L. (2014a). Biharmonic conformal immersions into three-dimensional manifolds, *Mediterr. J. Math.* DOI 10.1007/s00009-014-0420-3.

Ou, Y.-L. (2014b). On *f*-biharmonic maps and *f*-biharmonic submanifolds, arXiv:1306.3549.

Ou, Y.-L. and Lu, S. (2012). Biharmonic maps in two dimensions, *Ann. Mat. Pura Appl.* **192**, 127–144.

Ou, Y.-L. and Tang, L. (2012). On the generalized Chen's conjecture on biharmonic submanifolds, *Michigan Math. J.* **61**, 531–542.

Ou, Y.-L. and Wang, Z.-P. (2008). Linear biharmonic maps into Sol, Nil and Heisenberg spaces, *Mediterr. J. Math.* **5**, 379–394.

Ou, Y.-L. and Wang, Z.-P. (2011). Constant mean curvature and totally umbilical biharmonic surfaces in 3-dimensional geometries, *J. Geom. Phys.* **61**, 1845–1853.

Ozgür, C. and Güven, S. (2011). On some classes of biharmonic Legendre curves in generalized Sasakian space forms, *Collect. Math.* **65**, 203–218.

Papantoniou, V. J. and Petoumenos, K. (2012). Biharmonic hypersurfaces of type M_2^3 in $\mathbb{E}_2^4$, *Houston J. Math.* **38**, 93–114.

Pak, J. S. and Yoon, D. W. (2000). On null scrolls satisfying the condition $\Delta H = AH$, *Commun. Korean Math. Soc.* **15**, 533–540.

Park, J. (1994). Hypersurfaces satisfying the equation $\Delta x = Rx + b$, *Proc. Amer. Math. Soc.* **120**, 317–328.

Pashaie, F. and Kashani, S. M. B. (2013). Spacelike hypersurfaces in Riemannian or Lorentzian space forms satisfying $L_k x = Ax + b$, *Bull. Iranian Math. Soc.* **39**, 195–213.

Penrose, R. (1965). Gravitational collapse and space-time singularities, *Phys. Rev. Lett.* **14**, 57–59.

Perktas, S. Y., Kilic, E. and Keles, S. (2011). Biharmonic hypersurfaces of LP-Sasakian manifolds, *An. Stiint. Univ. Al. I. Cuza Iasi. Mat.* **57**, 387–408.

Peterson, K. M. (1853). *Über die Biegung der Flächen*, Doctoral Thesis (Dorpat University).

Petrović, M. (1984). On submanifolds of finite Chen type, *Zb. Rad.* **16**, 79–86.

Petrović, M. (2002). 3-type curves in the Euclidean space $\mathbb{E}^5$, *Kragujevac J. Math.* **24**, 355–365.

Petrović, M. (2004). 3-type curves in the Euclidean space $\mathbb{E}^6$, *Soochow J. Math.* **30**, 107–121.

Petrović-Torgašev, M. (2008). A property of closed finite type curves, *Bull. Austral. Math. Soc.* **77**, 145–149.

Petrović, M., Verstraelen, J. and Verstraelen, L. (2000). Principal normal spectral variations of space curves, *Proyecciones*, **19**, no. 2, 141–155.

Petrović, M. and Verstraelen, L. (1995). Curves of finite Chen type and k-minimal curves, *Geometry and Topology of Submanifolds* **VII**, 214–217.

Petrović, M. and Verstraelen, L. (1999). 3-type curves in the Euclidean space $\mathbb{E}^4$, *Novi Sad J. Math.* **29**, no. 3, 231–247.

Petrović, M., Verstraelen, L. and Vrancken, L. (1995). 3-type curves on ellipsoids of revolution, *Preprint Series, Dept. Math. KU Leuven*, **2**, 31–49.

Petrović, M., Verstraelen, L. and Vrancken, L. (1996). 3-type curves on hyperboloids and cones of revolution, *Publ. Inst. Math.* (Beograd), **59**, 138–152.

Pinkall, U. (1985). Hopf tori in S^3, *Invent. Math.* **81**, 379–386.

Pitts, J. T. (1981). *Existence and Regularity of Minimal Surfaces on Riemannian Manifolds*, Math. Notes **27**, Princeton Univ. Press (Princeton, N.J.).

Poincaré, H. (1895). Analysis situs, *J. Ecole polytech.* (2) **1**, 1–121.

Poisson, S. D. (1812). *Memoire sur les Surfaces Elastiques*, Mem. Cl. Sci. Math. Phys. (Inst. de France, Paris).

Ponge, R. and Reckziegel, H. (1993). Twisted products in pseudo-Riemannian geometry, *Geom. Dedicata*, **48**, 15–15.

Pyo, Y.-S. and Lee, J.-K. (1998). On finite type closed curves on the pseudo-hyperbolic space $H^3(-c^2)$, *Balkan J. Geom. Appl.* **3**, 111–120.

Reckziegel, H. (1985). Horizontal lifts of isometric immersions into the bundle space of a pseudo-Riemannian submersion, *Lecture Notes in Mathematics*, **1156**, 264–279.

Reilly, R. C. (1973). Variational properties of functions of the mean curvatures for hypersurfaces in space forms, *J. Diff. Geom.* **8**, 465–477.

Reilly, R. C. (1977). On the first eigenvalue of the Laplacian for compact submanifolds of Euclidean space, *Comm. Math. Helv.* **52**, 525–533.

Riemann, B. (1854). *Über die Hypothesen welche der Geometrie zu Grunde liegen*, Habili. Abhand. Königlichen Ges. Wissen. Göttingen, **13**.

Ros, A. (1983). Spectral geometry of CR-minimal submanifolds in the complex projective space, *Kodai Math. J.* **6**, 88–99.

Ros, A. (1984a). On spectral geometry of Kaehler submanifolds, *J. Math. Soc. Japan*, **36**, 433–447.

Ros, A. (1984b). Eigenvalue inequalities for minimal submanifolds and P-manifolds, *Math. Z.* **187**, 393–404.

Ros, A. (1999). The Willmore conjecture in the real projective space, *Math. Res. Lett.* **6**, 487–493.

Ros, A. (2001). The isoperimetric and Willmore problems, *Contemp. Math.* **288**, 149–161.

Rosca, R. (1972). On null hypersurfaces of a Lorentzian manifold. *Tensor (N.S.)* **23**, 66–74.

Roth, J. (2013). A note on biharmonic submanifolds of product spaces, *J. Geom.* **104**, 375–381.

Rotondaro, G. (1993). On total curvature of immersions and minimal submanifolds of spheres, *Comment. Math. Univ. Carolina*, **34**, 549–463.

Rouxel, B. (1982). *Sur quelques propriétés anallagmatiques de l'espace elliptique*, Mém. Collect., Acad. Roy. Belg. 1982 (128 pages).

Rouxel, B. (1993). Chen submanifolds, *Geometry and Topology of Submanifolds*, **VI**, 185–198.

Ruh, E. A. and Vilms, J. (1970). The tension of the Gauss map, *Trans. Amer. Math. Soc.* **149**, 569–573.

Sahin, B. (2011). Biharmonic Riemannian maps, *Ann. Polon. Math.* **102**, 39–49.

Sakai, T. (1971). On eigenvalues of Laplacian and curvature of Riemannian manifold, *Tohoku Math. J.* **23**, 589–603.

Sakamoto, K. (1977). Planar geodesic immersions, *Tohoku Math. J.* **29**, 25–56.

Sanini, A. (1983). Applicazioni tra varietà riemanniane con energia critica rispetto a deformazioni di metriche. *Rend. Mat.* **3**, 53–63.

Sard, A. (1942). The measure of the critical values of differentiable maps, *Bull. Amer. Math. Soc.* **48**, 883–890.

Sasahara, T. (2002). Submanifolds in a Sasakian manifold $R^{2n+1}(-3)$ whose ϕ-mean curvature vectors are eigenvectors, *J. Geom.* **75**, 166–178.

Sasahara, T. (2003a). Legendre surfaces whose mean curvature vectors are eigenvectors of the Laplace operator, *Note Mat.* **22**, 49–58.

Sasahara, T. (2003b). Spectral decomposition of the mean curvature vector field of surfaces in a Sasakian manifold $\mathbf{R}^{2n+1}(-3)$, *Results Math.* **43**, 168–180.

Sasahara, T. (2005a). Quasi-minimal Lagrangian surfaces whose mean curvature vectors are eigenvectors, *Demonstratio Math.* **38**, 185–196.

Sasahara, T. (2005b). Biminimal Legendrian surfaces in 5-dimensional Sasakian space forms, *Colloq. Math.* **108**, 297–304.

Sasahara, T. (2005c). Legendre surfaces with harmonic mean curvature vector field in the unit 5-sphere, *Rocky Mountain J. Math.* **40**, 313–320.

Sasahara, T. (2005d). Legendre surfaces in Sasakian space forms whose mean curvature vectors are eigenvectors, *Publ. Math. Debrecen*, **67**, 285–303.

Sasahara, T. (2007a). Biharmonic Lagrangian surfaces of constant mean curvature in complex space forms, *Glasg. Math. J.* **49**, 497–507.

Sasahara, T. (2007b). Biminimal Legendrian surfaces in 5-dimensional Sasakian space forms, *Colloq. Math.* **108**, 297–304.

Sasahara, T. (2008). Stability of biharmonic Legendrian submanifolds in Sasakian space forms, *Canad. Math. Bull.* **51**, 448–459.

Sasahara, T. (2009a). A short survey of biminimal Legendrian and Lagrangian submanifolds, *Bull. Hachinohe Inst. Tech.* **28**, 305–314.

Sasahara, T. (2009b). Biminimal Lagrangian surfaces of constant mean curvature in complex space forms, *Differential Geom. Appl.* **27**, 647–652.

Sasahara, T. (2010a). A classification result for biminimal Lagrangian surfaces in complex space forms, *J. Geom. Phys.* **60**, 884–895.

Sasahara, T. (2010b). Legendre surfaces with harmonic mean curvature vector field in the unit 5-sphere, *Rocky Mountain J. Math.* **40**, 313–320.

Sasahara, T. (2012a). Biharmonic submanifolds in nonflat Lorentz 3-space forms, *Bull. Aust. Math. Soc.* **85**, 422–432.

Sasahara, T. (2012b). Biminimal Lagrangian H-umbilical submanifolds in complex space forms, *Geom. Dedicata*, **160**, 185–193.

Sasahara, T. (2012c). Surfaces in Euclidean 3-space whose normal bundles are tangentially biharmonic, *Arch. Math.* **99**, 281–287.

Sasahara, T. (2013). Quasi-biharmonic Lagrangian surfaces in Lorentzian complex space forms, *Ann. Mat. Pura Appl.* **192**, 191–201.

Sasahara, T. (2014a). A class of biminimal Legendrian submanifolds in Sasaki space forms, *Math. Nachr.* **287**, 79–90.

Sasahara, T. (2014b). Classification results for λ-biminimal surfaces in 2-dimensional complex space forms, *Acta Math. Hungar.* (in press).

Schlafli, L. (1873). Nota alla Memoria del sig. Beltrami, *Ann. Mat. Pura Appl.* **5**, 178–193.

Scofield, P. D. (1995). Curves of constant precession, *Amer. Math. Monthly*, **288**, 765–790.

Seifert, H. (1932). Topologie dreidimensionaler gefaserter Räume, *Acta Math.* **60**, 147–238.

Shen, Y.-B. (1995). On spectral geometry of minimal surfaces in $\mathbb{C}P^n$, *Trans. Amer. Math. Soc.* **347**, 3873–3889.

Shiohama, K. and Takagi, R. (1970). A characterization of a standard torus in E^3, *J. Differential Geom.* **4**, 477–485.

Simon, L. (1993). Existence of surfaces minimizing the Willmore functional, *Comm. Anal. Geom.* **1**, 281–326.

Slebodziński, W. (1931). Sur les équations de Hamilton, *Bull. Acad. Roy. Belg.* **17**, 864–870.

Smale, S. (1959). The classification of immersions of spheres in Euclidean spaces, *Ann. Math.* **69**, 327–344.

Spivak, M. (1979). *A Comprehensive Introduction to Differential Geometry*, vol. IV, Publish or Perish (Berkeley, CA).

Sternberg, S. (1964). *Lectures on Differential Geometry*, (Prentice-Hall, NJ).

Suceavă, B. (2001). The Chen invariants of warped products of hyperbolic planes and their applications to immersibility problems, *Tsukuba J. Math.* **25**, 311–320.

Šucurović, E. (2000). A classification of 3-type curves in Minkowski 3-space $\mathbb{E}_1^3$, II. *Publ. Inst. Math.* (Beograd) (N.S.) **68(82)**, 117–132.

Šucurović, E. (2001). A classification of 2-type curves in the Minkowski space $\mathbb{E}_1^n$, *Publ. Inst. Math.* (Beograd) (N.S.) **70(84)**, 42–58.

Sunday, D. (1976). The total curvature of knotted spheres, *Bull. Amer. Math. Soc.* **82**, 140–142.

Tai, S.-S. (1968). Minimum imbed dings of compact symmetric spaces of rank one, *J. Differn. Geom.* **2**, 55–66.

Takagi, R. (1975). Real hypersurfaces in a complex projective space with constant principal curvatures I, II, *J. Math. Soc. Japan*, **27**, 43–53, 507–516.

Takagi, R. and Takeuchi, M. (1977). Degree of symmetric Kählerian submanifolds of a complex projective space, *Osaka J. Math.* **14**, 501–518.

Takahashi, T. (1966). Minimal immersions of Riemannian manifolds, *J. Math. Soc. Japan*, **18**, 380–385.

Takeuchi, M. (1981). Parallel submanifolds of space forms, *Manifolds and Lie Groups*, 429–447 (Birkhäuser, Boston).

Takeuchi, M. and Kobayashi, S. (1968). Minimal imbeddings of *R*-spaces, *J. Differential Geom.* **2**, 203–215.

Tazzioli, R. (1997). The role of differential parameters in Beltrami's work, *Historia Math.* **24**, 25–45.

Thomsen, G. (1923). Über konforme Geometrie I: Grundlagen der konformen Flächentheorie, *Abh. Math. Sem. Univ. Hamburg*, **3**, 31–56.

Turgay, N. C. (2014). On the marginally trapped surfaces in 4-dimensional space-times with finite type Gauss map, *Gen. Relativ. Gravit.* article 46:1621.

Turhan, E. and Körpinar, T. (2010). Biminimal general helix in the Heisenberg group Heis^3, *Kragujevac J. Math.* **34**, 51–60.

Turhan, E. and Körpinar, T. (2011). Null biminimal general helices in the Lorentzian Heisenberg group, *Thai J. Math.* **9**, 127–137.

Udagawa, S. (1986a). Einstein parallel Kaehler submanifolds in a complex projective space, *Tokyo J. Math.* **9**, 335–340.

Udagawa, S. (1986b). Spectral geometry of Kaehler submanifolds of a complex projective space, *J. Math. Soc. Japan*, **38**, 453–471.

Udagawa, S. (1986c). Spectral geometry of compact Hermitian symmetric submanifolds, *Math. Z.* **192**, 57–72.

Urakawa, H. (2011). The geometry of biharmonic maps, in: Harmonic maps and differential geometry, *Contemp. Math.* **542**, 159–175.

Veblen, O. and Whitehead, J. H. C. (1932). *The foundations of Differential Geometry*, Cambridge Tracts in Math. and Math. Phys. **29** (London, England).

Verheyen, P. (1985). Submanifolds with geodesic normal sections are helical, *Rend. Sem. Mat. Univ. Politecn. Torino*, **43**, 511–527.

Verstraelen, L. (1991). On submanifolds of finite Chen type and of restricted type, *Results Math.* **20**, 744–755.

Verstraelen, L. (1990). Curves and surfaces of finite Chen type, *Geometry and Topology of Submanifolds* **III**, 304–311.

Voigt, W. *Die fundamentalen physikalischen Eigenschaften der Krystalle in elementarer Darstellung*, (Leipzig, Germany, 1898).

Voss, K. (1956). Einige differentialgeometrische Kongruenzsätze für geschlossene Flächen und Hyperflächen, *Math. Ann.* **131**, 180–218.

Wallach, N. R. (1972). Minimal immersions of symmetric spaces into spheres, *Symmetric spaces*, Pure and Appl. Math. (M. Dekker), **8**, 1–40.

Wang, S. B. (1989). The first variation formula for k-harmonic mapping, *J. Nanchang Univ.* **13**, no. 1.

Wang, Z.-P. and Ou, Y.-L. (2011). Biharmonic Riemannian submersions from 3-manifolds, *Math. Z.* **269**, 917–925.

Wang, Z.-P., Ou, Y.-L. and Yang, H.-C. (2014). Biharmonic maps from a 2-sphere, *J. Geom. Phys.* **77**, 86–96.

Wei, S. W. (2008). p-harmonic geometry and related topics, *Bull. Transilv. Univ. Brasov*, Ser. III, **1(50)**, 415–453.

Weiner, J. (1978). On a problem of Chen, Willmore, et al., *Indiana Univ. Math. J.* **27**, 19–35.

Weingarten, J. (1861). Ueber eine Klasse auf einander abwickelbarer Flächen, *J. Reine Angew. Math.* **59**, 382–393.

Wettstein, B. (1978). *Congruence and existence of differentiable map*, Ph.D. Thesis (ETH Zürich, no. 6252).

Weyl, H. (1912). Das asymptotische Verteilungsgesetz der Eigenwerte linearer partieller Differentialgleichungen, *Math. Ann.* **71**, 411-479.

Weyl, H. (1913). *Die Idee der Riemannschen Fläche*, (Teubner, Stuttgart).

Weyl, H. (1918). Reine Infinitesimalgeometrie, *Math. Z.* **26**, 384–411.

Wheeler, G. (2013). Chen's conjecture and ϵ-superbiharmonic submanifolds of Riemannian manifolds, *Intern. J. Math.* **24**, no. 4, 1350028 (6 pages).

Whitney, H. (1936). Differentiable manifolds, *Ann. Math.* **37** 645–680.

Whitney, H. (1937). On regular closed curves in the plane, *Compositio Math.* **4**, 276–284.

Willmore, T. J. (1965). Note on embedded surfaces, *An. Sti. Univ. "Al. I. Cuza" Iasi, Sec. I. a Mat.* (N.S.) **11B**, 493–496.

Willmore, T. J. (1968). Mean curvature of immersed surfaces, *An. Sti. Univ. "Al. I. Cuza" Iasi, Sec. I. a Mat.* (N.S.) **14**, 99–103.

Willmore, T. J. (1971). Mean curvature of Riemannian immersion, *J. London Math. Soc.* **3**, 307–310.

Wintgen, P. (1978). On the total curvature of surfaces in $\mathbb{E}^4$, *Colloq. Math.* **39**, 289–296.

Wintgen, P. (1979). Sur l'inégalité de Chen-Willmore, *C. R. Acad. Sci. Paris Sér. A-B*, **288**, A993–A995.

Witt, E. (1941). Eine Identität zwischen Modulformen zweiten Grades, *Abh. Math. Sem. Univ. Hamburg*, **14**, 323–337.

Zafindratafa, G. (1995). On hypersurfaces of finite type, *Geometry and Topology of Submanifolds* **VI**, 289–292.

Zafindratafa, G. (1996). On finite type algebraic hypersurfaces of translation, *J. Geom.* **55**, 182–191.

Zhang, W. (2011). New examples of biharmonic submanifolds in CP^n and S^{2n+1}, *An. Stiint. Univ. Al. I. Cuza Iasi. Mat.* (N.S.) **57**, 207–218.

Zhang, Y. H. and Ma, Z. S. (1993). Spherical minimal submanifolds in which the second standard immersion of the sphere is of 3-type, *Sichuan Shifan Daxue Xuebao Ziran Kexue Ban*, **16**, 7–12.

Zhang, Y. Z. (2002). A Hopf-type theorem for 2-type topological two-spheres in S^7, *J. Zhejiang Univ. Sci. Ed.* **29**, 135–139.

Subject Index

Author Index

461

Printed in the United States
By Bookmasters